# FUNCTIONAL ANALYSIS

## an introduction

# PURE AND APPLIED MATHEMATICS

A Series of Monographs and Textbooks




1. K. YANO. Integral Formulas in Riemannian Geometry (1970)
2. S. KOBAYASHI. Hyperbolic Manifolds and Holomorphic Mappings (1970)
3. V. S. VLADIMIROV. Equations of Mathematical Physics (A. Jeffrey, editor; A. Littlewood, translator) (1970)
4. B. N. PSHENICHNYI. Necessary Conditions for an Extremum (L. Neustadt, translation editor; K. Makowski, translator) (1971)
5. L. NARICI, E. BECKENSTEIN, and G. BACHMAN. Functional Analysis and Valuation Theory (1971)
6. D. S. PASSMAN. Infinite Group Rings (1971)
7. L. DORNHOFF. Group Representation Theory (in two parts). Part A: Ordinary Representation Theory. Part B: Modular Representation Theory (1971, 1972)
8. W. BOOTHBY and G. L. WEISS (eds.). Symmetric Spaces: Short Courses Presented at Washington University (1972)
9. Y. MATSUSHIMA. Differentiable Manifolds (E. T. Kobayashi, translator) (1972)
10. L. E. WARD, JR. Topology: An Outline for a First Course (1972)
11. A. BABAKHANIAN. Cohomological Methods in Group Theory (1972)
12. R. GILMER. Multiplicative Ideal Theory (1972)
13. J. YEH. Stochastic Processes and the Wiener Integral (1973)
14. J. BARROS-NETO. Introduction to the Theory of Distributions (1973)
15. R. LARSEN. Functional Analysis: An Introduction (1973)

*In Preparation:*

B.-Y. CHEN. Geometry of Submanifolds
J. DIEUDONNE. Introduction to the Theory of Formal Groups
R. HERMANN. Geometry, Physics, and Systems
R. LARSEN. Banach Algebras: An Introduction
M. MARCUS. Finite-Dimensional Multilinear Algebra (in two parts)
C. PROCESI. Rings with Polynomial Identities
I. VAISMAN. Cohomology and Differential Forms
N. WALLACH. Harmonic Analysis on Homogeneous Spaces
K. YANO and S. ISHIHARA. Differential Geometry of Tangent and Cotangent Bundles

# FUNCTIONAL ANALYSIS

## an introduction

RONALD LARSEN
DEPARTMENT OF MATHEMATICS
WESLEYAN UNIVERSITY
MIDDLETOWN, CONNECTICUT

MARCEL DEKKER, INC. New York 1973

MARCEL DEKKER, INC.

95 Madison Avenue, New York, New York 10016

LIBRARY OF CONGRESS CATALOG CARD NUMBER: 72-90375

ISBN: 0-8247-6042-5

PRINTED IN THE UNITED STATES OF AMERICA

To

Karel deLeeuw

and

Ralph S. Phillips

with whom I began to learn functional analysis

# PREFACE

The exposition in the following pages is based on lectures I gave to second year mathematics graduate students at Wesleyan University during the academic years 1970 - 71 and 1971 - 72. The intent of the lectures was to provide the student with an introduction to functional analysis that not only presented the basic notions, theorems, and techniques of the subject, but also gave a modest sampling of the applications of functional analysis. This remains the goal of the present book. The choice of the topics and applications is clearly idiosyncratic, and I make no claim to a balanced, let alone definitive, treatment. I will consider the book a success if it but convinces the reader of the beauty, power, and utility of functional analysis.

An intelligent reading of the book presupposes at most the usual mathematical equipment possessed by second year mathematics graduate students. The main items required are some knowledge of point set topology, linear algebra, and elementary complex analysis, together with a good background in measure and integration theory, say, for example, as in Royden's book "Real Analysis" [Ry]. Results with which it is assumed the reader is familiar are frequently cited without further elaboration. However, in almost all such instances an appropriate reference is given.

The last section in each chapter consists of problems that hopefully are do-able with the material developed up to that point. An asterisk before a problem generally indicates that the problem may be of a more substantial nature or, in some cases, that it contains a result of particular importance. Most of the results in the body of the text that are formally stated without proof appear again in the problem sections and such problems are cross referenced accordingly.

The conclusion of proofs is indicated by the symbol □ at the right hand margin.

I would like to thank all of the graduate students at Wesleyan who passed through my course while this book was evolving for their comments and suggestions. In particular, I would like to thank David DeGeorge, Hans Engenes, Polly Moore Hemstead, and Michael Paul for their often perspicacious observations and questions that more than once kept me from foolish error. Those errors that remain, foolish or otherwise, are of course my own responsibility.

I am especially grateful to Polly Moore Hemstead, who not only passed through the course but also provided me with valuable editorial assistance and collected and organized the problem sets at the end of each chapter. Her efforts have greatly enhanced the final form of the book.

I would also like to thank Helen Diehl, who typed all of the original manuscript and a good deal of the final one.

Finally, thanks are due to the editors and staff of Marcel Dekker for their cheerful and expert cooperation during the production of the book.

Middletown, Connecticut
January, 1973

CONTENTS

"Atter og fram, det er like langt;
ut og inn, det er like trangt."

Peer Gynt

# FUNCTIONAL ANALYSIS
an introduction

# CHAPTER 1

# SEMINORMED AND NORMED LINEAR SPACES

__1.0. Introduction.__ In this chapter we develop some of the basic concepts and results concerning seminormed and normed linear spaces. We begin with the fundamental definitions and some elementary theorems about the construction of new normed linear spaces from given normed linear spaces. This is followed by a collection of examples, by no means exhaustive, of seminormed and normed linear spaces. The verifications are left to the reader and are, in some cases, nontrivial. References are cited at the end of the section.

Section 1.3 contains a discussion of finite-dimensional normed linear spaces in which it is shown that all such spaces are topologically ismorphic and that a normed linear space is finite-dimensional if and only if closed bounded sets are compact. In the next two sections we discuss the concept of the gauge of a convex balanced absorbing set, the relationship between gauges and seminorms, and the introduction of a topology into a seminormed linear space.

__1.1. Basic Definitions.__ We begin with several fundamental definitions. Throughout, $\Phi$ will denote either the real-number field or the complex-number field, $\mathbb{R}$ and $\mathbb{C}$ will stand for the real and complex fields, respectively, and $\mathbb{Z}$ will always denote the integers.

__Definition 1.1.1.__ A __linear space__ $V$ over a field $\Phi$ is a commutative group under a binary operation $+$, together with an operation of scalar multiplication from $\Phi \times V$ to $V$, denoted by juxtaposition, such that

(i) $a(x + y) = ax + ay$,

(ii) $(a + b)x = ax + bx$,

(iii) $a(bx) = (ab)x$,

(iv) $1x = x$ $\qquad (a,b \in \Phi;\ x,y \in V)$.

Of course, $1$ denotes the multiplicative identity in $\Phi$, and the operations of addition and multiplication in $\Phi$ have been indicated in the usual manner; $0$ will be used to denote the additive identity in both $V$ and $\Phi$. The context will make clear which is meant.

Definition 1.1.2. Let $V$ be a linear space over $\Phi$. Given $E \subset \Phi$; $a \in \Phi$; $A,B \subset V$; and $x_0 \in V$, we set

$$A + B = \{x + y \mid x \in A,\ y \in B\},$$
$$x_0 + B = \{x_0 + y \mid y \in B\},$$
$$EA = \{ax \mid a \in E,\ x \in A\},$$
$$aA = \{ax \mid x \in A\}.$$

Definition 1.1.3. Let $V$ be a linear space over $\Phi$ and let $W \subset V$. Then $W$ is a linear subspace of $V$ if $W + W \subset W$ and $\Phi W \subset W$. Furthermore, $W$ is convex if $ax + (1 - a)y \in W$ whenever $x,y \in W$ and $a \in \Phi$, $0 \le a \le 1$; $W$ is symmetric if $-1W = -W = W$; and $W$ is balanced if $aW \subset W$, $a \in \Phi$, $|a| \le 1$.

Clearly every balanced set is symmetric, but the converse need not be true. Moreover, a nonempty balanced set contains the origin.

It should be remarked that the terminology with regard to balanced sets is not universally the same. Many authors call a balanced set "circled" (for example, see [K, p. 176; KeNa, p. 14]), whereas others use the term "balanced" [Bb, p. 8; $E_1$, p. 50; T, p. 123; $W_1$, p. 22; Y, p. 24].

Definition 1.1.4. Let $V$ be a linear space over $\Phi$ and suppose that $p : V \to \mathbb{R}$. Then $p$ is a seminorm on $V$ if

(i) $p(x + y) \leq p(x) + p(y)$

(ii) $p(ax) = |a|p(x)$ $(x,y \in V;\ a \in \Phi)$.

Property (i) of $p$ is known as the triangle inequality, for obvious reasons. It is also referred to as the subadditivity of $p$.

We shall shortly see some examples of seminorms. First, however, we wish to prove the following proposition:

Proposition 1.1.1. Let $V$ be a linear space over $\Phi$ and let $p$ be a seminorm on $V$. Then

(i) $p(0) = 0$,

(ii) $p(x) \geq 0$,

(iii) $p(x - y) \geq |p(x) - p(y)|$ $(x,y \in V)$.

Proof. $p(0) = p(0x) = |0|p(x) = 0$, $x \in V$, and part (i) follows. Clearly part (ii) is a consequence of part (iii). But, for any $x,y \in V$, the subadditivity of $p$ reveals that

$$p(x) = p(x - y + y) \leq p(x - y) + p(y),$$

and hence $p(x) - p(y) \leq p(x - y)$. However, from the homogeneity property of $p$ we deduce that

$$p(x - y) = p[-(y - x)] = p(y - x) \geq p(y) - p(x),$$

and so $p(x - y) \geq |p(x) - p(y)|$. □

We could, of course, have included the results of Proposition 1.1.1 as part of the definition of a seminorm, and this is done by some writers.

Definition 1.1.5. Let $V$ be a linear space over $\Phi$ and let $p$ be a seminorm on $V$. If $p(x) = 0$ implies $x = 0$, then $p$ is said to be a norm on $V$.

Definition 1.1.6. Let $V$ be a linear space over $\Phi$. Then $V$ is said to be a seminormed linear space if there exists a family $P = \{p_\beta \mid \beta \in \Lambda\}$ of seminorms on $V$ such that $p_\beta(x) = 0$, $\beta \in \Lambda$, implies $x = 0$. If $V$ is a seminormed linear space and $P = \{p\}$, that is, $P$ is a singleton set, then $V$ is said to be a normed linear space.

Evidently the condition on the family of seminorms $P = \{p_\beta \mid \beta \in \Lambda\}$ required for $V$ to be a seminormed linear space is a replacement for the positive definiteness displayed by a norm. It will become clear in the sequel how this property of $\{p_\beta\}$ is utilized.

In the case that $V$ is a normed linear space, we shall generally call $p(x)$ the norm of $x$ and write $p(x) = \|x\|$. It is then easily seen that the equation $\rho(x,y) = \|x - y\|$, $x,y \in V$, defines a metric $\rho$ on $V$ and that a net $\{x_\alpha\} \subset V$ converges in this metric topology to $x \in V$ if and only if $\lim_\alpha \|x_\alpha - x\| = 0$. The details of these assertions are left to the reader. We shall generally refer to this metric topology as the norm topology. The next definition now clearly makes sense.

Definition 1.1.7. A normed linear space $V$ over $\Phi$ is said to be a Banach space if it is a complete metric space with the metric $\rho(x,y) = \|x - y\|$, $x,y \in V$.

Since there may be many families of seminorms under which a given linear space is a seminormed linear space, we shall write these spaces as pairs $(V,P)$, where $P$ is the relevant family of seminorms. When $P = \{p\}$, that is, $V$ is a normed linear space, we shall generally write $(V,\|\cdot\|)$.

Before we turn to some examples, we wish to state two theorems concerning normed linear spaces. The proofs are straightforward and are left to the reader.

Theorem 1.1.1. Let $(V,\|\cdot\|)$ be a normed linear space over $\Phi$.

(i) If $W \subset V$ is a linear subspace, then the closure of $W$ in $V$ is a linear subspace of $V$.

(ii) If $W \subset V$ is a linear subspace, then $(W,\|\cdot\|)$ is a normed linear space. If $(V,\|\cdot\|)$ is a Banach space and $W$ is a closed linear subspace, then $(W,\|\cdot\|)$ is a Banach space.

(iii) If $W \subset V$ is a closed linear subspace, then the quotient space $V/W$ is a normed linear space with the norm

$$|||x + W||| = \inf_{y \in W} \|x + y\| \qquad (x \in V).$$

If $(V,\|\cdot\|)$ is a Banach space and $W$ is a closed linear subspace, the $(V/W, |||\cdot|||)$ is a Banach space.

(iv) There exists a Banach space $(V_1,\|\cdot\|_1)$ over $\Phi$ and a mapping $\varphi : V \to V_1$ such that

(a) $\varphi$ is an isomorphism.

(b) $\varphi(V)$ is dense in $(V_1,\|\cdot\|_1)$.

(c) $\|\varphi(x)\|_1 = \|x\| \qquad (x \in V)$.

Moreover, if $(V_2,\|\cdot\|_2)$ is another Banach space that satisfies properties (a), (b), and (c), then there exists a mapping $\psi : V_1 \to V_2$ that is a surjective isomorphism such that $\|\psi(x)\|_2 = \|x\|_1$, $x \in V_1$.

The space $(V_1,\|\cdot\|_1)$ described in Theorem 1.1.1(iv) is, of course, called the completion of $V$. Clearly Theorem 1.1.1(i) also remains valid for any seminormed linear space $(V,P)$.

Theorem 1.1.2. Let $(V_1,\|\cdot\|_1),(V_2\|\cdot\|_2),\ldots,(V_n,\|\cdot\|_n)$ be normed linear spaces over $\Phi$ and let $V$ denote the topological product $V_1 \times V_2 \times\cdots\times V_n$ of the topological spaces $V_1,V_2,\ldots,V_n$, with the respective norm topologies and with linear space addition and scalar multiplication defined componentwise. Then $(V,\|\cdot\|)$ is a normed linear space over $\Phi$ such that the norm topology is equivalent to the product topology if $\|\cdot\|$ is defined as any of the following:

(i) $\|x\| = \sup_{k=1,2,\ldots,n} \|x_k\|_k$,

(ii) $\|x\| = \left[\sum_{k=1}^{n} (\|x_k\|_k)^p\right]^{1/p}$ $(1 \leq p < \infty)$,

where $x = (x_1, x_2, \ldots, x_n)$. If $(V_1, \|\cdot\|_1), (V_2, \|\cdot\|_2), \ldots, (V_n, \|\cdot\|_n)$ are Banach spaces, then so is $(V, \|\cdot\|)$.

It should be noted that the topological product of infinitely many normed linear spaces cannot be provided with a norm for which the norm topology is equivalent with the product topology (see, for example, [K, p. 150]).

1.2. Examples of Seminormed Linear Spaces. In this section we give a number of examples of seminormed and normed linear spaces. No proofs are provided for the various assertions. Some proofs will appear in later chapters, whereas others are left for the reader to establish.

Example 1.2.1. Let $X$ be a Hausdorff topological space. By $C(X)$, $C_o(X)$, and $C_c(X)$ we denote, respectively, the linear spaces over $\mathbb{C}$ of all continuous complex-valued functions on $X$ that are bounded, vanish at infinity, or have compact support. The linear space operations are the usual ones of pointwise addition and scalar multiplication. If $f$ is an element of any of these spaces, then we set

$$\|f\| = \|f\|_\infty = \sup_{t \in X} |f(t)|.$$

Then $(C(X), \|\cdot\|_\infty)$ and $(C_o(X), \|\cdot\|_\infty)$ are Banach spaces, whereas $(C_c(X), \|\cdot\|_\infty)$ is a normed linear space. Clearly if $X$ is compact, then $C(X) = C_o(X) = C_c(X)$, and $(C_c(X), \|\cdot\|_\infty)$ is a Banach space only in this case.

If $X$ is noncompact, then we denote by $C'(X)$ all continuous complex-valued functions on $X$. Then clearly $\|\cdot\|_\infty$ no longer defines a norm on $C'(X)$. However, if for each compact set $K \subset X$ we set

$$p_K(f) = \sup_{t \in K} |f(t)| \qquad (f \in C'(X)),$$

then $P = \{p_K \mid K \subset X, K \text{ compact}\}$ is a family of seminorms on $C'(X)$ for which $(C'(X),P)$ is a seminormed linear space. Moreover, it can be shown that $C'(X)$ is never a normed linear space when $X$ is noncompact.

We shall denote by $C^R(X)$, $C_o^R(X)$, and $C_c^R(X)$ the real parts of the functions in $C(X)$, $C_o(X)$, and $C_c(X)$, respectively. Obviously these are linear spaces over $\mathbb{R}$ with the same properties as the analogous spaces of complex functions. They are equivalently the spaces of continuous real-valued functions on $X$ that are bounded, vanish at infinity, or have compact support.

Example 1.2.2. Let $X$ be any set. We denote by $B(X)$ the linear space under pointwise operations of all bounded complex-valued functions defined on $X$. The linear space $B(X)$ is a Banach space with the norm

$$\|f\|_\infty = \sup_{t \in X} |f(t)| \qquad (f \in B(X)).$$

If $X$ is a locally compact Hausdorff topological space, then it is evident that $C(X)$ is a linear subspace of $B(X)$.

Example 1.2.3. Let $a < b$ and let $n$ be a nonnegative integer. We denote by $C^n([a,b])$ the linear space of n-times continuously differentiable real-valued functions on $[a,b]$. If we define

$$\|f\|_n = \sum_{k=0}^{n} \|f^{(k)}\|_\infty \qquad (f \in C^n([a,b])),$$

where $f^{(k)}$ denotes the kth derivative of $f$, then $(C^n([a,b]),\|\cdot\|_n)$ is a Banach space over $\mathbb{R}$.

Furthermore, set $C^\infty([a,b]) = \cap_{n=0}^\infty C^n([a,b])$. Then $C^\infty([a,b])$

is the linear space of real-valued functions on $[a,b]$ with derivatives of all orders. Set $p_n(f) = \|f\|_n$, $n = 0,1,2,\ldots,$ and $P = \{p_n \mid n = 0,1,2,\ldots\}$. Then $(C^\infty([a,b]), P)$ is a seminormed linear space. It cannot be made into a normed linear space. We shall prove this in Section 2.4. It is, however, the case that the seminorms $\{p_n\}$ can be used to define a complete metric on $C^\infty([a,b])$, namely, the metric $\rho$ defined as

$$\rho(f,g) = \sum_{n=0}^{\infty} \frac{p_n(f-g)}{2^n[1 + p_n(f-g)]} \qquad (f,g \in C^\infty([a,b])).$$

An alternative family of seminorms for $C^\infty([a,b])$ is given by $P = \{p_n \mid n = 0,1,2,\ldots\}$, where $p_n(f) = \|f^{(n)}\|_\infty$.

Example 1.2.4. Let $(X,S,\mu)$ be a positive-measure space. For $0 < p < \infty$, let $L_p(X,S,\mu) = L_p(X,\mu) = L_p(\mu)$ stand for the linear space of equivalence classes of $\mu$-measurable complex-valued functions on $X$ whose pth powers are integrable. The equivalence classes are, of course, determined by equality almost everywhere with respect to the measure $\mu$. If $1 \leq p < \infty$ and

$$\|f\|_p = [\int_X |f(t)|^p \, d\mu(t)]^{1/p} \qquad (f \in L_p(\mu)).$$

then $(L_p(\mu), \|\cdot\|_p)$ is a Banach space. For $0 < p < 1$, the spaces $L_p(\mu)$ are neither normed nor seminormed linear spaces. We shall prove this in Section 4.2.

By $L_\infty(X,S,\mu) = L_\infty(X,\mu)) = L_\infty(\mu)$ we shall denote the linear space of all equivalence classes of essentially bounded, $\mu$-measurable, complex-valued functions on $X$. If

$$\|f\|_\infty = \operatorname*{ess\,sup}_{t \in X} |f(t)| = \inf \{M \mid \mu\{t \mid |f(t)| > M\} = 0\}$$

then $(L_\infty(\mu), \|\cdot\|_\infty)$ is a Banach space.

Example 1.2.5. Suppose $X = \{n \mid n = 1,2,3,\ldots\}$, $S$ is the σ-algebra of all subsets of $X$, and $\mu$ is the counting measure; that is, $\mu(E)$ = number of points in $E$ if $E$ is finite, and $\mu(E) = \infty$ if $E$ is infinite. Then the spaces $L_p(\mu)$, $0 < p \leq \infty$, are linear spaces of complex sequences $\{a_n\}$ such that either $\Sigma_{n=1}^{\infty} |a_n|^p < \infty$ or $\sup_n |a_n| < \infty$. In this case we generally denote these spaces by $\ell_p$, $0 \leq p < \infty$. Clearly $\ell_p$, $1 \leq p \leq \infty$, are Banach spaces with the appropriate norms. Furthermore, it is evident that $\ell_\infty = C(X)$, and that $C_o(X)$ is a closed linear subspace of $\ell_\infty$ and hence a Banach space. It is obviously the space of all complex sequences that converge to zero, and we denote it by $c_o$. The collection of all convergent complex sequences is also easily seen to be a closed linear subspace of $\ell_\infty$ and hence a Banach space. It will be denoted by c.

Example 1.2.6. Let $X$ be a locally compact Hausdorff topological space. Then $M(X)$ denotes the linear space of all bounded, complex-valued, regular Borel measures on $X$. If for $\mu \in M(X)$ we set

$$\|\mu\| = |\mu|(X) = \sum_{k=1}^{n} |\mu(E_k)|,$$

where the sup is taken over all finite partitions of $X$ by Borel sets $E_k$, then $(M(X), \|\cdot\|)$ is a Banach space. $\|\mu\| = |\mu|(X)$ is also called the total variation of $\mu$.

Furthermore, we note that if $\nu$ is a given nonnegative regular Borel measure on $X$, not necessarily bounded, and $f \in L_1(X,\nu)$, then the formula

$$\mu_f(E) = \int_E f(t)d\nu(t),$$

where $E \subset X$ is a Borel set, defines an element $\mu_f \in M(X)$. Moreover, $\|\mu_f\| = \|f\|_1$.

It is necessary to note at this juncture that a complex-valued regular Borel measure need not be bounded; that is, $|\mu|(X)$ need not be finite. Indeed, for us such a measure $\mu$ is always of the form $\mu = \mu_1 + i\mu_2$, where $\mu_k$, $k = 1,2$, are signed regular Borel measures. Thus, for example, if $X = \mathbb{R}$ and $\nu$ denotes Lebesgue measure, then $i\nu$ is a complex-valued regular Borel measure that is not bounded. The reader should compare this with [$\mathrm{Ru}_2$, pp. 117-120].

Example 1.2.7. Let $E(\mathbb{C})$ denote the linear space of all entire functions. For each compact set $K \subset \mathbb{C}$, let

$$p_K(f) = \sup_{t \in K} |f(t)| \qquad (f \in E(\mathbb{C})).$$

If $P = \{p_K \mid K \subset \mathbb{C},\ K \text{ compact}\}$, then $(E(\mathbb{C}), P)$ is a seminormed linear space. It cannot be made into a normed linear space.

Example 1.2.8. Let $D$ be the closed unit disk in the complex plane and let $A(D)$ denote the linear space of all complex-valued functions on $D$ that are continuous on $D$ and analytic on the interior of $D$. Then $(A(D), \|\cdot\|)$ is a Banach space, where as usual $\|f\|_\infty = \sup_{t \in D} |f(t)|$.

Example 1.2.9. Let $V$ be either $\mathbb{R}^n$ or $\mathbb{C}^n$, that is, either real or complex n-dimensional space. For $1 \leq p < \infty$, set

$$\|x\|_p = \left[\sum_{k=1}^{n} |x_k|^p\right]^{1/p}$$

and for $p = \infty$

$$\|x\|_\infty = \sup_{k=1,2,\ldots,n} |x_k| \qquad (x = (x_1, x_2, \ldots, x_n) \in V).$$

Then $(V, \|\cdot\|_p)$ is a Banach space, $1 \leq p \leq \infty$. When $p = 2$ and $V = \mathbb{R}^n$, then, of course, we have the usual real Euclidean n-space. It is instructive to draw some sketches of the sets $\{x \mid \|x\|_p \leq 1\}$ in $\mathbb{R}^2$ for various values of $p$.

We shall discuss finite-dimensional normed linear spaces more fully in the next section.

Example 1.2.10. Let $V$ be a linear space over $\Phi$ and suppose that there is a mapping $\psi : V \times V \to \Phi$ for which

(i) $\psi(ax + by, z) = a\psi(x,z) + b\psi(y,z)$.

(ii) $\psi(x,y) = \overline{\psi(y,x)}$,

where the bar denotes complex conjugation.

(iii) $\psi(x,x) \geq 0$.

(iv) $\psi(x,x) = 0$ implies $x = 0$ $\qquad (x,y,z \in V;\ a,b \in \Phi)$.

Then, on setting $\|x\| = [\psi(x,x)]^{1/2}$, it can be shown that $(V, \|\cdot\|)$ is a normed linear space.

The mapping $\psi$ is called an inner, or scalar, product, and the spaces $V$ together with an inner product $\psi$ are called inner-product spaces. If these spaces, when normed as above, are Banach spaces, then they are said to be Hilbert spaces. We shall investigate such spaces in some detail in Chapter 13, and shall also refer to them in the intervening pages.

It is apparent that the spaces $\mathbb{R}^n$ and $\mathbb{C}^n$ are inner-product spaces with the usual inner product,

$$\psi(x,y) = \sum_{k=1}^{n} x_k \overline{y}_k,$$

$x = (x_1, x_2, \ldots, x_n)$, $y = (y_1, y_2, \ldots, y_n)$. The norm obtained from this inner product is clearly $\|\cdot\|_2$, and so the spaces are Hilbert spaces. An example of an inner-product space that is not a Banach space is obtained by letting $V = C([0,1])$ and defining

$$\psi(f,g) = \int_0^1 f(t)\overline{g(t)}\,dt \qquad (f,g \in C([0,1])).$$

For discussions and proofs of the foregoing material, as well as much more, the reader is referred to [BaNr, pp. 7-11; Da, pp. 28-31;

$DS_1$, Chapter IV; HSt, Chapter 4; K, pp. 130-144; Ry, pp. 111-119; T, pp. 87-94, 102-104, 106-109; Y, pp. 26-42].

1.3. Finite-Dimensional Normed Linear Spaces and a Theorem of Riesz. In this section we wish to prove some basic results about finite-dimensional normed linear spaces -- in particular, that they are all essentially either $(\mathbb{R}^n, \|\cdot\|_1)$ or $(\mathbb{C}^n, \|\cdot\|_1)$. We shall also see that in infinite-dimensional normed linear spaces there must always exist closed norm-bounded sets that are not compact in the metric topology induced by the norm of the space. To prove this we shall have need of a theorem of F. Riesz. We begin with a few basic definitions.

Definition 1.3.1. Let $(V_1, \|\cdot\|_1)$ and $(V_2, \|\cdot\|_2)$ be normed linear spaces. Then $(V_1, \|\cdot\|_1)$ and $(V_2, \|\cdot\|_2)$ are said to be topologically isomorphic if there exists a surjective isomorphism $\varphi : V_1 \to V_2$ for which there exist $M > 0$ and $m > 0$ such that

$$m\|x\|_1 \leq \|\varphi(x)\|_2 \leq M\|x\|_1 \qquad (x \in V_1).$$

If $V_1 = V_2$ and $(V_1, \|\cdot\|_1$ and $(V_2, \|\cdot\|_2)$ are topologically isomorphic under the identity mapping $\varphi(x) = x$, then $\|\cdot\|_1$ and $\|\cdot\|_2$ are said to be equivalent norms on $V_1 = V_2$.

We note that a topological isomorphism between two normed linear spaces is a mapping that is simultaneously an isomorphism of the linear spaces and a homeomorphism of the corresponding metric spaces. That this description of a topological isomorphism is equivalent to that given in Definition 1.3.1 will be apparent when we prove that a linear mapping between two normed linear spaces is continuous if and only if it is bounded (Theorem 3.2.1).

It is evident that the relation of topological isomorphism between normed linear spaces is reflexive, symmetric and transitive.

Theorem 1.3.1. Let $(V,\|\cdot\|)$ be an n-dimensional normed linear space over $\mathbb{R}(\mathbb{C})$. Then $(V,\|\cdot\|)$ is topologically isomorphic to $(\mathbb{R}^n,\|\cdot\|_1)$ $((\mathbb{C}^n,\|\cdot\|_1))$.

Proof. We shall give the proof only for the case of real scalars. Let $x_1,x_2,\ldots,x_n \in V$ be a basis for V. Then for every $x \in V$ there exist unique $a_1,a_2,\ldots,a_n \in \mathbb{R}$ such that $x = \Sigma_{k=1}^n a_k x_k$, and clearly every n-tuple $(a_1,a_2,\ldots,a_n) \in \mathbb{R}^n$ determines an element of V. Hence, with some elementary calculations, it is apparent that $\varphi[(a_1,a_2,\ldots,a_n)] = \Sigma_{k=1}^n a_k x_k$ defines a surjective isomorphism from $\mathbb{R}^n$ to V. Moreover, by the triangle inequality, it is clear that

$$\|\varphi[(a_1,a_2,\ldots,a_n)]\| \le M\left[\sum_{k=1}^n |a_k|\right] \qquad ((a_1,a_2,\ldots,a_n) \in \mathbb{R}^n),$$

where $M = \sup_{k=1,2,\ldots,n}\|x_k\|$. Thus to complete the proof it remains only to show the existence of an $m > 0$ such that

$$m\|(a_1,a_2,\ldots,a_n)\|_1 \le \left\|\sum_{k=1}^n a_k x_k\right\| \qquad ((a_1,a_2,\ldots,a_n) \in \mathbb{R}^n).$$

If $a_k = 0$, $k = 1,2,\ldots,n$, then this is clearly the case for any $m$, and we may assume without loss of generality that not all the $a_k$ terms are zero. Furthermore, suppose we have obtained some $m$ such that the desired inequality is valid for all $(a_1,a_2,\ldots,a_n) \in \mathbb{R}^n$ for which $\Sigma_{k=1}^n |a_k| = 1$. Then for any $(b_1,b_2,\ldots,b_n) \in \mathbb{R}^n$ such that $b = \Sigma_{k=1}^n |b_k| > 0$ we would have

$$\frac{m\left[\sum_{k=1}^n |b_k|\right]}{b} \le \left\|\frac{\sum_{k=1}^n b_k x_k}{b}\right\| = \left\|\frac{\sum_{k=1}^n b_k x_k}{b}\right\|$$

since $\Sigma_{k=1}^n |b_k/b| = 1$ and $\|\cdot\|$ is a seminorm. Hence

$$m\|(b_1,b_2,\ldots,b_n)\|_1 \le \|\varphi[(b_1,b_2,\ldots,b_n)]\|$$

for all $(b_1,b_2,\ldots,b_n) \in \mathbb{R}^n$.

Now to see that the desired estimate is indeed valid for some $m > 0$ and all $(a_1,a_2,\ldots,a_n) \in \mathbb{R}^n$ such that $\Sigma_{k=1}^n |a_k| = 1$, we consider the nonnegative function $f$ defined on $\mathbb{R}^n$ by $f[(a_1,a_2,\ldots,a_n)] = \|\Sigma_{k=1}^n a_k x_k\|$. From Proposition 1.1.1(iii) we see that, if $(a_1,a_2,\ldots,a_n) \in \mathbb{R}^n$ and $(b_1,b_2,\ldots,b_n) \in \mathbb{R}^n$, then

$$\begin{aligned} |f[(a_1,a_2,\ldots,a_n)] - f[(b_1,b_2,\ldots,b_n)]| &= \left| \left\| \sum_{k=1}^n a_k x_k \right\| - \left\| \sum_{k=1}^n b_k x_k \right\| \right| \\ &\leq \left\| \sum_{k=1}^n (a_k - b_k) x_k \right\| \\ &\leq M \sum_{k=1}^n |a_k - b_k|, \end{aligned}$$

from which we conclude that $f$ is continuous on $(\mathbb{R}^n, \|\cdot\|_1)$. However, elementary arguments (based, for example, on the compactness of the closed interval $[-1,1]$ in $\mathbb{R}$) reveal that

$$\{(a_1,a_2,\ldots,a_n) \mid \sum_{k=1}^n |a_k| = 1\}$$

is a compact subset of $(\mathbb{R}^n, \|\cdot\|_1)$. Therefore $f$, being a continuous function, must attain a minimum value $m \geq 0$ on this compact set. $m \neq 0$ because, if $m = 0$, then there would exist

$$(a_1,a_2,\ldots,a_n) \in \mathbb{R}^n, \quad \sum_{k=1}^n |a_k| = 1$$

such that $\Sigma_{k=1}^n a_k x_k = 0$, as $\|\cdot\|$ is a norm, thereby contradicting the choice of $x_1,x_2,\ldots,x_n$ as a basis for $V$. Consequently, $m > 0$, and

$$m\|(a_1,a_2,\ldots,a_n)\|_1 \leq \|\varphi[(a_1,a_2,\ldots,a_n)]\| \qquad ((a_1,a_2,\ldots,a_n) \in \mathbb{R}^n),$$

which completes the proof. □

<u>Corollary 1.3.1</u>. Let $(V_1,\|\cdot\|_1)$ and $(V_2,\|\cdot\|_2)$ be n-dimensional normed linear spaces over $\Phi$. Then $(V_1,\|\cdot\|_1)$ and $V_2,\|\cdot\|_2)$ are topologically isomorphic.

Proof. Both spaces are topologically isomorphic with either $(\mathbb{R}^n, \|\cdot\|_1)$ or $(\mathbb{C}^n, \|\cdot\|_1)$. □

Corollary 1.3.2. Let $V$ be any n-dimensional linear space over $\Phi$ and suppose the norms $\|\cdot\|_1$ and $\|\cdot\|_2$ are such that $(V, \|\cdot\|_1)$ and $(V, \|\cdot\|_2)$ are normed linear spaces. Then the norms $\|\cdot\|_1$ and $\|\cdot\|_2$ are equivalent.

Thus the metric topologies on finite-dimensional normed linear spaces that are determined by the norms are all equivalent. In particular, the norms $\|\cdot\|_p$, $1 \leq p \leq \infty$, on $\mathbb{R}^n$ or $\mathbb{C}^n$ are all equivalent.

Easy arguments, based on the preceding results and the analogous assertions for the spaces $\mathbb{R}^n$ and $\mathbb{C}^n$, yield the next corollary. The details are left to the reader.

Definition 1.3.2. Let $(V,P)$ be a seminormed linear space over $\Phi$. A subset $E$ of $V$ is said to be bounded if for each $p \in P$ there exists some $M_p > 0$ such that $p(x) \leq M_p$, $x \in E$.

Corollary 1.3.3. Let $(V, \|\cdot\|)$ be a finite-dimensional normed linear space over $\Phi$. Then

(i) $(V, \|\cdot\|)$ is a Banach space.

(ii) If $W$ is a linear subspace of $V$, then $(W, \|\cdot\|)$ is a closed linear subspace of $(V, \|\cdot\|)$.

(iii) If $E \subset V$ is a closed bounded set, then $E$ is compact.

The compactness referred to in Corollary 1.3.3(iii) is, of course, in the norm topology. It is of some interest to note that the compactness property of every closed bounded set characterizes the finite-dimensional normed linear spaces. We shall prove this shortly, but first we must establish a preliminary result of independent interest.

Theorem 1.3.2 (F. Riesz). Let $(V,\|\cdot\|)$ be a normed linear space over $\Phi$ and suppose $(W,\|\cdot\|)$ is a closed proper linear subspace of $(V,\|\cdot\|)$ and $0 < a < 1$. Then there exists some $x_a \in V$ such that $\|x_a\| = 1$ and $\inf_{x \in W}\|x - x_a\| \geq a$.

Proof. Since $W$ is a proper subspace of $V$, there is some $y \in V \sim W$. Let $d = \inf_{x \in W}\|x - y\|$. Then $d > 0$, as if $d = 0$, then there exists a sequence $\{x_n\} \subset W$ for which $\lim_n\|x_n - y\| = 0$. But since $W$ is closed, this implies that $y \in W$, contrary to the choice of $y$.

Now $d/a > d$, as $0 < a < 1$; thus by the definition of $d$ there exists some $x_o \in W$ such that $0 < \|x_o - y\| < d/a$. Define $x_a = (y - x_o)/\|x_o - y\|$. Clearly $\|x_a\| = 1$. Moreover, for each $x \in W$

$$\begin{aligned}\|x - x_a\| &= \left\|x - \frac{(y - x_o)}{\|x_o - y\|}\right\| \\ &= \frac{\|\,\|x_o - y\|x + x_o - y\|}{\|x_o - y\|} \\ &\geq \frac{d}{\|x_o - y\|} \\ &> a,\end{aligned}$$

since $\|x_o - y\|\, x + x_o \in W$, as $W$ is a linear space, and $\|x_o - y\| < d/a$.

Hence $\inf_{x \in W}\|x - x_a\| \geq a$. □

Geometrically the preceding result says that, given a proper closed linear subspace of a normed linear space, one can always find some point on the surface of the closed ball of unit radius about the origin whose distance from the subspace is less than unity but as close to unity as we please. It is not generally possible, even for Banach spaces, to find a point that is a unit distance from the given subspace; that is, one cannot generally take $a = 1$ in the statement of the theorem.

As an example let $V = \{f | f \in C^R([0,1]),\ f(0) = 0\}$. Then $(V, \|\cdot\|_\infty)$ is a Banach space over $\mathbb{R}$. Let $W = \{f | f \in V,\ \int_0^1 f(t)dt = 0\}$ It is easily seen that $(W, \|\cdot\|_\infty)$ is a proper closed linear subspace of $(V, \|\cdot\|)$. But suppose there is some $f_1 \in V$ such that $\|f_1\|_\infty = 1$ and $\inf_{f \in W} \|f - f_1\|_\infty \geq 1$. Then for each $g \in V \sim W$ define

$$b_g = \frac{\int_0^1 f_1(t)dt}{\int_0^1 g(t)dt} .$$

Clearly $f_1 - b_g g \in W$, and hence

$$\begin{aligned} 1 &\leq \|f_1 - (f_1 - b_g g)\|_\infty \\ &= \|b_g g\|_\infty . \end{aligned}$$

Thus, referring to the definition of $b_g$, we conclude that

$$\left|\int_0^1 g(t)dt\right| \leq \left|\int_0^1 f_1(t)dt\right| \|g\|_\infty \qquad (g \in V \sim W).$$

However, it is apparent that $g_n(t) = t^{1/n}$, $n = 1,2,3,\ldots,$ belongs to $V \sim W$, and so

$$\begin{aligned} \frac{n}{n+1} &= \left|\int_0^1 g_n(t)dt\right| \\ &\leq \left|\int_0^1 f_1(t)dt\right| \|g_n\|_\infty \\ &\leq \left|\int_0^1 f_1(t)dt\right| \qquad (n = 1,2,3,\ldots), \end{aligned}$$

from which it follows at once that $1 \leq \left|\int_0^1 f_1(t)dt\right|$. But, on the other hand, it is easily seen that $\|f_1\|_\infty = 1$, $f_1(0) = 0$, implies that $\left|\int_0^1 f_1(t)dt\right| < 1$.

This contradiction shows that an $f_1$ with the desired properties cannot exist.

Now we can establish the result indicated before Theorem 1.3.2.

Theorem 1.3.3. Let $(V, \|\cdot\|)$ be a normed linear space over $\Phi$. Then the following are equivalent:

(i) $(V, \|\cdot\|)$ is finite dimensional.

(ii) If $E \subset V$ is a closed bounded set, then $E$ is compact.

Proof. The implication from part (i) to part (ii) is Corollary 1.3.3(iii). Conversely, suppose that part (ii) holds and that $V$ is not finite dimensional. Then there exists an infinite sequence $\{y_k\} \subset V$ of linearly independent vectors. Let $W_k$ be the linear subspace of $V$ spanned by the vectors $y_1, y_2, \ldots, y_k$, $k = 1,2,3,\ldots$ . Then from Corollary 1.3.3(ii) we see that each $W_{k-1}$ is a proper closed linear subspace of $W_k$, $k = 2,3,4,\ldots,$ and hence by Theorem 1.3.2 there exists some $x_k \in W_k \sim W_{k-1}$ such that $\|x_k\| = 1$ and

$$\inf_{x \in W_{K-1}} \|x - x_k\| \geq 1/2, \qquad (k = 2,3,4,\ldots).$$

However, it is evident that $\{x \mid x \in V, \|x\| = 1\}$ is closed and bounded, and hence compact by (ii). Consequently $\{x_k\}$ must have a convergent subsequence. This, however, contradicts the fact that $\|x_n - x_k\| \geq 1/2$, $n \neq k$, and thus $V$ must be finite dimensional. □

We see then that in an infinite-dimensional normed linear space there must always exist some closed bounded set that is not compact in the norm topology. Thus the familiar Euclidean characterization of compact sets as closed bounded sets is not generally valid for normed linear spaces, nor is it valid for general metric spaces. In the Banach-Alaoglu Theorem (Theorem 9.4.1) we shall, however, obtain a substitute for the Euclidean characterization.

It is also worthwhile remarking that one need not assume that all closed bounded sets in a normed linear space $V$ are compact to conclude that the space is finite dimensional. A simple modification of the proof of Theorem 1.3.3 shows that it suffices to assume only that $\{x \mid x \in V, \|x\| \leq 1\}$ is compact.

1.4. Gauges and Seminorms. In this section we introduce the concept of the gauge of certain types of convex subsets of linear spaces and establish some basic relations between gauges, convex sets, and seminorms. Gauges will play an important role in the study of the connection between seminormed linear spaces and locally convex topological linear spaces, to be discussed in the next chapter. Once again we begin with some definitions.

Definition 1.4.1. Let $V$ be a linear space over $\Phi$. Then $E \subset V$ is said to be absorbing if for each $x \in V$ there exists some $a > 0$ such that $x \in aE$; equivalently there exists some $a' > 0$ such that $a'x \in E$.

Clearly, if $E$ is absorbing, then $0 \in E$.

Definition 1.4.2. Let $V$ be a linear space over $\Phi$ and let $p$ be a seminorm on $V$. For each $\lambda > 0$ we define

$$B_\lambda = \{x \mid x \in V, p(x) \leq \lambda\},$$

and

$$B_\lambda^o = \{x \mid x \in V, p(x) < \lambda\}$$

It is easily verified that $B_\lambda = \lambda B_1$ and $B_\lambda^o = \lambda B_1^o$. The sets $B_\lambda$ and $B_\lambda^o$ are examples of absorbing sets. More specifically we have the following proposition:

Proposition 1.4.1. Let $V$ be a linear space over $\Phi$ and let $p$ be a seminorm on $V$. Then for each $\lambda > 0$ the sets $B_\lambda$ and $B_\lambda^o$ are convex balanced absorbing sets.

Proof. The fact that $B_\lambda$ and $B_\lambda^o$ are convex balanced sets follows immediately from the definition of a seminorm. To show that $B_\lambda$ and $B_\lambda^o$ are absorbing it is thus sufficient to show that for each $x \in V$ there exists some $a > 0$ such that $ax \in B_\lambda$ or $ax \in B_\lambda^o$, respectively; that is, $p(ax) \leq \lambda$ or $p(ax) < \lambda$. Clearly, any $a > 0$ for which $1/a > p(x)/\lambda$ will suffice. $\square$

In particular, the sets $B_1$ and $B_1^o$, which one can think of as the closed and open unit balls about the origin of $V$ determined by $p$, are convex balanced absorbing sets. It would obviously be of interest to know whether, given a convex balanced absorbing set $B$ in a linear space $V$, there is some seminorm $p$ on $V$ for which either $B_1 = B$ or $B_1^o = B$. We shall see that this is a bit too much to ask for. Nevertheless, one can get reasonably close to such a result, as evidenced by Theorem 1.4.1. First, however, we define the gauge of a convex balanced absorbing set and discuss some of its properties.

*Definition 1.4.3.* Let $V$ be a linear space over $\Phi$ and let $B \subset V$ be a convex balanced absorbing set. For each $x \in V$ define

$$q(x) = \inf\{a \mid a > 0,\ x \in aB\}.$$

Then $q$ is said to be the *gauge* of $B$

The gauge $q$ is also called the *Minkowski functional* of $B$.

*Proposition 1.4.2.* Let $V$ be a linear space over $\Phi$ and let $B \subset V$ be a convex balanced absorbing set. Then the gauge of $B$ is a seminorm on $V$.

*Proof.* If the gauge of $B$ is $q$, then clearly $0 \leq q(x) < \infty$, $x \in V$, and $q(0) = 0$. Furthermore, for each $x \in V$ there is some $a > 0$ for which $x \in aB$, as $B$ is absorbing. Since $B$ is balanced, it follows readily that $baB$ is also balanced for each $b \in \Phi$.. In particular, we deduce that $baB = |b|aB$, $b \in \Phi$. Hence, since $bx \in baB = |b|aB$, we conclude from the definition of $q$ that $q(bx) \leq |b|a$, $b \in \Phi$. However, this estimate is valid for any $a > 0$ for which $x \in aB$, and so once again from the definition of $q$ we deduce that $q(bx) \leq |b|q(x)$, $b \in \Phi$. Utilizing this result, we see that for each $x \in V$

$$q(x) = q(\frac{bx}{b}) \leq \frac{q(bx)}{|b|} \qquad (b \in \Phi;\ b \neq 0),$$

whence $q(bx) = |b|q(x)$, $b \in \Phi$, $b \neq 0$. Since $q(0) = 0$, we can drop the restriction that $b \neq 0$ and conclude that $q$ satisfies the homogeneity property of a seminorm.

Finally, to show that $q$ is subadditive, we consider any $x,y \in V$. Since $B$ is absorbing, there are some $a > 0$ and $b > 0$ such that $x \in aB$ and $y \in bB$. Let us write $x = au$ and $y = bv$, where $u,v \in B$. Consequently, because $B$ is convex, we have

$$\frac{x + y}{a + b} = \frac{a}{a + b} u + \frac{b}{a + b} v$$

belongs to $B$, that is, $x + y \in (a + b)B$. Thus $q(x + y) \leq a + b$. Since this estimate also remains valid for any $a$ and $b$ such that $x \in aB$ and $y \in bB$, we again conclude that $q(x + y) \leq q(x) + q(y)$.

Therefore $q$ is a seminorm on $V$. □

Thus we see that every convex balanced absorbing set $B$ determines a seminorm $q$, the gauge of $B$. The best we can say about the connection between $B$ and the closed and open unit balls determined by its gauge is contained in the next theorem. We include Proposition 1.4.1 as a portion of the statement of this theorem for the purposes of comparison.

Theorem 1.4.1. Let $V$ be a linear space over $\Phi$.

(i) If $p$ is a seminorm on $V$, then $B_1$ and $B_1^o$ are convex balanced absorbing sets in $V$.

(ii) If $B \subset V$ is a convex balanced absorbing set, then there exists a unique seminorm $p$ on $V$ such that $B_1^o \subset B \subset B_1$.

Proof. We need only prove implication (ii). As the seminorm $p$ we take the gauge of $B$. Let $x \in B_1^o$. Then $p(x) < 1$, and so, by the definition of the gauge of $B$ and the fact that $B$ is balanced, there exists some $a$, $0 < a < 1$, such that $x \in aB \subset B$. Hence $B_1^o \subset B$. On the other hand, if $x \in B_1$, then

$$p(x) = \inf\{a \mid a > 0,\ x \in aB\} \leq 1,$$

and so $B \subset B_1$.

Finally, suppose $p_1$ is another seminorm on $V$ for which

$$\{x \mid p_1(x) < 1\} \subset B \subset \{x \mid p_1(x) \leq 1\}.$$

Then, in particular, we would have

$$\{x \mid p(x) < 1\} \subset \{x \mid p_1(x) \leq 1\}.$$

Suppose that there is some $x \in V$ for which $p(x) < p_1(x)$. Then there is some $b > 0$ such that $p(x) < b < p_1(x)$. Consequently $p(x/b) = p(x)/b < 1$, and hence $p_1(x/b) \leq 1$. But $p_1(x/b) = p_1(x)/b > 1$, a contradiction. Therefore, $p_1(x) \leq p(x)$, $x \in V$.

A similar argument, using the fact that

$$\{x \mid p_1(x) < 1\} \subset \{x \mid p(x) \leq 1\},$$

reveals that we also must have $p(x) \leq p_1(x)$, $x \in V$.

Thus $p = p_1$, and the theorem is proved. □

It is possible that the inclusions $B_1^o \subset B \subset B_1$ are both proper. For example, let $V = \mathbb{R}^2$ and take

$$B = \{(x,y) \mid x^2 + y^2 < 1\} \cup \{(1,0)\} \cup \{(-1,0)\}.$$

Then $B$ is convex, balanced, and absorbing, the gauge of $B$ is $p[(x,y)] = (x^2 + y^2)^{1/2} = \|(x,y)\|_2$, and the inclusions $B_1^o \subset B \subset B_1$ are proper.

On the other hand, if $B$ is a convex balanced absorbing set, the gauge of $B$ is a norm, and $B$ is closed in the topology determined by this norm, then $B = B_1$. The details are left to the reader. For this and other reasons the following result is of interest:

Theorem 1.4.2. Let $V$ be a linear space over $\Phi$, let $B \subset V$ be a convex balanced absorbing set, and let $q$ be the gauge of $B$. Then the following are equivalent:

(i) $q$ is a norm on $V$.
(ii) $B$ contains no positive-dimensional linear subspace of $V$.

Proof. Suppose $q$ is not a norm and let

$$W = \{x \mid x \in V, q(x) = 0\}.$$

Since $q$ is a seminorm, it follows at once that $W$ is a linear subspace of $V$ distinct from $\{0\}$. Clearly, $W \subset \{x \mid q(x) < 1\} \subset B$, by Theorem 1.4.1(ii), and so part (ii) of the theorem implies part (i).

Conversely, suppose $B$ does contain a positive-dimensional linear subspace $W$ of $V$. Then $W \neq \{0\}$ and, by Theorem 1.4.1(ii), $W \subset B \subset \{x \mid q(x) \leq 1\}$. If $x \in W$ and $q(x) \neq 0$, then for any positive integer $n$ we have $nx \in W$ and $q(nx) = nq(x) \neq 0$. But then, on the one hand, $\lim_n q(nx) = +\infty$, and, on the other hand, $q(nx) \leq 1$ because $W \subset B$, which is a contradiction. Consequently, if $x \in W$, then $q(x) = 0$, and $q$ is not a norm.

Therefore Theorem 1.4.2(i) implies Theorem 1.4.2(ii). □

1.5. Topology in Seminormed Linear Spaces. Given a seminormed linear space $(V,P)$ we wish, in a standard manner, to provide $V$ with a Hausdorff topology determined by the family of seminorms $P$. If $P = \{p\}$ is a singleton set, that is, $V$ is a normed linear space, then clearly the most natural candidate for the desired topology is the norm topology. It is apparent in this case that a net $\{x_\alpha\} \subset V$ converges in the norm topology to $x \in V$ if and only if $\lim_\alpha p(x_\alpha - x) = 0$. We also want the analog of this observation to be valid in an arbitrary seminormed linear space.

With these observations in mind we make the following definitions:

Definition 1.5.1. Let $(V,P)$ be a seminormed linear space over $\Phi$. For each $x \in V$, each $\epsilon > 0$, and each positive integer $n$ we set

$$U(x;\epsilon;p_1,p_2,\ldots,p_n) = \{y \mid y \in V;\ p_k(y - x) < \epsilon;\ k = 1,2,\ldots,n\},$$

where $p_1,p_2,\ldots,p_n$ are any $n$ seminorms in $P$. Furthermore, for each $x \in V$ we set

$$\mathcal{U}_P(x) = \{U(x;\epsilon;p_1,p_2,\ldots,p_n) \mid \epsilon > 0;\ n \in \mathbb{Z};\ n > 0;\ p_1,p_2,\ldots,p_n \in P\},$$

and

$$\mathcal{U}_P = \bigcup_{x \in V} \mathcal{U}_P(x).$$

The next proposition is then easily verified.

Proposition 1.5.1. Let $(V,P)$ be a seminormed linear space over $\Phi$. Then $\mathcal{U}_P$ is a base for a Hausdorff topology on $V$.

Proof. The routine arguments needed to show that $\mathcal{U}_P$ is a base for a topology are left to the reader. To see that this topology is Hausdorff suppose $x,y \in V$, $x \neq y$. Then there exists some $p \in P$ such that $p(x - y) > 0$, as $(V,P)$ is a seminormed linear space. A straightforward computation reveals that $U(x,\epsilon,p)$ and $U(y,\epsilon,p)$ are disjoint open neighborhoods of $x$ and $y$, respectively, if we take $\epsilon = p(x - y)/3$. □

The reader should note the use made here of the fact that not all the seminorms in $P$ vanish at any one nonzero vector in $V$.

Given a seminormed linear space $(V,P)$ we shall denote the topology generated by $\mathcal{U}_P$ as $\mathcal{T}_P$, and we shall denote $V$ with this topology by the pair $(V,\mathcal{T}_P)$. It is evident that $\mathcal{U}_P(x)$ is a base for $\mathcal{T}_P$ at $x$.

When $(V,P)$ is a normed linear space, the neighborhoods in $\mathcal{U}_P(x)$ are, of course, just open balls with center $x$, and the topology $\mathcal{T}_P$ coincides with the norm topology. In the general case of a seminormed linear space it is not sufficient, however, to use only neighborhoods of $x$ of the form $U(x,\epsilon,p)$, $p \in P$, as the elements of $\mathcal{U}_P(x)$ since the intersection of two such sets need not contain a third such set. Hence there are not enough such sets to form a base at $x$ for a topology on $V$. They do, of course, form a subbase at $x$. Concrete examples of this are easily constructed, for instance, in $(C^\infty([0,1]),\{p_n\})$, where $p_n(f) = \|f^{(n)}\|_\infty$, $n = 0,1,2\ldots$ .

The theorem alluded to at the beginning of this section can now be established.

<u>Theorem 1.5.1</u>. Let $(V,P)$ be a seminormed linear space over $\Phi$ and let $\{x_\alpha\} \subset V$ be a net, $x \in V$. Then the following are equivalent:

(i) The net $\{x_\alpha\}$ converges to $x$ in $(V,\mathcal{T}_P)$.
(ii) $\lim_\alpha p(x_\alpha - x) = 0$ for each $p \in P$.

<u>Proof</u>. If part (i) holds and $\epsilon > 0$, then for each $p \in P$ the neighborhood $U(x,\epsilon,p)$ is an open neighborhood of $x$ in $(V,\mathcal{T}_P)$, and so there exists some $\alpha_0$ such that, if $\alpha > \alpha_0$, then $x_\alpha \in U(x,\epsilon,p)$, that is, $p(x_\alpha - x) < \epsilon$. Thus $\lim_\alpha p(x_\alpha - x) = 0$, and part (i) implies part (ii).

Conversely, if Theorem 1.5.1(ii) holds, $\epsilon > 0$, and $p_1,p_2,\ldots p_n$ are in $P$, then there exist some $\alpha_k$, $k = 1,2,\ldots,n$, such that $p_k(x_\alpha - x) < \epsilon$ whenever $\alpha > \alpha_k$, $k = 1,2,\ldots,n$. Consequently $x_\alpha \in U(x;\epsilon;p_1,p_2,\ldots,p_n)$ whenever $\alpha > \alpha_0$, where $\alpha_0$ is chosen so that $\alpha_0 > \alpha_k$, $k = 1,2,\ldots,n$.

From this it follows at once that part (ii) of the theorem implies part (i).

□

Theorem 1.5.1 shows us that in any seminormed linear space $(V,P)$ it is possible to introduce a Hausdorff topology in which convergence of a net $\{x_\alpha\}$ is precisely the same as the convergence of each of the nets of numbers $\{p(x_\alpha)\}$, $p \in P$. An obvious question to raise is whether this process can be reversed: given a linear space $V$ with a Hausdorff topology $T$, can we define a family of seminorms $P$ on $V$ such that $T = T_P$? The answer is, in general, negative, even when the linear space operations and the topology "fit together" properly. However, for a large class of important spaces the process can be reversed, as will be seen in the next chapter. The notion of the gauge of a convex balanced absorbing set, introduced in Section 1.4, will play a crucial role in resolving this question.

## 1.6. Problems.

1. (Theorem 1.1.1) Let $(V,\|\cdot\|)$ be a normed linear space over $\Phi$.

(a) If $W \subset V$ is a linear subspace, prove that the closure of $W$ in $V$ is a linear subspace of $V$.

(b) If $W \subset V$ is a linear subspace, prove that $(W,\|\cdot\|)$ is a normed linear space. If $(V,\|\cdot\|)$ is a Banach space and $W$ is a closed linear subspace, prove that $(W,\|\cdot\|)$ is a Banach space.

(c) If $W \subset V$ is a closed linear subspace, prove that the quotient space $V/W$ is a normed linear space with the norm

$$|||x + W||| = \inf_{y \in W} \|x + y\| \qquad (x \in V).$$

If $(V,\|\cdot\|)$ is a Banach space and $W$ is a closed linear subspace, prove that $(V/W, |||\cdot|||)$ is a Banach space.

(d) Prove that there exists a Banach space $(V_1,\|\cdot\|_1)$ over $\Phi$ and a mapping $\varphi : V \to V_1$ such that

(i) $\varphi$ is an isomorphism.

(ii) $\varphi(V)$ is dense in $(V_1, \|\cdot\|_1)$.

(iii) $\|\varphi(x)\|_1 = \|x\|$. $(x \in V)$.

Moreover, if $(V_2, \|\cdot\|_2)$ is another Banach space that satisfies conditions (i) through (iii), prove the existence of a mapping $\psi : V_1 \to V_2$ that is a surjective isomorphism such that

$$\|\psi(x)\|_2 = \|x\|_1, \qquad (x \in V_1).$$

2. Let $(V, \|\cdot\|)$ be a normed linear space over $\Phi$ and let $W \subset V$ be a linear subspace that is complete. Prove that, if $(V/W, \||\cdot\||)$ is a Banach space, then $(V, \|\cdot\|)$ is also a Banach space.

3. (Theorem 1.1.2) Let $(V_1, \|\cdot\|_1), (V_2, \|\cdot\|_2), \ldots, (V_n, \|\cdot\|_n)$ be normed linear spaces over $\Phi$ and let $V$ denote the topological product $V_1 \times V_2 \times \cdots \times V_n$ of the topological spaces $V_1, V_2, \ldots, V_n$, with the respective norm topologies and with linear space addition and scalar multiplication defined componentwise. Prove that $(V, \|\cdot\|)$ is a normed linear space over $\Phi$ with the norm topology equivalent to the product topology if $\|\cdot\|$ is defined as either of the following:

$$\|x\| = \sup_{k=1,2,\ldots n} \|x_k\|_k,$$

$$\|x\| = \left[\sum_{k=1}^{n} (\|x_k\|_k)^p\right]^{1/p} \qquad (1 \le p < \infty),$$

where $x = (x_1, x_2, \ldots, x_n)$. Prove that, if $(V_1, \|\cdot\|_1), (V_2, \|\cdot\|_2), \ldots$ $\ldots, (V_n, \|\cdot\|_n)$ are Banach spaces, then so is $(V, \|\cdot\|)$.

4. Sketch the sets $\{x \mid \|x\|_p = 1\}$ in $\mathbb{R}^2$ for $p = 1,2,3$ and $p = \infty$. Note that the norms $\|\cdot\|_p$ on $\mathbb{R}^2$ are all equivalent, $1 \le p \le \infty$.

5. (Corollary 1.3.2) Let $V$ be any n-dimensional linear space over $\Phi$ and suppose $\|\cdot\|_1$ and $\|\cdot\|_2$ are norms such that $(V, \|\cdot\|_1)$ and $(V, \|\cdot\|_2)$ are normed linear spaces. Prove that the norms $\|\cdot\|_1$ and $\|\cdot\|_2$ are equivalent.

6. (Corollary 1.3.3) Let $(V, \|\cdot\|)$ be a finite-dimensional normed linear space over $\Phi$.

(a) Prove that $(V, \|\cdot\|)$ is a Banach space.

(b) If $W$ is a linear subspace of $V$, prove that $(W, \|\cdot\|)$ is a closed linear subspace of $(V, \|\cdot\|)$.

(c) If $E \subset V$ is a closed bounded set, prove that $E$ is compact.

7. (a) Prove that a finite-dimensional subspace of a normed linear space is closed.

(b) If $V$ is a normed linear space and $W \subset V$ is a closed linear subspace, prove that the quotient map $\varphi : V \to V/W$ defined by $\varphi(x) = x + W$ is continuous.

(c) If $V$ is a normed linear space, $W \subset V$ is a closed linear subspace, and $Y \subset V$ is a finite-dimensional linear subspace, prove that $W + Y$ is closed in $V$.

8. Prove that a Banach space is finite dimensional if and only if every linear subspace is closed.

*9. Give an example of a Banach space and a decreasing sequence of nonempty bounded closed sets whose intersection is empty. (Hint: Theorem 1.3.3 tells you where <u>not</u> to look for an example.)

10. Does $p(x + iy) = |x|$ define a seminorm on $\mathbb{C}$ ?

11. (a) Give an example of a set in the plane that is balanced in the real linear space $\mathbb{R}^2$, but not in the complex linear space $\mathbb{C}$.

(b) Prove that every set that is balanced in the complex linear space $\mathbb{C}$ is convex.

(c) Give an example in a complex linear space of a nonconvex balanced set.

12. (a) Give an example of a set in the plane that is absorbing, but not convex.

(b) Some authors define a set $A$ to be absorbing if for every $x \in V$ there exists some $\epsilon > 0$ such that $tx \in A$ whenever $t \in \Phi$ and $|t| \leq \epsilon$. Show that with this definition, if $A$ is absorbing, then $V = \bigcup_{n=1}^{\infty} nA$. Give an example of a set $A$ in a linear space $V$ that is absorbing by Definition 1.4.1 but for which $V \neq \bigcup_{n=1}^{\infty} nA$.

13. Given a convex balanced absorbing set $B$, prove that there exists a convex balanced absorbing set $A$ such that $A + A \subset B$.

*14. Let $U$ and $V$ be convex balanced absorbing sets with gauges $p$ and $q$, respectively.

(a) Prove that $U \cap V$ is a convex balanced absorbing set with gauge $r(x) = \max\{p(x), q(x)\}$.

(b) Prove that $U + V$ is a convex balanced absorbing set with gauge $s(x) = \inf\{p(v) + q(v - x) \mid v \in V\}$.

(c) Give an example to show that $U \cup V$ is not necessarily a convex balanced absorbing set.

Conclude that $r$ and $s$ define seminorms, but

$$t(x) = \min\{p(x), q(x)\}$$

is not necessarily a seminorm.

15. Let $A$ be the set of all sequences $\{x_n\}$ of complex numbers for which $|x_n| < 1/n$. Prove that $A$ is not absorbing in $c_o$. If $E$ denotes the space of all sequences $\{x_n\}$ such that $x_n \neq 0$ for only finitely many values of $n$, prove that $A \cap E$ is absorbing in $E$.

*16. An absorbing set $A \subset V$ is said to be open (closed) in rays if for each $x \in V$ the set $I_x = \{a \mid a > 0, x \in aA\}$ is open (closed) relative to the interval $I = (0, \infty)$. Let $A$ be a convex balanced absorbing set and $p$ its gauge. Use Theorem 1.4.1 to prove that, if $A$ is open in rays, then $A = B_1^o$, and if $A$ is closed in rays, then $A = B_1$, where, as usual, $B_1^o = \{x \mid x \in V, p(x) < 1\}$ and $B_1 = \{x \mid x \in V, p(x) \leq 1\}$.

*17. A series $\Sigma_{k=1}^{\infty} x_k$ in a normed linear space $(V, \|\cdot\|)$ is said to be <u>summable</u> to a sum $x$ if $x \in V$ and $\lim_n \|x - \Sigma_{k=1}^{n} x_k\| = 0$, that is, the sequence of partial sums converges to $x$. The series is said to be <u>absolutely summable</u> if $\Sigma_{k=1}^{\infty} \|x_k\| < \infty$. Prove that a normed linear space $(V, \|\cdot\|)$ is complete if and only if every absolutely summable series is summable.

# CHAPTER 2

# TOPOLOGICAL LINEAR SPACES

## 2.0. Introduction.

We have seen that, given a seminormed linear space (V,P) we can introduce a Hausdorff topology $\mathcal{T}_P$ into V that is intimately connected with the family of seminorms P. Now we wish to study such spaces -- that is, linear spaces equipped with a Hausdorff topology -- in their own right. In order to make these topological linear spaces interesting objects of investigation we shall demand that the linear space operations and the topology "fit together" properly, namely, that the operations be continuous. All seminormed linear spaces are examples of these topological linear spaces, but the converse is not true.

After introducing the basic notions and establishing some elementary results in Section 2.1, we shall look at finite-dimensional topological linear spaces in Section 2.2. There it will be seen that a topological linear space is finite dimensional if and only if its topology is locally compact. Those topological linear spaces that are seminormed linear spaces will be described in Section 2.3. They will be seen to be precisely those topological linear spaces whose topology has, at the origin, a neighborhood base consisting of convex open sets. The notion of the gauge of a convex balanced absorbing set will be instrumental in establishing this description. In Section 2.4 some properties of gauges in topological linear spaces will be given, and these results will then be used to characterize those topological linear spaces that are normable. Metrizable topological linear spaces will be discussed in the last section.

## 2.1. Topological Linear Spaces.

We begin this section with the definition of topological linear spaces.

<u>Definition 2.1.1</u>. A linear space V over $\Phi$ is said to be a <u>topological linear space over</u> $\Phi$ if there exists a Hausdorff topology $\mathcal{T}$ on V such that the following mappings are continuous:

(i) The mapping from $V \times V$, with the product topology, to V, defined by $(x,y) \to x + y$, $x,y \in V$.

(ii) The mappings from $\Phi \times V$, with the product topology, to V, defined by $(a,x) \to ax$, $a \in \Phi$, $x \in V$.

(iii) The mapping from V to V, defined by $x \to -x$, $x \in V$.

Thus, somewhat loosely speaking, a linear space V with a Hausdorff topology $\mathcal{T}$ is a topological linear space if the linear space operations of addition, inversion, and scalar multiplication are continuous. We shall generally denote a topological linear space by the pair $(V,\mathcal{T})$.

It should be noted that not all definitions of topological linear spaces include the assumption that the topolog. is Hausdorff. This assumption is not made, for example, in [$E_1$, KeNa, T, and $W_1$]. Moreover, most of the results we shall establish are also valid in this more general context. Nevertheless, we choose to include the hypothesis of Hausdorffness in our definition since all of the topological linear spaces we shall discuss in the sequel have natural Hausdorff topologies. For expositions without the Hausdorff assumption we refer the reader to [$E_1$, pp. 56-66; T, pp. 123-133; $W_1$, pp. 167-177.]. In addition, it should be remarked that rather weak additional restrictions on the topology of a topological linear space $(V,\mathcal{T})$ in place of the Hausdorff assumption are sufficient to imply that $\mathcal{T}$ is Hausdorff. For example, if $\mathcal{T}$ is a $T_0$-topology (i.e., given any two distinct points in V, at least one of them has an open neighborhood not containing the other), then $\mathcal{T}$ is a Hausdorff topology. In this case one can even prove that, besides being Hausdorff, $\mathcal{T}$ also has the property that any closed set and point not in the set have disjoint open neighborhoods. For some discussion of these results see [T, p. 126; $W_1$, pp. 175-176].

Examples of topological linear spaces are easy to come by, as shown by the next result.

Theorem 2.1.1. Let $(V,P)$ be a seminormed linear space over $\Phi$. Then $(V,\mathcal{T}_P)$ is a topological linear space over $\Phi$.

Proof. Since from Proposition 1.5.1 we see that $\mathcal{T}_P$ is a Hausdorff topology on $V$, it remains only to show that the linear space operations are continuous. For example, to see that addition is continuous, let $x_o, y_o \in V$, and $\epsilon > 0$, and consider the open neighborhood $U(x_o + y_o;\epsilon;p_1,p_2,\ldots,p_n)$ of $x_o + y_o$ in $\mathcal{T}_P$. If $U$ denotes the open neighborhood

$$U(x_o;\epsilon/2;p_1,p_2,\ldots,p_n) \times U(y_o;\epsilon/2;p_1,p_2,\ldots,p_n)$$

in the product topology on $V \times V$, then clearly $(x,y) \in U$ implies that

$$\begin{aligned} p_k[x + y - (x_o + y_o)] &\leq p_k(x - x_o) + p_k(y - y_o) \\ &< \epsilon \qquad\qquad (k = 1,2,\ldots,n), \end{aligned}$$

from which we conclude that addition is continuous. Similar arguments, whose details are omitted, establish the continuity of inversion and scalar multiplication. □

All the spaces described in Section 1.2 are topological linear spaces.

Now let us state some further results concerning topological linear spaces. The proofs are reasonably straightforward and are left to the reader.

Theorem 2.1.2. Let $(V,\mathcal{T})$ be a topological linear space over $\Phi$. Then for each $y \in V$ and each $a \in \Phi$, $a \neq 0$, the mappings $\varphi_y : V \to V$, defined by $\varphi_y(x) = x + y$, $x \in V$, and $\varphi_a : V \to V$, defined by $\varphi_a(x) = ax$, $x \in V$, are surjective homeomorphisms. Moreover, the image under $\varphi_y$ or $\varphi_a$ of a convex set is again a convex set, and the image under $\varphi_a$ of a linear subspace is again a linear subspace.

The fact that translation in a topological linear space $(V,\mathcal{T})$ is a homeomorphism is very useful, as it often allows us to reduce a "global" question to a "local" one. For example, suppose that $\mathcal{U}$ is some collection of open sets in $(V,\mathcal{T})$ that is invariant under translation; that is, $U \in \mathcal{U}$ implies that $U + x \in \mathcal{U}$, $x \in V$. To show that $\mathcal{U}$ is a base for the topology $\mathcal{T}$ it is then sufficient to demonstrate that $\mathcal{U}$ contains some subset that is a neighborhood base at the origin. Similarly we shall see in Chapter 3 that a linear transformation from $(V,\mathcal{T})$ to $(V,\mathcal{T})$ is continuous on $V$ if and only if it is continuous at the origin.

It is also worth remarking that the continuity of scalar multiplication in a topological linear space implies that every open neighborhood of the origin in $V$ is an absorbing set.

<u>Theorem</u> <u>2.1.3</u>. Let $(V,\mathcal{T})$ be a topological linear space over $\Phi$.

(i) If $W \subset V$ is a linear subspace, then the closure of $W$ is a linear subspace.

(ii) If $W \subset V$ is a linear subspace and $\mathcal{T}'$ is the relative topology on $W$ induced by $\mathcal{T}$, then $(W,\mathcal{T}')$ is a topological linear space.

(iii) If $W \subset V$ is a closed linear subspace, then $(V/W,\mathcal{T}')$ is a topological linear space if $\mathcal{T}'$ is the topology on $V/W$ whose open sets are sets of the form $\{x + W \mid x \in U\}$, $U \in \mathcal{T}$, that is, $\mathcal{T}'$, is the usual quotient topology.

<u>Theorem</u> <u>2.1.4</u>. Let $(V_\alpha,\mathcal{T}_\alpha)$, $\alpha \in \Lambda$, be topological linear spaces over $\Phi$ and let $V$ denote the product of the topological spaces $V_\alpha$, $\alpha \in \Lambda$, with the product topology $\mathcal{T}$ and with linear space operations defined componentwise. Then $(V,\mathcal{T})$ is a topological linear space over $\Phi$.

<u>2.2. Finite-Dimensional Topological Linear Spaces.</u> From Theorem 2.1.1 we see that the spaces $(\mathbb{R}^n, \|\cdot\|_p)$ and $(\mathbb{C}^n, \|\cdot\|_p)$, $1 \leq p \leq \infty$, are n-dimensional topological linear spaces over $\mathbb{R}$ and $\mathbb{C}$, respectively. As is perhaps not surprising, in view of the results on finite-dimensional normed linear spaces obtained in Section 1.3., all n-dimensional topological linear spaces are essentially either $(\mathbb{R}^n, \|\cdot\|_p)$ or $(\mathbb{C}^n, \|\cdot\|_p)$ for any $p$, $1 \leq p \leq \infty$. This is an immediate consequence of Theorem 2.2.1, whose proof is left to the reader.

<u>Definition 2.2.1.</u> Let $(V_1, T_1)$ and $(V_2, T_2)$ be topological linear spaces over $\Phi$. Then $(V_1, T_1)$ and $(V_2, T_2)$ are said to be <u>topologically isomorphic</u> if there exists a surjective isomorphism $\varphi : V_1 \to V_2$ that is a homeomorphism.

As in the case of normed linear spaces, it is apparent that the relation of topological isomorphism between topological linear spaces is reflexive, symmetric and transitive.

<u>Theorem 2.2.1.</u> Let $(V, T)$ be an n-dimensional topological linear space over $\mathbb{R}(\mathbb{C})$. Then $(V, T)$ is topologically isomorphic to $(\mathbb{R}^n, \| \ \|_1)((\mathbb{C}^n, \|\cdot\|_1))$.

<u>Corollary 2.2.1.</u> Let $(V_1, T_1)$ and $(V_2, T_2)$ be n-dimensional topological linear spaces over $\Phi$. Then $(V_1, T_1)$ and $(V_2, T_2)$ are topologically isomorphic.

<u>Corollary 2.2.2.</u> Let $(V, T)$ be a topological linear space over $\Phi$ and let $W$ be a finite-dimensional linear subspace of $V$. Then $W$ is a closed linear subspace of $(V, T)$.

A characterization of finite-dimensional topological linear spaces analogous to that given for finite-dimensional normed linear spaces in Theorem 1.3.3 is also valid. On the surface, the form of the characterization is somewhat different from that for finite-dimensional normed linear spaces since Theorem 1.3.3 is no longer meaningful in the context of topological linear spaces. The characterization should be compared with the remarks following Theorem 1.3.3.

Theorem 2.2.2. Let $(V,\mathcal{T})$ be a topological linear space over $\Phi$. Then the following are equivalent:

(i) $(V,\mathcal{T})$ is finite dimensional.

(ii) There exists a compact set $B \subset V$ with a nonempty interior such that the origin belongs to the interior of $B$.

Proof. Suppose $(V,\mathcal{T})$ is finite dimensional. Then, by Theorem 2.2.1, there exists some $n$ such that $(V,\mathcal{T})$ is topologically isomorphic to either $(\mathbb{R}^n, \|\cdot\|_1)$ or $(\mathbb{C}^n, \|\cdot\|_1)$. If $\varphi$ is this topological isomorphism from $V$ to either $\mathbb{R}^n$ or $\mathbb{C}^n$ then $B = \varphi^{-1}(B_1)$, where $B_1$ is the closed unit ball about the origin in either $\mathbb{R}^n$ or $\mathbb{C}^n$, satisfies condition (ii), which is thus implied by (i).

Conversely, suppose part (ii) holds. Since $B$ is compact and the interior of $B$ is a nonempty set that contains the origin, we deduce at once that there exist a finite number of points $x_1, x_2, \ldots, x_n$ in $B$ such that $B \subset \bigcup_{k=1}^n (x_k + (1/2)B)$. Let $W$ be the linear subspace of $V$ spanned by $x_1, x_2, \ldots, x_n$. Then $W$, being finite dimensional, is closed, and so $V/W$ is a topological linear space with the quotient topology. Let $\psi : V \to V/W$ denote the canonical homomorphism determined by $W$. Then from the definition of the quotient topology on $V/W$ we see that $\psi$ is continuous and that $\psi(B)$ is a compact subset of $V/W$ with a nonempty interior that contains the origin in $V/W$. Moreover, by the definition of $W$, we have $B \subset W + (1/2)B$, and so $\psi(B) \subset (1/2)\psi(B)$. By induction we then conclude that $2^k\psi(B) \subset \psi(B)$, $k = 1,2,3,\ldots$. However, the continuity of scalar multiplication in the topological linear space $V/W$, combined with the fact that the nonempty interior of $\psi(B)$ contains the origin, entails that each point in $V/W$ belongs to $2^k\psi(B)$ for some $k$. Hence $V/W = \psi(B)$, and so $V/W$ is compact.

Now, if $V \neq W$, then $V/W$ would not be the zero space and so, as is easily verified, it would contain a linear subspace isomorphic to the scalar field $\Phi$, thereby contradicting the compactness of $V/W$. Consequently $V = W$, and so $V$ is finite dimensional.

Therefore part (ii) of the theorem implies part (i). $\square$

An equivalent phrasing of this result is that $(V,\mathcal{T})$ is a finite-dimensional topological linear space if and only if $\mathcal{T}$ is a locally compact topology.

## 2.3. Locally Convex Topological Linear Spaces.

In the first section of this chapter we saw that, if $(V,P)$ is a seminormed linear space over $\Phi$, then $(V,\mathcal{T}_P)$ is a topological linear space over $\Phi$. We now wish to address our attention to the converse question: if $(V,\mathcal{T})$ is a topological linear space, does there exist a family of seminorms $P$ on $V$ such that $(V,P)$ is a seminormed linear space and $\mathcal{T}_P = \mathcal{T}$? The answer is, in general, negative.

For example, as noted in Example 1.2.4, the spaces $L_p([0,1],dt)$, $0 < p < 1$, where $dt$ is Lebesgue measure on $[0,1]$, are neither normed nor seminormed linear spaces. However, they are topological linear spaces if we use the topology $\mathcal{T}$ generated by the neighborhoods

$$U(f,\epsilon) = \{g \mid g \in L_p([0,1],dt),\ \|g - f\|_p < \epsilon\}$$

where $f \in L_p([0,1],dt)$ and $\epsilon > 0$ is arbitrary. (See, for example, [K, pp. 157-158]).

What additional restrictions are then necessary to ensure that a topological linear space is a seminormed linear space? The clue to the answer lies in examining the nature of open neighborhoods in seminormed linear spaces. To be more precise, we have the following proposition:

Proposition 2.3.1. Let $(V,P)$ be a seminormed linear space over $\Phi$. Then $\mathcal{T}_P$ contains a neighborhood base at each point in $V$ that consists of convex open sets.

Proof. An elementary argument shows that for each $x \in V$ the open neighborhoods $U(x;\epsilon;p_1,p_2,\ldots,p_n)$ are convex. □

We shall see that the existence, at each point of a topological linear space, of a neighborhood base consisting of convex open sets is sufficient to ensure that the topology comes from a family of seminorms. Since the existence of such a neighborhood base is of considerable importance, we make the following definition:

Definition 2.3.1. Let $(V,\mathcal{T})$ be a topological linear space over $\Phi$. Then $(V,\mathcal{T})$ is said to be locally convex if each point $x \in V$ has a neighborhood base consisting of convex open sets.

The spaces $L_p([0,1],dt)$, $0 < p < 1$, are not locally convex topological linear spaces, as will be shown in Section 4.2. On the other hand, the content of Proposition 2.3.1 is precisely that, if $(V,P)$ is a seminormed linear space, then $(V,\mathcal{T}_P)$ is locally convex. Before we establish the converse assertion we need to prove some preliminary results, which are also of independent interest.

Proposition 2.3.2. Let $(V,\mathcal{T})$ be a topological linear space over $\Phi$.

(i) Let $U$ be a (open) neighborhood of the origin in $V$. Then there exists a (open) neighborhood $N$ of the origin such that $aN \subset U$, $a \in \Phi$, $|a| \leq 1$.

(ii) Let $U$ be a (open) neighborhood of the origin in $V$. Then there exists a balanced (open) neighborhood $N_o$ of the origin such that $N_o \subset U$.

(iii) Let $U \subset V$ be a convex set. Then the closure of $U$ and the interior of $U$ are convex sets.

(iv) Let $U \subset V$ be a balanced set. Then the closure of $U$ is balanced, and the interior of $U$ is balanced provided the interior of $U$ contains the origin.

Proof. Let $U$ be a (open) neighborhood of the origin. Since scalar multiplication is continuous, there exists some $\epsilon > 0$ and a (open) neighborhood $U_1$ of the origin in $V$ such that $ax \in U$ whenever $|a| \leq \epsilon$ and $x \in U_1$. Set $N = \epsilon U_1$. Then for $x = \epsilon u \in N$ and $a \in \Phi$, $|a| \leq 1$, we have $ax \in U$ because $u \in U_1$ and

$|a\epsilon| = |a|\epsilon \leq \epsilon$. Hence $aN \subset U$ whenever $|a| \leq 1$ and $N$ is a (open) neighborhood of the origin. Thus part (i) is established.

To prove part (ii) we set $N_o = \cup_{|a|\leq 1} aN$, where $N$ is the neighborhood we have just constructed. Clearly $N_o$ is a (open) neighborhood of the origin in $V$ and $N_o \subset U$. Furthermore, if $b \in \Phi$, $|b| \leq 1$, then $bN_o \subset N_o$.

Parts (iii) and (iv) are left to the reader. □

If $V$ is a locally convex topological linear space, then Proposition 2.3.2 (ii) can be used to construct a neighborhood base at the origin consisting of convex balanced open sets.

Proposition 2.3.3. Let $(V,\mathcal{T})$ be a locally convex topological linear space over $\Phi$. Then there exists a neighborhood base at the origin in $V$ consisting of convex balanced open sets.

Proof. Let $\mathcal{U}$ be a neighborhood base at the origin consisting of convex open sets. If $U \in \mathcal{U}$, then by Proposition 2.3.2(ii) there exists a balanced open neighborhood $N$ of the origin such that $N \subset U$.

Let $N_1$ be the collection of all finite convex linear combinations of elements of $N$; that is, let $N_1$ consist of all elements of the form $\Sigma_{k=1}^{n} a_k x_k$ for some $n$, where $x_k \in N$, $0 \leq a_k \leq 1$, $k = 1,2,\ldots,n$, and $\Sigma_{k=1}^{n} a_k = 1$. Then clearly $N_1$ is a convex balanced neighborhood of the origin and $N_1 \subset U$, as $U$ is convex. By Proposition 2.3.2(iii) and (iv) we conclude that the interior of $N_1$, which is nonempty because $N_1 \supset N$, is a convex balanced open neighborhood of the origin that is contained in $U$.

Thus every $U \in \mathcal{U}$ contains a convex balanced open neighborhood of the origin and hence there exists a neighborhood base at the origin consisting of such sets. □

Now let us state and prove the theorem indicated earlier.

Theorem 2.3.1. Let $(V,\mathcal{T})$ be a topological linear space over $\Phi$. Then the following are equivalent:

(i) $(V,\mathcal{T})$ is a locally convex topological linear space over $\Phi$.

(ii) There exists a family of seminorms $P$ on $V$ such that $(V,P)$ is a seminormed linear space over $\Phi$ and $\mathcal{T}_P = \mathcal{T}$.

Proof. Proposition 2.3.1 shows that part (ii) implies part (i). Conversely, suppose that part (i) holds and let $\mathcal{U}$ be a neighborhood base at the origin in $V$ consisting of convex balanced open sets. The continuity of scalar multiplication implies that each $U \in \mathcal{U}$ is absorbing. For each $U \in \mathcal{U}$ define

$$p_U(x) = \inf\{a \mid a > 0,\ x \in aU\} \qquad (x \in V),$$

that is, $p_U$ is the gauge of the convex balanced absorbing set $U$. Set $P = \{p_U \mid U \in \mathcal{U}\}$. Then from Proposition 1.4.2 and Theorem 1.4.1 we see that $P$ is a family of seminorms on $V$ such that

$$\{x \mid p_U(x) < 1\} \subset U \subset \{x \mid p_U(x) \leq 1\}.$$

Moreover, since $\mathcal{U}$ is a neighborhood base at the origin in $V$, it follows at once that, if $p_U(x) = 0$, $U \in \mathcal{U}$, then $x \in U$ for all $U \in \mathcal{U}$, and hence $x = 0$ because $\mathcal{T}$ is a Hausdorff topology.

It is then evident that $(V,P)$ is a seminormed linear space and $\mathcal{T}_P = \mathcal{T}$. Consequently part (i) of the theorem implies part (ii). □

Thus we see that seminormed linear spaces and locally convex topological linear spaces are the same, and we have a means of moving from one type of space to the other, either through the construction of the topology $\mathcal{T}_P$ associated with a family of seminorms or through the use of the gauges of certain convex balanced absorbing sets. At times there are significant advantages in being able to choose which type of space one wishes to work with while investigating various questions. This will be apparent in the sequel.

2.4. Seminorms and Convex Balanced Absorbing Sets. We have already seen (Theorem 1.4.1) that, if $B$ is a convex balanced absorbing subset of a linear space $V$, then there exists a unique seminorm $p$ on $V$ such that

$$B_1^o = \{x \mid p(x) < 1\} \subset B \subset \{x \mid p(x) \leq 1\} = B_1.$$

The seminorm $p$ is, of course, the gauge of the set $B$. In the next theorem we shall investigate further the relationship between gauges and convex balanced absorbing sets in the context of topological linear spaces. Some preliminary definitions are desirable.

Definition 2.4.1. Let $(V,\mathcal{T})$ be a topological linear space over $\Phi$ and let $E \subset V$. Then the interior of $E$ will be denoted by $\text{int}(E)$ and the closure of $E$ by $\text{cl}(E)$. The set $E$ is said to be bounded if for each open neighborhood $U$ of the origin in $V$ there exists some $a > 0$ such that $E \subset aU$.

It is easily seen that the definition of boundedness in a topological linear space reduces in the case of seminormed linear spaces to that given in Definition 1.3.2.

Theorem 2.4.1. Let $(V,\mathcal{T})$ be a topological linear space over $\Phi$, let $B \subset V$ be a convex balanced absorbing set, and let $q$ be the gauge of $B$. Then

(i) $\text{int}(B) \subset B_1^o \subset B \subset B_1 \subset \text{cl}(B)$.
(ii) $B = B_1^o$ if $B$ is open.
(iii) $B = B_1$ if $B$ is closed.
(iv) If $q : V \to \mathbb{R}$ is continuous, then $B_1^o = \text{int}(B)$ and $B_1 = \text{cl}(B)$.
(v) $q : V \to \mathbb{R}$ is continuous if and only if $0 \in \text{int}(B)$.
(vi) If $B$ is bounded, then $q$ is a norm.

Proof. From Theorem 1.4.1 we know that $B_1^o \subset B \subset B_1$. Let $x \in \text{int}(B)$. Since $\text{int}(B)$ is open there exists an open neighborhood $U$ of $x$ such that $U \subset \text{int}(B)$. Moreover, by the continuity of scalar multiplication, there exists an $\epsilon > 0$ such that, if $|a - 1| < \epsilon$, then $ax \in U$. In particular, if $0 < a - 1 < \epsilon$, then $ax \in \text{int}(B) \subset B$, and so $x \in (1/a)B$. Thus $q(x) \leq 1/a < 1$, that is, $x \in B_1^o$. Hence $\text{int}(B) \subset B_1^o$.

On the other hand, let $x \in B_1$ so that $q(x) \leq 1$. If $q(x) < 1$, then $x \in B_1^o \subset \text{cl}(B)$. Therefore, without loss of generality, we can assume that $q(x) = 1$. But let $a_n = 1 - 1/n$, $n = 2,3,\ldots$ . Then

$$q(a_n x) = (1 - \frac{1}{n})q(x) = 1 - \frac{1}{n} < 1 \qquad (n = 2,3,\ldots),$$

and we conclude that $\{a_n x\} \subset B_1^o \subset B$, $n = 2,3,\ldots$ . However, since scalar multiplication is continuous, we deduce that $\lim_n a_n x = x \in \text{cl}(B)$. Therefore $B_1 \subset \text{cl}(B)$, and part (i) is proved.

Parts (ii) and (iii) follow immediately from part (i).

If $q : V \to \mathbb{R}$ is continuous, then $B_1^o$ is open, as it is the inverse image of $(-\infty,1)$. We conclude at once from $\text{int}(B) \subset B_1^o \subset B$ that $B_1^o = \text{int}(B)$. A similar argument shows that $B_1 = \text{cl}(B)$.

If $q$ is continuous, then $B_1^o = \text{int}(B)$ entails that $0 \in \text{int}(B)$, as $q(0) = 0 < 1$. Conversely, suppose $0 \in \text{int}(B)$, and let $U$ be an open neighborhood of $0$ such that $U \subset \text{int}(B)$. Since $\text{int}(B) \subset B_1^o$, we have $q(x) < 1$, $x \in U$. But then, for any $\epsilon > 0$, $\epsilon U$ is an open neighborhood of $0$ such that, if $x = \epsilon u \in \epsilon U$, then $q(x) = \epsilon q(u) < \epsilon$; that is, $q$ is continuous at $0$. The inequality

$$|q(x) - q(y)| \leq q(x - y) \qquad (x,y \in V),$$

which is valid for any seminorm by Proposition 1.1.1, then shows that $q$ is continuous, indeed even uniformly continuous, on $V$. This proves part (v) of the theorem.

Finally, suppose $B$ is bounded. If $x \neq 0$, then, since $\mathcal{T}$ is Hausdorff, there exists an open neighborhood $U$ of $0$ such that

$x \notin U$. By Proposition 2.3.2 we may assume that $U$ is also balanced. Because $B$ is bounded, there exists some $a > 0$ for which $B \subset aU$, and since $B$ is absorbing, there also exists some $b > 0$ such that $x \in bB$. Clearly then $x/ab \in U$. But this implies that $ab > 1$, as if $ab \leq 1$, then $ab(x/ab) = x \in abU \subset U$, as $U$ is balanced, contradicting the choice of $U$. Hence $b > 1/a$, and this holds for any $b$ such that $x \in bB$. By the definition of $q$ it then follows that $q(x) \geq 1/a > 0$.

Therefore if $x \neq 0$, then $q(x) \neq 0$, and so $q$ is a norm. This completes the proof of part (vi) and the theorem. $\square$

The last portion of Theorem 2.4.1 allows us to characterize those topological linear spaces that are normed linear spaces.

Theorem 2.4.2. Let $(V,\mathcal{T})$ be a topological linear space over $\Phi$. Then the following are equivalent:

(i) There exists a norm on $V$ such that the norm topology and $\mathcal{T}$ coincide.

(ii) $\mathcal{T}$ contains a bounded convex open neighborhood of the origin.

Proof. Clearly, if there exists a norm $\|\cdot\|$ on $V$ such that the norm topology coincides with $\mathcal{T}$, then $\{x \mid \|x\| < 1\}$ is a bounded convex open neighborhood of the origin. Hence part (i) implies part (ii).

Conversely, suppose $B \in \mathcal{T}$ is a bounded convex open neighborhood of the origin. By repeating the argument used in proving Proposition 2.3.3, we may assume without loss of generality that $B$ is also balanced. From Theorem 2.4.1 (vi) we conclude that the gauge $q$ of $B$ is a norm on $V$. It remains to prove that the topology generated by this norm coincides with $\mathcal{T}$.

First, from Theorems 2.4.1 (iv) and (v) we note that $B = \{x \mid q(x) < 1\}$, and hence for each $a > 0$ we have

$$aB = \{x \mid q(x) < a\}.$$

To show that the norm topology coincides with $T$ it suffices to show, in view of Theorem 2.1.2, that the family $\{aB \mid a > 0\}$ of open neighborhoods of the origin in the topology $T$ forms a neighborhood base at the origin. To accomplish this it suffices to prove that, if $U \in T$ is any open neighborhood of the origin, then there is some $a > 0$ for which $aB \subset U$. But if $U$ is such a neighborhood, then, since $B$ is bounded, there exists some $b > 0$ such that $B \subset bU$. Thus $aB \subset U$, where $a = 1/b > 0$.

Therefore part (ii) of the theorem implies part (i). □

Theorem 2.4.2 clearly provides us with a means of proving that certain topological linear spaces are not normed linear spaces. For example, consider the seminormed linear space $(C^{\infty}([0,2\pi]),P)$, where

$$P = \{p_n \mid p_n(f) = \sum_{k=0}^{n} \|f^{(k)}\|_{\infty},\ n = 0,1,2,\ldots\}.$$

To show that this is not a normed linear space it is sufficient to prove that there exists no norm on $C^{\infty}([0,2\pi])$ for which the norm topology and the topology $T_P$ coincide. In view of Theorem 2.4.2, this assertion can evidently be established by showing that $T_P$ contains no bounded open sets.

Thus suppose that $E \subset C^{\infty}([0,2\pi])$ is bounded and open -- that is, for each $n = 0,1,2,\ldots,$ there exists an $M_n > 0$ such that $p_n(f) \leq M_n$, $f \in E$, $n = 0,1,2,\ldots,$ and $E \in T_P$. Furthermore, let $h \in E$ be fixed. Then by the definition of $T_P$ there exists some $\epsilon > 0$ and $p_{k_1}, p_{k_2}, \ldots, p_{k_n}$ in $P$ such that

$$U(h;\epsilon;p_{k_1},p_{k_2},\ldots,p_{k_n}) \subset E.$$

Set $r = \sup_{j=1,2,\ldots,n} k_j$, and assume, without loss of generality, that $r > 0$. Clearly $p_{k_j}(f) \leq p_r(f)$, $f \in C^{\infty}([0,2\pi])$, $j = 1,2,\ldots,n$, and so $U(h,\epsilon,p_r) \subset E$.

Now for each $b > 1$ define $h_b(t) = h(t) + (\epsilon \sin bt)/2rb^r$, $0 \leq t \leq 2\pi$. Obviously $h_b \in C^{\infty}([0,2\pi])$, $b > 1$. Indeed, elementary

computations reveal that the kth derivative of $h_b$, depending on the value of $k$, is one of the following four functions:

$$h^{(k)}(t) \pm \frac{\epsilon \sin bt}{2rb^{r-k}},$$

$$h^{(k)}(t) \pm \frac{\epsilon \cos bt}{2rb^{r-k}}.$$

Consequently,

$$p_r(h - h_b) = \sum_{k=0}^{r} \|h^{(k)} - h_b^{(k)}\|_\infty$$

$$\leq \frac{\epsilon}{2r} \sum_{k=0}^{r} \frac{1}{b^{r-k}}$$

$$< \frac{\epsilon(r + 1)}{2r}$$

$$< \epsilon,$$

as $b > 1$.

In particular, we can then conclude that $h_b \in E$, $b > 1$, and so $\|h_b^{(n)}\|_\infty \leq p_n(h_b) \leq M_n$, $n = 0,1,2,\ldots,$ for all $b > 1$. But this clearly leads to a contradiction since, when $n > r$, $\|(\epsilon \sin bt)/2rb^{r-n}\|_\infty$ and $\|(\epsilon \cos bt)/2rb^{r-n}\|_\infty$ can be made as large as one chooses by a suitable choice of $b > 1$.

Therefore $E$ cannot be both bounded and open, and so $\mathcal{T}_P$ is not a norm topology.

It is, however, the case that the metric topology on $C^\infty([0,2\pi])$ determined by the metric

$$\rho(f,g) = \sum_{n=0}^{\infty} \frac{p_n(f - g)}{2^n[1 + p_n(f - g)]} \qquad (f,g \in C^\infty([0,2\pi]))$$

does coincide with $\mathcal{T}_P$. The details are left to the reader.

<u>2.5. Fréchet Spaces</u>. We begin this section with two definitions.

<u>Definition 2.5.1</u>. Let $(V,\mathcal{T})$ be a topological linear space over $\Phi$. Then $V$ is said to be <u>complete</u> (<u>sequentially complete</u>) if, whenever $\{x_\alpha\} \subset V$ is a Cauchy net (sequence) in $\mathcal{T}$, there exists some $x \in V$ such that $\lim_\alpha x_\alpha = x$.

<u>Definition 2.5.2</u>. Let $(V,\mathcal{T})$ be a locally convex topological linear space over $\Phi$. If $\mathcal{T}$ is metrizable and $(V,\mathcal{T})$ is complete, then $V$ is called a <u>Fréchet space</u>.

It is obvious that every Banach space is a Fréchet space. It will also be apparent from the following discussion that $(C^\infty([a,b]),\{p_n\})$, where $p_n(f) = \Sigma_{k=0}^n \|f^{(k)}\|_\infty$, $n = 0,1,2,\ldots$, is a Fréchet space, but, as we have seen in the preceding section, it is not a Banach space.

As we know from Theorem 2.3.1, if $(V,\mathcal{T})$ is a locally convex topological linear space, then there exists a family of seminorms $P$ on $V$ such that $(V,P)$ is a seminormed linear space and $\mathcal{T} = \mathcal{T}_P$. We shall see that the Fréchet spaces are precisely those complete locally convex topological linear spaces such that $P$ is countable. Before we establish this we wish to prove a lemma, which is of some independent interest.

<u>Lemma 2.5.1</u>. Let $(V,P)$ be a seminormed linear space over $\Phi$ such that $P = \{p_n\}$ is countable. Then there exists a countable family of seminorms $Q = \{q_m\}$ such that

(i) $(V,Q)$ is a seminormed linear space over $\Phi$.

(ii) $(V,P)$ and $(V,Q)$ are topologically isomorphic.

(iii) $q_1(x) \le q_2(x) \le \cdots \le q_m(x) \le q_{m+1}(x) \le \cdots$, $x \in V$.

<u>Proof</u>. For each $m = 1,2,3,\ldots$, define

$$q_m(x) = \sup\{p_1(x),\ldots,p_m(x)\} \qquad (x \in V).$$

We claim that each $q_m$ is a seminorm on $V$. Clearly $q_m(x) \ge 0$ and

$q_m(ax) = |a|q_m(x)$ where $x \in V$ and $a \in \Phi$. Moreover, if $x,y \in V$, then

$$\begin{aligned} q_m(x + y) &= \sup\{p_1(x + y),\ldots,p_m(x + y)\} \\ &\leq \sup\{p_1(x) + p_1(y),\ldots,p_m(x) + p_m(y)\} \\ &\leq \sup\{p_1(x),\ldots,p_m(x)\} + \sup\{p_1(y),\ldots,p_m(y)\} \\ &= q_m(x) + q_m(y). \end{aligned}$$

It is evident that (i) and (iii) hold. Moreover, we see that for each $\epsilon > 0$

$$\begin{aligned} \{x \mid p_1(x) < \epsilon/2,\ldots,p_m(x) < \epsilon/2\} &\subset \{x \mid q_m(x) < \epsilon\} \\ &\subset \{x \mid p_1(x) < \epsilon,\ldots,p_m(x) < \epsilon\}, \end{aligned}$$

and so the identity mapping on $V$ is obviously a topological isomorphism. □

We can now characterize the metrizable locally convex topological linear spaces.

<u>Theorem 2.5.1</u>. Let $(V,\mathcal{T})$ be a locally convex topological linear space over $\Phi$. Then the following are equivalent:

(i) $\mathcal{T}$ is metrizable.

(ii) There exists a countable family $P = \{p_n\}$ of seminorms on $V$ such that $(V,P)$ is a seminormed linear space and $\mathcal{T} = \mathcal{T}_P$.

<u>Proof</u>. If $\mathcal{T}$ is metrizable, then there exists a countable family of convex open sets $\{U_n\}$ that forms a neighborhood base at at the origin. Clearly we may assume, without loss of generality, that $U_n \supset U_{n+1}$, $n = 1,2,3,\ldots,$ and that each $U_n$ is a convex balanced absorbing set. The latter is possible because of Proposition 2.3.3. Let $p_n$ be the gauge of $U_n$, $n = 1,2,3,\ldots$ . Then from the argument used in the proof of Theorem 2.3.1 we see that $(V,P)$ is a seminormed linear space and $\mathcal{T} = \mathcal{T}_P$, where $P = \{p_n\}$.

Thus part (i) implies part (ii).

Conversely, suppose that part (ii) holds. Because of Lemma 2.5.1, we may assume that $P = \{p_n\}$ is such that $p_n(x) \leq p_{n+1}(x)$, $n = 1,2,3,\ldots$, $x \in V$. Evidently the sets $U_{n,k} = \{x \mid p_n(x) < 1/k\}$, $n,k = 1,2,3,\ldots$, form at the origin, a neighborhood base for $T$ consisting of convex balanced absorbing sets.

Define

$$\rho(x,y) = \sum_{n=1}^{\infty} \frac{p_n(x - y)}{2^n[1 + p_n(x - y)]} \qquad (x,y \in V).$$

It is easily verified that $\rho$ defines a metric on $V$ that is translation invariant, that is, $\rho(x - y,0) = \rho(x,y)$, $x,y \in V$. To see that the metric topology and $T$ agree it suffices, since $\rho$ is translation invariant, to examine what happens at the origin.

But let $n \geq 1$ and suppose $p_{n+1}(x) < 1/2^{n+1}$. Then since $p_1(x) \leq p_2(x) \leq \cdots \leq p_{n+1}(x) < 1/2^{n+1}$ and $p_k(x)/[1 + p_k(x)] \leq p_k(x)$, $k = 1,2,3,\ldots$, we see that

$$\begin{aligned}
\rho(x,0) &= \sum_{k=1}^{\infty} \frac{p_k(x)}{2^k[1 + p_k(x)]} \\
&= \sum_{k=1}^{n+1} \frac{p_k(x)}{2^k[1 + p_k(x)]} + \sum_{k=n+2}^{\infty} \frac{p_k(x)}{2^k[1 + p_k(x)]} \\
&\leq \frac{1}{2^{n+1}} \sum_{k=1}^{n+1} \frac{1}{2^k} + \sum_{k=n+2}^{\infty} \frac{1}{2^k} \\
&\leq \frac{1}{2^{n+1}} \left[ \sum_{k=1}^{\infty} \frac{1}{2^k} + \sum_{k=1}^{\infty} \frac{1}{2^k} \right] \\
&= \frac{1}{2^n}.
\end{aligned}$$

Hence $U_{n+1,1/2^{n+1}} \subset \{x \mid \rho(x,0) < 1/2^n\}$, $n = 1,2,3,\ldots$ .

On the other hand, suppose $\rho(x,0) < 1/2^{n+1}$, where $n \geq 1$. Write $n = m + k$, $m \geq 1$. Then $\rho(x,0) < 1/2^{m+k+1}$ implies that

$$\frac{p_m(x)}{2^m[1 + p_m(x)]} < \frac{1}{2^{m+k+1}}$$

and so

$$\frac{p_m(x)}{[1 + p_m(x)]} < \frac{1}{2^{k+1}} .$$

Solving for $p_m(x)$ we obtain

$$p_m(x) < \frac{1}{(2^{k+1} - 1)} \leq \frac{1}{2^k} .$$

Thus $\{x \mid \rho(x,0) < 1/2^{m+k+1}\} \subset U_{m,1/2^k}$, $m + k = 1,2,3,\ldots$ .

From these observations it is apparent that $\mathcal{T} = \mathcal{T}_P$ and the metric topology determined by $\rho$ coincide.

Therefore part (ii) of the theorem implies part (i). □

This is not the most general theorem about the metrization of topological linear spaces. Indeed, one can show that a topological linear space $(V,\mathcal{T})$ is metrizable if and only if there exists a countable neighborhood base at the origin for $\mathcal{T}$. The metric can also be constructed in this case so as to be translation invariant. For details the reader is referred to [K, pp. 162-164; KeNa, pp. 48-49].

## 2.6. Problems.

1. (Theorem 2.1.2) Let $(V,\mathcal{T})$ be a topological linear space over $\Phi$. Prove that for each $y \in V$ and each $a \in \Phi$, $a \neq 0$, the mappings $\varphi_y : V \to V$, defined by $\varphi_y(x) = x + y$, $x \in V$, and

$\varphi_a : V \to V$, defined by $\varphi_a(x) = ax$, $x \in V$, are surjective homeomorphisms. Prove that the image under $\varphi_y$ or $\varphi_a$ of a convex set is again a convex set, and that the image under $\varphi_a$ of a linear subspace is again a linear subspace.

2. Let $(V,\mathcal{T})$ be a topological linear space. Prove that every neighborhood of the origin in $V$ is absorbing.

3. Let $\mathbb{R}$ have the half-open interval topology; that is, let all intervals of the form $[a,b)$, where $a < b$, be a base for a topology $\mathcal{T}$. Prove that $(\mathbb{R},\mathcal{T})$ is <u>not</u> a topological linear space.

4. (Theorem 2.1.3) Let $(V,\mathcal{T})$ be a topological linear space over $\Phi$.

(a) If $W \subset V$ is a linear subspace, prove that the closure of $W$ is a linear subspace.

(b) If $W \subset V$ is a linear subspace and $\mathcal{T}'$ is the relative topology on $W$ induced by $\mathcal{T}$, prove that $(W,\mathcal{T}')$ is a topological linear space.

(c) If $W \subset V$ is a closed linear subspace, prove that $(V/W,\mathcal{T}')$ is a topological linear space, where $\mathcal{T}'$ is the topology on $V/W$ whose open sets are of the form $\{x + W \mid x \in U\}$, $U \in \mathcal{T}$, that is $\mathcal{T}'$ is the usual quotient topology.

5. (Theorem 2.1.4) Let $(V_\alpha,\mathcal{T}_\alpha)$, $\alpha \in \Lambda$, be topological linear spaces over $\Phi$ and let $V$ denote the product of the topological spaces $V_\alpha$, $\alpha \in \Lambda$, with the product topology $\mathcal{T}$ and with linear space operations defined componentwise. Prove that $(V,\mathcal{T})$ is a topological linear space over $\Phi$.

6. (Theorem 2.2.1) Let $(V,\mathcal{T})$ be an n-dimensional topological linear space over $\mathbb{R}(\mathbb{C})$. Prove that $(V,\mathcal{T})$ is topologically isomorphic to $(\mathbb{R}^n,\|\cdot\|_1)((\mathbb{C}^n,\|\cdot\|_1))$.

7. (Corollary 2.2.1) Let $(V_1, \mathcal{T}_1)$ and $(V_2, \mathcal{T}_2)$ be n-dimensional topological linear spaces over $\Phi$. Prove that $(V_1, \mathcal{T}_1)$ and $(V_2, \mathcal{T}_2)$ are topologically isomorphic.

8. (Corollary 2.2.2) Let $(V, \mathcal{T})$ be a topological linear space over $\Phi$ and let W be a finite-dimensional linear subspace of V. Prove that W is a closed linear subspace of $(V, \mathcal{T})$.

9. Letting $(V, \mathcal{T})$ be a topological linear space over $\Phi$ and letting A and B be any subsets of V, prove each of the following:

(a) $\mathrm{cl}(x + A) = x + \mathrm{cl}(A)$, for each $x \in V$.

(b) $\mathrm{cl}(aA) = a\,\mathrm{cl}(A)$, for each $a \in \Phi$, $a \neq 0$.

(c) $\mathrm{cl}(A) + \mathrm{cl}(B) \subset \mathrm{cl}(A + B)$.

(d) If U is open, then A + U is open.

(c) $A + \mathrm{int}(B) \subset \mathrm{int}(A + B)$.

(f) If $\mathcal{U}$ is a neighborhood base at the origin in V, then $\mathrm{cl}(A) = \bigcap\{A + U \mid U \in \mathcal{U}\}$.

*10. Let $(V, \mathcal{T})$ be a topological linear space over $\Phi$ and let A and B be any closed subsets of V.

(a) Give an example to show that A + B need not be closed.

(b) Prove that, if A is compact, then A + B is closed. Conclude that, when A is compact, $\mathrm{cl}(A) + \mathrm{cl}(B) = \mathrm{cl}(A + B)$.

(c) Prove that, if A is compact and $A \cap B = \emptyset$, then there exists a neighborhood U of the origin for which $(A + U) \cap (B + U) = \emptyset$.

11. (Proposition 2.3.2) Let $(V, \mathcal{T})$ be a topological linear space over $\Phi$.

(a) If $U \subset V$ is convex, prove that cl(U) and int(U) are convex.

(b) If $U \subset V$ is balanced, prove that cl(U) is balanced. If int(U) contains the origin, prove that int(U) is balanced.

12. Letting $V$ be a linear space over $\Phi$, prove each of the following:

(a) If $U_1,\ldots,U_n$ are convex subsets of $V$ and $a_1,\ldots,a_n$ are arbitrary scalars in $\Phi$, then $a_1U_1 +\cdots+ a_nU_n$ is a convex set.

(b) If $U$ is a convex subset of $V$ and $x_1,\ldots,x_n$ are elements of $U$, then $a_1x_1 +\cdots+ a_nx_n \in U$ when $a_1,\ldots,a_n$ are nonnegative real numbers such that $\Sigma_{k=1}^n a_k = 1$.

13. Letting $(V,T)$ be a topological linear space over $\Phi$ and letting $U$ be a convex subset of $V$ with a nonempty interior, prove each of the following:

(a) $cl(U) = cl[int(U)]$.

(b) $int(U) = int[cl(U)]$.

14. (a) If $B$ is balanced in the complex linear space $\mathbb{C}$, prove that $int(B)$ is also balanced.

(b) Give an example to show that, if $B$ is balanced in the real linear space $\mathbb{R}^2$, then $int(B)$ need not be balanced.

15. Let $(V,T)$ be a locally convex topological linear space and let $x_n \in V$, $n = 1,2,3,\ldots$ . If $\{x_n\}$ converges to $0$, prove that the sequence $\{(1/n)\ \Sigma_{k=1}^n x_k\}$ converges to $0$.

16. Let $(V,P)$ be a seminormed linear space over $\Phi$. We have defined a subset $E$ of $V$ to be bounded if for each $p \in P$ there exists some $M_p > 0$ such that $p(x) \leq M_p$, $x \in E$ (Definition 1.3.2). Prove that this is equivalent to the definition that $E$ is bounded if for every open neighborhood $U$ of the origin in $V$ there exists some $a > 0$ such that $E \subset aU$ (Definition 2.4.1).

17. Letting $(V,T)$ be a topological linear space over $\Phi$ and letting $B,B_1,\ldots,B_n$ be bounded subsets of $V$, prove each of the following:

(a) $\bigcup_{k=1}^{n} B_k$ is bounded.

(b) $B_1 + \cdots + B_n$ is bounded.

(c) $aB$ is bounded for each $a \in \Phi$.

(d) $x + B$ is bounded for each $x \in V$.

(e) $cl(B)$ is bounded.

18. Let $(V,\mathcal{T})$ be a topological linear space over $\Phi$. If $B$ is a compact subset of $V$, prove that $B$ is bounded. In particular, conclude that every convergent sequence in $V$ is bounded.

19. Let $(V,\mathcal{T})$ be a topological linear space over $\Phi$. Prove that a subset $E$ of $V$ is bounded if and only if, whenever $\{a_n\}$ is a sequence of scalars in $\Phi$ that converges to $0$ and $\{x_n\}$ is any sequence of elements of $E$, the sequence $\{a_n x_n\}$ converges to $0$.

20. Let $(V,\mathcal{T})$ be a topological linear space. Prove that every Cauchy sequence in $V$ is bounded.

21. Let $(V_\alpha,\mathcal{T}_\alpha)$ be a family of normable topological linear spaces. Prove that the topological product $V$ of the spaces $V_\alpha$ is normable -- that is, there exists a norm on $V$ such that the norm topology and the product topology $\mathcal{T}$ coincide -- if and only if the number of spaces $V_\alpha$ that are nonzero is finite.

*22. Letting $(V,\mathcal{T})$ be a topological linear space over $\Phi$ and letting $W \subset V$ be a closed linear subspace, prove each of the following:

(a) If $V$ is locally convex, then $V/W$ is locally convex. If the topology on $V$ is given by the family of seminorms $\{p_\alpha\}$, then the quotient topology on $V/W$ is given by the family of seminorms $\{\tilde{p}_\alpha\}$, where $\tilde{p}_\alpha(x + W) = \inf_{y\in W} p_\alpha(x + y)$, $x \in V$.

(b) If $V$ is a Fréchet space, then $V/W$ is a Fréchet space. Note that, if $V$ is an arbitrary, complete, locally convex space, then $V/W$ need not be complete.

23. Let $(V_\alpha, \mathcal{T}_\alpha)$ be a family of locally convex metrizable topological linear spaces over $\Phi$. Prove that the topological product $V$ of the spaces $V_\alpha$ is metrizable if and only if at most countably many of the spaces $V_\alpha$ are nonzero.

24. A topological linear space $(V, \mathcal{T})$ is said to be _locally bounded_ if there exists a bounded open neighborhood of the origin in $V$. Prove that every locally bounded topological linear space is metrizable.

*25. A locally convex topological linear space $(V, \mathcal{T})$ is said to be _bornological_ if, whenever $A$ is a balanced convex subset of $V$ that absorbs every bounded set in $V$ (that is, for every bounded $B \subset V$ there exists some $a_0 \in \Phi$ such that $B \subset aA$ whenever $|a| \geq |a_0|$), the origin is contained in the interior of $A$. Prove that every metrizable locally convex topological linear space is bornological.

CHAPTER 3

LINEAR TRANSFORMATIONS AND LINEAR FUNCTIONALS

3.0. Introduction. Linear transformations and linear functionals play a central role in functional analysis, and after defining these concepts, we shall examine a number of concrete examples. Then we shall discuss some basic results concerning linear transformations, the most important being the equivalence between the notions of continuity and boundedness, and the fact that the space of all continuous linear transformations from a normed linear space to a Banach space can be made into a Banach space in a natural manner. The last section of the chapter contains a few fundamental results pertaining to linear functionals. A characterization of continuity of linear functionals is established, and the question of the existence of sufficiently many nonzero continuous linear functionals on $V$ to separate the points of $V$ is considered. This will set the scene for the Hahn-Banach Theorem to be discussed in the next chapter.

3.1. Linear Transformations. Let us begin by defining linear transformations and linear functionals.

Definition 3.1.1. Let $V_1$ and $V_2$ be linear spaces over $\Phi$. A mapping $T : V_1 \to V_2$ is said to be a linear transformation if

$$T(ax + by) = aT(x) + bT(y) \qquad (x,y \in V;\ a,b \in \Phi).$$

The collection of all linear transformations from $V_1$ to $V_2$ will be denoted by $L'(V_1,V_2)$. If $(V_1,T_1)$ and $(V_2,T_2)$ are topological linear spaces, those $T \in L'(V_1,V_2)$ that are continuous will be denoted by $L(V_1,V_2)$.

Some authors use the term "<u>linear operator</u>" instead of "linear transformation".

It is readily seen that $L'(V_1,V_2)$ can be made into a linear space over $\Phi$ by defining

$$(T + S)(x) = T(x) + S(x)$$
$$(aT)(x) = aT(x) \qquad (T,S \in L'(V_1,V_2);\ x \in V_1;\ a \in \Phi).$$

We shall always assume that this has been done. Clearly, when $V_1$ and $V_2$ are topological linear spaces, $L(V_1,V_2)$ is a linear subspace of $L'(V_1,V_2)$.

Also it should be expressly noted that $T \in L'(V_1,V_2)$ is always defined on all of $V_1$. It is, however, possible to discuss linear transformations that are only defined on subsets of $V_1$, but this introduces additional difficulties that we prefer to avoid -- for example, the definition of linear space operations in $L'(V_1,V_2)$. It is evident though that the domain of a linear transformation must be a linear space, in any case.

When $V_1 = V_2 = V$, we shall generally write $L'(V)$ and $L(V)$ for $L'(V,V)$ and $L(V,V)$, respectively; and when $V_1 = V$ and $V_2 = \Phi$, we give the elements of $L'(V,\Phi)$ a special name.

<u>Definition 3.1.2</u>. Let V be a linear space over $\Phi$. Then the linear transformations in $L'(V,\Phi)$ are called <u>linear functionals</u> on V. If $(V,T)$ is a topological linear space and $\Phi$ has the usual absolute-value topology, then the elements of $L(V,\Phi)$ are called <u>continuous linear functionals</u> on V.

The linear space $L'(V,\Phi)$ will be denoted by $V'$ and a generic element in $V'$ by $x'$; similarly $L(V,\Phi)$ will be denoted by $V^*$ and its generic elements by $x^*$; $V^*$ will also be referred to as the <u>conjugate</u>, or <u>dual space</u> of V.

Obviously, since linear functionals are special instances of linear transformations, any results that we establish for the latter objects will automatically hold for the former.

Let us look at a few examples of linear transformations and functionals. As with the examples in Chapter 1, the verifications of the following assertions are left to the reader.

Example 3.1.1. Let $V_1 = \mathbb{R}^n$ and $V_2 = \mathbb{R}^m$, and consider the $m \times n$ real matrix $A = (a_{ij})$. Then

$$T(x) = A\begin{pmatrix} x_1 \\ \vdots \\ x_n \end{pmatrix} = \begin{pmatrix} \Sigma_{j=1}^n a_{1j}x_j \\ \vdots \\ \Sigma_{j=1}^n a_{mj}x_j \end{pmatrix} \qquad (x = (x_1,x_2,\ldots,x_n) \in \mathbb{R}^n)$$

defines a linear transformation in $L'(\mathbb{R}^n,\mathbb{R}^m)$. If we consider $\mathbb{R}^n$ and $\mathbb{R}^m$ with any topologies under which they are topological linear spaces (e.g., any of the norms $\|\cdot\|_p$, $1 \leq p \leq \infty$) then $T \in L(\mathbb{R}^n,\mathbb{R}^m)$. If $m = 1$, then $T \in (\mathbb{R}^n)^*$.

Example 3.1.2. Let $(V,\|\cdot\|) = (C([0,1]),\|\cdot\|_\infty)$. Then the transformation $T$ defined for each $f \in C([0,1])$ by

$$(Tf)(t) = \int_0^t f(s)\,ds \qquad (0 \leq t \leq 1).$$

belongs to $L(V)$.

Example 3.1.3. Let $(V,P) = (C^\infty([0,1]),\{p_n\})$, where

$$p_n(f) = \|f\|_n = \Sigma_{k=0}^n \|f^{(k)}\|_\infty \qquad (n = 0,1,2,\ldots).$$

Then $(V,P)$ is a seminormed linear space, and the transformation $T$ defined for each $f \in C^\infty([0,1])$ by

$$(Tf)(t) = f'(t) \qquad (0 \leq t \leq 1)$$

belongs to $L(V)$.

The elements of $C^{\infty}([0,1])^*$ are called distributions. Discussion of this important class of continuous linear functionals can be found, for example, in [$E_1$, pp. 297-418; $E_3$, pp. 46-132; Sh, Y, pp. 28-30, 46-52]. We, however, shall content ourselves with this passing mention.

Example 3.1.4. Let $X$ be a locally compact Hausdorff topological space and let $(V,\|\cdot\|_{\infty}) = (C_o(X),\|\cdot\|_{\infty})$.

For each $t_o \in X$ define

$$x^*(f) = f(t_o) \qquad (f \in C_o(X)).$$

Then $x^* \in C_o(X)^*$.

More generally, if $\mu \in M(X)$, then

$$x^*(f) = \int_X f(t)\, d\mu(t) \qquad (f \in C_o(X))$$

defines an element $x^* \in C_o(X)^*$. The converse of this last assertion is also true, but we shall not prove it in its full generality. We shall, however, establish it for a spacial case in the next chapter.

Example 3.1.5. Let $X$ be a locally compact Hausdorff space and let $(V,\|\cdot\|) = (M(X),\|\cdot\|)$, where as usual the norm in $M(X)$ is the total variation norm. For each $f \in C_o(X)$ define

$$x^*(\mu) = \int_X f(t)\, d\mu(t) \qquad (\mu \in M(X)).$$

Then $x^* \in M(X)^*$. However, it is not generally the case that all elements of $M(X)^*$ are obtained in this manner. For example, if $X$ is not compact, then the preceding formula defines an element of $M(X)^*$ for any $f \in C(X)$.

Example 3.1.6. Let $(V,\|\cdot\|) = (c_o,\|\cdot\|_{\infty})$ and let $\{a_n\} \in \ell_1$. Then

$$x^*(\{b_n\}) = \sum_{n=1}^{\infty} b_n a_n \qquad (\{b_n\} \in c_o)$$

defines an element of $c_o^*$. Moreover all of $c_o^*$ is obtained in this way. This is actually just a special case of Example 3.1.4 when $X = \{n \mid n = 1,2,3,\ldots\}$ since in this case $M(X) = \ell_1$.

Example 3.1.7. Let $(V_1,\|\cdot\|_1) = (L_1([-\pi,\pi],dt/2\pi),\|\cdot\|_1)$, where dt denotes Lebesgue measure on $[-\pi,\pi]$, and let $(V_2,\|\cdot\|_2) = (C_o(\mathbb{Z}),\|\cdot\|_\infty)$, where $\mathbb{Z}$ denotes the locally compact space of the integers with the discrete topology. For each integer n we define

$$\hat{f}(n) = \frac{1}{2\pi}\int_{-\pi}^{\pi} f(t)e^{-int}\,dt \qquad (f \in L_1([-\pi,\pi],dt/2\pi)).$$

Then $\hat{f} \in C_o(\mathbb{Z})$, and the transformation T, defined by $T(f) = \hat{f}$, $f \in L_1([-\pi,\pi],dt/2\pi)$, belongs to $L(V_1,V_2)$. That is, the Fourier transformation is a continuous linear transformation from $L_1([-\pi,\pi],dt/2\pi)$ to $C_o(\mathbb{Z})$. The fact that T maps $L_1([-\pi,\pi],dt/2\pi)$ into $C_o(\mathbb{Z})$ is nontrivial and is known as the Riemann-Lebesgue Lemma [$E_2$, p. 36].

Example 3.1.8. Let

$$(V,\|\cdot\|) = (L_p(X,S,\mu),\|\cdot\|_p) = (L_p(\mu),\|\cdot\|_p) \qquad (1 \le p \le \infty),$$

and let q, $1 \le q \le \infty$, be such that $1/p + 1/q = 1$, where $q = \infty$ if $p = 1$ and $q = 1$ if $p = \infty$. Then, if $g \in L_q(\mu)$, the formula

$$x^*(f) = \int_X f(t)\overline{g(t)}\,d\mu(t) \qquad (f \in L_p(\mu))$$

defines an element $x^* \in L_p(\mu)^*$. If $1 < p < \infty$, then the converse assertion is also valid; that is, every element of $L_p(\mu)^*$ is determined by an element of $L_q(\mu)$. We shall prove this in Section 8.3. If $p = 1$, then the converse assertion need not be valid. However, if $(X,S,\mu)$ is a measure space such that there exists a family $\{X_\alpha\} \subset S$ of disjoint subsets of X with the properties that $\mu$ is $\sigma$-finite when restricted to $X_\alpha$, and if $E \in S$ and $\mu(E) < \infty$, then

$E \cap X_\alpha \neq \emptyset$ for at most a countable number of $\alpha$, then every continuous linear functional on $L_1(\mu)$ is given by the preceding formula for some $g \in L_\infty(\mu)$. In particular this is true if $\mu$ is σ-finite (see, for example, [$DS_1$, pp. 289-290]). If X is a locally compact topological space and $\mu$ is a regular Borel measure, then it is always the case that every continuous linear functional on $L_1(\mu)$ is obtained from some $g \in L_\infty(\mu)$ [$E_1$, pp. 215-220,239-240].

When $p = \infty$, then, except in trivial cases, there always exist continuous linear functionals on $L_\infty(\mu)$ that are not given by the preceding formula for some $g \in L_1(\mu)$ (see [$DS_1$, p. 296]).

3.2. Some Basic Results Concerning Linear Transformations. The linearity property of linear transformations is of fundamental importance, and its consequences will be used repeatedly in the remainder of this volume. The reader should pay special note to the role linearity plays in the proofs of the results in this section, as they are typical of the importance of linearity. Subsequently the role of linearity may not always be as clearly displayed as it is here.

Proposition 3.2.1. Let $V_1$ and $V_2$ be linear spaces over $\Phi$ and let $T \in L'(V_1,V_2)$. Then

(i) $T(0) = 0$.

(ii) The range of T,

$$R(T) = \{y \mid y \in V_2,\ y = T(x) \text{ for some } x \in V_1\},$$

is a linear subspace of $V_2$.

(iii) T is injective if and only if $T(x) = 0$ implies $x = 0$.

(iv) If T is injective, then $T^{-1}$ exists and $T^{-1} \in L'(R(T),V_1)$.

Proof. Since the proofs are quite elementary, we give only that for part (iii). Clearly, if T is injective, then $T(x) = 0$ implies $x = 0$. Conversely, suppose that $x,y \in V_1$ and $T(x) = T(y)$. Then, since T is a linear mapping, we have $0 = T(x) - T(y) = T(x - y)$, and hence $x - y = 0$, that is, $x = y$. Consequently T is injective whenever $T(x) = 0$ implies $x = 0$.

□

Proposition 3.2.2. Let $(V_1, \mathcal{T}_1)$ and $(V_2, \mathcal{T}_2)$ be topological linear spaces over $\Phi$ and let $T \in \mathcal{L}'(V_1, V_2)$. Then the following are equivalent:

(i) There exists some $x_o \in V_1$ such that $T$ is continuous at $x_o$.

(ii) $T$ is continuous at the origin in $V_1$.

(iii) $T \in \mathcal{L}(V_1, V_2)$.

(iv) $T$ is uniformly continuous on $V_1$.

Proof. It is obvious that (iv) implies (iii) implies (ii) implies (i). Suppose that $T$ is continuous at $x_o$. If $U_2 \in \mathcal{T}_2$ is an open neighborhood of the origin in $V_2$, then from Theorem 2.1.2 we see that $W_2 = U_2 + T(x_o)$ is an open enighborhood of $T(x_o)$. Thus there exists some open neighborhood $W_1 \in \mathcal{T}_1$ of $x_o$ such that, if $x \in W_1$, then $T(x) \in W_2$. But then $U_1 = W_1 - x_o$ is an open neighborhood of the origin in $V_1$, and hence, if $y, z \in V_1$ are such that $y - z \in U_1$, then $T(y) - T(z) = T(y - z) \in W_2 - T(x_o) = U_2$. Consequently $T$ is uniformly continuous.

Therefore (i) implies (iv), and the proof is complete. □

One of the must useful results concerning the continuity of linear transformations between normed linear spaces is the relationship between 'boundedness' and continuity given in the next theorem.

Definition 3.2.1. Let $(V_1, \|\cdot\|_1)$ and $(V_2, \|\cdot\|_2)$ be normed linear spaces over $\Phi$ and let $T \in \mathcal{L}'(V_1, V_2)$. Then $T$ is said to be bounded if there exists some $M > 0$ such that

$$\|T(x)\|_2 \leq M\|x\|_1 \qquad (x \in V_1).$$

If $T \in \mathcal{L}'(V_1, V_2)$ is bounded, then we define

$$\|T\| = \inf\{M \mid \|T(x)\|_2 \leq M\|x\|_1,\ x \in V_1\}\ .$$

<u>Theorem 3.2.1</u>. Let $(V_1, \|\cdot\|_1)$ and $(V_2, \|\cdot\|_2)$ be normed linear spaces over $\Phi$ and let $T \in L'(V_1, V_2)$. Then the following are equivalent:

(i) $T \in L(V_1, V_2)$.
(ii) $T$ is bounded.

Moreover, if $T$ is bounded, then

$$\|T\| = \sup_{\substack{x \in V_1 \\ x \neq 0}} \frac{\|T(x)\|_2}{\|x\|_1} = \sup_{\substack{x \in V_1 \\ \|x\|_1 \leq 1}} \|T(x)\|_2 = \sup_{\substack{x \in V_1 \\ \|x\|_1 = 1}} \|T(x)\|_2 .$$

<u>Proof</u>. If $T$ is bounded, then the estimate

$$\|T(x) - T(y)\|_2 \leq M\|x - y\|_1 \qquad (x, y \in V_1)$$

immediately shows that $T$ is continuous. Conversely, suppose $T$ is continuous. Then, in particular, $T$ is continuous at the origin in $V_1$, and so there exists some $\epsilon > 0$ such that $\|x\|_1 < \epsilon$, implies $\|T(x)\|_2 < 1$. But then for any $y \in V_1$, $y \neq 0$, set $z = \epsilon y/2\|y\|_1$. Clearly $\|z\|_1 = \epsilon/2 < \epsilon$, and so $\|T(z)\|_2 < 1$. Simple computations then reveal that $\|T(y)\|_2 \leq 2\|y\|_1/\epsilon$. Obviously this estimate is also valid when $y = 0$, and therefore, taking $M = 2/\epsilon$, we conclude that $T$ is bounded.

The remainder of the proof is left to the reader. □

The notation $\|T\|$ is not haphazard, as $\|\cdot\|$ actually defines a norm on $L(V_1,V_2)$. Indeed, we have the following result:

Theorem 3.2.2. Let $(V_1,\|\cdot\|_1)$ and $(V_2,\|\cdot\|_2)$ be normed linear spaces over $\Phi$. Then

(i) $(L(V_1,V_2),\|\cdot\|)$ is a normed linear space over $\Phi$ where

$$\|T\| = \sup_{\substack{x\in V_1\\ \|x\|_1 = 1}} \|T(x)\|_2 \qquad (T \in L(V_1,V_2)).$$

(ii) $(L(V_1,V_2),\|\cdot\|)$ is a Banach space over $\Phi$ provided $(V_2,\|\cdot\|_2)$ is a Banach space over $\Phi$.

Proof. The proof of part (i) is routine and is omitted. Assuming that $(V_2,\|\cdot\|_2)$ is a Banach space, let $\{T_n\} \subset L(V_1,V_2)$ be a Cauchy sequence; that is, given $\epsilon > 0$, there exists some positive integer $N$ such that $n,m > N$ implies that $\|T_n - T_m\| < \epsilon$. Since

$$\|T_n(x) - T_m(x)\|_2 \le \|T_n - T_m\|\ \|x\|_1 \qquad (x \in V_1),$$

we see at once that $\{T_n(x)\}$ is a Cauchy sequence in $V_2$ for each $x \in V_1$. Thus, since $V_2$ is a Banach space, there exists for each $x \in V_1$ some element, call it $T(x)$, in $V_2$ such that $\lim_n\|T_n(x) - T(x)\|_2 = 0$. We claim that the mapping $x \to T(x)$ from $V_1$ to $V_2$ defines a linear transformation $T \in L(V_1,V_2)$.

To prove the linearity of $T$ we note that for each $x,y \in V_1$ and $a,b \in \Phi$ we have

$$\begin{aligned}\|T(ax + by) - aT(x) - bT(y)\|_2 &\le \|T(ax + by) - T_n(ax + by)\|_2\\ &\quad + \|aT_n(x) + bT_n(y) - aT(x) - bT(y)\|_2\\ &\le \|T(ax + by) - T_n(ax + by)\|_2\\ &\quad + |a|\,\|T_n(x) - T(x)\|_2\\ &\quad + |b|\,\|T_n(y) - T(y)\|_2.\end{aligned}$$

Since the right-hand side tends to zero as $n$ tends to infinity, we conclude that $T$ is linear.

The continuity of $T$ will be established, by Theorem 3.2.1, once we have shown that $T$ is bounded. But clearly, if $\{T_n\}$ is a Cauchy sequence in $L(V_1,V_2)$, then $\{\|T_n\|\}$ is a Cauchy sequence of nonnegative numbers, as $\|\cdot\|$ is a norm. Hence there exists some $M > 0$ for which $\|T_n\| \leq M$, $n = 1,2,3,\ldots$ . Thus for each $x \in V_1$ we have

$$\|T_n(x)\|_2 \leq M\|x\|_1 \qquad (n = 1,2,3,\ldots),$$

from which it follows that $\|T(x)\|_2 \leq M\|x\|_1$. Therefore $T \in L(V_1,V_2)$.

It remains only to show that $\{T_n\}$ converges to $T$ in $(L(V_1,V_2),\|\cdot\|)$.

But, as already noted, given $\epsilon > 0$ there exists $N$ such that, if $n,m > N$, then

$$\|T_n(x) - T_m(x)\|_2 \leq \|T_n - T_m\|\|x\|_1 < \epsilon\|x\|_1 \qquad (x \in V_1).$$

Holding $n$ fixed and letting $m$ tend to infinity, we deduce that for $n > N$,

$$\|T_n(x) - T(x)\|_2 \leq \epsilon\|x\|_1 \qquad (x \in V_1),$$

from which it follows by the definition of $\|\cdot\|$ that $\|T_n - T\| \leq \epsilon$, for $n > N$. That is, $\{T_n\}$ converges to $T$ in $(L(V_1,V_2),\|\cdot\|)$, and so $(L(V_1,V_2),\|\cdot\|)$ is a Banach space. □

Since the scalar field $\Phi$ can always be considered to be a Banach space, we have the following corollary:

<u>Corollary 3.2.1</u>. Let $(V,\|\cdot\|)$ be a normed linear space over $\Phi$. Then $(V^*,\|\cdot\|)$ is a Banach space over $\Phi$.

Note that the norm in $V^*$ here is, of course, that in $V^* = L(V,\Phi)$, and not the norm in $V$, despite the notation. On the surface this may appear confusing, but it will not be so in practice, since the context will make clear which norm is meant. Moreover, the method adopted helps to simplify notation.

It is perhaps worthwhile to write out explicitly the various expressions for $\|x^*\|$, $x^* \in V^*$, indicated in Theorem 3.2.1. Thus for $x^* \in V^*$ we have

$$\begin{aligned} \|x^*\| &= \sup_{x \neq 0} \frac{|x^*(x)|}{\|x\|} \\ &= \sup_{\|x\| \leq 1} |x^*(x)| \\ &= \sup_{\|x\| = 1} |x^*(x)|. \end{aligned}$$

We close this section with two further general results about linear transformations and normed linear spaces.

<u>Proposition 3.2.3</u>. Let $(V_1,\|\cdot\|_1)$ and $(V_2,\|\cdot\|_2)$ be normed linear spaces over $\Phi$ and let $T \in L(V_1,V_2)$. Then the following are equivalent:

(i) $T^{-1}$ exists and $T^{-1} \in L(R(T),V_1)$, where $R(T)$ is considered to be a linear subspace of $(V_2,\|\cdot\|_2)$.

(ii) There exists some $m > 0$ such that

$$m\|x\|_1 \leq \|T(x)\|_2 \qquad (x \in V_1).$$

<u>Proof</u>. If part (ii) holds, then clearly $T(x) = 0$ implies $x = 0$, and from Proposition 3.2.1(iii) and (iv) we see that $T^{-1}$ exists and belongs to $L'(R(T),V_1)$. Inequality (ii), however, says precisely that

$$\|T^{-1}(y)\|_1 \leq (1/m)\|y\|_2 \qquad (y \in R(T)),$$

that is, $T^{-1} \in L(R(T),V_1)$.

The converse implication is apparent. □

Clearly, we see from the preceding, that $\|T^{-1}\| \leq 1/m$.

Proposition 3.2.4. Let $(V_1,\|\cdot\|_1)$ and $(V_2,\|\cdot\|_2)$ be normed linear spaces over $\Phi$. If $(V_1,\|\cdot\|_1)$ is finite dimensional, then $L'(V_1,V_2) = L(V_1,V_2)$.

3.3. Some Basic Results Concerning Linear Functionals. If V is a linear space over $\mathbb{C}$, then it is easily seen that V can also be considered as a linear space over $\mathbb{R}$. Our first concern in this section is to investigate the connection between linear functionals on V as a linear space over $\mathbb{C}$ and linear functionals on V as a linear space over $\mathbb{R}$. This connection will be useful in discussing the Hahn-Banach Theorem. Second, we show that a linear functional on a topological linear space is continuous if and only if its kernel, that is, $\{x \mid x \in V, x'(x) = 0\}$, is a closed linear subspace of V. This is also a particularly useful result, as will be seen in Chapter 5. Finally, we shall give an example of a topological linear space that has no nonzero continuous linear functionals.

Definition 3.3.1. Let V be a linear space over $\mathbb{C}$. If x' is a linear functional on V considered as a linear space over $\mathbb{R}$, then x' is said to be a real linear functional on V.

Proposition 3.3.1. Let V be a linear space over $\mathbb{C}$ and let $x' \in V'$. If $x_1' = \mathrm{Re}(x')$ and $x_2' = \mathrm{Im}(x')$, that is, $x_1'$ and $x_2'$ are the real and imaginary parts of x', respectively, then $x_1'$ and $x_2'$ are real linear functionals on V. Moreover,

$$x'(x) = x_1'(x) - ix_1'(ix) = x_2'(ix) + ix_2'(x) \qquad (x \in V).$$

Proof. Let $a,b \in \mathbb{R}$ and $x,y \in V$. Then on the one hand,

$$\begin{aligned} x'(ax + by) &= ax_1'(x) + aix_2'(x) + bx_1'(y) + bix_2'(y) \\ &= ax_1'(x) + bx_1'(y) + i[ax_2'(x) + bx_2'(y)], \end{aligned}$$

while on the other hand,

$$x'(ax + by) = x_1'(ax + by) + ix_2'(ax + by).$$

Equating real and imaginary parts, we conclude at once that $x_1'$ and $x_2'$ are linear functionals over $\mathbb{R}$.

For the second part of the proposition we note first that for each $x \in V$ we have

$$x'(ix) = x_1'(ix) + ix_2'(ix).$$

But also, since $x' \in V'$,

$$\begin{aligned} x'(ix) &= ix'(x) = i[x_1'(x) + ix_2'(x)] \\ &= ix_1'(x) - x_2'(x). \end{aligned}$$

Again equating real and imaginary parts, we deduce that $x_1'(ix) = -x_2'(x)$, $x_2'(ix) = x_1'(x)$, $x \in V$, and so

$$\begin{aligned} x'(x) &= x_1'(x) + ix_2'(x) \\ &= x_1'(x) - ix_1'(ix) \\ &= x_2'(ix) + ix_2'(x). \end{aligned}$$

□

The last portion of this proposition shows us how to express a linear functional in terms of either its real or imaginary part alone. Conversely, the next proposition shows us how to define linear functionals over $\mathbb{C}$ by using real linear functionals. The details are straightforward and are left to the reader.

Proposition 3.3.2. Let $V$ be a linear space over $\mathbb{C}$ and let $x_1'$ be a real linear functional on $V$. If $x'(x) = x_1'(x) - ix_1'(ix)$ for each $x$ in $V$, then $x'$ is in $V'$.

The next theorem can be easily derived from the preceding two propositions, we omit the proof.

Theorem 3.3.1. Let $(V,\mathcal{T})$ be a topological linear space over $\mathbb{C}$ and let $x' \in V'$. Then the following are equivalent:

(i) $x' \in V^*$.

(ii) $\text{Re}(x')$ is a continuous real linear functional on $V$.

(iii) $\text{Im}(x')$ is a continuous real linear functional on $V$.

Now we turn to the second concern of this section. First we need a few definitions.

Definition 3.3.2. Let $V$ be a linear space over $\Phi$ and let $W$ be a proper linear subspace of $V$. Then $W$ is said to be a *maximal linear subspace* if, whenever $W \subset W_1$, either $W = W_1$ or $W_1 = V$.

It is easily seen that a proper linear subspace $W \subset V$ is maximal if and only if it is of *codimension one*, that is, for any $x_o \in V \sim W$ the linear subspace spanned by $\{x_o\} \cup W$ is all of $V$. Equivalently, $W$ is maximal if and only if $V/W$ is of dimension one.

Definition 3.3.3. Let $V$ be a linear space over $\Phi$ and let $x' \in V'$. Then $N(x') = \{x \mid x \in V, x'(x) = 0\}$ is called the *kernel* of $x'$.

Another common name for $N(x')$ is the *null space* of $x'$.

Proposition 3.3.3. Let $V$ be a linear space over $\Phi$ and let $x' \in V'$, $x' \neq 0$. Then $N(x')$ is a maximal linear subspace of $V$.

Proof. It is evident that $N(x')$ is a proper linear subspace. To prove that $N(x')$ is maximal it clearly suffices to show that, if $x_o \in V \sim N(x')$, then the linear span of $N(x') \cup \{x_o\}$ is all of $V$. But if $x \in V$, then, since $x'(x_o) \neq 0$, we see that

$$x - [x'(x)/x'(x_o)]x_o \in V$$

is such that $x'(x - [x'(x)/x'(x_o)]x_o) = 0$. Hence

$$x - [x'(x)/x'(x_o)]x_o \in N(x'),$$

and so $x \in N(x') + [x'(x)/x'(x_o)]x_o$.

Therefore the linear span of $N(x') \cup \{x_o\}$ is $V$. □

Corollary 3.3.1. Let $V$ be a linear space over $\Phi$ and let $x', x_1', x_2', \ldots, x_n'$ be in $V'$. Then the following are equivalent:

(i) $x'$ is linearly dependent on $x_1', x_2', \ldots, x_n'$.
(ii) $N(x') \supset \bigcap_{k=1}^{n} N(x_k')$.

The proof is left to the reader.

Next we establish the indicated characterization of continuous linear functionals.

Theorem 3.3.2. Let $(V, \mathcal{T})$ be a topological linear space over $\Phi$ and let $x' \in V'$. Then the following are equivalent:

(i) $x' \in V^*$.
(ii) $N(x')$ is closed.

Proof. Clearly (i) implies (ii), and if $x' = 0$, then obviously (ii) implies (i). On the other hand, suppose that $x' \neq 0$, that $N(x')$ is closed, and that $x'$ is not continuous. Then, by Proposition 3.2.2, $x'$ is not continuous at the origin, and so there exists some $\epsilon > 0$ such that every open neighborhood $U$ of the origin contains some point $x$ for which $|x'(x)| \geq \epsilon$.

Let $W = \{x \mid x \in V,\ x'(x) = \epsilon/2\}$. We claim that $W = N(x') + y_0$ for some $y_0 \in V$. Indeed, if $y \in V \sim N(x')$, then we see immediately that $x'(x + [\epsilon/2x'(y)]y) = \epsilon/2$, $x \in N(x')$, and so

$$N(x') + [\epsilon/2x'(y)]y \subset W.$$

Conversely, since $N(x')$ is of codimension one, if $z \in W$, then there exist some $a \in \Phi$, and $x \in N(x')$ such that $z = x + ay$, and so $x'(z) = ax'(y) = \epsilon/2$. Thus $a = \epsilon/2x'(y)$, and we conclude that $W = N(x') + y_0$, where $y_0 = [\epsilon/2x'(y)]y$.

Consequently, by Theorem 2.1.2, $W$ is a closed subset of $V$, as it is the translate of the closed subspace $N(x')$, and $0 \notin W$. Thus there exists an open neighborhood $U_0$ of $0$ such that

$U_0 \cap W = \emptyset$. Moreover, by Proposition 2.3.2, we may assume that $U_0$ is balanced. Now there exists some $x_0 \in U_0$ such that $|x'(x_0)| \geq \epsilon$. Let $a = \epsilon/2x'(x_0)$. Then clearly $|a| < 1$, and so $ax_0 \in U_0$. But $x'(ax_0) = x'([\epsilon/2x'(x_0)]x_0) = \epsilon/2$ implies that $ax_0 \in W$, contradicting the fact that $U_0 \cap W = \emptyset$.

Therefore $x'$ is continuous. □

If $x' \in V'$ and $x' \neq 0$, then Theorem 3.3.2 gives a simple and useful characterization of the continuity of $x'$. However, there is a more fundamental question concerning linear functionals that we have not yet faced: Do there exist any nonzero (continuous) linear functionals on a given (topological) linear space $V$? We shall see shortly that it is always the case, when $V \neq \{0\}$, that $V' \neq \{0\}$, while it may very well be the case that $V^* = \{0\}$. A related question is whether $V'$ and $V^*$ separate the points of $V$; that is, given $x,y \in V$, $x \neq y$, does there exist an $x' \in V'$ or $x^* \in V^*$ such that $x'(x) \neq x'(y)$ or $x^*(x) \neq x^*(y)$? Again we shall see that, when $V \neq \{0\}$, then $V'$ separates the points of $V$, but this need not be the case for $V^*$.

In the next chapter we shall establish general conditions on $V$ that ensure the existence of sufficiently many nonzero continuous linear functionals on $V$ to separate the points of $V$. Indeed, we shall see that this is always the case whenever $V$ is a locally convex topological linear space -- equivalently, a seminormed linear space. Somewhat different proofs of these facts will also be given in Chapter 5. In both cases the Hahn-Banach Theorem, in either its analytic or geometric form, will be instrumental in establishing the desired result. The remainder of this section is devoted to showing that $V'$ always contains sufficiently many nonzero elements to separate the points of $V$ and to giving an example of a particular linear space $V$ for which $V^*$ fails to have this property.

For the sake of completeness we make the following definition:

Definition 3.3.4. Let $V$ be a linear space over $\Phi$. Then a family of linear functionals $F \subset V'$ is said to separate points if, whenever $x,y \in V$, $x \neq y$, there exists some $x' \in F$ such that $x'(x) \neq x'(y)$.

Note, in particular, that, if $V \neq \{0\}$ and $F \subset V'$ separates points, then $V' \neq \{0\}$ and that $F \subset V'$ separates points if and only if $x'(x) = 0$, $x' \in F$, implies $x = 0$.

Theorem 3.3.3. Let $V$ be a linear space over $\Phi, V \neq \{0\}$. Then $V'$ separates points.

Proof. Let $x_o \in V$, $x_o \neq 0$. We claim that there exists a linear subspace $W_o \subset V$ of codimension one such that $x_o \notin W_o$.

Indeed, consider the family $\mathcal{W}$ of linear subspaces $W \subset V$ such that $x_o \notin W$. Clearly $\mathcal{W} \neq \emptyset$, as $\{0\} \in \mathcal{W}$. We introduce a partial ordering into $\mathcal{W}$ by setting $W_1 < W_2$ whenever $W_1 \subset W_2$, and $W_1, W_2 \in \mathcal{W}$. Suppose that $\mathcal{V} = \{W_\alpha\}_{\alpha \in \Lambda}$ is a linearly ordered subset of $\mathcal{W}$. Then, as is easily verified, $\cup_{\alpha \in \Lambda} W_\alpha$ is an upper bound for $\mathcal{V}$, that is, $\cup_{\alpha \in \Lambda} W_\alpha \in \mathcal{W}$ and $W_\alpha < \cup_{\alpha \in \Lambda} W_\alpha$, $\alpha \in \Lambda$. Consequently, by Zorn's Lemma [$DS_1$, p. 6], the family $\mathcal{W}$ has a maximal element, which we denote by $W_o$.

Moreover, $W_o$ is of codimension one, since if this is not the case, then the dimension of $V/W_o$ will be greater than one and so there will exist some $x_1 \in V \sim W_o$ such that $x_1$ will be linearly independent of $x_o$ and for which $x_o - ax_1 \notin W_o$, $a \in \Phi$. It is then easily verified that the linear span of $W_o \cup \{x_1\}$ is an element of $\mathcal{W}$ that properly contains $W_o$, thereby contradicting the maximality of $W_o$. Thus for each $x \in V$ there exists a unique $a \in \Phi$ and $y \in W_o$ such that $x = ax_o + y$. Define $x'(x) = a$. It is then evident that $x' \in V'$, $N(x') = W_o$, and $x'(x_o) = 1$.

Consequently, if $y_o, z_o \in V$, $y_o \neq z_o$, then, by the previous construction, if $x_o = y_o - z_o \neq 0$, we see that there exists some $x' \in V'$ such that $x'(x_o) = 1$; that is, $x'(y_o) - x'(z_o) = 1$.

Therefore $x'(y_o) \neq x'(z_o)$, and $V'$ separates points. □

As we have indicated, if $V \neq \{0\}$ is a topological linear space, then it may be the case that $V^* = \{0\}$, and so a fortiori $V^*$ does not separate points. A concrete example of this is provided by the topological linear spaces $L_p([0,1],dt)$, $0 < p < 1$, where $dt$ denotes Lebesgue measure on $[0,1]$.

Theorem 3.3.4. $L_p([0,1],dt)^* = \{0\}$, $0 < p < 1$.

Proof. Suppose that $x^* \in L_p([0,1],dt)^*$ and $x^* \neq 0$. Then there exists some $f \in L_p([0,1],dt)$ such that $|x^*(f)| \geq 1$. For each $s$, $0 \leq s \leq 1$, set $f_s^1 = \chi_{[0,s]}f$, where $\chi_{[0,s]}$ denotes the characteristic function of the interval $[0,s]$, and set $f_s^2 = f - f_s^1$. Clearly $f_s^1, f_s^2 \in L_p([0,1],dt)$, $0 \leq s \leq 1$, and

$$(\|f_s^1\|_p)^p = \int_0^s |f(t)|^p \, dt \qquad (0 \leq s \leq 1)$$

reveals that $(\|f_s^1\|_p)^p$ is a continuous monotone increasing function of $s$ and that

$$(\|f_0^1\|_p)^p = 0$$

and

$$(\|f_1^1\|_p)^p = (\|f\|_p)^p.$$

Consequently there exists some $s_o$, $0 < s_o < 1$, such that $(\|f_{s_o}^1\|_p)^p = (\|f\|_p)^p/2$. For this $s_o$ we also have

$$
\begin{aligned}
(\|f_{s_o}^2\|_p)^p &= \int_0^1 |f(t) - f_{s_o}^1(t)|^p \, dt \\
&= \int_{s_o}^1 |f(t)|^p \, dt \\
&= \int_0^1 |f(t)|^p \, dt - \int_0^{s_o} |f(t)|^p \, dt \\
&= (\|f\|_p)^p - (\|f_{s_o}^1\|_p)^p \\
&= \frac{(\|f\|_p)^p}{2} .
\end{aligned}
$$

Now since $|x^*(f)| \geq 1$, we deduce via the triangle inequality and the linearity of $x^*$ that either

$$|x^*(f_{s_o}^1)| \geq 1/2$$

or

$$|x^*(f_{s_o}^2)| \geq 1/2.$$

Define $f_1 = 2f_{s_o}^i$, where $i = 1$ or $2$ is such that

$$|x^*(f_{s_o}^i)| \geq 1/2.$$

It is then evident that $f_1 \in L_p([0,1],dt)$, $|x^*(f_1)| \geq 1$, and

$$\|f_1\|_p = 2^{(1 - 1/p)} \|f\|_p.$$

Repeating the argument with $f_1$ in place of $f$, we obtain some $f_2 \in L_p([0,1],dt)$ such that $|x^*(f_2)| \geq 1$ and

$$\|f_2\|_p = 2^{(1-1/p)}\|f_1\|_p = 2^{2(1-1/p)}\|f\|_p.$$

Continuing in this manner, we construct a sequence $\{f_n\} \subset L_p([0,1],dt)$ such that $|x^*(f_n)| \geq 1$ and $\|f_n\|_p = 2^{n(1-1/p)}\|f\|_p$, $n = 1,2,3,\ldots$ .

But since $0 < p < 1$, we have $1 - 1/p < 0$ and so $\lim_n \|f_n\|_p = 0$, while $|x^*(f_n)| \geq 1$, $n = 1,2,3,\ldots$, thereby contradicting the continuity of $x^*$.

Therefore $L_p([0,1],dt)^* = \{0\}$. □

## 3.4. Problems.

1. (Proposition 3.2.1) Let $V_1$ and $V_2$ be linear spaces over $\Phi$ and let $T \in L'(V_1,V_2)$. Prove each of the following:

(a) $T(0) = 0$.

(b) The range of $T$ is a linear subspace of $V_2$.

(c) If $T$ is injective, then $T^{-1}$ exists and

$$T^{-1} \in L'(R(T),V_1).$$

2. (Theorem 3.2.1) Let $(V_1,\|\cdot\|_1)$ and $(V_2,\|\cdot\|_2)$ be normed linear spaces over $\Phi$ and let $T \in L(V_1,V_2)$. Prove that

$$\begin{aligned}\|T\| &= \sup\{\|T(x)\|_2/\|x\|_1 \mid x \in V_1,\ x \neq 0\} \\ &= \sup\{\|T(x)\|_2 \mid x \in V_1,\ \|x\|_1 \leq 1\} \\ &= \sup\{\|T(x)\|_2 \mid x \in V_1,\ \|x\|_1 = 1\}.\end{aligned}$$

3. (Theorem 3.2.2) Let $(V_1,\|\cdot\|_1)$ and $(V_2,\|\cdot\|_2)$ be normed linear spaces over $\Phi$. Prove that $(L(V_1,V_2),\|\cdot\|)$ is a normed linear space over $\Phi$, where

$$\|T\| = \sup\{\|T(x)\|_2 \mid x \in V_1,\ \|x\|_1 = 1\}.$$

4. (Proposition 3.2.4) Let $(V_1,\|\cdot\|_1)$ and $(V_2,\|\cdot\|_2)$ be normed linear spaces over $\Phi$. Prove that, if $(V_1,\|\cdot\|_1)$ is finite dimensional, then $L'(V_1,V_2) = L(V_1,V_2)$.

5. Let $V_1$ and $V_2$ be linear spaces over $\Phi$, let $T : V_1 \to V_2$ be a linear transformation, and let $E \subset V_1$ be any subset of $V_1$.

(a) If $E$ is symmetric, prove that $T(E)$ is symmetric.

(b) If $E$ is balanced, prove that $T(E)$ is balanced.

(c) If $E$ is convex, prove that $T(E)$ is convex.

(d) Give an example to show that a nonconvex set may have a convex image.

6. Let $(V_1,\|\cdot\|_1)$ and $(V_2,\|\cdot\|_2)$ be normed linear spaces over $\Phi$ and let $T \in L'(V_1,V_2)$. Prove that $T \in L(V_1,V_2)$ if and only if $T$ maps bounded sets in $V_1$ into bounded sets in $V_2$.

7. Let $V_1$ and $V_2$ be linear spaces over $\Phi$ and let $T \in L'(V_1,V_2)$. Let $A$ be an absorbing subset of $V_1$ and suppose that $T$ is constant on $A$, that is, $T(x) = y$, $x \in A$, for some $y \in V_2$. Prove that $T$ is identically zero.

8. Let $(V_1,\|\cdot\|_1)$, $(V_2,\|\cdot\|_2)$ and $(V_3,\|\cdot\|_3)$ be normed linear spaces over $\Phi$. If $S \in L(V_1,V_2)$ and $T \in L(V_2,V_3)$, prove that the composition $TS \in L(V_1,V_3)$ and that

$$\|TS\| \leq \|T\|\|S\|.$$

9. Let $(V_1,\|\cdot\|_1)$ and $(V_2,\|\cdot\|_2)$ be normed linear spaces over $\Phi$. Let $T \in L(V_1,V_2)$ be an isomorphism onto $V_2$ and suppose $T^{-1} \in L(V_2,V_1)$.

(a) Prove that $\|T^{-1}\| \geq \|T\|^{-1}$.

(b) If $V_2$ is a Banach space, prove that $V_1$ is also a Banach space.

10. Let $(V_1,\|\cdot\|_1)$ and $(V_2,\|\cdot\|_2)$ be normed linear spaces over $\Phi$, with $V_1 \neq \{0\}$. Prove that, if $L(V_1,V_2)$ is complete, then $V_2$ is complete.

11. Let $(V_1, \|\cdot\|_1)$ and $(V_2, \|\cdot\|_2)$ be normed linear spaces over $\Phi$, let $T, T_n \in L(V_1, V_2)$, $n = 1, 2, 3, \ldots$, and let $x, x_n \in V_1$, $n = 1, 2, 3, \ldots$ . If $\{T_n\}$ converges to $T$ in $(L(V_1, V_2), \|\cdot\|)$ and $\{x_n\}$ converges to $x$ in $(V_1, \|\cdot\|_1)$, prove that $\{T_n(x_n)\}$ converges to $T(x)$ in $(V_2, \|\cdot\|_2)$.

12. Let $(V_1, P)$ and $(V_2, Q)$ be seminormed linear spaces over $\Phi$ and let $T \in L'(V_1, V_2)$. Prove that $T \in L(V_1, V_2)$ if and only if for every seminorm $q \in Q$ there exists a seminorm $p \in P$ and a positive number $c$ such that $q[T(x)] \leq c\, p(x)$ for all $x \in V_1$.

13. Let $T : \mathbb{R} \to \mathbb{R}$ be a mapping that is additive (that is, $T(x + y) = T(x) + T(y)$, where $x, y \in \mathbb{R}$) and continuous. Prove that $T$ is linear; that is, there exists some $c \in \mathbb{R}$ such that $T(x) = cx$.

*14. (Corollary 3.3.1) Let $V$ be a linear space over $\Phi$ and let $x', x_1', \ldots, x_n'$ be in $V'$. Prove that $x'$ is linearly dependent on $x_1', \ldots, x_n'$ if and only if $N(x') \supset \bigcap_{k=1}^{n} N(x_k')$.

15. (Proposition 3.3.2) Let $V$ be a linear space over $\mathbb{C}$ and let $x_1'$ be a real linear functional on $V$. If $x'(x) = x_1'(x) - i x_1'(ix)$, where $x \in V$, prove that $x' \in V'$.

16. (Theorem 3.3.1) Let $(V, \mathcal{T})$ be a topological linear space over $\mathbb{C}$ and let $x' \in V'$. Prove that the following are equivalent:

(a) $x' \in V^*$.

(b) $\mathrm{Re}(x')$ is a continuous real linear functional on $V$.

(c) $\mathrm{Im}(x')$ is a continuous real linear functional on $V$.

17. Let $V$ be a linear space over $\mathbb{C}$ and let $E$ be a balanced subset of $V$. If $x' \in V'$ and $\mathrm{Re}[x'(x)] \leq 1$ for all $x \in E$, prove that $|x'(x)| \leq 1$ for all $x \in E$.

18. Letting $V$ be a linear space over $\mathbb{C}$ and $x', y' \in V'$, prove each of the following:

(a) If $\mathrm{Re}[x'(x)] \leq \mathrm{Re}[y'(x)]$ for all $x \in V$, then $x' = y'$.

(b) If $|\mathrm{Re}[x'(x)]| \leq |\mathrm{Re}[y'(x)]|$ for all $x \in V$, then $x' = ry'$ for some $r \in \mathbb{R}$.

19. Let $(V, \|\cdot\|)$ be a normed linear space over $\mathbb{C}$. If $x^* \in V^*$, prove that $\|x^*\| = \|\mathrm{Re}(x^*)\|$.

20. Let $(V, \mathcal{T})$ be a topological linear space over $\Phi$ and let $x' \in V'$, $x' \neq 0$. Prove that the following are equivalent:

(a) $x' \in V^*$.

(b) $N(x')$ is not dense in $V$.

(c) $x'$ is bounded on some subset $U \subset V$ such that $\mathrm{int}(U)$ contains the origin.

(d) $x'(U)$ is a proper subset of $\Phi$ for some nonempty open subset $U \subset V$.

21. Let $(V, \|\cdot\|)$ be a Banach space over $\Phi$ and let $M$ and $N$ be closed linear subspaces of $V$. Suppose that each $x \in V$ has a unique representation in the form $x = y + z$, with $y \in M$, and $z \in N$. Prove that there exists a constant $K$ such that $\|y\| \leq K\|x\|$, $x \in V$.

22. Let $V$ be a linear space over $\Phi$ and let $x' \in V'$. If $W \subset V$ is a linear subspace of $V$ and $x'$ is bounded on $W$, that is, $x'(W)$ is a bounded subset of $\Phi$, prove that $W \subset N(x')$.

23. Let $V$ be a linear space over $\Phi$ and let $W$ be a maximal linear subspace of $V$. If $x \in V \sim W$, prove that there exists a unique $x' \in V'$ such that $W = N(x')$ and $x'(x) = 1$.

24. Let $(V, \|\cdot\|)$ be a normed linear space over $\Phi$ and let $(V_1, \|\cdot\|_1)$ be its completion (see Theorem 1.1.1). Prove that $V^*$ is isometrically isomorphic to $V_1^*$.

25. Let $(V,\mathcal{T})$ be an n-dimensional topological linear space over $\Phi$. Prove that $(V^*,\|\cdot\|)$ is also an n-dimensional topological linear space.

26. Let $(V,\|\cdot\|) = (\ell_1,\|\cdot\|_1)$. Give an example to show that $V^* \nsubseteq V'$.

27. Let $(V,\|\cdot\|) = (C([0,1]),\|\cdot\|_\infty)$.

(a) Let $T \in L(V)$ be defined by

$$T(f)(t) = tf(t)/(1 + t^2) \qquad (t \in [0,1];\ f \in C([0,1])).$$

Find $\|T\|$.

(b) Let $S \in L(V)$ be defined by

$$S(f)(t) = \int_0^1 K(t,s)f(s)\,ds \qquad (t \in [0,1]),$$

where $K(t,s)$ is continuous on $[0,1] \times [0,1]$ and $f \in C([0,1])$. Find $\|S\|$.

(c) Let $x^* \in V^*$ be defined by

$$x^*(f) = f(t_o)$$

for fixed $t_o \in [0,1]$ and all $f \in C([0,1])$. Find $\|x^*\|$.

(d) Let $y^* \in V^*$ be defined by

$$y^*(f) = \int_0^1 f(t)y(t)\,dt$$

for fixed $y \in C([0,1])$ and all $f \in C([0,1])$. Find $\|y^*\|$.

*28. Let $(V,\|\cdot\|) = (C([a,b]),\|\cdot\|_\infty)$ and let $W$ be the subspace of $V$ consisting of those functions $f$ that have continuous first and second derivatives on $[a,b]$ and are such that $f(a) = f(b) = 0$. Let $T \in L'(W,V)$ be defined by $T(f) = f''$, where $f''$ is the second derivative of $f$. Prove that $T^{-1}$ exists and that $R(T) = V$.

29. Let $(V,\|\cdot\|) = (C([0,1]),\|\cdot\|_\infty)$ and let $F \subset V^*$ be the family of linear functionals $F = \{\varphi_t \mid t \in [0,1]\}$, where $\varphi_t(f) = f(t)$, $f \in C([0,1])$. Prove that $F$ separates points.

*30. In Example 3.1.6 we saw that $c_0^* = \ell_1$. Prove that $c^* = \ell_1$ and exhibit the general form of an $x^* \in c^*$.

# CHAPTER 4

# THE HAHN-BANACH THEOREM: ANALYTIC FORM

4.0. Introduction. The Hahn-Banach Theorem is, together with the Uniform Boundedness and Open-Mapping Theorems, one of the most important theorems of functional analysis. In its analytic form, discussed in this chapter, the theorem assures us that a linear functional on a linear subspace of a linear space that is bounded by a seminorm can always be extended to the entire space in such a way that the seminorm boundedness is retained. The proof of the theorem and a number of its consequences will be given in Sections 4.1 and 4.2. Among these consequences is the fact that there exist sufficiently many continuous linear functionals on a nontrivial locally convex topological linear space to separate points. The remaining sections of this chapter are devoted to a sampling of various applications of the Hahn-Banach Theorem and its consequences, and to a proof, in the last section, of Helly's Theorem. Although the proof of the latter theorem makes no direct use of the Hahn-Banach Theorem as developed in this chapter, it does depend on a simple case of the geometric form of the theorem, and hence the proof provides a motivation for the exposition of the succeeding chapter.

4.1. The Hahn-Banach Theorem: Analytic Form. Suppose $V$ is a linear space over $\Phi$ and $W \subset V$ is a linear subspace. Suppose also that $p$ is a seminorm on $V$ and that $y'$ is a linear functional on $W$ such that

$$|y'(x)| \leq p(x) \qquad (x \in W).$$

Then can $y'$ be extended to a linear functional on all of $V$ that is also bounded by $p$? That is, does there exist some $x' \in V'$ such that

$$\text{(i)} \quad x'(x) = y'(x) \qquad (x \in W).$$

$$\text{(ii)} \quad |x'(x)| \leq p(x) \qquad (x \in V).$$

The Hahn-Banach Theorem tells us that such an extension is always possible, although it is, in general, not unique.

To prove this result we proceed in several stages. First, we show how to make such an extension for linear spaces over $\mathbb{R}$ when the codimension of $W$ is one and then apply Zorn's Lemma [$DS_1$, p. 6] to obtain the general result for arbitrary linear spaces over $\mathbb{R}$. The theorem when $\Phi = \mathbb{C}$ will then be established with the help of some of the results in Section 3.3.

Lemma 4.1.1. Let $V$ be a linear space over $\mathbb{R}$, let $p$ be a seminorm on $V$, and let $W \subset V$ be a linear subspace of codimension one. If $y' \in W'$ is such that $y'(x) \leq p(x)$, $x \in W$, then there exists some $x' \in V'$ such that

$$\text{(i)} \quad x'(x) = y'(x) \qquad (x \in W).$$

$$\text{(ii)} \quad x'(x) \leq p(x) \qquad (x \in V).$$

Proof. Let $x_o \in V \sim W$. Then, since $W$ is of codimension one, we see that $V$ is spanned by $W \cup \{x_o\}$, and thus every $x \in V$ can be uniquely expressed as $x = ax_o + y$ for some $a \in \mathbb{R}$ and $y \in W$. We define $x'$ as $x'(x) = a\gamma + y'(y)$ for some suitable choice of $\gamma$. Clearly such an $x'$ is a linear functional on $V$ whose restriction to $W$ agrees with $y'$. To show that $x'$ is also bounded by $p$ we need to look more closely at how one chooses $\gamma$.

Let $x_1, x_2 \in W$. Then

$$\begin{aligned} y'(x_1) - y'(x_2) = y'(x_1 - x_2) &\leq p(x_1 - x_2) \\ &= p(x_1 + x_o - x_2 - x_o) \\ &\leq p(x_1 + x_o) + p(-x_2 - x_o), \end{aligned}$$

and so we have

$$-p(-x_2 - x_o) - y'(x_2) \leq p(x_1 + x_o) - y'(x_1).$$

Thus for fixed $x_1 \in W$ we deduce that

$$\{-p(-x_2 - x_o) - y'(x_2) \mid x_2 \in W\} \subset \mathbb{R}$$

is bounded above and so has a least upper bound, call it $b_1$. Similarly $\{p(x_1 + x_o) - y'(x_1) \mid x_1 \in W\}$ has a greatest lower bound $b_2$. Clearly $b_1 \leq b_2$. Let $\gamma$ be any real number such that $b_1 \leq \gamma \leq b_2$. Note that by the choice of $\gamma$ we have

$$-p(-x - x_o) - y'(x) \leq \gamma \leq p(x + x_o) - y'(x) \qquad (x \in W).$$

To complete the proof we need to show that the $x'$ defined for the $\gamma$ chosen here is such that $x'(x) \leq p(x)$, $x \in V$. Suppose $x = ax_o + y$. We consider three cases.

Case 1. $a = 0$. Then $x'(x) = y'(y) \leq p(y) = p(x)$.

Case 2. $a > 0$. Then from $\gamma \leq p(y/a + x_o) - y'(y/a)$ we deduce that

$$\begin{aligned} a\gamma + y'(y) &\leq ap(y/a + x_o) \\ &= p(ax_o + y), \end{aligned}$$

and so $x'(x) \leq p(x)$.

Case 3. $a < 0$. Then from $-p(-y/a - x_o) - y'(y/a) \leq \gamma$ we deduce that

$$\begin{aligned} a\gamma + y'(y) &\leq -ap(-y/a - x_o) \\ &= p(ax_o + y), \end{aligned}$$

and again we have $x'(x) \leq p(x)$.

This completes the proof. □

A few remarks are in order before proceeding. First we note that the result is not more general than the one alluded to in our introductory remarks. Indeed, since $p$ is a seminorm, it is easily seen that $x'(x) \leq p(x)$ if and only if $|x'(x)| \leq p(x)$. The lemma was stated in this seemingly stronger form since it remains true, as given, and with the same proof, if we replace the assumption that $p$ is a seminorm with a slightly weaker assumption; namely, we need only assume that $p : V \to \mathbb{R}$ is such that

(1) $p(x) \geq 0$,

(2) $p(x + y) \leq p(x) + p(y)$,

(3) $p(ax) = ap(x)$ $\quad (a \in \mathbb{R};\ a \geq 0;\ x,y \in V)$.

These remarks apply also to the next theorem and will be used in proving a geometric form of the Hahn-Banach Theorem (Theorem 5.2.1) in the next chapter. However, when $\Phi = \mathbb{C}$, we need the hypothesis that $p$ is a seminorm.

The functional given by Lemma 4.1.1 is clearly not generally unique. Indeed, it is apparent that it is unique for a fixed $x_0 \in V \sim W$ if and only if

$$\inf_{x \in W} \{p(x + x_0) - y'(x)\} = \sup_{x \in W} \{-p(-x - x_0) - y'(x)\}.$$

<u>Theorem 4.1.1</u> (<u>Real Hahn-Banach Theorem</u>). Let $V$ be a linear space over $\mathbb{R}$, let $p$ be a seminorm on $V$, and let $W \subset V$ be a linear subspace. If $y' \in W'$ is such that $y'(x) \leq p(x)$, $x \in W$, then there exists some $x' \in V'$ such that

(i) $x'(x) = y'(x)$ $\quad (x \in W)$.

(ii) $x'(x) \leq p(x)$ $\quad (x \in V)$.

<u>Proof</u>. The proof is a standard application of Zorn's Lemma [$DS_1$, p. 6]. Let us call a pair $(u',U)$ an extension of $y'$ provided

(1) $U \subset V$ is a linear subspace such that $U \supset W$.

(2) $u'(x) = y'(x)$ $\quad (x \in W)$.

(3) $u'(x) \leq p(x)$ $\quad (x \in U)$.

Let $\mathcal{U}$ be the family of all extensions of $y'$. Clearly $\mathcal{U} \neq \phi$, as $(y',W) \in \mathcal{U}$. If $(u_1',U_1)$, $(u_2',U_2) \in \mathcal{U}$, then we write $(u_1',U_1) < (u_2',U_2)$ if $U_1 \subset U_2$ and $u_1'(x) = u_2'(x)$, $x \in U_1$. Then $<$ is a partial ordering on $\mathcal{U}$. Suppose $\{(u_\alpha',U_\alpha)\}$ is a linearly ordered subset of $\mathcal{U}$. Then it is easily verified that $(u',U)$ is an upper bound for $\{(u_\alpha',U_\alpha)\}$ where $U = \bigcup_\alpha U_\alpha$ and $u'(x) = u_\alpha'(x)$, $x \in U_\alpha$. Consequently, by Zorn's Lemma, $\mathcal{U}$ has a maximal element, call it $(x',V_o)$.

We claim that $V_o = V$. If this were not true, we would get the following: Let $x_o \in V \sim V_o$ and let $V_1$ be the linear space spanned by $V_o \cup \{x_o\}$. Clearly $V_o$ is a linear subspace of $V_1$ of codimension one, and so, by Lemma 4.1.1, there exists some $z' \in V_1'$ such that $(z',V_1)$ is an extension of $(x',V_o)$, that is, $(z',V_1) > (x',V_o)$. Since $V_o \neq V_1$, this contradicts the maximality of $(x',V_o)$.

Therefore $V_o = V$, and the theorem is proved. □

We can now combine these results with our knowledge of real linear functionals obtained in Section 3.3 to establish the Hahn-Banach Theorem for linear spaces over arbitrary $\Phi$.

Theorem 4.1.2 (Hahn-Banach Theorem). Let $V$ be a linear space over $\Phi$, let $p$ be a seminorm on $V$, and let $W \subset V$ be a linear subspace. If $y' \in W'$ is such that $|y'(x)| \leq p(x)$, $x \in W$, then there exists some $x' \in V'$ such that

(i) $x'(x) = y'(x)$ $(x \in W)$.

(ii) $|x'(x)| \leq p(x)$ $(x \in V)$.

Proof. If $\Phi = \mathbb{R}$, then the result is an immediate consequence of the Real Hahn-Banach Theorem and the remark immediately following Lemma 4.1.1. So assume that $\Phi = \mathbb{C}$ and let $y_1'(x) = \mathrm{Re}[y'(x)]$, $x \in W$. From Proposition 3.3.1 we see that $y_1'$ is a real linear functional on $W$, that is, a linear functional on $W$ considered as a linear space over $\mathbb{R}$. Moreover, we see that

$$y_1'(x) = \mathrm{Re}[y'(x)] \leq |y'(x)| \leq p(x) \qquad (x \in W).$$

Hence we may apply the Real Hahn-Banach Theorem and deduce the existence of a real linear functional $x_1'$ on $V$ such that $x_1'(x) = y_1'(x)$, $x \in W$, and $x_1'(x) \leq p(x)$, $x \in V$.

We define

$$x'(x) = x_1'(x) - ix_1'(ix) \qquad (x \in V).$$

Then from Proposition 3.3.2 we see that $x' \in V'$ and from Proposition 3.3.1 that $x'(x) = y'(x)$, $x \in W$. Finally, given $x \in V$, we see that

$$\begin{aligned} |x'(x)| &= \exp[-i \arg x'(x)]x'(x) \\ &= x'(\exp[-i \arg x'(x)]x) \\ &= x_1'(\exp[-i \arg x'(x)]x) \\ &\leq p(\exp[-i \arg x'(x)]x) \\ &= p(x) \end{aligned}$$

as $x'(\exp[-i \arg x'(x)]x) = |x'(x)| \in \mathbb{R}$, where $\exp(\cdot)$ denotes the usual exponential function. □

<u>4.2</u>. <u>Some Consequences of the Hahn-Banach Theorem</u>. We now consider many of the most useful and immediate consequences of the Hahn-Banach Theorem. In particular, we shall see that $V^*$ separates the points of $V$, and so $V^* \neq \{0\}$, whenever $V$ is a nontrivial seminormed linear space.

<u>Theorem 4.2.1</u>. Let $(V,P)$ be a seminormed linear space over $\Phi$ and let $W \subset V$ be a linear subspace. If $y^* \in W^*$, then there exists some $x^* \in V^*$ such that $x^*(x) = y^*(x)$, $x \in W$.

<u>Proof</u>. Since $y^*$ is continuous on $(W,\mathcal{T}_P)$, it is continuous at the origin, and so there exists an $\epsilon > 0$ and seminorms $p_1,p_2,\ldots,p_n$ in $P$ such that $|y^*(x)| < 1$ whenever

$p_k(x) < \epsilon$, $k = 1,2,\ldots,n$, $x \in W$. Define

$$p(x) = \frac{\sup\{p_1(x),\ldots,p_n(x)\}}{\epsilon} \qquad (x \in V).$$

Then, by essentially the same argument as that used in the proof of Lemma 2.5.1, we see that $p$ is a seminorm on $V$.

Moreover, given $x \in W$, let $a > p(x)$. Then from $p(x/a) < 1$ we deduce that $p_k(x/a) < \epsilon$, $k = 1,2,\ldots,n$, and so $|y^*(x/a)| < 1$; that is, $|y^*(x)| < a$. However, since this holds for each $a > p(x)$, we conclude that $|y^*(x)| \leq p(x)$, $x \in W$. Consequently by the Hahn-Banach Theorem (Theorem 4.1.2) there exists some $x' \in V'$ such that $x'(x) = y^*(x)$, $x \in W$, and $|x'(x)| \leq p(x)$, $x \in V$.

Furthermore, if $\delta > 0$ and $x \in V$ is such that $p_k(x) < \delta\epsilon$, $k = 1,2,\ldots,n$, then

$$p(x) = \frac{\sup\{p_1(x),\ldots,p_n(x)\}}{\epsilon} < \delta,$$

and so $|x'(x)| < \delta$. Thus $x'$ is continuous at the origin, and so, by Proposition 3.2.2, $x' = x^* \in V^*$. □

Corollary 4.2.1. Let $(V,P)$ be a seminormed linear space over $\Phi$ and let $V \neq \{0\}$. Then $V^*$ separates points.

Proof. Suppose $x,y \in V$, $x \neq y$. Then $x_o = x - y \neq 0$. Let $W$ be the one-dimensional linear subspace of $V$ spanned by $x_o$ and define $y^*(ax_o) = a$, $ax_o \in W$. Using Theorem 2.2.1 it is not difficult to verify that $y^* \in W^*$. Hence, by Theorem 4.2.1, there exists some $x^* \in V^*$ such that $x^*(x) = y^*(x)$, $x \in W$.

In particular, $x^*(x_o) = x^*(x - y) = 1 \neq 0$, and $V^*$ separates points. □

Corollary 4.2.2. Let $(V,P)$ be a seminormed linear space over $\Phi$ and let $V \neq \{0\}$. Then $V^* \neq \{0\}$.

Corollary 4.2.3. Let $(V,P)$ be a seminormed linear space over $\Phi$. If $x \in V$, $x \neq 0$, then there exists some $x^* \in V^*$ such that $x^*(x) = 1$.

Since, by Theorem 2.3.1, the concepts of seminormed linear spaces and locally convex topological linear spaces are equivalent, we conclude at once that Theorem 4.2.1 as well as Corollaries 4.2.1, 4.2.2, and 4.2.3 are all valid when $(V,\mathcal{T})$ is a locally convex topological linear space. In particular, $V^*$ separates points whenever $V$ is such a space and $V \neq \{0\}$. Thus we see that the topological linear spaces $L_p([0,1],dt)$, $0 < p < 1$, where $dt$ is Lebesgue measure on $[0,1]$, are not locally convex since, by Theorem 3.3.4, $L_p([0,1],dt)^* = \{0\}$, $0 < p < 1$.

These results are, of course, also valid in the special case in which $V$ is a normed linear space. Moreover, in this case several of these results can be sharpened.

Theorem 4.2.2. Let $(V,\|\cdot\|)$ be a normed linear space over $\Phi$ and let $W \subset V$ be a linear subspace. If $y^* \in W^*$, then there exists some $x^* \in V^*$ such that

(i) $x^*(x) = y^*(x) \qquad (x \in W)$.

(ii) $\|x^*\| = \|y^*\|$.

Proof. Let $p(x) = \|y^*\| \, \|x\|$, $x \in V$, where as usual

$$\|y^*\| = \sup_{\substack{\|x\| = 1 \\ x \in W}} |y^*(x)|.$$

Then $p$ is evidently a seminorm on $V$ and $|y^*(x)| \leq p(x)$, $x \in W$. Hence by the Hahn-Banach Theorem (Theorem 4.1.2) there exists some $x' \in V'$ such that $x'(x) = y^*(x)$, $x \in W$, and

$$|x'(x)| \leq p(x) = \|y^*\|\|x\| \qquad (x \in V).$$

Thus, by Theorem 3.2.1, $x' = x^* \in V^*$ and $\|x^*\| \leq \|y^*\|$

Moreover, from

$$\|y^*\| = \sup_{\substack{\|x\|=1 \\ x \in W}} |y^*(x)| = \sup_{\substack{\|x\|=1 \\ x \in W}} |x^*(x)|$$

$$\leq \sup_{\substack{\|x\|=1 \\ x \in V}} |x^*(x)|$$

$$= \|x^*\|,$$

we conclude that $\|x^*\| = \|y^*\|$. □

A substantial improvement of Corollary 4.2.3 is also available in normed linear spaces.

Theorem 4.2.3. Let $(V, \|\cdot\|)$ be a normed linear space over $\Phi$, let $W \subset V$ be a linear subspace, and let $x_o \in V \sim W$. If $d = \inf_{y \in W} \|x_o - y\| > 0$, then there exists some $x^* \in V^*$ such that

(i) $x^*(x) = 0$ $(x \in W)$.

(ii) $x^*(x_o) = d$.

(iii) $\|x^*\| = 1$.

Proof. Let $W_o \subset V$ be the linear subspace spanned by $W \cup \{x_o\}$ and define

$$y^*(ax_o + y) = ad \qquad (a \in \Phi;\ y \in W).$$

Clearly $y^*$ is a linear functional on $W_o$, $y^*(x_o) = d$, and $y^*(y) = 0$, $y \in W$. Moreover, we claim that $\|y^*\| = 1$.

Indeed, note first that if $x = ax_o + y$, $a \neq 0$, $y \in W$, then

$$\|x\| = \|ax_o + y\| = \|a[x_o - (-y/a)]\|$$

$$\geq |a|d.$$

Hence $|y^*(x)| = |a|d \leq \|x\|$; that is, $\|y^*\| \leq 1$. On the other hand, by the definition of $d$, given $\epsilon > 0$ there exists some $y \in W$ such that $\|x_o - y\| < d + \epsilon$. Set $z = (x_o - y)/\|x_o - y\|$. Then $z \in W_o, \|z\| = 1$, and

$$|y^*(z)| = \frac{|y^*(x_o - y)|}{\|x_o - y\|}$$

$$> \frac{d}{(d + \epsilon)} .$$

Since $\epsilon > 0$ is arbitrary, it follows from the definition of $\|y^*\|$ that $\|y^*\| \geq 1$, and thus $\|y^*\| = 1$.

An application of Theorem 4.2.2 completes the proof. □

Note, in particular, that Theorem 4.2.3 applies whenever $W$ is a closed linear subspace and $x_o \in V \sim W$.

A number of corollaries can be obtained from the preceding results. We leave the proofs to the reader.

Corollary 4.2.4. Let $(V, \|\cdot\|)$ be a normed linear space over $\Phi$, let $W \subset V$ be a linear subspace, and let $x_o \in V \sim W$. If $d = \inf_{y \in W} \|x_o - y\| > 0$, then there exists some $x^* \in V^*$ such that

(i) $x^*(x) = 0$ $(x \in W)$.
(ii) $x^*(x_o) = 1$.
(iii) $\|x^*\| = 1/d$.

Corollary 4.2.5. Let $(V, \|\cdot\|)$ be a normed linear space over $\Phi$. If $x_o \in V$, $x_o \neq 0$, then there exists some $x^* \in V^*$ such that

(i) $x^*(x_o) = \|x_o\|$.
(ii) $\|x^*\| = 1$.

Moreover,

$$\|x_o\| = \sup_{\substack{\|x^*\| = 1 \\ x^* \in V^*}} |x^*(x_o)| .$$

Corollary 4.2.6. Let $(V,\|\cdot\|)$ be a normed linear space over $\Phi$, and let $x_o \in V$. If $x^*(x_o) = 0$, $x^* \in V^*$, then $x_o = 0$.

Corollary 4.2.7. Let $(V,\|\cdot\|)$ be a normed linear space over $\Phi$, let $W \subset V$ be a closed linear subspace, and let $x_o \in V$. If $x^* \in V^*$ and $x^*(x) = 0$, $x \in W$, imply that $x^*(x_o) = 0$, then $x_o \in W$.

Corollary 4.2.8. Let $(V,\|\cdot\|)$ be a normed linear space over $\Phi$, and let $W \subset V$ be a linear subspace. Then the following are equivalent:

(i) $\mathrm{cl}(W) = V$.

(ii) If $x^* \in V^*$ is such that $x^*(x) = 0$, $x \in W$, then $x^* = 0$.

As the reader may suspect, a number of these results for normed linear spaces have valid analogs in the context of seminormed linear spaces, (i.e., locally convex topological linear spaces). We shall return to this in Section 5.3 after we obtain a geometric version of the Hahn-Banach Theorem.

In the remaining sections of this chapter we shall examine various applications of the Hahn-Banach Theorem and its consequences. Other uses of this important theorem will occur in subsequent chapters.

4.3. The Hahn-Banach Theorem and Abelian Semigroups of Transformations. We wish to establish an extension of the Hahn-Banach Theorem asserting that linear functionals on subspaces bounded by a seminorm can be extended to the entire space in such a way as to preserve not only this boundedness but also the action of the linear functionals with regard to certain families of linear transformations. The statement of the next theorem will make this rather vague assertion precise. First, however, we need a definition.

Definition 4.3.1. Let $V$ be a linear space over $\Phi$. Suppose that $\mathcal{G} \subset \mathcal{L}'(V)$ is a family of linear transformations such that

(i) $T,S \in \mathcal{G}$ implies $TS \in \mathcal{G}$.

(ii) $TS = ST$ $\qquad (S,T \in \mathcal{G})$.

Then $G$ is said to be an <u>Abelian semigroup of linear transformations</u> on $V$.

<u>Theorem 4.3.1</u>. Let $V$ be a linear space over $\Phi$, let $p$ be a seminorm on $V$, and let $G$ be an Abelian semigroup of linear transformations on $V$ such that $p[T(x)] \leq p(x)$, $x \in V$ and $T \in G$. Suppose $W \subset V$ is a linear subspace and $y' \in W'$ is such that

(a) $|y'(x)| \leq p(x)$ $(x \in W)$.

(b) $T(x) \in W$ $(x \in W;\ T \in G)$.

(c) $y'[T(x)] = y'(x)$ $(x \in W;\ T \in G)$.

Then there exists some $x' \in V'$ such that

(i) $x'(x) = y'(x)$ $(x \in W)$.

(ii) $|x'(x)| \leq p(x)$ $(x \in V)$.

(iii) $x'[T(x)] = x'(x)$ $(x \in V;\ T \in G)$.

<u>Proof</u>. As might be expected, the idea of the proof is to arrange things so that the Hahn-Banach Theorem (Theorem 4.1.2) can be applied. To this end we begin by defining a new seminorm on $V$: for each $x \in V$ we define

$$p_o(x) = \inf\left\{\frac{p[T_1(x)+\ldots+T_n(x)]}{n}\right\},$$

where the infimum is taken over all possible finite subsets $\{T_1,T_2,\ldots,T_n\} \subset G$. Since $p$ is a seminorm and $p[T(x)] \leq p(x)$, $T \in G$, we see at once that $0 \leq p_o(x) \leq p(x)$, $x \in V$. It is elementary to verify that $p_o(ax) = |a|p_o(x)$, $x \in \Phi$, as $p$ has this property.

Furthermore, let $x,y \in V$ and suppose $\epsilon > 0$. Then by the definition of $p_o$ there exist $T_1,\ldots,T_n$; $S_1,\ldots,S_m$ in $G$ for which

$$\frac{p[T_1(x)+\ldots+T_n(x)]}{n} < p_o(x) + \frac{\epsilon}{2}$$

$$\frac{p[S_1(y)+\ldots+S_m(y)]}{m} < p_o(y) + \frac{\epsilon}{2}.$$

Hence

$$
\begin{aligned}
p_o(x + y) &\leq \frac{p\left[\sum_{k=1}^{n} \sum_{j=1}^{m} T_k S_j(x + y)\right]}{nm} \\
&\leq \frac{p\left[\sum_{k=1}^{n} \sum_{j=1}^{m} T_k S_j(x)\right]}{nm} + \frac{p\left[\sum_{k=1}^{n} \sum_{j=1}^{m} T_k S_j(y)\right]}{nm} \\
&= \frac{p\left[\sum_{j=1}^{m} S_j\{\sum_{k=1}^{n} T_k(x)/n\}\right]}{m} + \frac{p\left[\sum_{k=1}^{n} T_k\{\sum_{j=1}^{m} S_j(y)/m\}\right]}{n} \\
&\leq \frac{\sum_{j=1}^{m} p\left[S_j\{\sum_{k=1}^{n} T_k(x)/n\}\right]}{m} + \frac{\sum_{k=1}^{n} p\left[T_k\{\sum_{j=1}^{m} S_j(y)/m\}\right]}{n} \\
&\leq \frac{\sum_{j=1}^{m} p\left[\sum_{k=1}^{n} T_k(x)/n\right]}{m} + \frac{\sum_{k=1}^{n} p\left[\sum_{j=1}^{m} S_j(y)/m\right]}{n} \\
&= \frac{p\left[\sum_{k=1}^{n} T_k(x)\right]}{n} + \frac{p\left[\sum_{j=1}^{m} S_j(y)\right]}{m} \\
&< p_o(x) + p_o(y) + \epsilon .
\end{aligned}
$$

Note that the commutativity and linearity of the elements of $G$, as well as the assumption that $p[T(x)] \leq p(x)$, $x \in V$, $T \in G$, are all used in establishing the preceding estimates. Since $\epsilon > 0$ is arbitrary, we conclude that $p_o(x + y) \leq p_o(x) + p_o(y)$, $x,y \in V$.

Thus $p_o$ is a seminorm on $V$ such that $p_o(x) \leq p(x)$, $x \in V$.

Moreover, if $x \in W$, then for any $T_1, T_2, \ldots, T_n$ in $G$ we have

$$|y'(x)| = \frac{|y'[T_1(x)] + \ldots + y'[T_n(x)]|}{n}$$

$$= \frac{\left|y'\left[\sum_{k=1}^{n} T_k(x)\right]\right|}{n}$$

$$\leq \frac{p\left[\sum_{k=1}^{n} T_k(x)\right]}{n}$$

since $y'[T(x)] = y'(x)$, $T \in G$. Because the $T_1, T_2, \ldots, T_n$ in $G$ are arbitrary, we can conclude that $|y'(x)| \leq p_o(x)$, $x \in W$.

Consequently, appealing to the Hahn-Banach Theorem (Theorem 4.1.2) with regard to the seminorm $p_o$, we see that there exists some $x' \in V'$ for which $x'(x) = y'(x)$, $x \in W$, and $|x'(x)| \leq p_o(x)$, $x \in V$. Furthermore we claim that $x'[T(x)] = x'(x)$, $x \in V$ and $T \in G$, on the following grounds: Suppose $x \in V$ and $T \in G$, then for each positive integer $n$

$$p_o[x - T(x)] \leq \frac{p\{T[x - T(x)] + T^2[x - T(x)] + \ldots + T^n[x - T(x)]\}}{n}$$

$$= \frac{p[T(x) - T^{n+1}(x)]}{n}$$

$$\leq \frac{p[(T(x)] + p[T^{n+1}(x)]}{n}$$

$$\leq \frac{2p(x)}{n},$$

from which we deduce at once that $p_o[x - T(x)] = 0$. However, since $|x'[x - T(x)]| \leq p_o[x - T(x)]$ we have $x'[T(x)] = x'(x)$. □

An examination of the proof shows that Theorem 4.3.1 also holds for linear spaces $V$ over $\mathbb{R}$ if one assumes only the following: $p(x) \geq 0$, $p(x + y) \leq p(x) + p(y)$, and $p(ax) = ap(x)$, $a \geq 0$ and $x,y \in V$. The reader should compare the remarks following Lemma 4.1.1.

Clearly, if $\mathcal{G}$ consists only of the identity transformation on $V$, then Theorem 4.3.1 reduces to the Hahn-Banach Theorem.

Let us look at an application of Theorem 4.3.1 to the existence of so called Banach limits. Consider the family $c$ of all convergent sequences of complex numbers. Then from well-known properties of convergent sequences we know that to every such sequence $\{a_k\}$ there corresponds a unique complex number $\lim_k a_k$ that has the following properties:

(i) If $c_k = aa_k + bb_k$, $k = 1,2,3,\ldots$, where $a,b \in \mathbb{C}$ and $\{a_k\},\{b_k\} \in c$, then $\{c_k\} \in c$ and $\lim_k c_k = a \lim_k a_k + b \lim_k b_k$.

(ii) If $e_k = 1$, $k = 1,2,3,\ldots$, then $\lim_k e_k = 1$.

(iii) For each $n = 1,2,3,\ldots$, if $c_k = a_{k+n}$, $k = 1,2,3,\ldots$, where $\{a_k\} \in c$, then $\{c_k\} \in c$ and $\lim_k c_k = \lim_k a_k$.

Is it possible to define a notion of "limit" for all bounded sequences of complex numbers that will satisfy all of these properties and will reduce to the ordinary limit if the sequence under consideration is convergent? A rephrasing of this question in functional analytic terms will indicate how we might proceed to answer it.

Note first that the collection of all bounded sequences of complex numbers is a Banach space $\ell_\infty$ with the norm

$$\|\{a_k\}\|_\infty = \sup_k |a_k| \qquad (\{a_k\} \in \ell_\infty)$$

and that $c$ is a closed linear subspace of $\ell_\infty$. This was mentioned in Example 1.2.4. The preceding discussion says, among other things, that $\lim_k a_k$ defines a linear functional on $c$, and, as is easily seen, $|\lim_k a_k| \leq \|\{a_k\}\|_\infty$, $\{a_k\} \in c$. Thus the question we have posed is equivalent to asking whether there exists a (continuous) linear functional on $\ell_\infty$ that satisfies the three properties of the limit functional on c indicated above and agrees with the ordinary limit functional on c.

Obviously a straightforward application of the Hahn-Banach Theorem (Theorem 4.1.2) with the seminorm $p(\{a_k\}) = \|\{a_k\}\|_\infty$ (or of Theorem 4.2.2) yields an extension to $\ell_\infty$ of the limit functional on c that satisfies properties (i) and (ii). In order to ensure that property (iii) will also hold for the extension obtained we appeal to Theorem 4.3.1.

To this end we define $T \in L(\ell_\infty)$ by $T(\{a_k\}) = \{c_k\}$, where $c_k = a_{k+1}$, $k = 1,2,3,\ldots$, $\{a_k\} \in \ell_\infty$, and set $G = \{T^n \mid n = 1,2,3,\ldots\}$. Clearly $G$ is an Abelian semigroup of linear transformations on $\ell_\infty$. Moreover, if $p(\{a_k\}) = \|\{a_k\}\|_\infty$, $\{a_k\} \in \ell_\infty$, then

$$\begin{aligned} p[T(\{a_k\})] &= \|T(\{a_k\})\|_\infty \\ &= \sup_{k \geq 2} |a_k| \\ &\leq \sup_{k \geq 1} |a_k| \\ &= \|\{a_k\}\|_\infty \\ &= p(\{a_k\}), \end{aligned}$$

and, from property (iii) of the limit functional on c, if $\{a_k\} \in c$, then $\lim_k T^n(\{a_k\}) = \lim_k a_k$, $n = 1,2,3,\ldots$ . Thus all the hypotheses of Theorem 4.3.1 are fulfilled, and an application of the theorem immediately gives the next result.

<u>Theorem 4.3.2</u>. There exists some $x^* \in \ell_\infty^*$ such that

(i) If $\{a_k\} \in c$, then $x^*(\{a_k\}) = \lim_k a_k$.

(ii) Given $n = 1,2,3,\ldots$, if $\{a_k\} \in \ell_\infty$ and $c_k = a_{k+n}$, $k = 1,2,3,\ldots$, then $x^*(\{c_k\}) = x^*(\{a_k\})$.

The continuous linear functional $x^*$ on $\ell_\infty$ obtained in Theorem 4.3.2 is generally called a <u>Banach limit</u>. We shall take a look at a different method of obtaining such functionals in Section 9.6.

<u>4.4. Adjoint Transformations</u>. Suppose $(V_1,P_1)$ and $(V_2,P_2)$ are nontrivial seminormed linear spaces over $\Phi$. Then from Corollary 4.2.2 we know that $V_k^* \neq \{0\}$, $k = 1,2$. There exists a simple, and quite useful, relationship between transformations $T \in L(V_1,V_2)$ and certain elements in $L'(V_2^*,V_1^*)$. To be precise we make the following definition:

<u>Definition 4.4.1</u>. Let $(V_1,P_1)$ and $(V_2,P_2)$ be seminormed linear spaces over $\Phi$, and let $V_k \neq \{0\}$, $k = 1,2$. If $T \in L(V_1,V_2)$, then we define a mapping $T^* \in L'(V_2^*,V_1^*)$ by setting

$$T^*(x_2^*)(x_1) = x_2^*[T(x_1)] \qquad (x_1 \in V_1;\ x_2^* \in V_2^*),$$

$T^*$ being called the <u>adjoint</u> of $T$.

It is easy to verify that the definition of $T^*$ indeed defines an element of $L'(V_2^*,V_1^*)$.

With the aid of the Hahn-Banach Theorem we can easily establish several results about the adjoint when the spaces $V_1$ and $V_2$ are normed linear spaces.

<u>Theorem 4.4.1</u>. Let $(V_1,\|\cdot\|_1)$ and $(V_2,\|\cdot\|_2)$ be normed linear spaces over $\Phi$ and suppose $T \in L(V_1,V_2)$. Then

(i) $T^* \in L(V_2^*,V_1^*)$ and $\|T^*\| = \|T\|$.

(ii) $T^*$ is injective if and only if $R(T)$ is dense in $(V_2,\|\cdot\|_2)$.

<u>Proof</u>. Let $x_1 \in V_1$ and $x^*_2 \in V_2^*$. Then

$$\begin{aligned}|T^*(x_2^*)(x_1)| &= |x_2^*[T(x_1)]| \\ &\leq \|x_2^*\|\|T\|\|x_1\|_1,\end{aligned}$$

and so $\|T^*(x_2^*)\| \leq \|T\| \, \|x_2^*\|$. Thus from Theorem 3.2.1 we see that $T^* \in L(V_2^*, V_1^*)$ and $\|T^*\| \leq \|T\|$. Moreover, suppose $\epsilon > 0$ and let $x_1 \in V_1$ be such that $\|x_1\| = 1$ and $\|T(x_1)\|_2 > \|T\| - \epsilon$, which is possible by the definition of $\|T\|$. Then by Corollary 4.2.5 there exists some $x_2^* \in V_2^*$ such that $\|x_2^*\| = 1$ and $x_2^*[T(x_1)] = \|T(x_1)\|_2$. But then we have

$$\begin{aligned} |T^*(x_2^*)(x_1)| &= |x_2^*[T(x_1)]| \\ &= \|T(x_1)\| \\ &> \|T\| - \epsilon, \end{aligned}$$

from which we conclude at once that $\|T^*(x_2^*)\| > \|T\| - \epsilon$. However, since $\|x_2^*\| = 1$, this entails also that $\|T^*\| > \|T\| - \epsilon$, and since $\epsilon > 0$ is arbitrary, we obtain $\|T^*\| \geq \|T\|$. Thus Theorem 4.4.1(i) is proved.

In order to establish the equivalence in Theorem 4.4.1(ii) let us first suppose that $T^*$ is injective. To show that $R(T)$ is dense it suffices, by Corollary 4.2.8, to prove that, if $x_2^* \in V_2^*$ is such that $x_2^*(x_2) = 0$, $x_2 \in R(T)$, then $x_2^* = 0$. However, given such an $x_2^* \in V_2^*$, we see that $T^*(x_2^*)(x_1) = x_2^*[T(x_1)] = 0$, $x_1 \in V_1$, and so $T^*(x_2^*) = 0$. Since $T^*$ is injective, it follows that $x_2^* = 0$.

Conversely, suppose $R(T)$ is dense in $(V_2, \|\cdot\|_2)$ and let $x_2^* \in V_2^*$ be such that $T^*(x_2^*) = 0$. Then $x_2^*[T(x_1)] = T^*(x_2^*)(x_1) = 0$, $x_1 \in V_1$, shows that $x_2^*$ vanishes on $R(T)$, from which we deduce, via the continuity of $x_2^*$ and the denseness of $R(T)$, that $x_2^* = 0$. Therefore $T^*$ is injective, and this completes the proof. □

Theorem 4.4.1 easily yields the following corollary. The details are left to the reader.

<u>Corollary 4.4.1</u>. Let $(V_1, \|\cdot\|_1)$ be a Banach space and $(V_2, \|\cdot\|_2)$ a normed linear space over $\Phi$, and suppose $T \in L(V_1, V_2)$ is such that $\|T(x)\|_2 \geq m\|x\|_1$, $x \in V_1$, for some $m > 0$. Then the following are equivalent:

(i) $T$ is surjective.

(ii) $T^*$ is injective.

We shall return to the notion of the adjoint in subsequent sections, particularly when we discuss Hilbert spaces in Chapter 13. A similar idea will also appear in the investigation of reflexivity of normed linear spaces in Section 8.1.

4.5. Separability of $V^*$. Here we wish to show that the separability of $V^*$ implies that of $V$, at least in the case that $V$ is a normed linear space. We recall the following definition:

Definition 4.5.1. Let $(V,\mathcal{T})$ be a topological linear space over $\Phi$. Then $V$ is said to be separable if it contains a countable dense subset.

Theorem 4.5.1. Let $(V,\|\cdot\|)$ be a normed linear space over $\Phi$. If $V^*$ is separable, then $V$ is separable.

Proof. Let $\{x_n^*\}$ be a countable dense subset of $V^*$ and let $\{x_n\} \subset V$ be so chosen, using the definition of $\|x_n^*\|$, that $\|x_n\| = 1$ and $|x_n^*(x_n)| \geq \|x_n^*\|/2$, $n = 1,2,3,\ldots$ . Let $D \subset V$ denote the family of all finite linear combinations with (complex) rational coefficients of $\{x_n\}$. Clearly $D$ is countable. Let $W$ denote the linear subspace of $V$ spanned by $\{x_n\}$. Obviously $D \subset W$ and $D$ is dense in $W$. Thus it is apparent that $D$ is dense in $V$ if and only if $W$ is dense in $V$.

Suppose then that $x^* \in V^*$ is such that $x^*(x) = 0$, $x \in W$. In particular, $x^*(x_n) = 0$, $n = 1,2,3,\ldots$ . Since $\{x_n^*\}$ is dense in $V^*$, there exists some sequence from the set $\{x_n^*\}$, call it $\{x_{n_k}^*\}$, such that $\lim_k \|x_{n_k}^* - x^*\| = 0$. However,

$$\begin{aligned}
\|x_{n_k}^* - x^*\| &= \sup_{\|x\|=1} |(x_{n_k}^* - x^*)(x)| \\
&\geq |(x_{n_k}^* - x^*)(x_{n_k})| \\
&= |x_{n_k}^*(x_{n_k})| \\
&\geq \|x_{n_k}^*\|/2 \qquad (k = 1,2,3,\ldots),
\end{aligned}$$

from which it follows that $\lim_k \|x^*_{n_k}\| = 0$. But $\lim_k \|x^*_{n_k} - x^*\| = 0$ implies that $\lim_k \|x^*_{n_k}\| = \|x^*\|$, and so $\|x^*\| = 0$, that is, $x^* = 0$. Consequently from Corollary 4.2.8 we conclude that W, and hence D, is dense in V. Therefore V is separable. □

It should be noted that the converse of this result need not be valid. For example, $\ell_1$ is a separable Banach space, whereas $\ell_1^* = \ell_\infty$ is not.

4.6. Annihilators. In this section we wish to introduce the notion of annihilators and to use it for describing the dual spaces of subspaces and quotient spaces of normed linear spaces. It will be apparent when we discuss Hilbert spaces in Chapter 13 that the concept of the annihilator is a natural generalization to normed linear spaces of the notion of an orthogonal complement in Euclidean spaces and in inner-product spaces in general.

Definition 4.6.1. Let $(V,\mathcal{T})$ be a topological linear space over $\Phi$. If $E \subset V$, then

$$E^\perp = \{x^* \mid x^* \in V^*,\ x^*(x) = 0,\ x \in E\}$$

is called the annihilator of E. If $E^* \subset V^*$, then

$$(E^*)_\perp = \{x \mid x \in V,\ x^*(x) = 0,\ x^* \in E^*\}$$

is called the annihilator of $E^*$.

Thus $E^\perp \subset V^*$ is the set of all continuous linear functionals on V that vanish identically on E, and $(E^*)_\perp \subset V$ is the set of common zeros of the continuous linear functionals on V that belong to $E^*$. It is evident that $E^\perp$ and $(E^*)_\perp$ may be rather trivial, for example, if $V^* = \{0\}$.

The proof of the next proposition is straightforward and is left to the reader.

Proposition 4.6.1. Let $(V,\mathcal{T})$ be a topological linear space over $\Phi$, $E \subset V$, and $E^* \subset V^*$. Then

(i) $E^{\perp} \subset V^*$ is a linear subspace.

(ii) If $V$ is a normed linear space, then $E^{\perp} \subset V^*$ is a closed linear subspace.

(iii) $(E^*)_{\perp} \subset V$ is a closed linear subspace.

(iv) $E \subset (E^{\perp})_{\perp}$.

(v) If $V$ is a normed linear space, $V \neq \{0\}$, and $E \subset V$ is a proper linear subspace, then $E^{\perp} \neq \{0\}$.

Among other things, the preceding proposition says that $E \subset (E^{\perp})_{\perp}$ whenever $E \subset V$. The next result gives us a condition under which the containment is equality.

Theorem 4.6.1. Let $(V,\|\cdot\|)$ be a normed linear space over $\Phi$. If $W \subset V$ is a linear subspace, then $\mathrm{cl}(W) = (W^{\perp})_{\perp}$.

Proof. Since $W \subset (W^{\perp})_{\perp}$ and $(W^{\perp})_{\perp}$ is closed, we see that $\mathrm{cl}(W) \subset (W^{\perp})_{\perp}$. But if $x_0 \in (W^{\perp})_{\perp} \sim \mathrm{cl}(W)$, then, since $\mathrm{cl}(W)$ is a closed linear subspace, we see from Theorem 4.2.3 that there exists some $x^* \in V^*$ such that $x^*(x_0) \neq 0$ and $x^*(x) = 0$, $x \in \mathrm{cl}(W)$. In particular, $x^* \in W^{\perp}$. However, $x_0 \in (W^{\perp})_{\perp}$ implies that $x^*(x_0) = 0$, thereby contradicting the choice of $x^*$.

Therefore $\mathrm{cl}(W) = (W^{\perp})_{\perp}$. □

Corollary 4.6.1. Let $(V,\|\cdot\|)$ be a normed linear space over $\Phi$ and let $W \subset V$ be a linear subspace. Then the following are equivalent:

(i) $W$ is a closed linear subspace.

(ii) $W = (W^{\perp})_{\perp}$.

Next we shall use the concept of an annihilator to describe the spaces of continuous linear functionals on the subspaces and quotient spaces of normed linear spaces.

First we make a general definition.

<u>Definition 4.6.2</u>. Let $(V_1, \|\cdot\|_1)$ and $(V_2, \|\cdot\|_2)$ be normed linear spaces over $\Phi$. A mapping $\varphi : V_1 \to V_2$ is said to be an <u>isometry</u> if $\|\varphi(x)\|_2 = \|x\|_1$, $x \in V_1$. Such a mapping is also said to be <u>isometric</u>.

<u>Theorem 4.6.2</u>. Let $(V, \|\cdot\|)$ be a normed linear space over $\Phi$ and let $W \subset V$ be a closed linear subspace. Then there exist surjective isometric isomorphisms between $V^*/W^\perp$ and $W^*$, and between $(V/W)^*$ and $W^\perp$.

<u>Proof</u>. If $x^* + W^\perp \in V^*/W^\perp$, then clearly the equation $(x^* + W^\perp)(x) = x^*(x)$, $x \in W$, defines a continuous linear functional on $W$. Thus the mapping $\varphi : V^*/W^\perp \to W^*$, defined by $\varphi(x^* + W^\perp)(x) = x^*(x)$, $x \in W$, is well defined. It is easily seen that $\varphi$ is an isomorphism. Moreover, suppose $y^* \in W^*$. Then, by Theorem 4.2.2, there exists some $x^* \in V^*$ such that $\|x^*\| = \|y^*\|$ and $x^*(x) = y^*(x)$, $x \in W$. Clearly, $\varphi(x^* + W^\perp) = y^*$, and so $\varphi$ is surjective.

Furthermore, the preceding argument, combined with the definition of the norm in $V^*/W^\perp$ (Theorem 1.1.1), reveals that, if $y^* \in W^*$ and $\varphi(x^* + W^\perp) = y^*$, where $x^* \in V^*$ is that chosen as in the preceding paragraph, then

$$\|y^*\| = \|x^*\| \geq |||x^* + W^\perp|||,$$

that is, $\|\varphi(x^* + W^\perp)\| \geq |||x^* + W^\perp|||$. On the other hand, if $y^* = \varphi(x^* + W^\perp)$, then

$$\begin{aligned} \|y^*\| &= \sup_{\substack{x \in W \\ \|x\| = 1}} |y^*(x)| \\ &= \sup_{\substack{x \in W \\ \|x\| = 1}} |(x^* + z^*)(x)| \\ &\leq \|x^* + z^*\| \qquad (z^* \in W^\perp), \end{aligned}$$

and hence $\|y^*\| = \|\varphi(x^* + W^\perp)\| \leq |||x^* + W^\perp|||$.

Therefore $\varphi$ is an isometry, and the first assertion of the theorem is proved.

The second assertion is even easier to prove and is left to the reader. □

Analogous results are also valid for locally convex topological linear spaces (see, for example, [K, pp. 276-279]).

4.7. Ideals in $L_1(\mathbb{R},dt)$. As noted in Example 1.2.4, the space $(L_1(\mathbb{R},dt),\|\cdot\|_1)$ is a Banach space, where as usual $dt$ denotes Lebesgue measure on $\mathbb{R}$. It is also possible to introduce an operation of multiplication into $L_1(\mathbb{R},dt)$ so that $L_1(\mathbb{R},dt)$ becomes an algebra. The main purpose of this section is to use the Hahn-Banach Theorem to give a description of the closed ideals in this algebra.

First, however, we must see what the multiplication in $L_1(\mathbb{R},dt)$ is to be. The following definition and proposition are more general than needed for our present purposes, but we shall have use for these more general results in subsequent sections.

Definition 4.7.1. Let $f$ and $h$ be Lebesgue-measurable functions defined on $\mathbb{R}$. Then the convolution $f * h$ of $f$ and $h$ is defined formally for almost all $t \in \mathbb{R}$ by the expression

$$f * h(t) = \int_{\mathbb{R}} f(t - s)h(s)\, ds.$$

Various properties of convolution are collected in the next proposition.

Proposition 4.7.1. Let $f,g \in L_1(\mathbb{R},dt)$ and $h \in L_p(\mathbb{R},dt)$, $1 \leq p \leq \infty$. Then

(i) $f * h \in L_p(\mathbb{R},dt)$.

(ii) $\|f * h\|_p \leq \|f\|_1\ \|h\|_p$.

(iii) $f * (g * h) = (f * g) * h$.

(iv) $(f + g) * h = f * h + g * h$.

(v) $a(f * h) = (af) * h = f * (ah)$ $(a \in \mathbb{C})$.

(vi) $f * h = h * f$.

Moreover, if $f \in L_p(\mathbb{R},dt)$ and $h \in L_q(\mathbb{R},dt)$, $1/p + 1/q = 1$, $1 \leq p \leq \infty$, then $f * g \in C(\mathbb{R},dt)$ and $\|f * h\|_\infty \leq \|f\|_p \|h\|_q$.

We do not give the proof of this result, but instead refer the reader to [HS, pp. 396-399]. The proof of parts (i) and (ii) of the proposition is nontrivial and is obtained essentially as a consequence of Fubini's Theorem [Ry, p. 269]. Parts (i) and (iii) through (vi) of the proposition say precisely that $L_1(\mathbb{R},dt)$ is a commutative algebra over $\mathbb{C}$ with convolution as multiplication. Part (ii) says, incidentally, that the algebra $L_1(\mathbb{R},dt)$ is a Banach algebra, but this does not concern us here.

A few more preliminary notions are required.

Definition 4.7.2. For each $s \in \mathbb{R}$, if $f$ is a function on $\mathbb{R}$, then we define $\tau_s f$ by the equation

$$(\tau_s f)(t) = f(t - s) \qquad (t \in \mathbb{R}).$$

If one makes the usual agreement about identifying Lebesgue measurable functions that are equal almost everywhere, then it is easily seen that $\tau_s \in \mathcal{L}(L_p(\mathbb{R}, dt))$, $s \in \mathbb{R}$, $1 \leq p \leq \infty$, and that $\tau_s(f * g) = (\tau_s f) * g = f * (\tau_s g)$, $f \in L_1(\mathbb{R},dt)$, $g \in L_p(\mathbb{R},dt)$, $1 \leq p \leq \infty$.

We can now state the theorem alluded to previously.

Theorem 4.7.1. Let $W \subset L_1(\mathbb{R},dt)$ be a linear subspace. Then the following are equivalent:

(i) $W$ is a closed translation invariant subspace.

(ii) $W$ is a closed ideal.

Proof. Suppose $W$ is a closed translation invariant subspace. To see that $W$ is a closed ideal it suffices to show that $f * g \in W$ whenever $f \in L_1(\mathbb{R},dt)$ and $g \in W$. So let $f \in L_1(\mathbb{R},dt)$ and $g \in W$.

We recall from Example 3.1.8 that the elements $x^*$ of $L_1(\mathbb{R},dt)^*$ can be identified with the elements $\varphi$ of $L_\infty(\mathbb{R},dt)$ by the formula

$$x^*(f) = \int_{\mathbb{R}} f(t)\overline{\varphi(t)}\,dt \qquad (f \in L_1(\mathbb{R},dt)).$$

Thus, if $\varphi \in L_\infty(\mathbb{R},dt)$ is such that

$$\int_{\mathbb{R}} g(t)\overline{\varphi(t)}\,dt = 0 \qquad (g \in W),$$

then after some simple computations we see that $g * \varphi^*(0) = 0$, $g \in W$, where $\varphi^*(t) = \overline{\varphi(-t)}$, the bar denoting complex conjugation. Since $W$ is translation invariant, we deduce further that $(f * g) * \varphi^* = f * (g * \varphi^*) = 0$, as $g * \varphi^* = 0$. Hence

$$\int_{\mathbb{R}} f * g(t)\overline{\varphi(t)}\,dt = 0 \quad (f \in L_1(\mathbb{R},dt);\ g \in W),$$

and so from Corollary 4.2.7, we conclude that $f * g \in W$, $f \in L_1(\mathbb{R},dt)$; that is, $W$ is an ideal. Thus part (i) implies part (ii).

Conversely, suppose $W$ is a closed ideal and $\varphi \in L_\infty(\mathbb{R},dt)$ is as before. If $g \in W$, then $f * g \in W$, $f \in L_1(\mathbb{R},dt)$, as $W$ is an ideal, and so $(f * g) * \varphi^*(0) = f * (g * \varphi^*)(0) = 0$, $f \in L_1(\mathbb{R},dt)$ by the choice of $\varphi$. However, from Proposition 4.7.1 we see that $g * \varphi^* \in L_\infty(\mathbb{R},dt)$, and so $g * \varphi^*$ defines a continuous linear functional on $L_1(\mathbb{R},dt)$ that vanishes on all of $L_1(\mathbb{R},dt)$ and hence is the zero functional. However, it is easily seen that

$$\begin{aligned} f * (g * \varphi^*) &= \int_{\mathbb{R}} f(t)g * \varphi^*(-t)\,dt \\ &= \int_{\mathbb{R}} f(t)\ \left[\int_{\mathbb{R}} g(-t - u)\overline{\varphi(-u)}\,du\right] dt \\ &= \int_{\mathbb{R}} f(t)\ \left[\int_{\mathbb{R}} g(-t + u)\overline{\varphi(u)}\,du\right] dt \end{aligned}$$

$$= \int_{\mathbb{R}} f(t) \{\overline{\int_{\mathbb{R}} g[-t - u)]\varphi(u)\, du}\}\, dt$$
$$= \int_{\mathbb{R}} f(t)\overline{g^* * \varphi(t)}\, dt \qquad (f \in L_1(\mathbb{R},dt)).$$

Consequently we conclude that $g^* * \varphi = 0$, and hence $g * \varphi^* = (g^* * \varphi)^* = 0$. Since $g * \varphi^*$ is continuous, by Proposition 4.7.1, we see that for each $s \in \mathbb{R}$.

$$0 = g * \varphi^*(-s) = \tau_s(g * \varphi^*)(0)$$
$$= (\tau_s g) * \varphi^*(0),$$

and this holds for each $g \in W$.

Thus we have shown that any continuous linear functional on $L_1(\mathbb{R},dt)$ that vanishes on $W$ must also vanish at each $\tau_s g$ when $s \in \mathbb{R}$ and $g \in W$. Consequently, since $W$ is closed, by Corollary 4.2.7, $\tau_s g \in W$ whenever $s \in \mathbb{R}$ and $g \in W$; that is, $W$ is translation invariant.

Therefore part (ii) of Theorem 4.7.1 implies part (i), and the proof is complete. □

## 4.8. Continuous Linear Functionals on C([0,1]).

In Example 3.1.5 we asserted that every continuous linear functional $x^*$ on $C([0,1])$ could be realized as

$$x^*(f) = \int_0^1 f(t)\, d\mu(t) \qquad (f \in C([0,1])),$$

where $\mu$ is an appropriate bounded, regular, and complex-valued Borel measure on $[0,1]$. We now wish to use the Hahn-Banach Theorem to prove this result. For the sake of completeness we include the next definition.

*Definition 4.8.1.* A complex-valued function $g$ defined on $[0,1]$ is said to be of *bounded variation* if

$$V(g) = \sup \sum_{k=1}^{n} |g(t_k) - g(t_{k-1})| < \infty,$$

where the supremum is taken over all partitions of $[0,1]$ of the form $0 = t_0 < t_1 < \cdots < t_n = 1$, and $n$ is an arbitrary positive integer.

Theorem 4.8.1. The following are equivalent:

(i) $x^* \in C([0,1])^*$.

(ii) There exists some complex-valued function $g$ of bounded variation defined on $[0,1]$ such that

$$x^*(f) = \int_0^1 f(t)\, dg(t) \qquad (f \in C([0,1])),$$

where the integral is the Riemann-Stieltjes integral.

Moreover, if $x^* \in C([0,1])^*$, then $\|x^*\| = V(g)$.

Proof: It is left to the reader to show that part (ii) of the theorem implies part (i). Conversely, let $x^* \in C([0,1])^*$. Then $x^*$ is a continuous linear functional on the linear subspace $C([0,1])$ of $B([0,1])$, the normed linear space of bounded complex-valued functions on $[0,1]$ with the usual supremum norm $\|\cdot\|_\infty$. Hence, by Theorem 4.2.2, there exists some $z^* \in B([0,1])^*$ such that $\|x^*\| = \|z^*\|$ and $z^*(f) = x^*(f)$, $f \in C([0,1])$.

Now for each $0 < s \leq 1$ the characteristic function of the interval $[0,s]$ belongs to $B([0,1])$. Denote this characteristic function by $\chi_s$, set $\chi_0 = 0$, and define $g(s) = z^*(\chi_s)$, $0 \leq s \leq 1$. We claim that $g$ is of bounded variation.

Indeed, suppose $0 = t_0 < t_1 < \cdots < t_n = 1$ is any partition of $[0,1]$. Define for each $k = 1,2,\ldots,n$

$$a_k = \exp\{-i \arg[g(t_k) - g(t_{k-1})]\} \quad \text{for} \quad g(t_k) \neq g(t_{k-1}),$$

$$a_k = 0 \quad \text{for} \quad g(t_k) = g(t_{k-1}),$$

and set

$$h(t) = a_1 \quad \text{for} \quad t_0 \leq t \leq t_1,$$

$$h(t) = a_k \quad \text{for} \quad t_{k-1} < t \leq t_k;\ k = 2,3,\ldots,n.$$

Clearly $h \in B([0,1])$ and $\|h\|_\infty \leq 1$. Moreover, straightforward arguments reveal that $h = \Sigma_{k=1}^{n} a_k(\chi_{t_k} - \chi_{t_{k-1}})$.

Consequently

$$z^*(h) = \sum_{k=1}^{n} a_k[z^*(\chi_{t_k}) - z^*(\chi_{t_{k-1}})]$$

$$= \sum_{k=1}^{n} a_k[g(t_k) - g(t_{k-1})]$$

$$= \sum_{k=1}^{n} |g(t_k) - g(t_{k-1})|.$$

From this identity we immediately deduce that

$$\sum_{k=1}^{n} |g(t_k) - g(t_{k-1})| \leq \|z^*\|\|h\|_\infty \leq \|x^*\|.$$

hence $g$ is of bounded variation, and $V(g) \leq \|x^*\|$, as the partition $0 = t_0 < t_1 < \cdots < t_n = 1$ is arbitrary.

Next, let $f \in C([0,1])$ and let $0 = t_0 < t_1 < \cdots < t_n = 1$ be a partition of $[0,1]$. Then set $h_1 = \Sigma_{k=1}^{n} f(t_{k-1})(\chi_{t_k} - \chi_{t_{k-1}})$. As before, $h_1 \in B([0,1])$ and

$$z^*(h_1) = \sum_{k=1}^{n} f(t_{k-1})[g(t_k) - g(t_{k-1})].$$

However, given $\epsilon > 0$, it is easily seen that $\|h_1 - f\|_\infty < \epsilon/2\|z^*\|$ and

$$\left| \sum_{k=1}^{n} f(t_{k-1})[g(t_k) - g(t_{k-1})] - \int_0^1 f(t)\, dg(t) \right| < \frac{\epsilon}{2}$$

whenever $\sup_k|t_k - t_{k-1}|$ is sufficiently small. The Riemann-Stieltjes integral exists since $f \in C([0,1])$ and $g$ is of bounded variation. Consequently we conclude by the continuity of $z^*$ and the arbitrariness of $\epsilon > 0$ that

$$z^*(f) = \int_0^1 f(t)\, dg(t).$$

But $z^*(f) = x^*(f)$, and hence part (ii) of the theorem is proved.

Therefore parts (i) and (ii) are equivalent.

The fact that $\|x^*\| = V(g)$ follows at once from the estimate $V(g) \le \|x^*\|$ obtained here and the well-known estimate for Riemann-Stieltjes integrals,

$$\left|\int_0^1 f(t)\, dg(t)\right| \le \|f\|_\infty V(g) \qquad (f \in C([0,1])). \square$$

It needs to be noted that the correspondence between the elements of $C([0,1])^*$ and functions of bounded variation, although linear, is not injective. For example, any two functions $g_1$ and $g_2$ of bounded variation such that

$$\int_0^1 f(t)\, dg_1(t) = \int_0^1 f(t)\, dg_2(t) \qquad (f \in C([0,1]))$$

obviously correspond to the same element of $C([0,1])^*$. One can, however, use this observation to intoduce an equivalence relation into the space of functions of bounded variation and then obtain an injective mapping from $C([0,1])^*$ to the space of equivalence classes. This is done, for instance, in [BaNr, pp. 217-227; K, pp. 197-201; T, pp. 195-201]. An alternative way to approach this difficulty is to recognize that it is the measure $dg$ determined by $g$, and not the function itself, that corresponds to the continuous linear functional. This observation also allows one to envision the generalization of Theorem 4.8.1 to $C_0(X)$, where $X$ is a locally compact Hausdorff space.

We shall not prove the theorem in this generality, but we shall state it precisely and use it in subsequent chapters. The reader is referred to [$DS_1$, pp. 261-265; KeNa, p. 127; Ry, pp. 308-311; $Ru_2$, pp. 130-133].

<u>Theorem 4.8.2</u>. (<u>Riesz Representation Theorem</u>). Let $X$ be a locally compact Hausdorff topological space. Then the following are equivalent:

(i) $x^* \in C_o(X)^*$.

(ii) There exists a unique $\mu \in M(X)$ such that

$$x^*(f) = \int_X f(t)\, d\mu(t) \qquad (f \in C_o(X)).$$

Moreover, the correspondence between $x^*$ and $\mu$ defines a surjective isometric isomorphism between $C_o(X)^*$ and $M(X)$.

The reader is referred to Examples 1.2.1 and 1.2.6 for the definitions of $C_o(X)$ and $M(X)$, respectively.

<u>4.9. A Moment Problem</u>. One version of the moment problem in classical mathematics asks the following question: Given a sequence of real numbers $c_0, c_1, c_2, \ldots,$ when does there exist a real-valued function $g$ defined on $[0,1]$ that is of bounded variation and such that

$$\int_0^1 t^n\, dg(t) = c_n \qquad (n = 0,1,2,\ldots).$$

The integrals are, of course, Riemann-Stieltjes integrals. This moment problem also has a probabilistic interpretation. Essentially, it asks: When do the noncentral moments of a probability distribution determine the distribution?

In view of the discussion in the preceding section we see at once that this moment problem is equivalent to asking: When does there exist a continuous linear functional $x^* \in C([0,1])^*$ such that $x^*(t^n) = c_n$, $n = 0,1,2,\ldots$ ? This observation prompts us to pose a general functional-analytic moment problem: Suppose $(V,\mathcal{T})$ is a topological linear space over $\Phi$ and suppose $\{c_\alpha\}_{\alpha \in \Lambda} \subset \Phi$ and $\{x_\alpha\}_{\alpha \in \Lambda} \subset V$ are given. When does there exist some $x^* \in V^*$ such that $x^*(x_\alpha) = c_\alpha$, $\alpha \in \Lambda$?

As is to be expected, this moment problem, in general, does not have a solution. But the Hahn-Banach Theorem does provide us with an appropriate tool with which to obtain rather simple necessary and sufficient conditions for the existence of solutions in the case of normed linear spaces.

Theorem 4.9.1. Let $(V,\|\cdot\|)$ be a normed linear space over $\Phi$ and suppose $\{c_\alpha\}_{\alpha\in\Lambda} \subset \Phi$ and $\{x_\alpha\}_{\alpha\in\Lambda} \subset V$. Then the following are equivalent:

(i) There exists some $x^* \in V^*$ such that $x^*(x_\alpha) = c_\alpha$, $\alpha \in \Lambda$.

(ii) There exists some $M > 0$ such that

$$\left|\sum_\alpha a_\alpha c_\alpha\right| \leq M\left\|\sum_\alpha a_\alpha x_\alpha\right\|,$$

the sum being taken over all possible subsets $\{a_\alpha\}_{\alpha\in\Lambda} \subset \Phi$, where at most a finite number of the $a_\alpha$ are not zero.

Proof. If $x^* \in V^*$ is such that $x^*(x_\alpha) = c_\alpha$, $\alpha \in \Lambda$, then clearly for any set $\{a_\alpha\}_{\alpha\in\Lambda} \subset \Phi$ with only finitely many nonzero $a_\alpha$ we have

$$\begin{aligned}\left|\sum_\alpha a_\alpha c_\alpha\right| &= \left|\sum_\alpha a_\alpha x^*(x_\alpha)\right| \\ &= \left|x^*\left(\sum_\alpha a_\alpha x_\alpha\right)\right| \\ &\leq \|x^*\|\left\|\sum_\alpha a_\alpha x_\alpha\right\|.\end{aligned}$$

Hence part (i) of the theorem implies part (ii).

Conversely, suppose part (ii) holds and let $W \subset V$ be the linear subspace spanned by $\{x_\alpha\}_{\alpha\in\Lambda}$. For each $\Sigma_\alpha a_\alpha x_\alpha \in W$ define

$$y^*\left(\sum_\alpha a_\alpha x_\alpha\right) = \sum_\alpha a_\alpha c_\alpha.$$

We claim that $y^*$ is a well-defined continuous linear functional on $W$.

To see that $y^*$ is well defined suppose that $\Sigma_\alpha a_\alpha x_\alpha = \Sigma_\beta b_\beta x_\beta \in W$ and define $\{a'_\gamma\},\{b'_\gamma\} \subset \Phi$, for $\gamma \in \Lambda$ as follows:

$$a'_\gamma = a_\gamma \quad \text{for} \quad a_\gamma \neq 0,$$
$$a'_\gamma = 0 \quad \text{for} \quad a_\gamma = 0.$$

$$b'_\gamma = b_\gamma \quad \text{for} \quad b_\gamma \neq 0,$$
$$b'_\gamma = 0 \quad \text{for} \quad b_\gamma = 0.$$

Clearly at most a finite number of the $a'_\gamma$ and $b'_\gamma$ are nonzero, and by part (ii) of the theorem we have

$$\begin{aligned} \left|\sum_\alpha a_\alpha c_\alpha - \sum_\beta b_\beta c_\beta\right| &= \left|\sum_\gamma (a'_\gamma - b'_\gamma)c_\gamma\right| \\ &\leq M \left\|\sum_\gamma (a'_\gamma - b'_\gamma)x_\gamma\right\| \\ &= M \left\|\sum_\alpha a_\alpha x_\alpha - \sum_\beta b_\beta x_\beta\right\| \\ &= 0, \end{aligned}$$

and so $y^*(\Sigma_\alpha a_\alpha x_\alpha) = y^*(\Sigma_\beta b_\beta x_\beta)$. Thus $y^*$ is well-defined. It is evident from the definition of $y^*$ and part (ii) of the theorem that $y^* \in W^*$ and $\|y^*\| \leq M$. Consequently by Theorem 4.2.2 there exists some $x^* \in V^*$ such that $\|x^*\| = \|y^*\|$ and $x^*(x) = y^*(x)$, $x \in W$. In particular, $x^*(x_\alpha) = c_\alpha$, $\alpha \in \Lambda$. Therefore part (ii) of the theorem implies part (i). □

It should be noted that the functional $x^*$ in part (i) of Theorem 4.9.1, whose existence is implied by part (ii), is such that $\|x^*\| \leq M$.

<u>4.10. Helly's Theorem</u>. In the preceding section we considered the problem of the existence of $x^* \in V^*$ such that $x^*(x_\alpha) = c_\alpha$, $\alpha \in \Lambda$, where $\{x_\alpha\}_{\alpha\in\Lambda} \subset \Phi$ are given. A superficially similar, but more difficult, problem is the existence of an $x \in V$ such that $x_k^*(x) = c_k$, $k = 1,2,\ldots,n$, where $\{x_k^*\} \subset V^*$ and $\{c_k\} \subset \Phi$ are given.

The theorem we are about to prove, which is due to E. Helly [Hy], provides necessary and sufficient conditions for the existence of a solution to such a problem. The theorem and its corollary are very useful results, as will be apparent from the development of various sections in subsequent chapters.

The proof we give relies on a result concerning convex sets and hyperplanes in Euclidean space, which we use without further proof. This result asserts that in Euclidean space a convex set with a nonempty interior and a point not in the set can always be separated from each other by a hyperplane, that is, by a translate of some linear subspace of codimension one. In actuality this result is just the Hahn-Banach Theorem for Euclidean spaces in disguise. Such a geometric version of the Hahn-Banach Theorem will be discussed in the next chapter, and its equivalence with our analytic form of the theorem will be established.

The proof of Helly's Theorem will also contain a special case of the Open-Mapping Theorem, another of the most fundamental theorems of functional analysis. The Open-Mapping Theorem will be discussed in Chapter 7.

<u>Theorem 4.10.1 (Helly's Theorem)</u>. Let $(V,\|\cdot\|)$ be a normed linear space over $\Phi$, let $M > 0$, and suppose $x_1^*,x_2^*,\ldots,x_n^*$ are in $V^*$ and $c_1,c_2,\ldots,c_n$ are in $\Phi$. Then the following are equivalent:

(i) For each $\epsilon > 0$ there exists some $x_\epsilon \in V$ such that $\|x_\epsilon\| \leq M + \epsilon$ and $x_k^*(x_\epsilon) = c_k$, $k = 1,2,\ldots,n$.

(ii) $|\Sigma_{k=1}^n a_k c_k| \leq M\,\|\Sigma_{k=1}^n a_k x_k^*\|$ for all choices of $a_1,a_2,\ldots,a_n$ in $\Phi$.

<u>Proof</u>. If part (i) holds, then for each $\epsilon > 0$ there exists some $x_\epsilon \in V$, $\|x_\epsilon\| \leq M + \epsilon$, for which $x_k^*(x_\epsilon) = c_k$, $k = 1,2,\ldots,n$. It then follows easily that for any $a_1,a_2,\ldots,a_n$ in $\Phi$ we have

$$\left|\sum_{k=1}^{n} a_k c_k\right| = \left|\left(\sum_{k=1}^{n} a_k x_k^*\right)(x_\epsilon)\right|$$

$$\leq \|\sum_{k=1}^{n} a_k x_k^*\| \|x_\epsilon\|$$

$$\leq (M + \epsilon) \|\sum_{k=1}^{n} a_k x_k^*\|.$$

Since $\epsilon > 0$ is arbitrary, we conclude at once that part (i) implies part (ii).

Conversely, suppose part (ii) holds. If $c_k = 0$, $k = 1,2,\ldots,n$, then part (i) is trivial on taking $x_\epsilon = 0$. Thus without loss of generality we may assume that some $c_k \neq 0$. Moreover, we claim that we may assume that $x_1^*, x_2^*, \ldots, x_n^*$ are linearly independent on the following grounds: Suppose they are linearly dependent and let $x_1^*, x_2^*, \ldots, x_m^*$, $m \leq n$, be a maximal linearly independent subset of $x_1^*, x_2^*, \ldots, x_n^*$, where we have for convenience possibly renumbered the $x_k^*$ and $c_k$. Then, given $\epsilon > 0$, suppose $x_\epsilon \in V$ is such that $\|x_\epsilon\| \leq M + \epsilon$ and $x_k^*(x_\epsilon) = c_k$, $k = 1,2,\ldots,m$. Now, if $m < k \leq n$, then there exist some $b_{kj} \in \Phi$ such that

$$x_k^* = \sum_{j=1}^{m} b_{kj} x_j^*$$

and hence

$$x_k^*(x_\epsilon) = \sum_{j=1}^{m} b_{kj} x_j^*(x_\epsilon) = \sum_{j=1}^{m} b_{kj} c_j.$$

However, from part (ii) of the theorem we see that

$$|c_k - \sum_{j=1}^{m} b_{kj} c_j| \leq M \|x_k^* - \sum_{j=1}^{m} b_{kj} x_j^*\| = 0,$$

and so $x_k^*(x_\epsilon) = c_k$, $k = m + 1,\ldots,n$.

This shows that we may assume, without loss of generality, that $x_1^*, x_2^*, \ldots, x_n^*$ are linearly independent.

Finally, we assume for the purposes of the proof that $\Phi = \mathbb{R}$. This is purely a matter of convenience, and the same arguments _mutatis mutandis_ as given below apply in the case $\Phi = \mathbb{C}$.

We define a mapping $T : V \to \mathbb{R}^n$ by setting

$$T(x) = (x_1^*(x), x_2^*(x), \ldots, x_n^*(x)) \qquad (x \in V).$$

Clearly $T$ is linear. Moreover, we claim that $T$ is surjective by the following argument: If $T(V) \neq \mathbb{R}^n$, then $T(V)$, being a proper linear subspace of $\mathbb{R}^n$, is contained in some linear subspace of $\mathbb{R}^n$ of codimension one. Thus there exist $a_1, a_2, \ldots, a_n$ in $\mathbb{R}$, not all of which are zero, such that $\Sigma_{k=1}^n a_k x_k^*(x) = 0$, $x \in V$. But this clearly contradicts the linear independence of $x_1^*, x_2^*, \ldots, x_n^*$ and hence $T$ is surjective.

Denote the standard basis vectors in $\mathbb{R}^n$ by $e_k$; that is, $e_k = (e_{k1}, e_{k2}, \ldots, e_{kn})$, where $e_{kj} = 0$ if $j \neq k$, $e_{kk} = 1$, $k = 1,2,..,n$. Let $x_k \in V$ be such that $T(x_k) = e_k$, $k = 1,2,\ldots,n$. Given $\epsilon > 0$, set $K_\epsilon = \{x \mid x \in V, \|x\| \leq M + \epsilon\}$. Obviously $K_\epsilon$, and so $T(K_\epsilon)$, is convex and balanced. Furthermore, we claim that $T(K_\epsilon)$ has a nonempty interior that contains the origin in $\mathbb{R}^n$.

Indeed, let $\delta_\epsilon = (M + \epsilon)/(n \sup_{k=1,2,\ldots,n} \|x_k\|)$. Then, if $|a_k| < \delta_\epsilon$, $k = 1,2,\ldots,n$, we have

$$\left\| \sum_{k=1}^{n} a_k x_k \right\| \leq \sum_{k=1}^{n} |a_k| \, \|x_k\| \leq M + \epsilon.$$

Thus, if $|a_k| < \delta_\epsilon$, $k = 1,2,\ldots,n$, then $\Sigma_{k=1}^n a_k x_k \in K_\epsilon$, and so $T(\Sigma_{k=1}^n a_k x_k) = \Sigma_{k=1}^n a_k e_k \in T(K_\epsilon)$. Consequently $T(K_\epsilon)$ contains the open ball about the origin of radius $\delta_\epsilon$ in $(\mathbb{R}^n, \|\cdot\|_\infty)$, and so $T(K_\epsilon)$ has a nonempty interior that contains the origin.

Parenthetically we remark that $T(K_\epsilon)$ having a nonempty interior follows at once from the Open-Mapping Theorem (Theorem 7.2.1) on noting that $T$ is continuous.

Recall now that we wish to prove the existence of some $x_\epsilon \in K_\epsilon$ such that $x_k^*(x_\epsilon) = c_k$, $k = 1,2,\ldots,n$; that is, $T(x_\epsilon) = c = (c_1,c_2,\ldots,c_n)$. Suppose no such $x_\epsilon$ exists. Then $c \notin T(K_\epsilon)$. Since $T(K_\epsilon)$ is a convex balanced set with a nonempty interior that contains the origin and $c \notin T(K_\epsilon)$, there exist $b_1,b_2,\ldots,b_n$ in $\mathbb{R}$ such that

(a) $\displaystyle\sum_{k=1}^{n} b_k c_k > 0.$

(b) $\displaystyle\left|\sum_{k=1}^{n} b_k x_k^*(x)\right| \leq \sum_{k=1}^{n} b_k c_k \qquad (x \in K_\epsilon),$

that is, there exists a hyperplane that separates $T(K_\epsilon)$ and $c$. Here, incidentally, is the geometric form of the Hahn-Banach Theorem. But

$$\sup_{x \in K_\epsilon} \left|\sum_{k=1}^{n} b_k x_k^*(x)\right| = \sup_{x \in K_\epsilon} \frac{\left|\left(\sum_{k=1}^{n} b_k x_k^*\right)(x)\right|(M + \epsilon)}{M + \epsilon}$$

$$= \sup_{\|x\| \leq 1} \left|\left(\sum_{k=1}^{n} b_k x_k^*\right)(x)\right|(M + \epsilon)$$

$$= (M + \epsilon)\left\|\sum_{k=1}^{n} b_k x_k^*\right\|.$$

Consequently we conclude that

$$(M + \epsilon)\left\|\sum_{k=1}^{n} b_k x_k^*\right\| \leq \sum_{k=1}^{n} b_k c_k,$$

thereby contradicting part (ii) of the theorem.

Therefore $c \in T(K_\epsilon)$, and part (ii) implies part (i). □

<u>Corollary 4.10.1</u>. Let $(V,\|\cdot\|)$ be a Banach space over $\Phi$. If $x^{**}$ is a continuous linear functional on the Banach space $(V^*,\|\cdot\|)$, then for any $x_1^*,x_2^*,\ldots,x_n^*$ in $V^*$ and any $\epsilon > 0$ there exists some $x_\epsilon \in V$ such that

(i) $\|x_\epsilon\| \leq \|x^{**}\| + \epsilon$.

(ii) $x_k^*(x_\epsilon) = x^{**}(x_k^*)$ $(k = 1,2,\ldots,n)$.

<u>Proof</u>. Apply Helly's Theorem with $c_k = x^{**}(x_k^*)$, $k = 1,2,\ldots,n$, and $M = \|x^{**}\|$. □

Corollary 4.10.1 will be instrumental in proving that uniformly convex Banach spaces are reflexive (Theorem 8.2.1).

Incidentally, the separation property for hyperplanes in $\mathbb{R}^n$ that was utilized in the last portion of the proof of Helly's Theorem will be proved in the next chapter (Corollary 5.2.1) for arbitrary topological linear spaces.

## 4.11. Problems.

1. (Corollary 4.2.3) Let $(V,P)$ be a seminormed linear space over $\Phi$. Prove that, if $x \in V$, $x \neq 0$, then there exists some $x^* \in V^*$ such that $x^*(x) = 1$.

2. (Corollary 4.2.4) Let $(V,\|\cdot\|)$ be a normed linear space over $\Phi$, let $W \subset V$ be a linear subspace, and let $x_o \in V \sim W$. Prove that, if $d = \inf_{y \in W} \|x_o - y\| > 0$, then there exists some $x^* \in V^*$ such that

(a) $x^*(x) = 0$ $(x \in W)$.

(b) $x^*(x_o) = 1$.

(c) $\|x^*\| = 1/d$.

3. (Corollary 4.2.5) Let $(V,\|\cdot\|)$ be a normed linear space over $\Phi$. Prove that, if $x_o \in V$, $x_o \neq 0$, then there exists some $x^* \in V^*$ such that

(a) $x^*(x_o) = \|x_o\|$.
(b) $\|x^*\| = 1$.

Moreover, prove that

$$\|x_o\| = \sup_{\substack{\|x^*\| = 1 \\ x^* \in V^*}} |x^*(x_o)|.$$

*4. Prove that the dual result to Problem 3 is not valid; that is, give an example of a normed linear space $V$ and an $x_o^* \in V^*$, $x_o^* \neq 0$, such that for no $x \in V$, $\|x\| = 1$, is it true that $x_o^*(x) = \|x_o^*\|$.

5. Let $(V,\|\cdot\|)$ be a normed linear space over $\Phi$, let $W \subset V$ be a linear subspace, and let $x_o \in V$. Prove that

$$\inf_{y \in W} \|x_o - y\| = \sup\{|x^*(x_o)| \mid x^* \in V^*, \|x^*\| = 1, \quad W \subset N(x^*)\}.$$

6. Let $(V,T)$ be a topological linear space over $\Phi$, let $x_o \in V$, and let $p$ be a continuous seminorm on $V$. Prove that there exists on $x^* \in V^*$ such that $x^*(x_o) = p(x_o)$ and $|x^*(x)| \leq p(x)$, $x \in V$. Use this to give another proof of Corollary 4.2.1.

7. Let $(V,P)$ be a seminormed linear space over $\Phi$ and let $E$ be a closed convex balanced subset of $V$. Prove that for any $x_o \notin E$ there exists an $x^* \in V^*$ such that $x^*(x_o) > 1$ and $|x^*(x)| \leq 1$, $x \in E$.

8. Let $(V,P)$ be a seminormed linear space over $\Phi$ and let $U$ be a convex balanced neighborhood of the origin in $V$, that is, $0 \in \operatorname{int}(U)$. Prove that for any $x_o \notin U$ there exists an $x^* \in V^*$ such that

$$x^*(x_o) \geq \sup_{x \in U} |x^*(x)|.$$

9. (Corollary 4.2.6) Let $(V,\|\cdot\|)$ be a normed linear space over $\Phi$, and let $x_o \in V$. Prove that, if $x^*(x_o) = 0$ for all $x^* \in V^*$, then $x_o = 0$.

10. (Corollary 4.2.7) Let $(V,\|\cdot\|)$ be a normed linear space over $\Phi$, let $W \subset V$ be a closed linear subspace, and let $x_o \in V$. Prove that, if $x^* \in V^*$ and $x^*(x) = 0$, $x \in W$, implies that $x^*(x_o) = 0$, then $x_o \in W$.

11. (Corollary 4.2.8) Let $(V,\|\cdot\|)$ be a normed linear space over $\Phi$ and let $W \subset V$ be a linear subspace. Prove that the following are equivalent:

(a) $\mathrm{cl}(W) = V$.
(b) If $x^* \in V^*$ is such that $x^*(x) = 0$, $x \in W$, then $x^* = 0$.

12. Let $(V,\|\cdot\|) = (\mathbb{R}^3,\|\cdot\|_1)$ and let $W = \{(x,y,z) \mid x + 2y = z = 0\}$. Define $x^* \in W^*$ by $x^*[(x,y,z)] = x$. Give two different extensions of $x^*$ to all of $\mathbb{R}^3$ with the same norm as $x^*$.

13. Let $(V,\|\cdot\|)$ be a normed linear space over $\Phi$ and let $B_1 = \{x \mid x \in V,\ \|x\| \leq 1\}$. Furthermore, let $f$ be a mapping from $B_1$ into $\mathbb{R}$ such that $f(ax + by) = af(x) + bf(y)$ whenever $x$, $y$, and $ax + by$ are in $B_1$. Prove that $f$ can be extended to all of $V$ so that $f \in V'$.

*14 Let $V$ be a linear space over $\mathbb{R}$ and let $P \subset V$ be such that

(a) $x,y \in P$ and $a,b \geq 0$ imply $ax + by \in P$.
(b) $x \in P$ and $-x \in P$ imply $x = 0$.

Then $P$ is called a convex cone. For $x,y \in V$ define $x \leq y$ to mean that $y - x \in P$. Prove that $\leq$ is a partial ordering on $V$.

Let $W$ be a linear subspace of $V$ such that for all $x \in V$, we have $(x + W) \cap P \neq \emptyset$ if and only if $(-x + W) \cap P \neq \emptyset$. Suppose $y' \in W'$ is such that $x \in W$ and $x \geq 0$ imply $y'(x) \geq 0$. Prove that $y'$ can be extended to $x' \in V'$ such that $x \in V$ and $x \geq 0$ imply $x'(x) \geq 0$.

*15. Let $(V, \|\cdot\|)$ be a normed linear space over $\Phi$. Without using the embedding of $V$ in $V^{**}$, prove that there is a set $X$ such that $V$ is isometrically isomorphic with a subspace of the Banach space $(B(X), \|\cdot\|_\infty)$, the bounded functions on $X$ (see Example 1.2.2). Conclude that, if $(V, \|\cdot\|)$ is a separable normed linear space, then $V$ is isometrically isomorphic to a subspace of $(\ell_\infty, \|\cdot\|_\infty)$.

16. Let $(V, \|\cdot\|)$ be a Banach space over $\Phi$. If $x$ is a mapping from $\mathbb{C}$ into $V$, then $x$ is said to be analytic on a domain $\Omega \subset \mathbb{C}$ if

$$\lim_{h \to 0} \left\| \frac{x(z + h) - x(z)}{h} \right\|$$

exists for every $z \in \Omega$, $z + h \in \Omega$. If $x^* \in V^*$ and $x$ is analytic on $\Omega$, then the mapping $x^*(x) : \mathbb{C} \to \mathbb{C}$ defined by $[x^*(x)](z) = x^*[x(z)]$ is analytic on $\Omega$ in the usual complex-variable sense. The mapping $x$ is entire if $x$ is analytic on $\mathbb{C}$, and $x$ is bounded if $\|x(z)\| \leq M$ for all $z \in \mathbb{C}$ and some $M > 0$. Prove that, if $x : \mathbb{C} \to V$ is a bounded entire function, then $x$ is a constant.

*17. Let $D$ denote the closed unit disk in $\mathbb{C}$, let $\Gamma$ denote the unit circle, and let $U = \mathrm{int}(D)$. If $f \in C(D)$ and $E \subset D$, define $\|f\|_E = \sup\{|f(z)| \mid z \in E\}$.

(a) Let $f(z) = \Sigma_{k=0}^n a_k z^k$, where $a_0, \ldots, a_n$ are in $\mathbb{C}$, be a polynomial. Prove that $\|f\|_U = \|f\|_\Gamma$. [Note that $\|f\|_U = \|f\|_D$ for any $f \in C(D)$.]

(b) Let $W$ be any linear subspace of $C(D)$ that contains all polynomials and is such that $\|f\|_U = \|f\|_\Gamma$ for all $f \in W$. Prove that for each $z \in U$ there exists some $\mu_z \in M(\Gamma)$ such that

$$f(z) = \int_\Gamma f(\zeta)\, d\mu_z(\zeta) \qquad (f \in W).$$

[Hint: consider the subspace $W_\Gamma$ of $C(\Gamma)$ consisting of all functions on $\Gamma$ that are restrictions to $\Gamma$ of members of $W$, and the functional $y' \in W'_\Gamma$ defined by $y'(f) = f(z)$ for fixed $z \in U$. Use Theorems 4.1.2 and 4.8.2.]

(c) For each $n \in Z$ define $u_n(\zeta) = \zeta^n$, $\zeta \in D$. Prove that

$$\int_\Gamma u_n(\zeta)\, d\mu_z(\zeta) = r^{|n|} e^{in\theta},$$

where $z = re^{i\theta} \in U$. (Note that $u_n \in W$, $n = 0,1,2,\ldots$ .)

(d) For $0 \leq r < 1$ define

$$P_r(s) = \sum_{n=-\infty}^{\infty} r^{|n|} e^{ins} \qquad (s \in \mathbb{R}).$$

Compute

$$\frac{1}{2\pi}\int_{-\pi}^{\pi} P_r(\theta - t) e^{int}\, dt \qquad (\theta \in \mathbb{R};\ n \in \mathbb{Z}),$$

and compare the answer with

$$\int_\Gamma u_n(\zeta)\, d\mu_z(\zeta) \qquad (n \in \mathbb{Z}),$$

where $z = re^{i\theta} \in U$.

(e) Since, as seen in (d), we have

$$\int_\Gamma f(\zeta)\, d\mu_z(\zeta) = \frac{1}{2\pi}\int_{-\pi}^{\pi} f(e^{it}) P_r(\theta - t)\, dt$$

whenever $f = u_n$, $n \in \mathbb{Z}$, and $z = re^{i\theta} \in U$, and since every $f \in C(\Gamma)$ may be approximated uniformly by trigonometric polynomials, it can be shown that

$$\begin{aligned} f(z) &= \int_\Gamma f(\zeta)\, d\mu_z(\zeta) \\ &= \frac{1}{2\pi}\int_{-\pi}^{\pi} f(e^{it}) P_r(\theta - t)\, dt \qquad (f \in W), \end{aligned}$$

where $z = re^{i\theta} \in U$. Prove that

$$P_r(\theta - t) = \frac{1 - r^2}{1 - 2r\cos(\theta - t) + r^2}.$$

$P_r$ is called the Poisson kernel. To summarize, we have proved that, if $W$ is a linear subspace of $C(D)$ that contains all polynomials and is such that $\|f\|_U = \|f\|_\Gamma$, $f \in W$, then for each $z = re^{i\theta} \in U$ the Poisson integral representation

$$f(z) = \frac{1}{2\pi}\int_{-\pi}^{\pi} \frac{1 - r^2}{1 - 2r\cos(\theta - t) + r^2} f(e^{it})\, dt$$

is valid for each $f \in W$.

18. Let $\mathcal{B}(\mathbb{R})$ denote all the bounded subsets of $\mathbb{R}$. Use Theorem 4.3.1 to prove the existence of a set function $\mu$ defined for all $E \in \mathcal{B}(\mathbb{R})$ with the following properties:

(a) If $E,F \in \mathcal{B}(\mathbb{R})$ and $E \cap F = \phi$, then $\mu(E \cup F) = \mu(E) + \mu(F)$.

(b) If $E \in \mathcal{B}(\mathbb{R})$, then $\mu(E + t) = \mu(E)$, $t \in \mathbb{R}$.

(c) If $E,F \in \mathcal{B}(\mathbb{R})$ and $E \subset F$, then $\mu(E) \leq \mu(F)$.

(d) If $E \in \mathcal{B}(\mathbb{R})$ is Lebesgue measurable, then $\mu(E)$ is the Lebesgue measure of $E$.

19. (Corollary 4.4.1) Let $(V_1,\|\cdot\|_1)$ be a Banach space and $(V_2,\|\cdot\|_2)$ a normed linear space over $\Phi$ and suppose $T \in L(V_1,V_2)$ is such that $\|T(x)\|_2 \geq m\|x\|_1$, $x \in V_1$, for some $m > 0$. Prove that the following are equivalent:

(a) $T$ is surjective.

(b) $T^*$ is injective.

20. Let $(V_1,\|\cdot\|_1)$ be a Banach space and $(V_2,\|\cdot\|_2)$ be a normed linear space over $\Phi$ and suppose $T \in L(V_1,V_2)$ is such that $\|T(x)\|_2 \geq m\|x\|_1$, $x \in V_1$, for some $m > 0$. Prove that $(R(T),\|\cdot\|_2)$ is a Banach space.

21. Letting $(V_1,\|\cdot\|_1)$ and $(V_2,\|\cdot\|_2)$ be Banach spaces over $\Phi$ and $S,T \in L(V_1,V_2)$, prove each of the following:

(a) $(aS)^* = \bar{a}S^*$, $a \in \Phi$.

(b) $(S + T)^* = S^* + T^*$.

(c) If $T^{-1}$ exists and $\mathcal{R}(T)$ is dense in $(V_2,\|\cdot\|_2)$, then $(T^{-1})^* = (T^*)^{-1}$.

(d) If $(V_3, \|\cdot\|_3)$ is also a Banach space over $\Phi$ and $A \in \mathcal{L}(V_2,V_3)$, then $(AT)^* = T^*A^*$.

(e) If $\varphi(T)$ is the mapping from $\mathcal{L}(V_1,V_2)$ to $\mathcal{L}(V_2^*,V_1^*)$, defined by $\varphi(T) = T^*$, then $\varphi$ is an isometric isomorphism.

22. Let $(V_1,\|\cdot\|_1)$ and $(V_2,\|\cdot\|_2)$ be normed linear spaces over $\Phi$ and let $T \in \mathcal{L}(V_1,V_2)$. Prove that $T$ is injective if and only if the family $T^*(V_2^*) \subset V_1^*$ separates the points of $V_1$.

23. Let $(V,\|\cdot\|)$ be a normed linear space over $\Phi$ and let $T \in \mathcal{L}(V)$. Suppose that for some $x \in V$ and $x^* \in V^*$ we have $T(x) = ax$ and $T^*(x^*) = bx^*$, where $a,b \in \Phi$, $a \neq b$. Prove that $x^*(x) = 0$.

24. Let $V_1 = \mathbb{R}^n$, $V_2 = \mathbb{R}^m$, and $T \in \mathcal{L}(V_1,V_2)$. Then $T$ can be represented by an $m \times n$ real matrix $(a_{ij})$ (see Example 3.1.1). Prove that $T^*$ is represented by the matrix $(a_{ji})$, the transpose of $(a_{ij})$.

25. Let $(V,\|\cdot\|) = (c_o,\|\cdot\|_\infty)$ and let $T \in \mathcal{L}(V)$ be the mapping defined by $T(\{a_n\}) = \{a_n/n\}$, $\{a_n\} \in c_o$. Prove that $\mathcal{R}(T)$ is dense and find an expression for $T^*$.

26. (Proposition 4.6.1) Let $(V,\mathcal{T})$ be a topological linear space over $\Phi$, let $E \subset V$, and let $E^* \subset V^*$. Prove each of the following:

(a) $E^\perp \subset V^*$ is a linear subspace.

(b) If $V$ is a normed linear space, then $E^\perp \subset V^*$ is a closed linear subspace.

(c) $(E^*)_\perp \subset V$ is a closed linear subspace.

(d) $E \subset (E^\perp)_\perp$.

(e) If $V$ is a normed linear space, $V \neq \{0\}$, and $E \subset V$ is a proper linear subspace, then $E^\perp \neq \{0\}$.

*27. (Theorem 4.6.2) Let $(V,\|\cdot\|)$ be a normed linear space over $\Phi$ and let $W \subset V$ be a closed linear subspace. Prove that there exists a surjective isometric isomorphism between $(V/W)^*$ and $W^{\perp}$.

*28. Let $(V,\mathcal{T})$ be a topological linear space over $\Phi$. Let $E,F \subset V$ and $E^*,F^* \subset V^*$.

(a) Prove that, if $E \subset F$, then $F^{\perp} \subset E^{\perp}$.

(b) Prove that, if $E^* \subset F^*$, then $(F^*)_{\perp} \subset (E^*)_{\perp}$.

(c) If $E$ and $F$ are closed linear subspaces and $E^{\perp} = F^{\perp}$, prove that $E = F$.

(d) Give an example of a space $V$ and a subset $E^* \subset V^*$ for which $[(E^*)_{\perp}]^{\perp} \nsubseteq E^*$.

29. Let $(V_1,\|\cdot\|_1)$ and $(V_2,\|\cdot\|_2)$ be normed linear spaces over $\Phi$. Let $T \in L(V_1,V_2)$, let $T^* \in L(V_2^*,V_1^*)$ be the adjoint of $T$, let $N(T) = \{x \mid x \in V_1,\ T(x) = 0\}$ and $N(T^*) = \{x^* \mid x^* \in V_2^*,\ T^*(x^*) = 0\}$. $N(T)$ and $N(T^*)$ are called the <u>kernel</u> or <u>null space</u> of $T$ and $T^*$. Prove each of the following:

(a) $\{\mathrm{cl}[R(T)]\}^{\perp} = N(T^*)$.

(b) $\mathrm{cl}[R(T)] = [N(T^*)]_{\perp}$.

(c) $\{\mathrm{cl}[R(T^*)]\}_{\perp} = N(T)$.

(d) $\mathrm{cl}[R(T^*)] \subset [N(T)]^{\perp}$, and if $V_1$ and $R(T)$ are Banach spaces, then $\mathrm{cl}[R(T^*)] = [N(T)]^{\perp}$.

*30. (Proposition 4.7.1) Let $f,g \in L_1(\mathbb{R},dt)$ and $h \in L_p(\mathbb{R},dt)$, $1 \le p \le \infty$. Prove each of the following:

(a) $f * h \in L_p(\mathbb{R},dt)$.

(b) $\|f * h\|_p \le \|f\|_1 \|h\|_p$.

(c) $f * (g * h) = (f * g) * h$.

(d) $(f + g) * h = f * h + g * h$.

(e) $a(f * h) = (af) * h = f * (ah)$, $a \in \mathbb{C}$.

(f) $f * h = h * f$.

Moreover, prove that, if $f \in L_p(\mathbb{R},dt)$ and $h \in L_q(\mathbb{R},dt)$, $1/p + 1/q = 1$, $1 \le p \le \infty$, then $f * h \in C(\mathbb{R})$ and $\|f * h\|_{\infty} \le \|f\|_p\|h\|_q$.

Finally, if $f \in L_1(\mathbb{R},dt)$ and $h \in L_p(\mathbb{R},dt)$, $1 \leq p \leq \infty$, or $f \in L_p(\mathbb{R},dt)$ and $h \in L_q(\mathbb{R},dt)$, $1/p + 1/q = 1$, prove that $\tau_s(f) * h = f * \tau_s(h) = \tau_s(f * h)$, $s \in \mathbb{R}$, where $\tau_s(g)(t) = g(t - s)$, $t \in \mathbb{R}$.

31. Let $g$ be a complex-valued function of bounded variation defined on $[0,1]$ and define

$$x^*(f) = \int_0^1 f(t)\, dg(t) \qquad (f \in C([0,1])).$$

Prove that $x^* \in (C([0,1]))^*$.

*32. (Theorem 4.8.2) Let $X$ be a locally compact Hausdorff space. Prove that the following are equivalent:

(a) $x^* \in C_o(X)^*$.

(b) There exists a unique $\mu \in M(X)$ such that

$$x^*(f) = \int_X f(t)\, d\mu(t) \qquad (f \in C_o(x)).$$

Prove that the correspondence between $x^*$ and $\mu$ defines a surjective isometric isomorphism between $C_o(X)^*$ and $M(X)$.

33. Let $(V,\|\cdot\|)$ be a normed linear space over $\Phi$ and let $\{x_k\}$ be a sequence in $V$ such that $\lim_k \|x_k\| = 0$. Let $\{a_k\}$ be a sequence in $\Phi$ that does not converge to zero. Prove that, given any $M > 0$, there exists some finite set $\{b_1,\ldots,b_n\}$ of scalars such that

$$\left|\sum_{k=1}^{n} b_k a_k\right| > M\left\|\sum_{k=1}^{n} b_k x_k\right\|.$$

*34. Let $(V,\mathcal{T})$ be a topological linear space over $\Phi$ and let $E$ be a convex balanced compact subset of $V$. Let $F \subset V^*$ and let $\{c_f : f \in F\}$ be a family of scalars. Prove that there is some $x_o \in E$ such that $f(x_o) = c_f$ for all $f \in F$ if and only if

$$\left|\sum_{k=1}^{n} a_k c_k\right| \le \sup\left\{\left|\sum_{k=1}^{n} a_k f_k(x)\right| \;\middle|\; x \in E\right\}$$

for all choices of $f_1,\ldots,f_n$ in $F$ and scalars $a_1,\ldots,a_n$ in $\Phi$, where $c_k = c_{f_k}$.

# CHAPTER 5

# THE HAHN-BANACH THEOREM: GEOMETRIC FORM

5.0. Introduction. In the preceding chapter we discussed at some length the Hahn-Banach Theorem, as well as its consequences and applications. The form of the theorem we proved was an analytic one, involving the possibility of extending linear functionals. However, the Hahn-Banach Theorem also enables us to obtain a considerable amount of geometric information about topological linear spaces, much of which is generalizations of well-known theorems concerning Euclidean spaces. We now wish to turn to an exposition of this geometric aspect of the Hahn-Banach Theorem. Our treatment here will not be as long nor as detailed as the preceding one.

After introducing the concepts of linear variety and hyperplane, and establishing some basic results concerning such objects, we shall prove a geometric form of the Hahn-Banach Theorem. This theorem will assert that, if $K$ is a convex absorbing set in a linear space $V$ and $L_o$ is a linear variety disjoint from $K$, then there exists a real hyperplane $L$ that contains $L_o$ and is such that $K$ lies on one side of $L$. We shall use the Real Hahn-Banach Theorem (Theorem 4.1.1) to prove this theorem and shall show that the two theorems are equivalent.

Subsequently we shall use the geometric form of the Hahn-Banach Theorem to reestablish some of the consequences of the Hahn-Banach Theorem discussed in Section 4.2, and to extend some of those results, which we have proved only for normed linear spaces, to locally convex topological linear spaces. We conclude the chapter in Section 5.4 with the statement, without proof, of several additional results of a geometric nature.

5.1. Linear Varieties and Hyperplanes. To begin with we wish to introduce the notions of linear variety and hyperplane in a linear space, and to give some elementary results concerning such objects. We recall that a linear space $V$ over $\Phi$ can always be considered a linear space over $\mathbb{R}$. Throughout this chapter, whenever we speak of, for example, a real linear subspace of $V$ or a real linear functional on $V$, we mean a linear subspace of $V$ considered as a linear space over $\mathbb{R}$ or a linear functional on $V$ considered as a linear space over $\mathbb{R}$. Thus, for instance, $V = \mathbb{C}$ can be thought of as a linear space over $\mathbb{R}$, and it is then the same space as $\mathbb{R}^2$. The linear space $W = \mathbb{R}$ over $\mathbb{R}$ is clearly a real linear subspace of $V = \mathbb{C}$.

Definition 5.1.1. Let $V$ be a linear space over $\Phi$. A set $L \subset V$ is said to be a (real) linear variety if there exists a (real) linear subspace $W \subset V$ and some $x_0 \in V$ such that $L = x_0 + W$. A (real) linear variety $L$ is said to be a (real) hyperplane if $W$ is a maximal (real) linear subspace.

Thus a (real) linear variety is a translate of a (real) linear subspace, and a (real) hyperplane is a translate of a maximal (real) linear subspace. Clearly in the case that $\Phi = \mathbb{R}$ the notions of hyperplane and real hyperplane are identical. Recall also that a maximal linear subspace is by definition (Definition 3.3.2) a proper linear subspace.

It is easily seen that every linear variety is a real linear variety, but not conversely. For example, $\mathbb{R}$ is a real linear subspace, and hence a real linear variety, in $\mathbb{C}$, but $\mathbb{R}$ is not a linear variety in $\mathbb{C}$. Also, $\mathbb{R} \subset \mathbb{C}$ is an example of a real hyperplane that is not a hyperplane. An example of a hyperplane that is not a real hyperplane is provided, for instance, by $\mathbb{C} \subset \mathbb{C}^2$.

There is an intimate connection between hyperplanes and linear functionals, as shown by the following proposition:

Proposition 5.1.1. Let $V$ be a linear space over $\Phi$ and $L \subset V$. Then the following are equivalent:

(i) $L$ is a (real) hyperplane.

(ii) There exists a (real) linear functional $x'$ on $V, x' \neq 0$, and some (real) number $a \in \Phi$ such that $L = \{x \mid x \in V, x'(x) = a\}$.

Moreover, the (real) hyperplane $L$ contains the zero vector if and only if $a = 0$.

Proof. Suppose $L$ is a hyperplane, say $L = x_0 + W$, where $W$ is a linear subspace of codimension one. Let $x_1 \in V \sim W$. Then $\{x_1\} \cup W$ spans $V$, and the formula

$$x'(cx_1 + y) = c \qquad (c \in \Phi;\ y \in W)$$

defines a linear functional on $V$ and $x' \neq 0$. If $a = x'(x_0)$, then obviously $L = \{x \mid x'(x) = a\}$, and $a = 0$ if and only if $x_0 \in W$, that is, $W = L$ contains the origin.

Conversely, suppose $L = \{x \mid x'(x) = a\}$, where $x' \in V'$, $x' \neq 0$. Then there exists some $x_1 \in V$ such that $x'(x_1) = 1$. Let $x_0 = ax_1$ and $W = N(x')$, the kernel of $x'$. From Proposition 3.3.3 we know that $N(x')$ is a maximal linear subspace. We claim that $L = x_0 + N(x')$. Clearly, if $x = x_0 + y$, $y \in N(x')$, then $x'(x) = ax'(x_1) = a$, and so $x \in L$. Thus $x_0 + N(x') \subset L$. On the other hand, if $x \in L$, then $x'(-x_0 + x) = -x'(x_0) + x'(x) = -a + a = 0$ shows that $-x_0 + x \in N(x')$; that is, $x \in x_0 + N(x')$. Hence $L = x_0 + N(x')$, and $L$ is a hyperplane.

The arguments when $L$ is a real hyperplane or $x'$ is a real linear functional are essentially the same and are left to the reader.$\square$

The description of hyperplanes in terms of linear functionals will be quite useful in the succeeding development.

Some other elementary properties of linear varieties and of convex sets are collected in the next two propositions. Their proofs are left to the reader.

Proposition 5.1.2. Let $(V,\mathcal{T})$ be a topological linear space over $\Phi$. Then

(i) If $L \subset V$ is a (real) linear variety, then $\mathrm{cl}(L)$ is a (real) linear variety.

(ii) If $L \subset V$ is a (real) hyperplane, then either $L$ is closed or $\mathrm{cl}(L) = V$.

Clearly, in the case of real linear varieties and real hyperplanes we are considering $(V,\mathcal{T})$ as a topological linear space over $\mathbb{R}$.

Proposition 5.1.3. Let $V$ be a linear space over $\Phi$. Then

(i) If $K \subset V$ is convex, then the following sets are convex:

$aK$, $a \in \Phi$; $x + K$, $x \in V$; $T(K)$, $\{x \mid T(x) \in K\}$, $T \in L'(V)$.

(ii) If $K_1, K_2 \subset V$ are convex, then $K_1 + K_2$ and $K_1 - K_2$ are convex.

(iii) If $\{K_\alpha\}_{\alpha \in \Lambda}$ is a family of convex sets in $V$, then $\bigcap_\alpha K_\alpha$ is convex.

(iv) If $E \subset V$, then $\bigcap_{\substack{K \supset E \\ K \text{ convex}}} K$ is convex, and if $K_1 \subset V$ is convex and $K_1 \supset E$, then $K_1 \supset \bigcap_{\substack{K \supset E \\ K \text{ convex}}} K$.

The last portion of the proposition says that the intersection of all the convex sets that contain a set $E$ is the smallest convex set that contains $E$.

Note also that a (real) linear variety is a convex set.

Before continuing, another definition is necessary.

Definition 5.1.2. Let $V$ be a linear space over $\Phi$ and let $x'$ be a real linear functional on $V$, $x' \neq 0$. Then for each $a \in \mathbb{R}$ the sets $\{x \mid x'(x) > a\}$, $\{x \mid x'(x) < a\}$, $\{x \mid x'(x) \geq a\}$, and $\{x \mid x'(x) \leq a\}$ are called half-spaces determined by the real hyperplane $L = \{x \mid x'(x) = a\}$. A set $E \subset V$ is said to lie on one side of the real hyperplane $L$ if $E$ is contained in any one of the half-spaces determined by $L$, and $E$ is said to lie strictly on one side of $L$ if $E$ lies on one side of $L$ and $E \cap L = \emptyset$.

Obviously half-spaces are convex sets. If $V$ is a topological linear space and $L = \{x \mid x^*(x) = a\}$ is a real hyperplane, $x^*$ being a real continuous linear functional, then the first two half-spaces indicated in the preceding paragraph are open sets and the latter two are closed.

The following proposition is evident from the definition:

Proposition 5.1.4. Let $V$ be a linear space over $\Phi$, let $x'$ be a real linear functional on $V$, $x' \neq 0$, let $L = \{x \mid x'(x) = a\}$ for some $a \in \mathbb{R}$ and let $E \subset V$. Then

(i) $E$ lies on one side of $L$ if and only if either $E \subset \{x \mid x'(x) \leq a\}$ or $E \subset \{x \mid x'(x) \geq a\}$.

(ii) $E$ lies strictly on one side of $L$ if and only if either $E \subset \{x \mid x'(x) < a\}$ or $E \subset \{x \mid x'(x) > a\}$.

A necessary and sufficient condition for a convex set to lie strictly on one side of a real hyperplane is provided by the next theorem.

Theorem 5.1.1. Let $V$ be a linear space over $\Phi$, let $L \subset V$ be a real hyperplane, and let $K \subset V$ be convex. Then the following are equivalent:

(i) $K$ lies strictly on one side of $L$.

(ii) $K \cap L = \emptyset$.

Proof. The validity of part (i) implying part (ii) is established by definition. Conversely, suppose $K \cap L = \emptyset$, let $x'$ be a real linear functional on $V$, let $x' \neq 0$, and let $a \in \mathbb{R}$ be such that $L = \{x \mid x'(x) = a\}$. If $K$ does not lie strictly on one side of $L$, then from Proposition 5.1.4 and $K \cap L = \emptyset$ we see that there exist some $x_1, x_2 \in K$ for which $x'(x_1) < a < x'(x_2)$.

However, since $K$ is convex, $bx_1 + (1 - b)x_2 \in K$, $0 \leq b \leq 1$, and so $x'[bx_1 + (1 - b)x_2] = bx'(x_1) + (1 - b)x'(x_2)$ is clearly a continuous function of $b$ on $[0,1]$, taking the values $x'(x_1)$ at $b = 1$ and $x'(x_2)$ at $b = 0$. Consequently there exists some $b_o$, $0 < b_o < 1$, such that $x'[b_o x_1 + (1 - b_o)x_2] = a$. Thus $b_o x_1 + (1 - b_o)x_2 \in K \cap L$, contrary to assumption.

Therefore part (ii) of the theorem implies part (i), and the proof is complete. □

We conclude this section with the following proposition:

Proposition 5.1.5. Let $(V,\mathcal{T})$ be a topological linear space over $\Phi$ and let $L \subset V$ be a real hyperplane. If $E \subset V$ lies on one side of $L$ and $\text{int}(E) \neq \emptyset$, then

(i) $L$ is closed.
(ii) $\text{int}(E)$ lies strictly on one side of $L$.
(iii) $\text{cl}(E)$ lies on one side of $L$.

Proof. Using Proposition 5.1.1, let $x'$ be a real linear functional on $V$, let $x' \neq 0$, and let $a \in \mathbb{R}$ be such that $L = \{x \mid x'(x) = a\}$. Without loss of generality, we may suppose that $E \subset \{x \mid x'(x) \leq a\}$, as otherwise we merely need replace $x'$ by $-x'$ and $a$ by $-a$.

If $L$ is not closed, then by Theorem 2.1.2 the hyperplane $L_1 = \{x \mid x'(x) = a + 1\}$ is also not closed, and from Proposition 5.1.2 we see that $L_1$ is dense in $V$. Since $\text{int}(E) \neq \emptyset$, it follows at once that $\text{int}(E) \cap L_1 \neq \emptyset$, contradicting the fact that $\text{int}(E) \subset \{x \mid x'(x) \leq a\}$. Hence $L$ is closed.

Moreover, from the proof of Proposition 5.1.1 we see that the real hyperplane $L$ is of the form $x_o + N(x')$ for some suitable choice of $x_o$. Consequently, since translation in $V$ is a homeomorphism, we see that $N(x') = -x_o + L$ is closed, and so $x'$ is continuous by Theorem 3.3.2. Thus $\{x \mid x'(x) < a\}$ is an open set, and $\text{int}(E) \subset \{x \mid x'(x) < a\}$ shows that $\text{int}(E)$ lies strictly on one side of $L$.

Part (iii) of this proposition is now apparent, as $\{x \mid x'(x) \leq a\}$ is closed. □

5.2. The Hahn-Banach Theorem: Geometric Form. The geometric content of the Hahn-Banach Theorem is that certain kinds of sets can be separated from one another by real hyperplanes. The precise meaning of this will become clear in the statement of the theorem and in some further geometric consequences to be mentioned in Section 5.4. There are various other equivalent geometric formulations of the theorem in addition to the one presented here (see, for example, [$DS_1$, pp. 412, 417, and 418; $E_1$, pp. 116-118; KeNa, pp. 22 and 23; T, pp. 142 and 151; $W_1$, pp. 46-51, 219 and 220]).

Before we continue to the statement and proof of the Hahn-Banach Theorem in its geometric form we wish to introduce what could be called the gauge of a convex absorbing set. Since, however, we have already used the term "gauge" in another context (Section 1.4) in dealing with convex balanced absorbing sets, we shall not use the term again here. We content ourselves with the following definition:

Definition 5.2.1. Let $V$ be a linear space over $\Phi$ and let $K \subset V$ be a convex absorbing set. Then for each $x \in V$ we set

$$r_K(x) = \inf\{a \mid a > 0,\ x \in aK\}.$$

If $K$ were a convex balanced absorbing set, then $r_K$ would be just the gauge of $K$, and it would be a seminorm on $V$ by Proposition 1.4.2. In general, though, $r_K$ need not be a seminorm.

However, it is easily seen by the same arguments as those used in the proof of Proposition 1.4.2 that $r_K$ has the following properties:

(1) $r_K(x) \geq 0$ $\quad (x \in V)$.

(2) $r_K(0) = 0$.

(3) $r_K(x + y) \leq r_K(x) + r_K(y)$ $\quad (x,y \in V)$.

(4) $r_K(ax) = ar_K(x)$ $\quad (a \in \Phi;\ a \geq 0)$.

We shall use the next lemma in the proof of the Hahn-Banach Theorem.

Lemma 5.2.1. Let $V$ be a linear space over $\Phi$ and let $K \subset V$ be a convex absorbing set. If $L \subset V$ is a real hyperplane such that $K \cap L = \emptyset$ then there exists a real linear functional $x'$ on $V$ such that

(i) $L = \{x \mid x'(x) = 1\}$.

(ii) $-r_K(-x) \leq x'(x) \leq r_K(x)$ $\quad (x \in V)$.

Proof. Since $L$ is a real hyperplane, there exists, by Proposition 5.1.1, a real linear functional $y'$ on $V$, $y' \neq 0$, and an $a \in \mathbb{R}$ such that $L = \{x \mid y'(x) = a\}$. Since $K \cap L = \emptyset$ and $0 \in K$, as $K$ is absorbing, we see that $a \neq 0$. Hence $x' = y'/a$ is a real linear functional on $V$ such that $L = \{x \mid x'(x) = 1\}$.

Since $K \cap L = \emptyset$, from Theorem 5.1.1 we see that $K$ lies strictly on one side of $L$. Moreover, $K \subset \{x \mid x'(x) < 1\}$, as $0 \in K$ and $x'(0) = 0$. Let $x \in V$ and suppose $b > 0$ is such that $x \in bK$. Then $x/b \in K$, and hence $x'(x/b) < 1$; that is, $x'(x) < b$. Since this holds for any such $b$, we conclude that $x'(x) \leq r_K(x)$, $x \in V$.

Consequently for each $x \in V$ we also have $x'(-x) \leq r_K(-x)$, and thus $-r_K(-x) \leq -x'(-x) = x'(x)$, $x \in V$. □

We can now state and prove a geometric form of the Hahn-Banach Theorem.

Theorem 5.2.1. Let $V$ be a linear space over $\Phi$ and let $K \subset V$ be a convex absorbing set. If $L_o \subset V$ is a linear variety such that $K \cap L_o = \emptyset$, then there exists a real hyperplane $L = \{x \mid x'(x) = 1\}$, where $x'$ is a real linear functional on $V$, such that

(i) $L_o \subset L$.

(ii) $x'(x) \leq r_K(x)$ $\quad (x \in V)$.

(iii) $K \subset \{x \mid x'(x) \leq 1\}$.

Proof. Since $L_o$ is a linear variety, it is a real linear variety, and so there exists a real linear subspace $W_o \subset V$ and some $x_o \in V$ such that $L_o = x_o + W_o$. Moreover, $x_o \notin W_o$, since if $x_o \in W_o$, then $L_o = W_o$ and so $K \cap L_o \neq \emptyset$, contrary to hypothesis. Let $W \subset V$ be the real linear subspace spanned by $\{x_o\} \cup W_o$. Clearly $L_o \subset W$ and $W_o$ is a maximal real linear subspace of $W$. Thus $L_o$ is a real hyperplane in $W$.

Now consider $K \cap W$. It is easily seen that $K \cap W$, as a subset of $W$, is a convex absorbing set. Furthermore, we claim that $r_{K \cap W}$, computed on the linear space $W$, is precisely $r_K$ restricted to $W$. This is evident from

$$\begin{aligned} r_{K \cap W}(x) &= \inf\{a \mid a > 0,\ x \in a(K \cap W)\} \\ &= \inf\{a \mid a > 0,\ x \in aK \cap W\} \\ &= \inf\{a \mid a > 0,\ x \in aK\} \\ &= r_K(x) \qquad (x \in W). \end{aligned}$$

The second of these equalities is valid since $W$ is a real linear subspace. Finally, we note that $(K \cap W) \cap L_o = \emptyset$, as $K \cap L_o = \emptyset$.

Hence, by Lemma 5.2.1, there exists a real linear functional $y'$ on $W$ such that $L_o = \{x \mid x \in W,\ y'(x) = 1\}$ and $y'(x) \leq r_K(x)$, $x \in W$.

Next, appealing to the Real Hahn-Banach Theorem (Theorem 4.1.1) and the remarks following Lemma 4.1.1, we deduce the existence of a real linear functional $x'$ on $V$ such that $x'(x) = y'(x)$, $x \in W$, and $x'(x) \leq r_K(x)$, $x \in V$.

Let $L = \{x \mid x'(x) = 1\}$. Then $L$ is a real hyperplane in $V$, and $L_o \subset L$. Moreover, if $x \in K$, then $r_K(x) \leq 1$, and so $x'(x) \leq 1$; that is, $K \subset \{x \mid x'(x) \leq 1\}$. □

Thus, when $K$ is a convex absorbing set and $L_o$ is a linear variety such that $K \cap L_o = \emptyset$, there exists a real hyperplane $L$ that contains $L_o$ and is such that $K$ lies on one side of $L$. The improvement on this provided by the next corollary is precisely what was used in the last portion of the proof of Helly's Theorem (Theorem 4.10.1).

<u>Corollary 5.2.1</u>. Let $(V,\mathcal{T})$ be a topological linear space over $\Phi$ and let $K \subset V$ be a convex absorbing set for which $\text{int}(K) \neq \emptyset$, then there exists a closed real hyperplane $L = \{x \mid x^*(x) = 1\}$, where $x^*$ is a continuous real linear functional on $V$, such that

(i) $L_o \subset L$.
(ii) $\text{int}(K) \subset \{x \mid x^*(x) < 1\}$.
(iii) $\text{cl}(K) \subset \{x \mid x^*(x) \leq 1\}$.

<u>Proof</u>. Immediate from Proposition 5.1.5 and Theorem 5.2.1. □

Note, in particular, that, if $K$ is an open convex absorbing set, then $K$ lies strictly on one side of the real hyperplane $L$.

Incidentally, it is not necessary to use the analytic form of the Hahn-Banach Theorem to prove the geometric form, as we just did. But we prefer doing it this way, rather than take the time to develop the machinery necessary for an independent proof. A proof without the use of the analytic form of the theorem can be found, for instance, in [$W_1$, pp. 46-51, 219, and 220].

We have alluded at several points to the fact that the geometric form of the Hahn-Banach Theorem (Theorem 5.2.1) is equivalent to the analytic form. We shall now show that this is indeed the case when $\Phi = \mathbb{R}$. The equivalence of the analytic form of the Hahn-Banach Theorem over an arbitrary field $\Phi$ (Theorem 4.1.2) and Theorem 5.2.1 is an easy consequence of the theorem we present next.

Since we shall only consider $\Phi = \mathbb{R}$, we shall drop the adjective "real" for hyperplanes in the statement and proof of the theorem.

Theorem 5.2.2. Let $V$ be a linear space over $\mathbb{R}$. Then the following are equivalent:

(i) Suppose $p : V \to \mathbb{R}$ is such that

(a) $p(x) \geq 0$ $\quad (x \in V)$,

(b) $p(x + y) \leq p(x) + p(y)$ $\quad (x,y \in V)$,

(c) $p(ax) = ap(x)$ $\quad (a \in \mathbb{R};\ a \geq 0)$,

and let $W \subset V$ be a linear subspace. If $y' \in W'$ is such that $y'(x) \leq p(x)$, $x \in W$, then there exists some $x' \in V'$ such that $x'(x) = y'(x)$, $x \in W$, and $x'(x) \leq p(x)$, $x \in V$.

(ii) Suppose $K \subset V$ is a convex absorbing set. If $L_o \subset V$ is a linear variety such that $K \cap L_o = \phi$, then there exists a hyperplane $L = \{x \mid x'(x) = 1\}$, $x' \in V'$, such that $L_o \subset L$, $K \subset \{x \mid x'(x) \leq 1\}$, and $x'(x) \leq r_K(x)$, $x \in V$.

Proof. The proof of Theorem 5.2.1 shows immediately that part (i) implies (ii). Conversely, suppose the implication in part (ii) is valid, let $W \subset V$ be a linear subspace, and let $y' \in W'$ be such that $y'(x) \leq p(x)$, $x \in W$. If $y' = 0$, then the conclusion of part (i) holds trivially with $x' = 0$. Thus we may assume that $y' \neq 0$.

Let $L_o = \{x \mid x \in W,\ y'(x) = 1\}$. Then, by Proposition 5.1.1, $L_o$ is a hyperplane in $W$, and so a linear variety in $V$. If $K = \{x \mid x \in V,\ p(x) < 1\}$, then, by the same arguments as those used in proving Proposition 1.4.1, we see that $K$ is a convex absorbing set in $V$. Since $y'(x) \leq p(x)$, $x \in W$, we have $K \cap L_o = \phi$. Hence, applying part (ii), we deduce the existence of a hyperplane $L = \{x \mid x \in V,\ x'(x) = 1\}$ in $V$, where $x'$ is a linear functional on $V$, such that $L_o \subset L$, $K \subset \{x \mid x'(x) \leq 1\}$, and $x'(x) \leq r_K(x)$, $x \in V$.

Now, if $x \in V$, then from the definition of $K$ we see that

$$\begin{aligned} x'(x) \leq r_K(x) &= \inf\{a \mid a > 0,\ x \in aK\} \\ &= \inf\{a \mid a > 0,\ x/a \in K\} \\ &= \inf\{a \mid a > 0,\ p(x/a) < 1\} \\ &= \inf\{a \mid a > 0,\ p(x) < a\} \\ &= p(x). \end{aligned}$$

Thus $x'(x) \leq p(x)$, $x \in V$, and, moreover, this also shows that $K \cap L = \emptyset$.

The proof will be completed once we show that $x'(x) = y'(x)$, $x \in W$. Recall that $L_o = \{x \mid x \in W,\ y'(x) = 1\}$ is a hyperplane in $W$, as is $L \cap W = \{x \mid x \in W,\ x'(x) = 1\}$. The latter is obvious since the restriction of $x'$ to $W$ is a linear functional on $W$. Clearly $L_o \subset L \cap W$, as $L_o \subset L$. We claim that $L_o = L \cap W$.

Indeed, since $L_o$ and $L \cap W$ are hyperplanes, then, as can be seen, for example, from the proof of Proposition 5.1.1, there exist $x_1, x_2 \in W$ such that $L_o = x_1 + N(y')$ and $L \cap W = x_2 + N(x')$, where, as before, we consider $x'$ to be a linear functional on $W$ and compute $N(x')$ in $W$. Now $x_1 + N(y') = L_o \subset L \cap W = x_2 + N(x')$ shows that $N(y') \subset -x_1 + x_2 + N(x')$, from which we deduce at once, since $N(y')$ and $N(x')$ are linear subspaces, that $-x_1 + x_2 \in N(x')$, and so $-x_1 + x_2 + N(x') = N(x')$. But then $N(y') = N(x')$ since $N(y') \subset N(x')$, both spaces are maximal linear subspaces of $W$, and $x'$ is not identically zero on $W$. Hence $L_o = x_1 + N(y') \subset x_2 + N(y') = L \cap W$. Thus to show the opposite containment it clearly suffices to prove that $y'(x_2) = 1$.

Suppose $y'(x_2) = a$. If $a = 0$, then clearly $x_2 \in N(y')$ and so $x_1 + N(y') \subset x_2 + N(y') = N(y')$, from which we conclude that $x_1 \in N(y')$. But this contradicts the fact that $y'(x_1) = 1$. Thus $a \neq 0$.

But then $x_2/a \in W$, and $y'(x_2/a) = 1$ implies that $x_2/a \in L_0 \subset L \cap W$. Consequently $x'(x_2/a) = 1/a = 1$; that is, $a = 1$. Therefore $y'(x_2) = 1$ and $L_0 = L \cap W$.

Next consider the linear functional $z' = x' - y'$ on $W$. Since $L_0 = L \cap W$, we see at once that $z'(x) = x'(x) - y'(x) = 1 - 1 = 0$, $x \in L_0$; that is, $L_0 \subset N(z')$. Moreover, since $0 \notin L_0$ and $0 \in N(z')$, we see that $L_0 \neq N(z')$. Furthermore, we claim that $N(z') = W$. Indeed, on the one hand we have

$$\begin{aligned} L_0 - L_0 &= x_1 + N(y') - [x_1 + N(y')] \\ &= N(y') - N(y') \\ &= N(y'), \end{aligned}$$

and on the other hand,

$$\begin{aligned} L_0 - L_0 &\subset N(z') - N(z') \\ &= N(z'). \end{aligned}$$

Hence $N(y') \subset N(z')$. However, since $x_1 \in L_0 \subset N(z')$ and $x_1 \notin N(y')$, as $y'(x_1) = 1$, we see that $N(y') \neq N(z')$. Since the kernels are maximal linear subspaces of $W$, we must then conclude that $N(z') = W$.

But this last assertion says precisely that $x'(x) = y'(x)$, $x \in W$, so part (ii) implies part (i). This completes the proof.

□

The argument used in the preceding proof to show that $x'$ and $y'$ agreed on the space $W$ is worthwhile noticing for itself. Its content is precisely the next proposition.

Proposition 5.2.1. Let $V$ be a linear space over $\Phi$ and let $L \subset V$ be a hyperplane that does not contain the zero vector. If $x', y' \in V'$ are such that $x'(x) = y'(x)$, $x \in L$, then $x' = y'$.

## 5.3. Some Consequences of the Hahn-Banach Theorem Revisited.

In section 4.2 we discussed many consequences of the Hahn-Banach Theorem and used the analytic form of the theorem (Theorem 4.1.2) to prove them. Now we wish to reconsider some of these results and show how the geometric form of the Hahn-Banach Theorem (Theorem 5.2.1) can be used in their proofs.

Theorem 5.3.1. Let $(V,\mathcal{T})$ be a locally convex topological linear space over $\Phi$ and $V \neq \{0\}$. Then $V^*$ separates points.

Proof. Because of linearity, it is clearly sufficient to show that, if $x_o \in V$, $x_o \neq 0$, then there exists some $x^* \in V^*$ such that $x^*(x_o) \neq 0$. But, if $x_o \neq 0$, then, since $V$ is locally convex, there exists a convex open neighborhood $K$ of $0$ such that $x_o \notin K$. As always, such a neighborhood is absorbing since scalar multiplication is continuous. Thus $K$ is a convex absorbing set such that $\text{int}(K) = K \neq \emptyset$, and $K \cap L_o = \emptyset$, where $L_o = \{x_o\}$. Therefore, by Corollary 5.2.1, there exists a closed real hyperplane $L = \{x \mid y^*(x) = 1\} = x_o + N(y^*)$, where $y^*$ is some continuous real linear functional on $V$, such that $K \cap L = \emptyset$. However, by Proposition 3.3.2 and Theorem 3.3.1, if

$$x^*(x) = y^*(x) - iy^*(ix) \qquad (x \in V),$$

then $x^* \in V^*$ and

$$x^*(x_o) = y^*(x_o) - iy^*(ix_o) \neq 0. \qquad \square$$

Corollary 5.3.1. Let $(V,\mathcal{T})$ be a locally convex topological linear space over $\Phi$. If $V \neq \{0\}$, then $V^* \neq \{0\}$.

Corollary 5.3.2. Let $(V,\mathcal{T})$ be a locally convex topological linear space over $\Phi$, and let $x_o \in V$. If $x^*(x_o) = 0$, $x^* \in V^*$, then $x_o = 0$.

Corollary 5.3.3. Let $(V,\mathcal{T})$ be a locally convex topological linear space over $\Phi$, let $W \subset V$ be a closed linear subspace, and let $x_o \in V$. If $x^* \in V^*$ and $x^*(x) = 0$, $x \in W$, imply that $x^*(x_o) = 0$, then $x_o \in W$.

Proof. Suppose that $x_o \notin W$. Then, since $W$ is closed, there exists a convex open neighborhood $U$ of $x_o$ such that $U \cap W = \emptyset$. Let $K = -x_o + U$ and $L_o = -x_o + W$. Then $K$ is an open convex absorbing set, and $L_o$ is a linear variety such that $K \cap L_o = \emptyset$. Consequently, by Corollary 5.2.1, there exists a closed real hyperplane $L = \{x \mid y^*(x) = 1\}$, where $y^*$ is a continuous real linear functional on $V$, such that $L_o \subset L$ and $K \cap L = \emptyset$. In particular, $y^*(-x_o) = 1$, and so $y^*(x_o) = -1$, as $-x_o \in L_o = -x_o + W \subset L$.

But, if $x \in W$, then $-x_o + x \in L_o$ and

$$y^*(x) = y^*(-x_o + x + x_o) = y^*(-x_o + x) + y^*(x_o) = 1 - 1 = 0;$$

that is, $W \subset N(y^*)$. Consequently, if $x_o \notin W$, then we have constructed a real continuous linear functional $y^*$ on $V$ such that $y^*(x) = 0$, $x \in W$, and $y^*(x_o) \neq 0$. This, however, contradicts the hypotheses of the corollary.

Indeed, define $x^* \in V^*$ by $x^*(x) = y^*(x) - iy^*(ix)$, $x \in V$. Since $W$ is a linear subspace of $V$ and $y^*$ vanishes on $W$, we see at once that $x^*(x) = 0$, $x \in W$. But $x^*(x_o) = y^*(x_o) - iy^*(ix_o) \neq 0$, contrary to assumption.

Therefore $x_o \in W$. □

Corollary 5.3.4. Let $(V,\mathcal{T})$ be a locally convex topological linear space over $\Phi$ and let $W \subset V$ be a linear subspace. Then the following are equivalent:

(i) $\mathrm{cl}(W) = V$.

(ii) If $x^* \in V^*$ is such that $x^*(x) = 0$, $x \in W$, then $x^* = 0$.

## 5.4. Some Further Geometric Consequences of the Hahn-Banach Theorem.

As already mentioned, the geometric form of the Hahn-Banach Theorem also provides us with some information about when various sorts of sets in a topological linear space can be separated from one another by a real hyperplane. We would now like to state a sampling of such results. The proofs, which can be based on our previous work in this chapter, are left to the reader. Details and further results can also be found in [$DS_1$, pp. 409-418; $E_1$, pp. 116-144; K, pp. 186-188, and 243-245; KeNa, pp. 22 and 23; $W_1$, pp. 46-51, 219 and 220].

Definition 5.4.1. Let $V$ be a linear space over $\Phi$ and let $E_1, E_2 \subset V$. A real hyperplane $L \subset V$ is said to separate $E_1$ and $E_2$ if $E_1$ and $E_2$ lie on different sides of $L$, and it is said to strictly separate $E_1$ and $E_2$ if $E_1$ and $E_2$ lie strictly on different sides of $L$.

Thus, in view of Propositions 5.1.1 and 5.1.4, we see that $E_1$ and $E_2$ are separated by $L = \{x \mid x'(x) = a\}$, where $x'$ is some real linear functional on $V$, if $E_1 \subset \{x \mid x'(x) \leq a\}$ and $E_2 \subset \{x \mid x'(x) \geq a\}$, or vice versa. Similarly $L$ strictly separates $E_1$ and $E_2$ if $E_1 \subset \{x \mid x'(x) < a\}$ and $E_2 \subset \{x \mid x'(x) > a\}$, or vice versa. The geometric form of the Hahn-Banach Theorem (Theorem 5.2.1) says that, if $K$ is a convex absorbing set and $L_o$ is a linear variety disjoint from $K$, then there exists a real hyperplane $L$ that separates $K$ and $L_o$. Corollary 5.2.1 says that, if $\text{int}(K) \neq \phi$, then there exists a closed real hyperplane that separates $K$ and $L_o$.

One can also separate pairs of disjoint convex sets, provided they meet certain requirements.

Theorem 5.4.1. Let $(V, T)$ be a topological linear space over $\Phi$ and let $K_1, K_2 \subset V$ be nonempty disjoint convex sets.

(i) If $\text{int}(K_1) \neq \phi$, then there exists a closed real hyperplane $L$ that separates $K_1$ and $K_2$.

(ii) If $K_1$ and $K_2$ are open, then there exists a closed real hyperplane $L$ that strictly separates $K_1$ and $K_2$.

Corollary 5.4.1. Let $(V, T)$ be a locally convex topological linear space over $\Phi$, let $K \subset V$ be a nonempty closed convex set, and let $x_o \in V$. If $x_o \notin K$, then there exists a closed real hyperplane L that strictly separates K and $\{x_o\}$.

Corollary 5.4.2. Let $(V,\mathcal{T})$ be a locally convex topological linear space over $\Phi$ and let $K \subset V$ be a nonempty closed convex set. Then K is the intersection of all the closed half-spaces that contain K.

A final result, analogous to Proposition 5.1.5, is the next proposition.

Proposition 5.4.1. Let $(V,\mathcal{T})$ be a topological linear space over $\Phi$ and let $K_1, K_2 \subset V$ be nonempty disjoint convex sets such that $\text{int}(K_1) \neq \emptyset$. If L is a real hyperplane that separates $K_1$ and $K_2$, then L is closed.

We shall use some of the results from this section when we discuss the Krein-Mil'man Theorem in Section 11.2.

## 5.5. Problems.

1. (Proposition 5.1.2) Letting $(V,\mathcal{T})$ be a topological linear space over $\Phi$, prove each of the following:

(a) If $L \subset V$ is a (real) linear variety, then $\text{cl}(L)$ is a (real) linear variety.

(b) If $L \subset V$ is a (real) hyperplane, then either L is closed or $\text{cl}(L) = V$.

2. (Proposition 5.1.3) Letting V be a linear space over $\Phi$, prove each of the following:

(a) If $K \subset V$ is convex and $T \in \mathcal{L}'(V)$, then $\{x \mid T(x) \in K\}$ is convex.

(b) If $\{K_\alpha\}$ is a family of convex sets in V, then $\bigcap_\alpha K_\alpha$ is convex.

(c) If $E \subset V$, let $co(E) = \bigcap_{\substack{K \supset E \\ K \text{ convex}}} K$. Prove that $co(E)$ is convex, and if $K_1 \subset V$ is convex and $K_1 \supset E$, then $K_1 \supset co(E)$.

3. Let $(V, \mathcal{T})$ be a topological linear space over $\Phi$. If $x^*$ is a continuous real linear functional on $V$ and $a \in \mathbb{R}$, let $H = \{x \mid x^*(x) \leq a\}$ and $U = \{x \mid x^*(x) < a\}$ and prove that

(a) $cl(U) = H$.
(b) $int(H) = U$.

4. Let $(V, \|\cdot\|)$ be a normed linear space over $\Phi$, let $x^* \in V^*$, $x^* \neq 0$, and let $L = \{x \mid x^*(x) = 1\}$. If $d$ is the distance of the hyperplane $L$ from the origin, prove that $d = 1/\|x^*\|$.

5. (Corollary 5.2.1) Let $(V, \mathcal{T})$ be a topological linear space over $\Phi$ and $K \subset V$ a convex absorbing set for which $int(K) \neq \emptyset$. If $L_o \subset V$ is a linear variety such that $K \cap L_o = \emptyset$, prove that there exists a closed real hyperplane $L = \{x \mid x^*(x) = 1\}$, where $x^*$ is a continuous real linear functional on $V$, such that

(a) $L_o \subset L$
(b) $int(K) \subset \{x \mid x^*(x) < 1\}$
(c) $cl(K) \subset \{x \mid x^*(x) \leq 1\}$.

6. (Proposition 5.2.1) Let $V$ be a linear space over $\Phi$ and let $L \subset V$ be a hyperplane that does not contain the zero vector. Prove that, if $x', y' \in V'$ are such that $x'(x) = y'(x)$, $x \in L$, then $x' = y'$.

*7. (Theorem 5.4.1) Letting $(V, \mathcal{T})$ be a topological linear space over $\Phi$, and $K_1, K_2 \subset V$ be nonempty disjoint convex sets prove each of the following:

(a) If $int(K_1) \neq \emptyset$, then there exists a closed real hyperplane $L$ that separates $K_1$ and $K_2$.

(b) If $K_1$ and $K_2$ are open, then there exists a closed real hyperplane $L$ that strictly separates $K_1$ and $K_2$.

8. (Corollary 5.4.1) Let $(V,\mathcal{T})$ be a locally convex topological linear space over $\Phi$, $K \subset V$ be a nonempty closed convex set, and let $x_o \in V$. Prove that, if $x_o \notin K$, then there exists a closed real hyperplane $L$ that strictly separates $K$ and $\{x_o\}$

9. Give an example in the plane of two nonempty disjoint closed convex sets that can be separated, but not strictly separated.

10. (Corollary 5.4.2) Let $(V,\mathcal{T})$ be a locally convex topological linear space over $\Phi$ and let $K \subset V$ be a nonempty closed convex set. Prove that $K$ is the intersection of all the closed half-spaces that contain $K$.

11. (Proposition 5.4.1) Let $(V,\mathcal{T})$ be a topological linear space over $\Phi$ and let $K_1, K_2 \subset V$ be nonempty disjoint convex sets such that $\mathrm{int}(K_1) \neq \emptyset$. Prove that if $L$ is a real hyperplane that separates $K_1$ and $K_2$, then $L$ is closed.

12. Let $V$ be a linear space over $\Phi$ and let $E_1, E_2 \subset V$. Prove that $E_1$ and $E_2$ can be separated (strictly separated) if and only if $E_1 - E_2$ and $\{0\}$ can be separated (strictly separated).

*13. Let $V$ be a linear space over $\Phi$ and let $E_1, E_2 \subset V$ be convex subsets of $V$. Prove that $E_1$ and $E_2$ can be strictly separated if and only if there exists a convex absorbing set $U$ such that $(E_1 + U) \cap E_2 = \emptyset$.

14. Let $(V,\mathcal{T})$ be a locally convex topological linear space over $\Phi$, let $E \subset V$ be a closed convex subset, and let $K \subset V$ be a compact convex subset. Prove that, if $E \cap K = \emptyset$, then there exists a closed real hyperplane $L$ that strictly separates $E$ and $K$. [Hint: see Chapter 2, Problem 10(c).]

15. Let $(V,\mathcal{T})$ be a topological linear space over $\Phi$. Prove that the following are equivalent:

(a) Every subspace of finite codimension is dense in $V$.

(b) There exist no closed hyperplanes in $V$.

# CHAPTER 6

# THE UNIFORM BOUNDEDNESS THEOREM

6.0. Introduction. We turn our attention in this chapter to another fundamental theorem of functional analysis, the Uniform Boundedness Theorem. It asserts, for example, that a family of continuous linear transformations between two Banach spaces that is pointwise bounded is actually uniformly bounded. The proof of this theorem, which is an application of the Baire Category Theorem, together with the proof of the related Banach-Steinhaus Theorem, will be given in Section 6.2. In Section 6.1 we recall the notion of category and the Baire Category Theorem, and use the latter to prove Osgood's Theorem. This result will, in turn, provide us with the key step in the proof of the Uniform Boundedness Theorem. The remaining sections of the chapter are devoted to some applications of the Uniform Boundedness and Banach-Steinhaus Theorems. We shall see additional applications in subsequent chapters.

6.1. The Baire Category Theorem and Osgood's Theorem. The purpose of this section is to recall, without proof, some facts about the concept of category in metric spaces and to prove Osgood's Theorem. An application of the latter in the next section will give us the Uniform Boundedness Theorem.

For the sake of completeness we make the following definitions:

Definition 6.1.1. Let $X$ be a topological space and $E \subset X$. Then $E$ is said to be nowhere dense if $\text{int}[\text{cl}(E)] = \emptyset$. Furthermore, $E$ is said to be of category I if $E$ is the countable union of nowhere dense subsets of $X$; $E$ is said to be of category II if $E$ is not of category I.

If $X = \mathbb{R}$, then, for example, the Cantor ternary set in $[0,1]$ is nowhere dense, the rationals are of category I, and the irrationals are of category II.

The details of the next proposition are left to the reader.

Proposition 6.1.1. Let $X$ be a topological space and $E \subset X$. Then the following are equivalent:

(i) $E$ is nowhere dense.

(ii) If $\mathcal{O} \subset X$ is open, then there exists some open set $U \subset \mathcal{O}$ such that $U \cap E = \emptyset$.

(iii) $\mathrm{cl}[X \sim \mathrm{cl}(E)] = X$.

Moreover, if $E$ is a closed set of category II, then $\mathrm{int}(E) \neq \emptyset$.

We are primarily interested in the notion of category in the context of metric spaces. The most important result in this connection is the next theorem.

Theorem 6.1.1 (Baire Category Theorem). Every complete metric space is of category II in itself.

Discussions and proofs of this theorem can be found, for example, in [BaNr, pp. 76-80; $DS_1$, p. 20; Ry, pp. 139 and 140; $W_2$, pp. 178-180].

Osgood's Theorem, to be proved here, is a classic example of the application of the Baire Category Theorem. Before we state and prove the theorem we require one further definition.

Definition 6.1.2. Let $X$ be a topological space and $f : X \to \mathbb{R}$. The function $f$ is said to be lower semicontinuous if $\{t \mid t \in \mathbb{R}, f(t) \leq a\}$ is a closed subset of $X$ for each $a \in \mathbb{R}$.

Clearly, if $f$ is continuous, it is lower semicontinuous. The converse need not be the case.

Theorem 6.1.2 (Osgood's Theorem). Let $X$ be a topological space and suppose $\{f_\alpha\}_{\alpha \in \Lambda}$ is a family of real-valued lower semicontinuous functions defined on $X$. If $E \subset X$ is a set of category II such that for each $t \in E$ there exists some $M_t > 0$ for which

$$\sup_{\alpha \in \Lambda} f_\alpha(t) \leq M_t,$$

then there exists some open set $\mathcal{O} \subset X$ and some $M > 0$ such that

$$\sup_{\substack{\alpha \in \Lambda \\ t \in \mathcal{O}}} f_\alpha(t) \leq M.$$

Proof. For each positive integer $n$ define

$$E_n = \{t \mid f_\alpha(t) \leq n,\ \alpha \in \Lambda\}.$$

Since the functions $f_\alpha$ are lower semicontinuous, it is apparent that each $E_n$ is closed, being the intersection of closed sets. Moreover, evidently $E \subset \bigcup_{n=1}^{\infty} E_n$. Thus, since $E$, and hence $\bigcup_{n=1}^{\infty} E_n$, is of category II, we conclude that there exists some $n_o$ for which $E_{n_o}$ is not nowhere dense, that is, $\operatorname{int}(E_{n_o}) \neq \phi$. Clearly, if $\mathcal{O} = \operatorname{int}(E_{n_o})$, then

$$\sup_{\substack{\alpha \in \Lambda \\ t \in \mathcal{O}}} f_\alpha(t) \leq n_o = M.$$

□

Another phrasing of Osgood's Theorem is worthwhile noting: if a family of real-valued lower semicontinuous functions on $X$ is bounded above at each point of some set of category II in $X$, then the family is uniformly bounded above on some open subset of $X$. When we apply this in the particular context of normed linear spaces and linear transformations, we shall see that the open set can be replaced by the whole space. This is essentially the point of the Uniform Boundedness Theorem.

6.2. The Uniform Boundedness Theorem and the Banach-Steinhaus Theorem. We begin at once with the statement and proof of the main theorem.

Theorem 6.2.1 (Uniform Boundedness Theorem). Let $(V_1, \|\cdot\|_1)$ and $(V_2, \|\cdot\|_2)$ be normed linear spaces over $\Phi$ and suppose $\{T_\alpha\}_{\alpha \in \Lambda} \subset L(V_1, V_2)$. If $E \subset V_1$ is a set of category II such that for each $x \in E$ there exists some $M_x > 0$ for which

$$\sup_{\alpha \in \Lambda} \|T_\alpha(x)\|_2 \le M_x,$$

then there exists some $M > 0$ such that

$$\sup_{\alpha} \|T_\alpha\| \le M.$$

Proof. It is evident that for each $\alpha \in \Lambda$ the function $f_\alpha(x) = \|T_\alpha(x)\|_2$, $x \in V_1$, is a continuous, and hence lower semicontinuous, real-valued function on $V_1$. For each $x \in E$, $\sup_{\alpha \in \Lambda} f_\alpha(x) \le M_x$. Thus by Osgood's Theorem (Theorem 6.1.2) there exists some $M' > 0$ and an open set $\mathcal{O} \subset V_1$ such that

$$\sup_{\substack{\alpha \in \Lambda \\ x \in \mathcal{O}}} \|T_\alpha(x)\|_2 \le M'.$$

In particular, there exist some $x_0 \in \mathcal{O}$ and some $\delta > 0$ such that $B(x_0, \delta) = \{x \mid \|x - x_0\|_1 < \delta\} \subset \mathcal{O}$, and

$$\sup_{\substack{\alpha \in \Lambda \\ x \in B(x_0,\delta)}} \|T_\alpha(x)\|_2 \le M'.$$

Now, if $y \in V_1$, $\|y\|_1 < \delta$, then clearly $y + x_0 \in B(x_0, \delta)$, and so

$$\begin{aligned} \|T_\alpha(y)\|_2 &\le \|T_\alpha(y + x_0)\|_2 + \|T_\alpha(x_0)\|_2 \\ &\le 2M' \qquad (\alpha \in \Lambda). \end{aligned}$$

Finally, if $z \in V_1$, $z \neq 0$, we set $y = (\delta/2\|z\|_1)z$. Clearly $\|y\|_1 < \delta$, and so from the preceding observation we conclude that $\|T_\alpha(y)\|_2 \le 2M'$, $\alpha \in \Lambda$. Consequently

$$\|T_\alpha(z)\|_2 = \|T_\alpha[\frac{2\|z\|_1}{\delta} y]\|_2$$

$$= \frac{2\|z\|_1}{\delta} \|T_\alpha(y)\|_2$$

$$\leq \frac{4M'}{\delta} \|z\|_1 \qquad (\alpha \in \Lambda).$$

Thus, setting $M = 4M'/\delta$, we see that $\|T_\alpha(z)\|_2 \leq M\|z\|_1$, $\alpha \in \Lambda$, $z \in V_1$; that is,

$$\sup_{\alpha \in \Lambda} \|T_\alpha\| \leq M.$$

□

If the space $V_1$ is of category II in itself, then the Uniform Boundedness Theorem asserts that, if $\{T_\alpha\}_{\alpha \in \Lambda}$ is norm bounded at each point of $V_1$, then it is uniformly norm bounded on $V_1$. The next corollary is a special case of this observation.

Corollary 6.2.1. Let $(V_1, \|\cdot\|_1)$ be a Banach space and let $(V_2, \|\cdot\|_2)$ be a normed linear space over $\Phi$ and suppose $\{T_\alpha\}_{\alpha \in \Lambda} \subset L(V_1, V_2)$. If for each $x \in V_1$ there exists some $M_x$ such that

$$\sup_{\alpha \in \Lambda} \|T_\alpha(x)\|_2 \leq M_x,$$

then there exists some $M > 0$ such that

$$\sup_{\alpha} \|T_\alpha\| \leq M.$$

One of the most useful consequences of the Uniform Boundedness Theorem is the next result.

Theorem 6.2.2 (Banach-Steinhaus Theorem). Let $(V_1, \|\cdot\|_1)$ be a Banach space and let $(V_2, \|\cdot\|_2)$ be a normed linear space over $\Phi$. If $\{T_n\} \subset L(V_1, V_2)$ is a sequence such that for each $x \in V_1$ there exists some $z_x \in V_2$ for which $\lim_n T_n(x) = z_x$, then the formula $T(x) = z_x$, $x \in V_1$, defines a unique element $T \in L(V_1, V_2)$ such that $\|T\| \leq \liminf_n \|T_n\|$.

Proof. It is easily verified that the formula $T(x) = z_x$, $x \in V_1$, defines a unique linear transformation from $V_1$ to $V_2$. Moreover, since $\{T_n(x)\}$ is a Cauchy sequence in $V_2$ for each $x \in V_1$, it follows from the triangle inequality that $\{\|T_n(x)\|_2\}$ is a Cauchy sequence of numbers that converges to $\|z_x\|_2 = \|T(x)\|_2$. Hence for each $x \in V_1$ there exists some $M_x > 0$ such that

$$\sup_n \|T_n(x)\|_2 \leq M_x.$$

Consequently from the Uniform Boundedness Theorem (Theorem 6.2.1) there exists some $M > 0$ for which

$$\sup_n \|T_n\| \leq M.$$

Then the estimates

$$\begin{aligned} \|T(x)\|_2 &= \lim_n \|T_n(x)\|_2 \\ &\leq \liminf_n \|T_n\| \, \|x\|_1 \\ &\leq M \, \|x\|_1 \qquad (x \in V_1) \end{aligned}$$

reveal that $T \in L(V_1,V_2)$ and $\|T\| \leq \liminf_n \|T_n\|$. □

By employing Theorems 6.2.1 and 6.2.2 it is easy to prove the next corollary. The details are left to the reader.

Corollary 6.2.2. Let $(V_1,\|\cdot\|_1)$ and $(V_2,\|\cdot\|_2)$ be Banach spaces over $\Phi$. Suppose $\{T_n\} \subset L(V_1,V_2)$ is a sequence such that

(i) For each $x \in V_1$ there exists some $M_x > 0$ such that $\sup_n \|T_n(x)\|_2 \leq M_x$.

(ii) There exists some $x_0 \in V_1$, some $\delta > 0$, and a set $E \subset B(x_0,\delta) = \{x \mid \|x - x_0\|_1 < \delta\}$ such that $E$ is dense in $B(x_0,\delta)$ and such that for each $x \in E$ there exists some $z_x \in V_2$ for which $\lim_n T_n(x) = z_x$.

Then there exists a unique element $T \in L(V_1,V_2)$ such that

(a) $T(x) = \lim_n T_n(x) \qquad (x \in V_1)$.

(b) $\|T\| \leq \liminf_n \|T_n\|$.

In particular, the conclusions of the corollary hold if the sequence $\{T_n(x)\}$ converges for each $x$ in some dense subset of $V_1$ and hypothesis (i) is satisfied.

The Banach-Steinhaus Theorem is not generally valid if the sequence $\{T_n\}$ is replaced by a net $\{T_\alpha\} \subset L(V_1,V_2)$. This is the case since a Cauchy net need not be norm bounded. Corollary 6.2.2 does, however, remain valid if one replaces the sequence by a net. The details are left to the reader.

It is also worthwhile remarking that some sort of category assumption is required in order to ensure the validity of the Uniform Boundedness Theorem. As an example of this, consider the linear space $V$ of all complex-valued polynomials defined on $\mathbb{R}$ with the usual pointwise operations. We note that $x(t) \in V$ if and only if $x(t) = \Sigma_{k=0}^{\infty} a_k t^k$, $a_k \in \mathbb{C}$, where there exists some integer $N(x) \geq 0$ for which $a_k = 0$, $k \geq N(x)$. If we set

$$\|x(t)\| = \sup_k |a_k| \qquad (x(t) \in V),$$

then it is easily verified that $(V,\|\cdot\|)$ is a normed linear space. It is not, however, a Banach space because, for example, it is apparent that $\Sigma_{k=1}^{\infty} k^{-2}t^k \notin V$, but it is approximated in the norm of $V$ by polynomials.

For each positive integer $n$ define $x_n^* \in V^*$ by

$$x_n^*[x(t)] = x_n^*\left(\sum_{k=0}^{\infty} a_k t^k\right) = \sum_{k=0}^{n-1} a_k \qquad (x(t) \in V).$$

Clearly $x_n^*$ is a linear functional on $V$, and

$$|x_n^*[x(t)]| = \left|\sum_{k=0}^{n-1} a_k\right| \leq n \sup_k |a_k| = n\|x(t)\|$$

shows that $x_n^* \in V^*$ and $\|x_n^*\| \leq n$. Actually $\|x_n^*\| = n$ since, if

$y(t) = \Sigma_{k=0}^{n-1} t^k$, then $\|y\| = 1$ and $x_n^*[y(t)] = n$. Thus we see that $\{x_n^*\} \in V^*$ and $\{\|x_n^*\|\}$ is unbounded.

However, if $x(t) \in V$, then there does exist some $M_x > 0$ for which

$$\sup_n |x_n^*[x(t)]| \leq M_x.$$

Indeed, if $x(t) = \Sigma_{k=0}^{m} a_k t^k$ and $n \leq m$, then

$$\begin{aligned} |x_n^*[x(t)]| &= \left|\sum_{k=0}^{n-1} a_k\right| \leq n \sup_k |a_k| \\ &= n \|x(t)\| \\ &\leq (m+1)\|x(t)\|, \end{aligned}$$

whereas if $n > m$, then

$$|x_n^*[x(t)]| = \left|\sum_{k=0}^{m} a_k\right| \leq (m+1)\|x(t)\|.$$

Hence, if $M_x = (m+1)\|x(t)\|$, we see that

$$\sup_n |x_n^*[x(t)]| \leq M_x.$$

Consequently we conclude that the Uniform Boundedness Theorem fails to hold in this case. The reason for the failure, of course, lies in the fact that $V$ is not of category II in itself.

We have not given the most general form of the Uniform Boundedness and Banach-Steinhaus Theorems, preferring instead to concentrate only on these results in the context of normed linear spaces. This will be sufficient for our needs. These theorems and their corollaries do, however, have valid analogs in a more general context. We state two of them here without proof and refer the reader for further discussion to [$DS_1$, pp. 51-55; $E_1$, pp. 462-466, 476-477 and 480-483; K, pp. 168-170; $W_1$, pp. 116-118 and 223-225].

Theorem 6.2.3. Let $(V_1, \mathcal{T}_1)$ and $(V_2, \mathcal{T}_2)$ be Fréchet spaces over $\Phi$ and suppose $\rho_k$ is the metric on $V_k$ that defines the topology $\mathcal{T}_k$, $k = 1,2$. If $\{T_\alpha\}_{\alpha \in \Lambda} \subset L(V_1, V_2)$ and if for each $x \in V_1$ there exists some $M_x > 0$ for which

$$\sup_{\alpha \in \Lambda} \rho_2(T_\alpha(x), 0) \leq M_x,$$

then, given $\epsilon > 0$, there exists a $\delta > 0$ such that

$$\sup_{\substack{x \in V_1 \\ \rho_1(x,0) < \delta}} \rho_2(T_\alpha(x), 0) \leq \epsilon \qquad (\alpha \in \Lambda).$$

In other words, the conclusion is that $T_\alpha(x)$ tends to 0 in $V_2$ uniformly in $\alpha$ as $x$ tends to 0 in $V_1$.

Theorem 6.2.4. Let $(V_1, \mathcal{T}_1)$ and $(V_2, \mathcal{T}_2)$ be Fréchet spaces over $\Phi$ and suppose $\rho_k$ is the metric on $V_k$ that defines the topology $\mathcal{T}_k$, $k = 1,2$. Suppose $\{T_\alpha\}_{\alpha \in \Lambda}$ is a net in $L(V_1, V_2)$ such that

(i) For each $x \in V_1$ there exists some $M_x > 0$ for which $\sup_\alpha \rho_2(T_\alpha(x), 0) \leq M_x$.

(ii) There exists some $x_o \in V_1$, some $\delta > 0$, and a set $E \subset B(x_o, \delta) = \{x \mid \rho_1(x, x_o) < \delta\}$ such that $E$ is dense in $B(x_o, \delta)$ and such that for each $x \in E$ there exists some $z_x \in V_2$ for which $\lim_\alpha T_\alpha(x) = z_x$.

Then there exists a unique element $T \in L(V_1, V_2)$ such that $T(x) = \lim_\alpha T_\alpha(x)$, $x \in V_1$.

6.3. The Strong Operator Topology. We have already seen (Theorem 3.2.2) that $(L(V_1, V_2), \|\cdot\|)$ is a Banach space provided that $(V_2, \|\cdot\|_2)$ is a Banach space. The norm in $L(V_1, V_2)$ then, of course, defines a natural topology in $L(V_1, V_2)$ under which $L(V_1, V_2)$ is a locally convex topological linear space. There are many other ways, however, to introduce a topology into $L(V_1, V_2)$ to the same effect.

Some of these topologies will be considered in greater detail in Chapters 9 and 10. Here we concern ourselves with only one such topology, called the strong operator topology.

Definition 6.3.1. Let $(V_1,\|\cdot\|_1)$ and $(V_2,\|\cdot\|_2)$ be normed linear spaces over $\Phi$. For each $T \in L(V_1,V_2)$, each $\epsilon > 0$, and each positive integer $n$ we set

$$U(T;\epsilon;x_1,x_2,\ldots,x_n) = \{S \mid S \in L(V_1,V_2),\ \|S(x_k) - T(x_k)\|_2 < \epsilon,\ k = 1,2,\ldots,n\},$$

where $x_1,x_2,\ldots,x_n$ are in $V_1$.

Furthermore, for each $T \in L(V_1,V_2)$ we set

$$\mathcal{U}_{so}(T) = \{U(T;\epsilon;x_1,x_2,\ldots,x_n) \mid \epsilon > 0,\ n \in Z,\ n > 0,\ x_1,x_2,\ldots,x_n \text{ in } V_1\}$$

and

$$\mathcal{U}_{so} = \bigcup_{T \in L(V_1,V_2)} \mathcal{U}_{so}(T).$$

The proof of the next proposition is left to the reader.

Proposition 6.3.1. Let $(V_1,\|\cdot\|_1)$ and $(V_2,\|\cdot\|_2)$ be normed linear spaces over $\Phi$. Then

(i) $\mathcal{U}_{so}$ is a base for a Hausdorff topology $so(\mathcal{T})$ on $L(V_1,V_2)$ such that $(L(V_1,V_2),so(\mathcal{T}))$ is a locally convex topological linear space over $\Phi$.

(ii) A net $\{T_\alpha\} \subset L(V_1,V_2)$ converges to $T \in L(V_1,V_2)$ in the topology $so(\mathcal{T})$ if and only if $\lim_\alpha \|T_\alpha(x) - T(x)\|_2 = 0$ for each $x \in V_1$.

As indicated, the topology generated by $\mathcal{U}_{so}$ will be called the strong operator topology for $L(V_1,V_2)$ and denoted by $so(\mathcal{T})$.

If both $V_1$ and $V_2$ are Banach spaces, then we can obtain a completeness result with the strong operator topology.

<u>Theorem 6.3.1</u>. Let $(V_1, \|\cdot\|_1)$ and $(V_2, \|\cdot\|_2)$ be Banach spaces over $\Phi$. Then $(L(V_1, V_2), so(T))$ is sequentially complete.

<u>Proof</u>. Suppose $\{T_n\} \subset L(V_1, V_2)$ is a Cauchy sequence in $so(T)$. Then from Proposition 6.3.1(ii) we see that $\{T_n(x)\}$ is a Cauchy sequence in $(V_2, \|\cdot\|_2)$ for each $x \in V_1$, and hence there exists some $z_x \in V_2$ such that $\lim_n \|T_n(x) - z_x\|_2 = 0$, $x \in V_1$.

Since $(V_1, \|\cdot\|_1)$ is a Banach space, we may apply the Banach-Steinhaus Theorem (Theorem 6.2.2) to conclude that the formula

$$T(x) = z_x = \lim_n T_n(x) \qquad (x \in V_1)$$

defines an element $T \in L(V_1, V_2)$. It is apparent from the definition that $\{T_n\}$ converges to $T$ in $so(T)$.

Therefore $(L(V_1, V_2), so(T))$ is sequentially complete. □

An application of the Uniform Boundedness Theorem (Theorem 6.2.1) proves the next result.

<u>Theorem 6.3.2</u>. Let $(V_1, \|\cdot\|_1)$ be a Banach space and $(V_2, \|\cdot\|_2)$ a normed linear space over $\Phi$. If $\{T_n\} \subset L(V_1, V_2)$ is a Cauchy sequence in $so(T)$, then there exists some $M > 0$ such that

$$\sup_n \|T_n\| \leq M.$$

<u>Proof</u>. Since $\{T_n\}$ is a Cauchy sequence in $so(T)$, we see that $\{T_n(x)\}$ is a Cauchy sequence in $(V_2, \|\cdot\|_2)$ for each $x \in V_1$ and hence norm bounded. Thus for each $x \in V_1$ there exists some $M_x > 0$ for which $\sup_n \|T_n(x)\|_2 \leq M_x$. An appeal to the Uniform Boundedness Theorem completes the proof. □

Recall that, if $a \in \mathbb{R}$ is such that $|1 - a| < 1$, then clearly $a \neq 0$, and the inverse of $a$ can be computed by the geometric series $\Sigma_{k=0}^{\infty} (1 - a)^k$. We now wish to apply the sequential completeness of $(L(V), so(T))$, when $V$ is a Banach space, to obtain the analog of this fact in $L(V)$.

Theorem 6.3.3. Let $(V, \|\cdot\|)$ be a Banach space over $\Phi$ and suppose $T \in L(V)$. If $\|I - T\| < 1$, where $I(x) = x$, $x \in V$, then

(i) $T$ is bijective.
(ii) $T^{-1} \in L(V)$.
(iii) $\{\Sigma_{k=0}^{n} (I - T)^k\}$ converges to $T^{-1}$ in $so(T)$.

Proof. Let $\epsilon > 0$. Then, since $\|I - T\| < 1$, we see at once that there exists an $N$ such that, if $n \geq m > N$, then $\Sigma_{k=m}^{n} \|I - T\|^k < \epsilon$. Hence, if $n \geq m > N$,

$$\begin{aligned} \left\| \sum_{k=m}^{n} (I - T)^k(x) \right\| &\leq \sum_{k=m}^{n} \|(I - T)^k(x)\| \\ &\leq \sum_{k=m}^{n} \|I - T\|^k \|x\| \\ &< \epsilon\|x\| \qquad (x \in V). \end{aligned}$$

Consequently it follows easily that $\{\Sigma_{k=0}^{n} (I - T)^k\} \subset L(V)$ is a Cauchy sequence in $so(T)$ and hence converges to some element $S \in L(V)$ in $so(T)$ by Theorem 6.3.1. Moreover, if $x \in V$, then

$$\begin{aligned} T[S(x)] &= T[\lim_n \sum_{k=0}^{n} (I - T)^k(x)] \\ &= \lim_n [T[\sum_{k=0}^{n} (I - T)^k](x)] \\ &= \lim_n ([I - (I - T)][\sum_{k=0}^{n} (I - T)^k](x)) \\ &= \lim_n ([\sum_{k=0}^{n} (I - T)^k - \sum_{k=1}^{n+1} (I - T)^k](x)) \end{aligned}$$

$$= \lim_n ([I - (I - T)^{n+1}](x))$$

$$= x - \lim_n (I - T)^{n+1}(x)$$

$$= x.$$

The last equality is valid since $\|(I - T)^{n+1}(x)\| \leq \|I - T\|^{n+1}\|x\|$ and $\|I - T\| < 1$. Hence $T$ is surjective. By a similar computation we obtain $S[T(x)] = x$, $x \in V$, from which it follows at once that $T$ is injective.

Therefore $T$ is bijective, and $S = T^{-1}$. □

The geometric series $\Sigma_{k=1}^{\infty}(I - T)^k$ is called the <u>Neumann series</u>. If we set $m = 1 - \|I - T\| > 0$, then from $\|x - T(x)\| \leq \|I - T\|\ \|x\|$, $x \in V$, we immediately deduce that $\|T(x)\| \geq \|x\| - \|I - T\|\ \|x\| = m\|x\|$. Proposition 3.2.3 would then ensure that $T^{-1}$ exists as an element of $L(R(T),V)$. As we have seen, however, a good deal more can be deduced.

An examination of the proof also reveals that the sequence $\{\Sigma_{k=0}^{n}(I - T)^k\}$ is even a Cauchy sequence in the norm topology on $L(V)$, and so, by the same arguments as those used in the proof, $T^{-1} = \Sigma_{k=0}^{\infty}(I - T)^k$. This form of the result will be used in Chapter 13 when we discuss the spectral theory of bounded linear transformations on Hilbert spaces.

The theorem was stated in this particular form merely to indicate an application of the sequential completeness of the strong operator topology.

<u>6.4. Local Membership in $L_q(\mathbb{R},dt)$</u>. First we make precise what we mean by local membership.

<u>Definition 6.4.1</u>. Let $g$ be a Lebesgue measurable complex-valued function on $\mathbb{R}$ and $1 \leq q \leq \infty$. Then $g$ is said to <u>belong locally to</u> $L_q(\mathbb{R},dt)$ if for every $a,b \in \mathbb{R}$, $-\infty < a < b < \infty$, the function

$\chi_{[a,b]} g \in L_q(\mathbb{R}, dt)$, where $\chi_{[a,b]}$ denotes the characteristic function of $[a,b]$.

Obviously, if $g \in L_q(\mathbb{R}, dt)$, then $g$ belongs locally to $L_q(\mathbb{R}, dt)$, but the converse need not be valid. The next result will give sufficient conditions under which local membership implies membership.

<u>Theorem 6.4.1</u>. Let $1 < q < \infty$ and $1/p + 1/q = 1$. Suppose $g$ is a Lebesgue measurable complex-valued function on $\mathbb{R}$ such that

(i) $g$ belongs locally to $L_q(\mathbb{R}, dt)$.

(ii) $\int_{\mathbb{R}} |f(t)g(t)|\, dt < \infty \qquad (f \in L_p(\mathbb{R}, dt)).$

Then $g \in L_q(\mathbb{R}, dt)$.

<u>Proof</u>. If $g(t) = 0$ for almost all $t$, then the conclusion is trivial. So we may assume that this is not the case. For each positive integer $n$ define

$$x_n^*(f) = \int_{-n}^{n} f(t)g(t)\, dt \qquad (f \in L_p(\mathbb{R}, dt).$$

It is apparent that $x_n^*$ is a well-defined linear functional on $L_p(\mathbb{R}, dt)$, and an application of Hölder's inequality reveals that

$$|x_n^*(f)| \le \int_{\mathbb{R}} |f(t)|\, |\chi_{[-n,n]} g(t)|\, dt$$
$$\le \|f\|_p \|\chi_{[-n,n]} g\|_q \qquad (f \in L_p(\mathbb{R}, dt);\ n = 1,2,3,\ldots).$$

Hence $x_n^* \in L_p(\mathbb{R}, dt)^*$, $n = 1,2,3,\ldots$ . Moreover, from the Lebesgue Dominated Convergence Theorem [Ry, p. 229], we see that

$$\lim_n x_n^*(f) = \int_{\mathbb{R}} f(t)g(t)\, dt \qquad (f \in L_p(\mathbb{R}, dt)).$$

Consequently, appealing to the Banach-Steinhaus Theorem (Theorem 6.2.2), we conclude that there exists some $x^* \in L_p(\mathbb{R}, dt)^*$ such that

$$x^*(f) = \int_{\mathbb{R}} f(t)g(t)\, dt \qquad (f \in L_p(\mathbb{R}, dt)).$$

In particular, we have

$$\left|\int_{\mathbb{R}} f(t)g(t)\,dt\right| \le \|x^*\|\,\|f\|_p \qquad (f \in L_p(\mathbb{R},dt)).$$

Now for each positive integer $n$ define

$$f_n(t) = \chi_{[-n,n]}(t)\,\exp[-i \arg g(t)]\,|g(t)|^{q-1}$$

with the usual conventions about $\arg g(t)$ if $g(t) = 0$. We claim that $f_n \in L_p(\mathbb{R},dt)$, $n = 1,2,3,\ldots$ . Indeed, since $1/p + 1/q = 1$, it follows that $(q-1)p = q$, and so

$$\begin{aligned}
\|f_n\|_p &= \left(\int_{\mathbb{R}} |f_n(t)|^p\,dt\right)^{1/p} \\
&= \left(\int_{\mathbb{R}} |\chi_{[-n,n]}g(t)|^{(q-1)p}\,dt\right)^{1/p} \\
&= \left(\int_{\mathbb{R}} |\chi_{[-n,n]}g(t)|^{q}\,dt\right)^{1/p} \\
&= \left(\|\chi_{[-n,n]}g\|_q\right)^{q/p} \\
&< \infty \qquad\qquad (n = 1,2,3,\ldots),
\end{aligned}$$

as $g$ belongs locally to $L_q(\mathbb{R},dt)$.

But then

$$\begin{aligned}
x^*(f_n) &= \int_{\mathbb{R}} f_n(t)g(t)\,dt \\
&= \int_{\mathbb{R}} \chi_{[-n,n]}(t)|g(t)|^q\,dt \\
&\le \|x^*\|\,\|f_n\|_p \\
&= \|x^*\|\left(\int_{\mathbb{R}} \chi_{[-n,n]}(t)|g(t)|^q\,dt\right)^{1/p},
\end{aligned}$$

from which it follows that

$$\left(\int_{\mathbb{R}} \chi_{[-n,n]}(t)|g(t)|^q\,dt\right)^{1/q} \le \|x^*\| \qquad (n = 1,2,3,\ldots),$$

because $1 - 1/p = 1/q$.

An appeal to the Monotone Convergence Theorem [Ry, p. 227] then allows us to conclude that

$$(\int_{\mathbb{R}} |g(t)|^q \, dt)^{1/q} \leq \|x^*\|,$$

that is, $g \in L_q(\mathbb{R}, dt)$. □

6.5. A Result in the Theory of Summability. Consider an infinite matrix $A = (a_{nk})_{n,k=1}^{\infty}$, $a_{nk} \in \mathbb{C}$, such that, whenever $\{x_k\}$ is a sequence of complex numbers, $\Sigma_{k=1}^{\infty} a_{nk} x_k$ converges, $n = 1,2,3,\ldots$ . It is easily seen that such a matrix $A$ defines a linear mapping $T_A$ from the space of all complex sequences to itself if we set $T_A(\{x_k\}) = \{y_n\}$, where $y_n = \Sigma_{k=1}^{\infty} a_{nk} x_k$, $n = 1,2,3,\ldots$ . We wish to discuss certain matrices that map the convergent sequences into themselves in this manner. Recall that the space $c$ of all convergent complex-valued sequences is a Banach space with the norm

$$\|\{x_k\}\|_\infty = \sup_k |x_k| \qquad (\{x_k\} \in c).$$

Definition 6.5.1. Suppose $A = (a_{nk})_{n,k=1}^{\infty}$ is a matrix of complex numbers such that

(i) If $\{x_k\} \in c$, then $\{\Sigma_{k=1}^{\infty} a_{nk} x_k\} \in c$,
(ii) If $\{x_k\} \in c$ and $\lim_k x_k = a$, then $\lim_n \Sigma_{k=1}^{\infty} a_{nk} x_k = a$.

Then $A$ is called a Toeplitz matrix.

For example, suppose $A = (a_{nk})_{n,k=1}^{\infty}$ is such that for each $n = 1,2,3,\ldots, a_{nk} = 1/n$, $k = 1,2,\ldots,n$; $a_{nk} = 0$, $k > n$. Then $\{\Sigma_{k=1}^{\infty} a_{nk} x_k\} = \{(\Sigma_{k=1}^{n} x_k)/n\}$ is the sequence whose nth term is the arithmetic average of the first $n$ terms of the sequence $\{x_k\}$. In this case it is well known that $\{\Sigma_{k=1}^{\infty} a_{nk} x_k\}$ converges whenever $\{x_k\}$ converges, and to the same limit. Hence $A$ is a Toeplitz matrix.

The next theorem provides a complete characterization of such matrices.

Theorem 6.5.1 (Silverman-Toeplitz Theorem). Let $A = (a_{nk})_{n,k=1}^{\infty}$ be a matrix of complex numbers. Then the following are equivalent:

(i) $A$ is a Toeplitz matrix.

(ii) $A$ is such that

(a) $\sup_n \sum_{k=1}^{\infty} |a_{nk}| < \infty$,

(b) $\lim_n a_{nk} = 0$, $k = 1,2,3,\ldots$,

(c) $\lim_n \sum_{k=1}^{\infty} a_{nk} = 1$.

Proof. Suppose $A$ satisfies the conditions in part (ii). From condition (a) it is apparent that $\Sigma_{k=1}^{\infty} a_{nk}$ and $\Sigma_{k=1}^{\infty} a_{nk}x_k$, $\{x_k\} \in c$, are absolutely convergent series for each $n$. Thus, if $\{x_k\} \in c$ and $\lim_k x_k = a$, we can write

$$\sum_{k=1}^{\infty} a_{nk}x_k = \sum_{k=1}^{\infty} a_{nk}(x_k - a) + a\sum_{k=1}^{\infty} a_{nk} \qquad (n = 1,2,3,\ldots).$$

Since $\lim_n \Sigma_{k=1}^{\infty} a_{nk} = 1$, it is evident that $\lim_n a\Sigma_{k=1}^{\infty} a_{nk} = a$.

Furthermore, let $\epsilon > 0$ and write

$$\sum_{k=1}^{\infty} a_{nk}(x_k - a) = \sum_{k=1}^{m} a_{nk}(x_k - a) + \sum_{k=m+1}^{\infty} a_{nk}(x_k - a).$$

We choose $m$ so large that for $k > m$

$$|x_k - a| < \frac{\epsilon}{2M}$$

where $M = \sup_n \Sigma_{k=1}^{\infty} |a_{nk}|$. Such a choice of $m$ is possible, as $\lim_k x_k = a$. Keeping $m$ fixed, we next choose $N$ such that, if $n > N$, then

$$\sum_{k=1}^{m} |a_{nk}|\ |x_k - a| < \frac{\epsilon}{2}.$$

This choice of $N$ is possible since $\lim_n a_{nk} = 0$, $k = 1,2,3,\ldots$ . Consequently, combining these estimates, we see that for $n > N$

$$\left|\sum_{k=1}^{\infty} a_{nk}(x_k - a)\right| \leq \sum_{k=1}^{m} |a_{nk}||x_k - a| + \sum_{k=m+1}^{\infty} |a_{nk}||x_k - a|$$

$$< \frac{\epsilon}{2} + \frac{M\epsilon}{2M}$$

$$= \epsilon,$$

that is, $\lim_n \Sigma_{k=1}^{\infty} a_{nk}x_k = a$, and $A$ is a Toeplitz matrix. Therefore part (ii) of the theorem implies part (i).

Conversely, suppose $A$ is a Toeplitz matrix. Given a fixed $m = 1,2,3,\ldots,$ we define $\{x_k^m\} \in c$ by

$$x_k^m = 1 \quad \text{for} \quad k = m,$$

$$x_k^m = 0 \quad \text{for} \quad k \neq m.$$

Clearly $\lim_k x_k^m = 0$, and since $A$ is a Toeplitz matrix, we have $\lim_n \Sigma_{k=1}^{\infty} a_{nk}x_k^m = \lim_n a_{nm} = 0$. Similarly, applying $A$ to the sequence consisting only of ones, we deduce that $\lim_n \Sigma_{k=1}^{\infty} a_{nk} = 1$. Hence $A$ satisfies conditions (b) and (c).

Finally, for each positive integer $n$ we define

$$x_n^*(\{x_k\}) = \sum_{k=1}^{\infty} a_{nk}x_k \qquad (\{x_k\} \in c).$$

Clearly each $x_n^*$ is a linear functional on $c$. We claim that it is continuous. To see this it suffices to prove that for each $n = 1,2,3,\ldots,\Sigma_{k=1}^{\infty}|a_{nk}| < \infty$. Suppose that this is not the case for a fixed $n$. Then there exists a sequence $\{k_j\}$ of positive integers such that $k_j < k_{j+1}$ and

$$\sum_{k=k_j+1}^{k_{j+1}} |a_{nk}| > j \qquad (j = 1,2,3,\ldots).$$

Next define a sequence $\{x_k\}$ of complex numbers by

$$x_k = 0 \quad \text{for} \quad 1 \le k \le k_1$$

$$x_k = \frac{\exp(-i \arg a_{nk})}{j} \quad \text{for} \quad k_j < k \le k_{j+1};\ j = 1,2,3,\ldots .$$

Clearly $\{x_k\} \in c$, indeed, $\lim_k x_k = 0$. But

$$\sum_{k=1}^{\infty} a_{nk}x_k = \sum_{j=1}^{\infty} \left[ \sum_{k=k_j+1}^{k_{j+1}} \frac{|a_{nk}|}{j} \right]$$

$$\ge \sum_{j=1}^{\infty} 1,$$

which contradicts the assumption that $\Sigma_{k=1}^{\infty} a_{nk}x_k$ converges.

Thus $x_n^* \in c^*$, and $\|x_n^*\| \le \Sigma_{k=1}^{\infty} |a_{nk}|$, $n = 1,2,3,\ldots$ . Actually the last inequality is an equality as we see from the following argument: For any $m = 1,2,3,\ldots,$ let $\{y_k^m\} \in c$ be defined as

$$y_k^m = \exp(-i \arg a_{nk}) \quad \text{for} \quad 1 \le k \le m,$$

$$y_k^m = 0 \quad \text{for} \quad k > m.$$

Then $\|\{y_k^m\}\|_\infty = 1$ and $x_n^*(\{y_k^m\}) = \Sigma_{k=1}^{m} |a_{nk}|$. It follows at once from this that $\|x_n^*\| = \Sigma_{k=1}^{\infty} |a_{nk}|$, $n = 1,2,3,\ldots$ .

But, since $A$ is a Toeplitz matrix, we have

$$\lim_n x_n^*(\{x_k\}) = \lim_n \sum_{k=1}^{\infty} a_{nk}x_k$$

$$= \lim_k x_k \qquad (\{x_k\} \in c),$$

and from the Uniform Boundedness Theorem (Theorem 6.2.1) we conclude that

$$\sup_n \|x_n^*\| = \sup_n \sum_{k=1}^{\infty} |a_{nk}| < \infty.$$

Therefore part (i) implies part (ii), and the proof is complete. □

6.6. Divergent Fourier Series. Our main purpose in this section is to prove the existence of a continuous function $f$, defined on $[-\pi,\pi]$, whose Fourier series diverges at the origin. In order to accomplish this we use some results, without proof, from the theory of Fourier series.

In Example 3.1.7 we defined the Fourier transform $\hat{f}$ of $f \in L_1([-\pi,\pi],dt/2\pi)$ by

$$\hat{f}(k) = \frac{1}{2\pi}\int_{-\pi}^{\pi} f(t)e^{-ikt}\,dt \qquad (k \in \mathbb{Z}).$$

The Fourier series of such an $f$ is, formally, the expression

$$\sum_{k=-\infty}^{\infty} \hat{f}(k)e^{iks}.$$

The symmetric partial sums of the Fourier series can be expressed in an integral form: for each nonnegative integer $n$

$$\sum_{k=-n}^{n} \hat{f}(k)e^{iks} = \frac{1}{2\pi}\int_{-\pi}^{\pi} f(t)D_n(s-t) \qquad (f \in L_1([-\pi,\pi],dt/2\pi)),$$

where

$$D_n(u) = \sum_{k=-n}^{n} e^{iku} = \frac{\sin[(n+1/2)u]}{\sin(u/2)} \quad (n = 0,1,2,\ldots).$$

Note also that we are here tacitly assuming that $f$ and $D_n$ have been extended periodically to all of $\mathbb{R}$, in order to ensure that the previous integral expression is meaningful. In similar circumstances to be discussed in succeeding chapters we shall always make this assumption without further comment. The function $D_n$ is generally referred to as the Dirichlet kernel. A discussion of this result can be found in [$E_2$, pp. 78 and 79].

With these observations, we can turn to the indicated theorem.

Theorem 6.6.1. There exists some $f \in C([-\pi,\pi])$ such that the Fourier series of $f$ diverges at $t = 0$.

Proof. To prove this it suffices to show the existence of some $f \in C([-\pi,\pi])$ such that the symmetric partial sums of the Fourier series of $f$ at $t = 0$ are unbounded -- that is, to show that

$$\sup_n \left| \sum_{k=-n}^{n} \hat{f}(k) \right| = \sup_n \left| \frac{1}{2\pi} \int_{-\pi}^{\pi} f(t) D_n(-t)\, dt \right| = \infty$$

for some $f \in C([-\pi,\pi])$.

Now clearly each $D_n \in C([-\pi,\pi])$ and $D_n(-u) = D_n(u)$, $n = 0,1,2,\ldots$ . We define for each nonnegative integer $n$ the linear functional $x_n^*$ on $C([-\pi,\pi])$ by

$$x_n^*(f) = \frac{1}{2\pi} \int_{-\pi}^{\pi} f(t) D_n(t)\, dt \qquad (f \in C([-\pi,\pi])).$$

The estimate

$$|x_n^*(f)| \le \left[\frac{1}{2\pi} \int_{-\pi}^{\pi} |D_n(t)|\, dt\right] \|f\|_\infty \qquad (f \in C([-\pi,\pi]))$$

shows that $x_n^* \in C([-\pi,\pi])^*$ and

$$\|x_n^*\| \leq \frac{1}{2\pi}\int_{-\pi}^{\pi}|D_n(t)|\ dt$$
$$= \|D_n\|_1 \qquad (n = 0,1,2,\ldots).$$

Moreover, from the Riesz Representation Theorem (Theorem 4.8.2) and Example 1.2.6 we see that $\|x_n^*\| = \|D_n\|_1$, $n = 0,1,2,\ldots$ .

Now, if the Fourier series of every $f \in C([-\pi,\pi])$ converged at $t = 0$, then we would have for each $f \in C([-\pi,\pi])$

$$\sup_n |x_n^*(f)| < \infty.$$

An appeal to the Uniform Boundedness Theorem (Theorem 6.2.1) would then allow us to conclude that

$$\sup_n \|x_n^*\| < \infty.$$

Consequently to deduce the existence of some $f \in C([-\pi,\pi])$ whose Fourier series diverges at $t = 0$ we need only show that $\sup_n\|x_n^*\| = \sup_n\|D_n\|_1 = \infty$. However, by an elementary change of variable, we obtain

$$\|D_n\|_1 = \frac{1}{2\pi}\int_{-\pi}^{\pi}\left|\frac{\sin(n + 1/2)t}{\sin t/2}\right| dt$$
$$= \frac{2}{\pi}\int_0^{\pi/2}\left|\frac{\sin(2n + 1)t}{\sin t}\right| dt$$
$$\geq \frac{2}{\pi}\int_0^{\pi/2}\left|\frac{\sin(2n + 1)t}{t}\right| dt \quad (n = 0,1,2,\ldots).$$

The inequality is valid since $0 \leq \sin t \leq t$, for $0 \leq t \leq \pi/2$. But

$$\int_0^{\pi/2}\left|\frac{\sin(2n + 1)t}{t}\right| dt = \sum_{k=0}^{2n}\int_{\frac{k\pi}{2(2n+1)}}^{\frac{(k+1)\pi}{2(2n+1)}}\left|\frac{\sin(2n + 1)t}{t}\right| dt$$

$$\geq \sum_{k=0}^{2n} \frac{2(2n+1)}{(k+1)\pi} \int_{\frac{k\pi}{2(2n+1)}}^{\frac{(k+1)\pi}{2(2n+1)}} |\sin(2n+1)t| \, dt$$

$$\geq \sum_{k=0}^{2n} \frac{2}{(k+1)\pi} \int_{\frac{k\pi}{2}}^{\frac{(k+1)\pi}{2}} |\sin t| \, dt$$

$$= \frac{2}{\pi} \sum_{k=0}^{2n} \left(\frac{1}{k+1}\right) \int_0^{\pi/2} \sin t \, d$$

$$= \frac{2}{\pi} \sum_{k=0}^{2n} \frac{1}{(k+1)},$$

where again an elementary change of variable has been effected. Hence

$$\|D_n\|_1 \geq \frac{4}{\pi^2} \sum_{k=0}^{2n} \frac{1}{k+1} \qquad (n = 0,1,2,\ldots),$$

from which it follows at once that $\sup_n \|D_n\|_1 = \infty$. □

Not only do there exist $f \in C([-\pi,\pi])$ whose Fourier series diverge at $t = 0$, but the set of all such $f$ is of category II in $C([-\pi,\pi]$. Even more can be said, as is shown by the following corollary:

Corollary 6.6.1. Let $E$ denote the set of $f \in C([-\pi,\pi]$ such that

$$\sup_n \left| \sum_{k=-n}^{n} \hat{f}(k) \right| < \infty.$$

Then $E$ is of category I in $C([-\pi,\pi])$.

<u>Proof</u>. If this corollary is not true then $E$ is of category II and

$$\sup_n \left| \sum_{k=-n}^{n} \hat{f}(k) \right| = \sup_n |x_n^*(f)| < \infty \qquad (f \in E).$$

Thus from the Uniform Boundedness Theorem (Theorem 6.2.1) we would have

$$\sup_n \|x_n^*\| = \sup_n \|D_n\|_1 < \infty,$$

contrary to the result, obtained in the proof of Theorem 6.6.1, that $\sup_n \|D_n\|_1 = \infty$.

Therefore $E$ is of category I. □

An explicit example of a continuous function whose Fourier series diverges at the origin can be found in [$E_2$, pp. 154 and 155].

## <u>6.7</u>. <u>Problems</u>.

1. (Proposition 6.1.1) Let $X$ be a topological space and $E \subset X$. Prove that the following are equivalent:

(a) $E$ is nowhere dense.

(b) If $\mathcal{O} \subset X$ is open, then there exists some open set $U \subset \mathcal{O}$ such that $U \cap E = \emptyset$.

(c) $\mathrm{cl}[X \sim \mathrm{cl}(E)] = X$.

Prove that, if $E$ is a closed set of category II, then $\mathrm{int}(E) \neq \emptyset$.

*2. (Theorem 6.1.1) Prove that every complete metric space is of category II in itself.

3. Let $X$ be a topological space and prove the following:

(a) If $E \subset X$ is of category I and $A \subset E$, then $A$ is also of Category I.

(b) If $\{E_n\}$ is a sequence of subsets of $X$ that are of category I, then $\bigcup_{n=1}^{\infty} E_n$ is also of category I.

(c) If $X$ is a complete metric space and $E \subset X$ is of category I, then $\mathrm{cl}[X \sim E] = X$.

4. Let $(V, \|\cdot\|)$ be a normed linear space over $\Phi$ and let $W \subset V$ be a proper closed linear subspace of $V$. Prove that $W$ is nowhere dense in $V$.

5. Prove that $c$ is nowhere dense in $\ell_\infty$.

6. Let $BV([0,1])$ denote the functions of bounded variation on $[0,1]$, with the metric

$$\rho(f,g) = |f(0) - g(0)| + V(f - g),$$

where $V(f - g)$ denotes the total variation of $f - g$. [see section 4.8 for a definition of $V(f - g)$]. Prove that the continuous functions of bounded variation are of category I in $BV([0,1])$.

*7. Let $C_1([0,1])$ denote the subset of $C([0,1])$ consisting of all continuous functions $f$ on $[0,1]$ for which $f(0) = f(1)$. Extend these functions periodically to the whole line and define

$$\rho(f,g) = \max\{|f(x) - g(x)| \mid x \in [0,1]\}.$$

Let $\mathcal{C}$ denote the set of extensions of the functions in $C_1([0,1])$.

(a) Prove that $(\mathcal{C}, \rho)$ is a complete metric space.

(b) Prove that the set of functions in $\mathcal{C}$ that are somewhere differentiable is of category I in $\mathcal{C}$. [Hint: consider the sets

$$K_n = \{f \in \mathcal{C} \mid \text{for some } x,\ \left|\frac{f(x + h) - f(x)}{h}\right| \le n \text{ for all } h > 0\}$$

and let $K = \bigcup_{n=1}^{\infty} K_n$. Prove that $K$ is of category I.]

(c) Conclude that there exist continuous nowhere differentiable functions.

*8. Let $B([0,1])$ denote the space of all bounded real-valued functions on $[0,1]$ with the usual supremum norm. The Baire subspace $B_n$ of $B([0,1])$ of type $n$ is defined to be all the bounded functions on $[0,1]$ that are pointwise limits of sequences of functions in the Baire subspace $B_{n-1}$, $n = 1,2,3,\ldots,$ and $B_0 = C([0,1])$; $B = \bigcup_{n=0}^{\infty} B_n$ is all the bounded functions on $[0,1]$ of finite Baire type. If $\omega$ denotes the ordinal number of the nonnegative integers, then $B_\omega$ denotes all the bounded functions on $[0,1]$ that are pointwise limits of sequences of functions in $B$. It can be shown that $B_n$, $n = 0,1,2,\ldots,$ and $B_\omega$ are closed linear subspaces of $B([0,1])$.

(a) Prove that there exists a sequence $\{f_n\} \subset B$ that converges uniformly to $f \in B([0,1])$ such that $f \notin B$. Clearly then $f \in B_\omega \sim B$.

(b) Prove that not every $f \in B_\omega$ is the uniform limit of a sequence $\{f_n\} \subset B$. (Hint: use Theorem 1.3.2.)

9. (Corollary 6.2.2) Let $(V_1, \|\cdot\|_1)$ and $(V_2, \|\cdot\|_2)$ be Banach spaces over $\Phi$. Suppose $\{T_n\} \subset L(V_1,V_2)$ is a sequence such that

(a) For each $x \in V_1$ there exists some $M_x > 0$ such that $\sup_n \|T_n(x)\|_2 \leq M_x$.

(b) There exists some $x_0 \in V_1$, some $\delta > 0$, and a set $E \subset B(x_0,\delta) = \{x \mid \|x - x_0\|_1 < \delta\}$ such that $E$ is dense in $B(x_0,\delta)$ and such that for each $x \in E$ there exists some $z_x \in V_2$ for which $\lim_n T_n(x) = z_x$.

Prove that there exists a unique element $T \in L(V_1,V_2)$ such that

(i) $T(x) = \lim_n T_n(x)$, $x \in V_1$.
(ii) $\|T\| \leq \liminf_n \|T_n\|$.

*10. (Theorem 6.2.3) Let $(V_1,\mathcal{T}_1)$ and $(V_2,\mathcal{T}_2)$ be Fréchet spaces over $\Phi$ and suppose $\rho_k$ is the metric on $V_k$ that defines the topology $\mathcal{T}_k$, $k = 1,2$. Suppose $\{T_\alpha\}_{\alpha\in\Lambda} \subset L(V_1,V_2)$ and suppose that for each $x \in V_1$ there exists some $M_x > 0$ for which

$$\sup_{\alpha\in\Lambda} \rho_2(T_\alpha(x),0) \leq M_x.$$

Prove that, given $\epsilon > 0$, there exists some $\delta > 0$ such that

$$\sup_{\substack{x\in V_1\\ \rho_1(x,0)<\delta}} \rho_2(T_\alpha(x),0) \leq \epsilon \qquad (\alpha \in \Lambda).$$

*11 (Theorem 6.2.4) Let $(V_1,\mathcal{T}_1)$ and $(V_2,\mathcal{T}_2)$ be Fréchet spaces over $\Phi$ and suppose $\rho_k$ is the metric on $V_k$ that defines the topology $\mathcal{T}_k$, $k = 1,2$. Suppose $\{T_\alpha\}_{\alpha\in\Lambda}$ is a net in $L(V_1,V_2)$ such that

(a) For each $x \in V_1$ there exists some $M_x > 0$ for which $\sup_\alpha \rho_2(T_\alpha(x),0) \leq M_x$.

(b) There exists some $x_o \in V_1$, some $\delta > 0$, and a set $E \subset B(x_o,\delta) = \{x \mid \rho_1(x,x_o) < \delta\}$ such that $E$ is dense in $B(x_o,\delta)$ and such that for each $x \in E$ there exists some $z_x \in V_2$ for which $\lim_\alpha = T_\alpha(x) = z_x$.

Prove that there exists a unique element $T \in L(V_1,V_2)$ such that $T(x) = \lim_\alpha T_\alpha(x)$, $x \in V_1$.

12. Let $(V,\mathcal{T})$ be a Fréchet space over $\Phi$. Prove that any lower semicontinuous seminorm on $V$ is continuous.

13. Suppose $\{a_k\}$ is a sequence of complex numbers such that $\Sigma_{k=1}^\infty a_k b_k$ converges for each sequence $\{b_k\} \in c_o$. Prove that $\Sigma_{k=1}^\infty |a_k| < \infty$.

14. Let $(V,\mathcal{T})$ be a locally convex topological linear space over $\Phi$. Suppose $E \subset V$ is weakly bounded, that is,

$$\sup_{x \in E} |x^*(x)| < \infty$$

for each $x^* \in V^*$. Prove that $E$ is bounded.

15. (a) Let $(V,\|\cdot\|)$ be a normed linear space over $\Phi$ and let $\{x_n\}$ be a sequence in $V$ such that $\lim_n x^*(x_n)$ exists for each $x^* \in V^*$. Prove that $\{\|x_n\| \mid n = 1,2,\ldots\}$ is bounded.

(b) Let $(V,\|\cdot\|)$ be a Banach space over $\Phi$ and let $\{x_n^*\}$ be a sequence in $V^*$ such that $\lim_n x_n^*(x)$ exists for each $x \in V$. Prove that $\{\|x_n^*\| \mid n = 1,2,\ldots\}$ is bounded.

16. Let $\{f_n\}$ be a sequence of continuous real-valued functions on $\mathbb{R}$. Prove that there is an interval $I$ and a number $M > 0$ such that $|f_n(x)| < M$ for every $x \in I$ and $n = 1,2,3,\ldots$ .

17. Let $V$ be a linear space over $\Phi$. Suppose a translation invariant metric $\rho$ is defined on $V$, that is, $\rho(x,y) = \rho(x - y,0)$, and let $\mathcal{T}$ be the topology determined by $\rho$. Furthermore, suppose $(V,\mathcal{T})$ satisfies the following three conditions:

(a) If $\{a_n\} \subset \Phi$ and $\lim_n a_n = 0$, then $\lim_n a_n x = 0$, $x \in V$.
(b) If $\{x_n\} \subset V$ and $\lim_n x_n = 0$, then $\lim_n a x_n = 0$, $a \in \Phi$.
(c) $(V,\mathcal{T})$ is complete.

Prove that $(V,\mathcal{T})$ is a complete metrizable topological linear space.

*18. Let $(V_1,\|\cdot\|_1)$ be a Banach space and $(V_2,\|\cdot\|_2)$ a normed linear space over $\Phi$. For each $m = 1,2,3,\ldots$, let $\{T_n^m\}$ be a sequence in $L(V_1,V_2)$ such that for every $m$ there is some $x_m \in V_1$ for which

$$\lim_n \sup \|T_n^m(x_m)\|_2 = \infty.$$

Prove that there is an $x \in V_1$ for which

$$\lim_n \sup \|T_n^m(x)\|_2 = \infty \qquad (m = 1,2,3,\ldots).$$

(This is known as the principle of condensation of singularities.)

19. Prove that there exists some $f \in C([-\pi,\pi])$ such that the Fourier series of $f$ diverges at all $t \in [-\pi,\pi]$ except for those $t \in E$ where $E$ is some set of category I in $[-\pi,\pi]$.

20. (Proposition 6.3.1) Letting $(V_1,\|\cdot\|_1)$ and $(V_2,\|\cdot\|_2)$ be normed linear spaces over $\Phi$, prove each of the following:

(a) $\mathcal{U}_{so}$ is a base for a Hausdorff topology $so(\mathcal{T})$ on $L(V_1,V_2)$ such that $(L(V_1,V_2),so(\mathcal{T}))$ is a locally convex topological linear space over $\Phi$.

(b) A net $\{T_\alpha\} \subset L(V_1,V_2)$ converges to $T \in L(V_1,V_2)$ in the topology $so(\mathcal{T})$ if and only if $\lim_\alpha \|T_\alpha(x) - T(x)\|_2 = 0$, $x \in V_1$.

21. Let $(V,\|\cdot\|)$ be a Banach space over $\Phi$ and let $\{x_n^*\} \subset V^*$ be such that $\|x_n^*\| = 1$ for each $n$ and $\bigcup_{k=1}^\infty \bigcap_{n=k}^\infty N(x_n^*)$ is dense in V. Prove that $\{x_n^*\}$ converges to zero in the strong operator topology on $V^*$.

22. Let $(V_1,\|\cdot\|_1)$ be a normed linear space over $\Phi$ and $(V_2,\|\cdot\|_2)$ a Banach space over $\Phi$. Let $\{T_\alpha\} \subset L(V_1,V_2)$ be a net such that $\lim \sup_\alpha \|T_\alpha\| < \infty$. Let $E = \{x \mid x \in V_1, \lim_\alpha T_\alpha(x) \text{ exists}\}$. Prove that $E$ is a closed linear subspace of $V_1$. Conclude that, if $\{T_\alpha(x)\}$ converges for all $x$ in some dense subset $E$ of $V_1$, then $\{T_\alpha(x)\}$ converges for all $x \in V_1$.

23. Let $(V_1,\mathcal{T}_1)$, $(V_2,\mathcal{T}_2)$, and $(V_3,\mathcal{T}_3)$ be topological linear spaces over $\Phi$. A mapping $f(x,y) : V_1 \times V_2 \to V_3$ is called <u>bilinear</u> if for each $y \in V_2$ the mapping $x \to f(x,y)$ is linear from $V_1$ to $V_3$, and for each $x \in V_1$ the mapping $y \to f(x,y)$ is linear from $V_2$ to $V_3$. A bilinear map is continuous if it is continuous as a mapping from $(V_1 \times V_2,\mathcal{T}_1 \times \mathcal{T}_2)$ to $(V_3,\mathcal{T}_3)$; it is separately continuous if all the partial mappings $x \to f(x,y)$ and $y \to f(x,y)$ are continuous.

(a) Prove that a bilinear mapping $f$ is continuous if and only if it is continuous at the origin $(0,0) \in V_1 \times V_2$.

(b) If $V_1$ and $V_2$ are complete and metrizable, prove that every separately continuous bilinear mapping $f(x,y) : V_1 \times V_2 \to V_3$ is continuous.

(c) Generalize problem (b) to a family $F$ of bilinear mappings; that is, with the hypotheses of problem (b), let $F$ be a family of bilinear mappings $f(x,y) : V_1 \times V_2 \to V_3$ such that each $f \in F$ is separately continuous, and for each fixed $(x,y) \in V_1 \times V_2$, $\{f(x,y) \mid f \in F\}$ is a bounded set in $V_3$. Prove that $F$ is equicontinuous -- that is, for each neighborhood $U_1$ of the origin in $V_3$ and point $(x,y) \in V_1 \times V_2$ there exists a neighborhood $U_2$ of $(x,y)$ such that $f(x_1,y_1) - f(x,y) \in U_1$ for all $(x_1,y_1)$ in $U_2$ and all $f \in F$.

*24. Prove that each of the following matrices $(a_{nk})_{n,k=0}^{\infty}$ is a Toeplitz matrix:

(a) Arithmetic means:

$$a_{nk} = \frac{1}{n+1} \quad \text{for} \quad 0 \leq k \leq n,$$
$$a_{nk} = 0 \quad \text{for} \quad k > n.$$

(b) Cesàro means: For each fixed nonnegative integer $r$ let

$$A_0^r = 1,$$
$$A_n^r = \frac{(r+1)(r+2)\cdots(r+n)}{n!} \quad \text{for} \quad n \geq 1.$$

Define

$$a_{nk} = \frac{A_{n-k}^{r-1}}{A_n^r} \quad \text{for} \quad 0 \leq k \leq n,$$

$$a_{nk} = 0 \quad \text{for} \quad k > n.$$

(c) Euler-Knopp matrix: Let $0 < r < 1$ and let

$$a_{nk} = \binom{n}{k} r^k (1 - r)^{n-k} \quad \text{for} \quad 0 \leq k \leq n,$$

$$a_{nk} = 0 \quad \text{for} \quad k > n.$$

where $\binom{n}{k} = n!/k!(n - k)!$.

(d) Nörlund mean: Let $p_0 > 0$, $p_n \geq 0$, $n \geq 1$, $P_n = p_0 + \dots + p_n$, and suppose $\lim_n p_n/P_n = 0$. Define

$$a_{nk} = \frac{p_{n-k}}{P_n} \quad \text{for} \quad 0 \leq k \leq n,$$

$$a_{nk} = 0 \quad \text{for} \quad k > n.$$

(Actually, $(a_{nk})_{n,k=0}^{\infty}$ is a Toeplitz matrix if and only if $\lim_n p_n/P_n = 0$.)

25. Let $A = (a_{nk})_{n,k=1}^{\infty}$ be a Toeplitz matrix. A sequence $x = \{x_k\}$ is said to have the A-limit $A(x)$ if for each $n = 1,2,3,\dots$, the sum $A_n(x) = \Sigma_{k=1}^{\infty} a_{nk}x_k$ converges and $\lim_n A_n(x) = A(x)$. Since $A = (a_{nk})$ is a Toeplitz matrix, if $x \in c$, then the A-limit of $x$ exists and equals the limit of the sequence $x$. Prove that there exists some $x \in \ell_\infty$ whose A-limit does not exist.

26. If $f \in L_1([-\pi,\pi],dt/2\pi)$, let $s_n(f)$ denote the nth partial sum of the Fourier series of $f$, that is,

$$s_n(f)(t) = \sum_{k=-n}^{n} \hat{f}(k)e^{ikt} \qquad (t \in [-\pi,\pi]).$$

Prove that there exists some $f \in L_1([-\pi,\pi],dt/2\pi)$ such that $\{\|s_n(f)\|_1 \mid n = 0,1,2,\dots\}$ is unbounded. In particular, there exists some $f \in L_1([-\pi,\pi],dt/2\pi)$ such that $s_n(f)$ does not converge to $f$ in the $L_1$-norm.

# CHAPTER 7

# THE OPEN MAPPING AND THE CLOSED GRAPH THEOREMS

7.0. Introduction. Our concern in this chapter is a pair of fundamental theorems, known generally as the Open Mapping and the Closed Graph Theorems. These theorems give powerful results about linear transformations whose graphs are closed sets. Indeed, we shall see that, if $T$ is such a transformation between two Fréchet spaces, then it must be a continuous transformation, this being the Closed Graph Theorem, whereas if the range of $T$ is of category II, then $T$ is surjective and maps open sets onto open sets, this being the Open Mapping Theorem.

We begin with the definition and some properties of a closed mapping between topological spaces in Section 7.1 and then shift our attention to linear closed mappings in the next two sections. Here we shall prove the Open Mapping and Closed Graph Theorems, as well as some of their immediate consequences. The remainder of the chapter, as usual, is devoted to some applications.

7.1. Closed Mappings. In this section we wish to discuss the concept of a closed mapping between two topological spaces. The results we obtain in this context will provide some motivation for the development in the succeeding two sections.

Definition 7.1.1. Let $X$ and $Y$ be arbitrary sets and suppose $f : X \to Y$. Then the graph of $f$, denoted by $G(f)$, is defined as $G(f) = \{(t,f(t)) \mid t \in X\}$.

Clearly $G(f) \subset X \times Y$, and $(t,u) \in G(f)$ if and only if $f(t) = u$.

Definition 7.1.2. Let $(X,\mathcal{T})$ and $(Y,\mathcal{U})$ be topological spaces and suppose $f : X \to Y$. $f$ is said to be closed if $G(f)$ is a closed subset of the topological product space $(X \times Y, \mathcal{T} \times \mathcal{U})$.

It is perhaps necessary to remark at once that, if $f : X \to Y$ is closed, then it does not necessarily follow that $f(E) \subset Y$ is a closed set whenever $E \subset X$ is a closed set. We shall, however, examine the connection between these two properties a bit further in Proposition 7.1.2.

Our first result provides an alternative, and useful, characterization of closed mappings.

Proposition 7.1.1. Let $(X,\mathcal{T})$ and $(Y,\mathcal{U})$ be topological spaces and suppose $f : X \to Y$. Then the following are equivalent:

(i) $f$ is closed.

(ii) If $\{t_\alpha\} \subset X$ is a net such that there exist some $t \in X$ and $u \in Y$ for which $\lim_\alpha t_\alpha = t$ and $\lim_\alpha f(t_\alpha) = u$, then $f(t) = u$.

Proof. Suppose $f$ is closed, $\lim_\alpha f(t_\alpha) = u$. Then in the topological product space $(X \times Y, \mathcal{T} \times \mathcal{U})$ we have $\lim_\alpha (t_\alpha, f(t_\alpha)) = (t,u)$. But the net $\{(t_\alpha, f(t_\alpha))\} \subset G(f)$, and $G(f)$ is a closed subset of $X \times Y$. Hence $(t,u) \in G(f)$, that is, $f(t) = u$, and part (i) implies part (ii).

Conversely, suppose $f$ satisfies implication (ii) and let $\{(t_\alpha, f(t_\alpha))\} \subset G(f)$ be a net such that $\lim_\alpha (t_\alpha, f(t_\alpha)) = (t,u)$. It then follows from the definition of the product topology that $\lim_\alpha t_\alpha = t$ and $\lim_\alpha f(t_\alpha) = u$, whence we conclude by part (ii), that $f(t) = u$. Thus $(t,u) \in G(f)$, and $G(f)$ is closed.

Therefore part (ii) of the theorem implies part (i). □

Corollary 7.1.1. Let $(X,\mathcal{T})$ and $(Y,\mathcal{U})$ be topological spaces, suppose that $f : X \to Y$ and that $\mathcal{U}$ is a Hausdorff topology. If $f$ is continuous, then $f$ is closed.

The proof is left to the reader.

Without the assumption that $\mathcal{U}$ is a Hausdorff topology the corollary may fail. For instance, if $X = Y$ and $\mathcal{T} = \mathcal{U}$ is not a Hausdorff topology, then $f : X \to X$ defined by $f(t) = t$ is continuous, but not closed.

An example of a closed mapping that is not continuous is also perhaps in order. Letting $X = [0,1]$ and $Y = \mathbb{R}$, with the usual topologies, define

$$f(t) = 1/t \quad \text{for} \quad 0 < t \leq 1,$$

$$f(t) = 0 \quad \text{for} \quad t = 0.$$

It is easily verified that $f$ is closed, but not continuous.

Under certain conditions, however, it is the case that a closed mapping is continuous, as is shown by part (v) of the next proposition.

Proposition 7.1.2. Let $(X,\mathcal{T})$ and $(Y,\mathcal{U})$ be topological spaces and suppose that $f : X \to Y$ is closed.

(i) If $f$ is bijective, then the inverse function $f^{-1}$ exists and $f^{-1} : Y \to X$ is closed.

(ii) If $K \subset X$ is compact, then $f(K) \subset X$ is closed.

(iii) If $(X,\mathcal{T})$ is compact, then $f(E) \subset X$ is closed whenever $E \subset X$ is closed.

(iv) If $K \subset Y$ is compact, then $\{t \mid f(t) \in K\}$ is closed.

(v) If $(Y,\mathcal{U})$ is compact, then $f$ is continuous.

(vi) If $(X,\mathcal{T})$ is compact and $f$ is bijective, then the inverse function $f^{-1}$ exists and is continuous.

(vii) If $(X,\mathcal{T})$ is compact and $f$ is bijective, then $f(E) \subset Y$ is open whenever $E \subset X$ is open.

Proof. Part (i) is obvious since the mapping $(t,u) \to (u,t)$ is a homeomorphism of $X \times Y$ onto $Y \times X$.

To prove part (ii) suppose $\{u_\alpha\} = \{f(t_\alpha)\}$ is a net in $f(K)$ and $\lim_\alpha f(t_\alpha) = u$. Since $K$ is compact, $\{t_\alpha\} \subset K$ has a subnet $\{t_\beta\}$ that converges to some $t \in K$. Thus $\lim_\beta t_\beta = t$ and $\lim_\beta f(t_\beta) = u$, from which we conclude, by Proposition 7.1.1, that $f(t) = u$. Hence $f(K)$ is closed.

Part (iii) follows at once from part (ii); part (iv) is established by an argument similar to the one just used.

If $(Y,\mathcal{U})$ is compact, then each closed subset $K \subset Y$ is compact, and thus from part (iv) we see that $\{t \mid f(t) \in K\} \subset X$ is closed. Thus $f$ is continuous.

Part (vi) follows at once from parts (i) and (v), whereas part (vii) is immediate from part (vi). □

Those of the preceding results that will be of most interest to us in the next two sections when we discuss linear closed mappings are parts (v) and (vii). These results assert that, under certain conditions, a closed mapping is continuous or open -- that is, maps open sets onto open sets. We shall be interested in seeing when such assertions are valid for linear closed mappings. The results will be the Closed Graph and Open Mapping Theorems, respectively.

## 7.2. The Open Mapping Theorem.

In this section we prove the Open Mapping Theorem for closed linear transformations between Fréchet spaces.

First, however, let us note again the relevant results about closed mappings. Suppose $(V_1,\mathcal{T}_1)$ and $(V_2,\mathcal{T}_2)$ are topological linear spaces over $\Phi$ and $T \in \mathcal{L}'(V_1,V_2)$. Then from the appropriate parts of Propositions 7.1.1 and 7.1.2 we see that

(i) $T$ is closed if and only if $T(x) = y$ whenever a net $\{x_\alpha\} \subset V_1$, $x \in V_1$, and $y \in V_2$ are such that $\lim_\alpha x_\alpha = x$ and $\lim_\alpha T(x_\alpha) = y$.

(ii) If $T$ is closed and bijective, then $T^{-1} \in L'(V_2,V_1)$ is closed.

(iii) If $T$ is continuous, then $T$ is closed.

At this point it is also perhaps worthwhile to give an example of a linear transformation that is closed, but not continuous. The previous example of a closed discontinuous mapping is clearly not linear.

Let $V_1 = V_2$ be the linear space over $\mathbb{R}$ of all real-valued polynomials defined on $[0,1]$ and set

$$\|f\|_1 = \|f\|_\infty,$$
$$\|f\|_2 = \|f\|_\infty + \|f'\|_\infty \qquad (f \in V_1 = V_2),$$

where $f'$, of course, denotes the derivative of $f$. Then $(V_1,\|\cdot\|_1)$ and $(V_2,\|\cdot\|_2)$ are normed linear spaces over $\mathbb{R}$.

Suppose $T \in L'(V_1,V_2)$ is defined by $T(f) = f'$, $f \in V_1$. Then it is easily seen that there exists no $M > 0$ such that

$$\|T(f)\|_2 = \|f'\|_\infty + \|f''\|_\infty \leq M\|f\|_\infty = M\|f\|_1 \qquad (f \in V_1),$$

and so $T$ is not continuous. On the other hand, suppose $\lim_n\|f_n - f\|_1 = 0$ and $\lim_n\|T(f_n) - g\|_2 = 0$, where $\{f_n\} \subset V_1$, $f \in V_1$, and $g \in V_2$. Then by the definitions of $\|\cdot\|_1$ and $\|\cdot\|_2$ we see that $\lim_n\|f_n - f\|_\infty = 0$ and $\lim_n\|f_n' - g\|_\infty = 0$. But clearly for each $n$

$$f_n(t) - f_n(0) = \int_0^t f_n'(s)\,ds \qquad (0 \leq t \leq 1),$$

from which we conclude at once that

$$f(t) - f(0) = \int_0^t g(s)\,ds \qquad (0 \leq t \leq 1).$$

Since $g$ is continuous, we see that $T(f) = f' = g$. Therefore $T$ is closed.

As will be seen when we come to the Closed Graph Theorem (Theorem 7.3.1), the underlying reason for $T$ not being continuous is that $(V_k, \|\cdot\|_k)$, $k = 1,2$, are not complete normed linear spaces.

Let us examine another example in the spaces $(V_k, \|\cdot\|_k)$, $k = 1,2$. This time consider the identity transformation I defined by $I(f) = f$, $f \in V_1$. Then $I$ is linear, but not continuous, as there exists no $M > 0$ for which

$$\|I(f)\|_2 = \|f\|_\infty + \|f'\|_\infty \leq M\|f\|_\infty = M\|f\|_1 \qquad (f \in V_1).$$

Moreover, if $\lim_n \|f_n - f\|_1 = 0$ and $\lim_n \|I(f_n) - g\|_2 = 0$, then $\lim_n \|f_n - f\|_\infty = \lim_n \|f_n - g\|_\infty = 0$, and so $I(f) = f = g$. Thus $I$ is closed.

Furthermore, we note that $I$ is clearly bijective, and so $I^{-1} \in L'(V_2, V_1)$ is closed. Indeed $I^{-1}$ is even continuous, as

$$\|I^{-1}(f)\|_1 = \|f\|_\infty \leq \|f\|_\infty + \|f'\|_\infty = \|f\|_2 \qquad (f \in V_2).$$

Hence $I^{-1} : V_2 \to V_1$ is an example of a closed, actually continuous, linear transformation that is invertible, but whose inverse, in this case $I$, is not continuous.

Again, as we shall see from the second corollary (Corollary 7.2.2) of the Open Mapping Theorem, this occurs essentially because the spaces $(V_k, \|\cdot\|_k)$, $k = 1,2$, are not complete.

Now let us turn to the proof of the Open Mapping Theorem. In order to reduce the length of the proof of the theorem itself we first establish a number of rather technical lemmas. We prove the lemmas and the theorem for Fréchet spaces. In view of the discussion in Section 2.5 and in particular because of Lemma 2.5.1, we denote here a Fréchet space by $(V, \{p_n\})$, assuming without loss of generality that $\{p_n\}$ is a countable family of seminorms such that $p_n(x) \leq p_{n+1}(x)$, $x \in V$, $n = 1,2,3,\ldots$ .

<u>Lemma 7.2.1</u>. Let $(V_1,\{p_k\})$ and $(V_2,\{q_m\})$ be Fréchet spaces over $\Phi$ and suppose $T \in L'(V_1,V_2)$ is such that $T(V_1)$ is of category II in $V_2$. If $U_n = \{x \mid x \in V_1,\ p_n(x) < 1/n^2\}$, $n = 1,2,3,\ldots$, then there exists a sequence $\{\epsilon_n\}$, $\epsilon_n > 0$, and a sequence of positive integers $\{k_n\}$ such that

(i) $\lim_n \epsilon_n = 0$.
(ii) $k_n \geq n$, $n = 1,2,3,\ldots$ .
(iii) $\mathrm{cl}[T(U_n)] \supset \{y \mid y \in V_2,\ q_{k_n}(y) < \epsilon_n\}$ $(n = 1,2,3,\ldots)$.

<u>Proof</u>. Let $n$ be a positive integer. Clearly $V_1 = \bigcup_{m=1}^{\infty} mU_n$, and so $T(V_1) = \bigcup_{n=1}^{\infty} mT(U_n)$. Since $T(V_1)$ is of category II in $V_2$, it follows that there exists some $m_o$ for which $\mathrm{int}\{\mathrm{cl}[T(U_n)]\} \neq \phi$.

We claim, moreover, that $\mathrm{cl}[T(U_n)]$ contains an open neighborhood of the origin, arguing as follows: Suppose $y \in \mathrm{int}\{\mathrm{cl}[T(U_n)]\}$. Since $p_n$ is a seminorm, it is evident that $U_n$ is a convex balanced set. Thus from Propositions 2.3.2(iii) and (iv), since $T$ is linear, we see that $\mathrm{int}\{\mathrm{cl}[T(U_n)]\}$ is convex and balanced. Therefore $-y \in \mathrm{int}\{\mathrm{cl}[T(U_n)]\}$, and so $0 = y/2 + (-y)/2 \in \mathrm{int}\{\mathrm{cl}[T(U_n)]\}$, from which it is apparent that $\mathrm{cl}[T(U_n)]$ contains an open neighborhood of the origin.

Consequently, by the definition of the topology in $V_2$, there exist some $\epsilon_n > 0$ and some positive integer $k_n$ such that

$$\begin{aligned}\mathrm{cl}[T(U_n)] &\supset \{y \mid y \in V_2,\ q_1(y) < \epsilon_n,\ldots,q_{k_n}(y) < \epsilon_n\}\\ &= \{y \mid y \in V_2,\ q_{k_n}(y) < \epsilon_n\},\end{aligned}$$

as $q_m(y) \leq q_{m+1}(y)$, $y \in V_2$, $m = 1,2,3,\ldots$ . It is, moreover, evident that we can choose $\epsilon_n$ and $k_n$ successively in this fashion so that $\lim_n \epsilon_n = 0$ and $k_n \geq n$. □

<u>Lemma 7.2.2</u>. Let $(V_1,\{p_k\})$ and $(V_2,\{q_m\})$ be Fréchet spaces over $\Phi$ and suppose $T \in L'(V_1,V_2)$ is such that $T(V_1)$ is of category II in $V_2$. For each positive integer $n$ choose $\epsilon_n > 0$

and a positive integer $k_n \geq n$ such that

(i) $\lim_n \epsilon_n = 0$.

(ii) $\mathrm{cl}[T(U_n)] \supset \{y \mid y \in V_2,\ q_{k_n}(y) < \epsilon_n\} = W_n$, where $U_n = \{x \mid x \in V_1,\ p_n(x) < 1/n^2\}$, $n = 1,2,3,\ldots$ .

If $T$ is a closed mapping, then $T(2U_n) \supset W_n$, $n = 1,2,3,\ldots$ .

Proof. The existence of the sequences $\{\epsilon_n\}$ and $\{k_n\}$ is of course ensured by Lemma 7.2.1. We give the argument only for the case $n = 1$. The other cases are proved in a similar manner.

Let $y \in W_1$. We need to show the existence of some $x \in 2U_1$ for which $T(x) = y$. Now from Lemma 7.2.1 we know that $\mathrm{cl}[T(U_1)] \supset W_1$, and hence there exists some $x_1 \in U_1$ such that $q_{k_2}[y - T(x_1)] < \epsilon_2$. From the definition of $W_2$ we see that $y - T(x_1) \in W_2$. Hence, since $\mathrm{cl}[T(U_2)] \supset W_2$, there exists some $x_2 \in U_2$ such that $q_{k_3}[y - T(x_1) - T(x_2)] < \epsilon_3$. Continuing in this fashion, we obtain, by repeated applications of Lemma 7.2.1, a sequence $\{x_n\} \subset V_1$ such that

(a) $x_n \in U_n$.

(b) $q_{k_{n+1}}[y - T(\Sigma_{k=1}^n x_k)] < \epsilon_{n+1}$ $\qquad (n = 1,2,3,\ldots)$.

We claim that $\{z_n\} = \{\Sigma_{k=1}^n x_k\}$ is a Cauchy sequence in the seminormed linear space $(V_1, \{p_k\})$.

Indeed, given any seminorm $p_j$, if $n > m > j$, then

$$\begin{aligned}
p_j(z_n - z_m) &= p_j\Big(\sum_{k=m+1}^{n} x_k\Big) \\
&\leq \sum_{k=m+1}^{n} p_j(x_k) \\
&\leq \sum_{k=m+1}^{n} p_k(x_k) \\
&< \sum_{k=m+1}^{n} \frac{1}{k^2},
\end{aligned}$$

as $p_k(x) \leq p_{k+1}(x)$, $x \in V_1$, $k = 1,2,3,\ldots$. Since $\Sigma_{k=1}^{\infty} 1/k^2$ is a convergent series, it follows at once that $\{z_n\}$ is a Cauchy sequence.

However, $(V_1,\{p_k\})$ is a Fréchet space, and hence $\{z_n\}$ converges in $V_1$, say to $x$. Moreover, the sequence $\{T(z_n)\} \subset V_2$ converges to $y$ as we see from the following: if $q_m$ is any seminorm and $n$ is chosen so large that $k_{n+1} \geq m$, then

$$q_m[y - T(z_n)] \leq q_{k_{n+1}}[y - T(z_n)] < \epsilon_{n+1}.$$

Since $\lim_n \epsilon_n = 0$, we deduce immediately that $\{T(z_n)\}$ converges to $y$ in the seminormed linear space $(V_2,\{q_m\})$.

However, since $T$ is a closed mapping, from Proposition 7.1.1 we see that $T(x) = y$. Furthermore, we observe that

$$\begin{aligned} p_1(x) &= p_1(\lim_n \sum_{k=1}^{n} x_k) \\ &\leq \limsup_n \sum_{k=1}^{n} p_1(x_k) \\ &\leq \limsup_n \sum_{k=1}^{n} p_k(x_k) \\ &\leq \sum_{k=1}^{\infty} \frac{1}{k^2} \\ &= \frac{\pi^2}{6} \\ &< 2. \end{aligned}$$

Therefore $x \in 2U_1$, and the proof is complete. □

We can now state and prove the Open Mapping Theorem.

Theorem 7.2.1' (Open Mapping Theorem). Let $(V_1,\{p_k\})$ and $(V_2,\{q_m\})$ be Fréchet spaces over $\Phi$ and suppose $T \in L'(V_1,V_2)$ is such that $T(V_1)$ is of category II in $V_2$. If $T$ is a closed mapping, then

(i) $T(U) \subset V_2$ is open whenever $U \subset V_1$ is open.
(ii) $T$ is surjective.

Proof. Let $\epsilon_n > 0$ and positive integers $k_n \geq n$, $n = 1,2,3,\ldots$, be so chosen that $T(2U_n) \supset W_n = \{y \mid y \in V_2, q_{k_n}(y) < \epsilon_n\}$, where $U_n = \{x \mid x \in V_1, p_n(x) < 1/n^2\}$, $n = 1,2,3,\ldots$ . Such choices are possible by Lemmas 7.2.1 and 7.2.2.

Since $T$ is linear, to extablish part (i) it clearly suffices to show that, if $U$ is an open neighborhood of the origin in $V_1$, then $T(U)$ is an open set in $V_2$. But if $x \in U$, then there exists some positive integer $n$ and some $\epsilon > 0$ such that $x + \epsilon U_n \subset U$, as $U$ is open. Consequently

$$T(x + \epsilon U_n) = T(x) + (\tfrac{\epsilon}{2})T(2U_n) \subset T(U).$$

However, $T(2U_n) \supset W_n$, and so $T(U) \supset T(x) + (\epsilon/2)W_n$. Since $(\epsilon/2)W_n$ is an open neighborhood of the origin in $V_2$, we conclude that $T(U)$ is open.

Next suppose $y \in V_2$. If $y = 0$, then $T(0) = y$, and if $y \neq 0$, then there exists some $k_n$ for which $q_{k_n}(y) \neq 0$. If the latter were not the case, that is, if $q_{k_n}(y) = 0$, $n = 1,2,3,\ldots$, then we would have $q_m(y) = 0$, $m = 1,2,3,\ldots$, as $k_n \geq n$ and $q_m(y) \leq q_{m+1}(y)$, and so $y = 0$, contrary to assumption. Set $z = [\epsilon_n/2q_{k_n}(y)]y$. Then $q_{k_n}(z) = \epsilon_n/2 < \epsilon_n$, and so $z \in W_n \subset T(2U_n)$. Thus there exists some $x \in 2U_n$ such that $T(x) = z$. From this we deduce that $T([2q_{k_n}(y)/\epsilon_n]x) = y$. Therefore $T$ is surjective. □

It is instructive at this point for the reader to refer back to the proof of Helly's Theorem (Theorem 4.10.1) and see how the Open Mapping Theorem could be used to shorten it.

Let us give two rather easy corollaries of the theorem.

Corollary 7.2.1. Let $(V_1,\{p_k\})$ and $(V_2,\{q_m\})$ be Fréchet spaces over $\Phi$ and suppose $T \in L'(V_1,V_2)$ is a closed mapping. Then the following are equivalent:

(i) The range of $T$ is of category II in $V_2$.

(ii) $T$ is surjective.

In particular, if $T$ is a closed linear transformation between Fréchet spaces, then either $T$ is surjective or its range is of category I. Moreover, the Open Mapping Theorem says that, if $T$ is surjective, then it is an open mapping. This immediately gives us the next corollary.

Corollary 7.2.2. Let $(V_1,\{p_k\})$ and $(V_2,\{q_m\})$ be Fréchet spaces over $\Phi$. If $T \in L'(V_1,V_2)$ is a closed bijective mapping, then $T^{-1} \in L(V_2,V_1)$.

As a corollary to the corollary we have the following useful result:

Corollary 7.2.3. Let $V$ be a linear space over $\Phi$ and suppose $T_1$ and $T_2$ are Hausdorff topologies on $V$ such that $(V,T_1)$ and $(V,T_2)$ are Fréchet spaces over $\Phi$. If $T_1 \supset T_2$, then $T_1 = T_2$.

Proof. Consider the identity transformation $I : V \to V$ defined by $I(x) = x$, $x \in V$. Clearly $I$ is linear, bijective, and continuous from $(V,T_1)$ to $(V,T_2)$. Hence, by Corollary 7.2.2, $I^{-1} = I$ is linear, bijective, and continuous from $(V,T_2)$ to $(V,T_1)$, and so $T_1 \subset T_2$. □

7.3. The Closed Graph Theorem. If $T$ is a closed bijective linear transformation between two Fréchet spaces, then Corollary 7.2.3 asserts that $T^{-1}$ is continuous. In particular, $T^{-1}$ is also a closed bijective linear transformation between Fréchet spaces, and hence it must have a continuous inverse -- that is, $T$ is continuous.

Thus we see that every closed bijective linear transformation between Fréchet spaces is continuous. Actually much more than this is true: every closed linear transformation between Fréchet spaces is continuous. This is the content of the Closed Graph Theorem, one of the most important consequences of the Open Mapping Theorem.

Theorem 7.3.1 (Closed Graph Theorem). Let $(V_1,\{p_k\})$ and $(V_2,\{q_m\})$ be Fréchet spaces over $\Phi$. If $T \in L'(V_1,V_2)$ is a closed mapping, then $T \in L(V_1,V_2)$.

Proof. Since $T$ is a closed mapping, $G(T)$, the graph of $T$, is a closed linear subspace in the topological product space $V_1 \times V_2$. It is not difficult to verify that this product space is obtained from the seminormed linear space $(V_1 \times V_2,\{r_{km}\})$, where for each pair $(x,y) \in V_1 \times V_2$ we have $r_{km}[(x,y)] = p_k(x) + q_m(y)$, $k,m = 1,2,3,\ldots$ . Moreover, $(V_1 \times V_2,\{r_{km}\})$ is complete -- that is, it is a Fréchet space, and so $G(T)$ is a Fréchet space. The details are left to the reader.

Define the mapping $S : G(T) \to V_1$ by $S[(x,T(x))] = x$, $x \in V_1$. Clearly $S$ is linear and surjective. Suppose that $\{(x_n,T(x_n))\} \subset G(T)$ is a sequence that converges to $(x,T(x))$ in $G(T)$. Then for each $k,m = 1,2,3,\ldots,$ we would have

$$\lim_n \{p_k(x_n - x) + q_m[T(x_n) - T(x)]\} = 0,$$

from which it is apparent that

$$\lim_n p_k\{S[(x_n,T(x_n))] - S[(x,T(x))]\} = \lim_n p_k(x_n - x) = 0$$

for each $k = 1,2,3,\ldots$ . Hence from Theorem 1.5.1 we conclude that $S \in L(G(T),V_1)$. An appeal to the Open Mapping Theorem (Theorem 7.2.1) then reveals that $S(U) \subset V_1$ is open whenever $U \subset G(T)$ is open. Note that in the preceding argument it suffices to consider sequences, rather than nets, since the spaces $G(T)$ and $V_1$ are Fréchet spaces and hence complete metric spaces.

Now suppose $W$ is any open neighborhood of the origin in $V_2$ and consider the open neighborhood $U$ of the origin in $G(T)$ given by $U = \{(x,T(x)) \mid x \in V_1, T(x) \in W\}$. Then $S(U)$ is an open neighborhood of the origin in $V_1$. Moreover, if $z \in S(U)$, then there exists some $(x,T(x)) \in U$ such that $S[(x,T(x))] = z$, and so $T(z) = T(x) \in W$.

Therefore $T[S(U)] \subset W$, and $T$ is continuous. □

The example of a closed discontinuous mapping in Section 7.1 demonstrates that the assumption of linearity is necessary for the validity of the Closed Graph Theorem.

Our most frequent use of the Closed Graph Theorem will be in the context of Banach spaces. Because of this, and because the proof is also instructive, we wish to restate and reprove the theorem in this setting.

Theorem 7.3.2 (Closed Graph Theorem). Let $(V_1,\|\cdot\|_1)$ and $(V_2,\|\cdot\|_2)$ be Banach spaces over $\Phi$. If $T \in L'(V_1,V_2)$ is a closed mapping, then $T \in L(V_1,V_2)$.

Proof. As before, define $S : G(T) \to V_1$ by $S[(x,T(x))] = x$, $x \in V_1$, where $G(T)$, the graph of $T$, is considered to be a closed linear subspace of the Banach space $V_1 \times V_2$ with the norm $\|(x,T(x))\| = \|x\|_1 + \|T(x)\|_2$, $x \in V_1$. Clearly $S$ is linear and bijective, and the estimate

$$\|S[(x,T(x))]\|_1 = \|x\|_1 \le \|x\|_1 + \|T(x)\|_2 = \|(x,T(x))\| \qquad (x \in V_1)$$

shows that $S \in L(G(T),V_1)$. Consequently, by Corollary 7.2.2, $S^{-1} \in L(V_1,G(T))$.

Thus, in particular, we see that

$$\|x\|_1 + \|T(x)\|_2 = \|(x,T(x))\|$$

$$= \|S^{-1}(x)\|$$
$$\le \|S^{-1}\| \|x\|_1 \qquad (x \in V_1),$$

from which we deduce that

$$\|T(x)\|_2 \le (\|S^{-1}\| - 1)\|x\|_1 \qquad (x \in V_1).$$

Therefore $T$ is continuous. □

For similar reasons we also want to look again at Corollary 7.2.3 in the context of Banach spaces.

Theorem 7.3.3 (Two Norm Theorem). Let $V$ be a linear space over $\Phi$ and suppose $\|\cdot\|_1$ and $\|\cdot\|_2$ are norms on $V$ such that $(V,\|\cdot\|_1)$ and $(V,\|\cdot\|_2)$ are Banach spaces over $\Phi$. If there exists some $M > 0$ such that $\|x\|_1 \le M\|x\|_2$, $x \in V$, then there exists some $m > 0$ such that $\|x\|_2 \le m\|x\|_1$, $x \in V$.

Proof. Consider the identity mapping $I(x) = x$, $x \in V$. Clearly $I$ is a linear mapping from $(V,\|\cdot\|_1)$ onto $(V,\|\cdot\|_2)$. We need to show that $I$ is continuous. We shall accomplish this by using the Closed Graph Theorem.

So suppose that $\{x_n\} \subset V$, $x,y \in V$, are such that $\lim_n \|x_n - x\|_1 = 0$ and $\lim_n \|I(x_n) - y\|_2 = 0$. Then for each $n$ we have

$$\|I(x) - y\|_1 = \|x - y\|_1$$
$$\le \|x - x_n\|_1 + \|x_n - y\|_1$$
$$\le \|x - x_n\|_1 + M\|I(x_n) - y\|_2,$$

from which it follows that $I(x) = y$.

Thus we see that $I$ is a closed mapping and from the Closed Graph Theorem (Theorem 7.3.2) conclude that $I$ is continuous. □

In particular, if $\|x\|_1 \leq M\|x\|_2$, $x \in V$, then $\|\cdot\|_1$ and $\|\cdot\|_2$ are equivalent norms on $V$.

As in our treatment of the Uniform Boundedness Theorem we have not discussed the most general forms of the Open Mapping and Closed Graph Theorems. For other developments of these two fundamental results the reader is referred to [$DS_1$, pp. 55-58; $E_1$, pp. 419-458; K, pp. 166-168; KeNa, pp. 97-100]. Also it should be remarked that, although we employed the Open Mapping Theorem to obtain the Closed Graph Theorem, it is possible to do the opposite: one can first prove the Closed Graph Theorem and then use it to deduce the Open Mapping Theorem. This is done, for example, in [KeNa, pp. 97 -100].

<u>7.4. A Uniform Boundedness Theorem for Continuous Linear Functionals.</u> As a first application of the Closed Graph Theorem we prove a version of the Uniform Boundedness Theorem (Theorem 6.2.1) for continuous linear functionals on Fréchet spaces.

<u>Theorem 7.4.1.</u> Let $(V,\{p_k\})$ be a Fréchet space over $\Phi$ and suppose $\{x^*_\alpha\}_{\alpha\in\Lambda} \subset V^*$. If for each $x \in V$ there exists some $M_x > 0$ such that $\sup_{\alpha\in\Lambda}|x^*_\alpha(x)| \leq M_x$, then for each $\epsilon > 0$ there exists some open neighborhood $U$ of the origin in $V$ such that $\sup_{\alpha\in\Lambda}|x^*_\alpha(x)| < \epsilon$, $x \in U$.

<u>Proof.</u> Consider the space $B(\Lambda)$ of all bounded $\Phi$-valued functions on $\Lambda$ with the norm $\|f\|_\infty = \sup_{\alpha\in\Lambda}|f(\alpha)|$, $f \in B(\Lambda)$. As indicated in Example 1.2.2, $(B(\Lambda),\|\cdot\|_\infty)$ is a Banach space over $\Phi$. For each $x \in V$ define $T(x)$ on $\Lambda$ by $T(x)(\alpha) = x^*_\alpha(x)$, $\alpha \in \Lambda$. Then $T(x) \in B(\Lambda)$ since

$$\sup_{\alpha\in\Lambda}|T(x)(\alpha)| = \sup_{\alpha\in\Lambda}|x^*_\alpha(x)| \leq M_x.$$

Clearly, the correspondence $x \to T(x)$ defines an element $T \in L'(V,B(\Lambda))$. Moreover, we claim that $T$ is continuous. To prove this we appeal to the Closed Graph Theorem (Theorem 7.3.1).

Indeed, suppose $\{x_n\} \subset V$, $x \in V$, and $f \in B(\Lambda)$ are such that $\{x_n\}$ converges to $x$ in $V$ and $\{T(x_n)\}$ converges to $f$ in $B(\Lambda)$. The latter assumption asserts that

$$\begin{aligned}\lim_n \|T(x_n) - f\|_\infty &= \lim_n \sup_{\alpha \in \Lambda} |T(x_n)(\alpha) - f(\alpha)| \\ &= \lim_n \sup_{\alpha \in \Lambda} |x^*_\alpha(x_n) - f(\alpha)| \\ &= 0.\end{aligned}$$

Thus, in particular, for each $\alpha \in \Lambda$, $\lim_n x^*_\alpha(x_n) = f(\alpha)$. But $x^*_\alpha \in V^*$, and so $\lim_n x^*_\alpha(x_n) = x^*_\alpha(x)$ for each $\alpha \in \Lambda$. Consequently $T(x)(\alpha) = f(\alpha)$, $\alpha \in \Lambda$, that is, $T(x) = f$, and so $T$ is a closed mapping.

Hence the Closed Graph Theorem (Theorem 7.3.1) leads us to conclude that $T \in L(V,B(\Lambda))$. Therefore, given $\epsilon > 0$, there exists some $\delta > 0$ and some seminorm $p_n$ such that, if $p_n(x) < \delta$, then $\|T(x)\|_\infty < \epsilon$. We need only one seminorm because of our standing assumption that $p_k(x) \le p_{k+1}(x)$, $x \in V$, $k = 1,2,3,\ldots$ . Setting $U = \{x \mid p_n(x) < \delta\}$, we see that

$$\begin{aligned}\sup_{\alpha \in \Lambda} |x^*_\alpha(x)| &= \sup_{\alpha \in \Lambda} |T(x)(\alpha)| \\ &= \|T(x)\|_\infty \\ &< \epsilon \qquad (x \in U),\end{aligned}$$

which completes the proof. □

7.5. <u>Some Results on Norms in $C([0,1])$</u>. We know from Section 1.2 that $C([0,1])$ is a Banach space under the usual supremum norm

$$\|f\|_\infty = \sup_{t \in [0,1]} |f(t)| \qquad (f \in C([0,1])).$$

In this section we wish to give some results concerning other possible norms on $C([0,1])$.

Theorem 7.5.1. Let $\|\cdot\|$ be a norm on $C([0,1])$ such that $(C([0,1]),\|\cdot\|)$ is a Banach space over $\mathbb{C}$ and suppose that, whenever $\{f_n\} \subset C([0,1])$ and $f \in C([0,1])$ are such that $\lim_n\|f_n - f\| = 0$, it is the case that $\lim_n f_n(t) = f(t)$, $0 \leq t \leq 1$. Then $\|\cdot\|$ and $\|\cdot\|_\infty$ are equivalent norms on $C([0,1])$.

Proof. Consider the identity mapping $I$ from $(C([0,1]),\|\cdot\|_\infty)$ to $(C([0,1]),\|\cdot\|)$. Clearly $I$ is linear and bijective. Moreover, if $\{f_n\} \subset C([0,1])$ and $f,g \in C([0,1])$ are such that $\lim_n\|f_n - f\|_\infty = 0$ and $\lim_n\|I(f_n) - g\| = 0$, then for each $t$, $0 \leq t \leq 1$, we have

$$|f(t) - g(t)| \leq |f(t) - f_n(t)| + |f_n(t) - g(t)|$$
$$\leq \|f - f_n\|_\infty + |f_n(t) - g(t)|.$$

Hence $I(f) = f = g$, as $\lim_n\|I(f_n) - g\| = 0$ implies that $\lim_n I(f_n)(t) = \lim_n f_n(t) = g(t)$, $0 \leq t \leq 1$. Thus $I$ is a closed mapping and hence continuous by the Closed Graph Theorem (Theorem 7.3.2).

Consequently there exists some $M > 0$ such that $\|f\| \leq M\|f\|_\infty$, $f \in C([0,1])$, from which it follows at once, by the Two Norm Theorem (Theorem 7.3.3), that $\|\cdot\|$ and $\|\cdot\|_\infty$ are equivalent. $\square$

The last step of the argument could have appealed to the Open Mapping Theorem, rather than the Two Norm Theorem. To be precise, since $I$ is continuous, linear, and bijective, we see from Corollary 7.2.2 that $I^{-1}$ is continuous, and hence $I$ is a homeomorphism; that is, $\|\cdot\|$ and $\|\cdot\|_\infty$ are equivalent.

There are, of course, many norms that one can introduce into $C([0,1])$. Theorem 7.5.1 says that any such complete norm whose notion of convergence is at least as strong as pointwise convergence must be equivalent to the supremum norm.

An example of a norm on $C([0,1])$ under which $C([0,1])$ is not a Banach space is given by the $L_1$-norm:

$$\|f\|_1 = \int_0^1 |f(t)|\ dt \qquad (f \in C([0,1])).$$

It is easy to verify that $\|\cdot\|_1$ is a norm on $C([0,1])$ and that $\|f\|_1 \leq \|f\|_\infty$, $f \in C([0,1])$. If $(C([0,1]),\|\cdot\|_1)$ were a Banach space, then by the Two Norm Theorem (Theorem 7.3.3) there would exist some $m > 0$ such that $\|f\|_\infty \leq m\|f\|_1$, $f \in C([0,1])$. However, consider for each positive integer $n$ the function $f_n \in C([0,1])$, defined by

$$f_n(t) = \left(\frac{-n^2}{2}\right)t + n \qquad \left(0 \leq t \leq \frac{2}{n}\right),$$

$$f_n(t) = 0 \qquad \left(\frac{2}{n} < t \leq 1\right).$$

Then simple computations reveal that $\|f_n\|_\infty = n$ and $\|f_n\|_1 = 1$, $n = 1,2,3,\ldots,$ in contradiction to the previous estimates.

Consequently $(C([0,1]),\|\cdot\|_1)$ is not a Banach space.

<u>7.6. A Criterion for the Continuity of Linear Transformations on $\ell_2$.</u> As should be evident by now, one of the chief uses of the Closed Graph Theorem is to establish the continuity of a given linear transformation. In the preceding examples the proof of continuity was auxiliary to some other purpose, but now we wish to discuss a situation in which it is the central concern. Namely, we seek conditions on $T \in L'(\ell_2)$ that will ensure the continuity of $T$. We shall see that this is the case whenever $T$ commutes with a certain shift transformation on $\ell_2$. To be precise, we make the following definition:

<u>Definition 7.6.1.</u> Let $S \in L'(\ell_2)$ denote the linear transformation defined by $S(\{a_k\}) = \{b_k\}$, where $b_1 = 0$ and $b_k = a_{k-1}$, $k = 2,3,4,\ldots,$ for each $\{a_k\} \in \ell_2$.

Thus $S$ shifts all the components of a sequence $\{a_k\} \in \ell_2$ one position to the right and puts a zero in the first position.

We have the following criterion for the continuity of $T \in L'(\ell_2)$:

<u>Theorem 7.6.1</u>. If $T \in L'(\ell_2)$ and $TS = ST$, then $T \in L(\ell_2)$.

<u>Proof</u>. To begin with we wish to establish some notational conventions that will simplify the proof. For each $m = 1,2,3,\ldots,$ we denote by $e^m = \{e_k^m\} \in \ell_2$ the sequence such that $e_k^m = 0$, $k \neq m$, $e_m^m = 1$. If $x = \{a_k\} \in \ell_2$, then we set $(x)_m = (\{a_k\})_m = a_m$, $m = 1,2,3,\ldots$; that is, $(x)_m$ is the mth component of the sequence x. It is immediately evident that $|(x)_m| \le \|x\|_2$, $x \in \ell_2$, for each $m = 1,2,3,\ldots$ .

Consequently, for each $n,m = 1,2,3,\ldots,$ we claim that the mapping $S_{n,m}$ defined by $S_{n,m}(x) = (a_n T[e^n])_m$, $x = \{a_k\} \in \ell_2$, belongs to $\ell_2^*$. Clearly each $S_{n,m}$ is linear and

$$\begin{aligned} |S_{n,m}(x)| &= |(a_n T[e^n])_m| \\ &= |a_n (T[e^n])_m| \\ &= |(T[e^n])_m| \, |a_n| \\ &\le |(T[e^n])_m| \, \|x\|_2 \qquad (x \in \ell_2) \end{aligned}$$

shows that $S_{n,m}$ is continuous from $\ell_2$ to $\mathbb{C}$.

Now we claim that, if $T \in L'(\ell_2)$ and $TS = ST$, then $T$ is a closed mapping. To see this, since $T$ is linear, it suffices to show that, if $\{x^j\}$ is a sequence in $\ell_2$ and $y \in \ell_2$ are such that $\lim_j \|x^j\|_2 = 0$ and $\lim_j \|T(x^j) - y\|_2 = 0$, then $y = 0$; that is, $(y)_m = 0$, $m = 1,2,3,\ldots$ .

But it is easily seen that for each $m = 1,2,3,\ldots,$

$$x^j = \{a_k^j\} = \sum_{k=1}^{m} a_k^j e^k + S^m(z_m^j) \qquad (j = 1,2,3,\ldots),$$

where $(z_m^j)_k = a_{m+k}^j$, $k = 1,2,3,\ldots$ . Since $TS = ST$, it follows at once that $TS^m = S^m T$, $m = 1,2,3,\ldots,$ and so

$$(T[x^j])_m = \sum_{k=1}^{m} (a_k^j T[e^k])_m + (S^m T[z_m^j])_m$$

$$= \sum_{k=1}^{m} S_{k,m}(x^j) \qquad (m,j = 1,2,3,\ldots).$$

The equations are valid since $(S^m T[z_m^j])_m = 0$ because $S^m$ introduces zeros in the first $m$ positions when applied to any sequence.

Thus, since each $S_{k,m}$, $k = 1,2,\ldots,m$; $m = 1,2,3,\ldots,$ is a continuous linear functional on $\ell_2$, we conclude that for each $m = 1,2,3,\ldots,$

$$\lim_j (T[x^j])_m = \lim_j \sum_{k=1}^{m} S_{k,m}(x^j)$$

$$= \sum_{k=1}^{m} S_{k,m}(\{0\})$$

$$= 0.$$

However, $\lim_j \|T(x^j) - y\|_2 = 0$ implies that $\lim_j |(T[x^j] - y)_m| = 0$, $m = 1,2,3,\ldots,$ and hence $y = 0$.

Therefore $T$ is a closed mapping, and so, by the Closed Graph Theorem (Theorem 7.3.1), $T$ is continuous. □

7.7. Separable Banach Spaces. In the preceding three sections we have given some applications of the Closed Graph Theorem. These are, of course, indirectly also applications of the Open Mapping Theorem, since the former was derived from the latter. In this and the next section we shall give two instances of direct applications of the Open Mapping Theorem.

Theorem 7.7.1. If $(V,\|\cdot\|)$ is a separable Banach space over $\mathbb{C}$, then there exists a closed linear subspace $W$ of $(\ell_1,\|\cdot\|_1)$ such that $V$ is topologically isomorphic to $\ell_1/W$, where the quotient space is taken with the usual quotient norm

$$|||x + W||| = \inf_{y \in W} \|x + y\|_1 \qquad (x \in \ell_1).$$

Proof. Since $V$ is separable, there exists a sequence $\{z_k\} \subset B_1 = \{z \mid z \in V, \|z\| \leq 1\}$ that is dense in $B_1$. For each $x = \{a_k\} \in \ell_1$ we define $T_1(x) = \Sigma_{k=1}^{\infty} a_k z_k$. The series converges in $(V, \|\cdot\|)$ since $\Sigma_{k=1}^{\infty} \|a_k z_k\| \leq \Sigma_{k=1}^{\infty} |a_k| = \|x\|_1$. Clearly then $T_1 \in L(\ell_1, V)$. Let $W = \{x \mid x \in \ell_1, T_1(x) = 0\}$. Then it is readily verified that $W$ is a closed linear subspace of $\ell_1$, and so $\ell_1/W$ is a Banach space with the quotient norm (Theorem 1.1.1(iii)).

Now define $T : \ell_1/W \to V$ by setting $T(x + W) = T_1(x)$, $x \in \ell_1$. $T$ is evidently a well-defined injective continuous linear transformation from $\ell_1/W$ to $V$, as $W = \{x \mid x \in \ell_1, T_1(x) = 0\}$. Moreover, we claim that $T$ is surjective. To see this it suffices to show that $T_1$ is surjective.

Suppose $z \in V$ and $\|z\| \leq 1$. Then there exists some positive integer $k_1$ such that $\|z - z_{k_1}\| < 1/2$. Furthermore, $\|2(z - z_{k_1})\| < 1$, so there exists some positive integer $k_2 > k_1$ such that $\|2(z - z_{k_1}) - z_{k_2}\| < 1/2$; that is, $\|z - z_{k_1} - z_{k_2}/2\| < 1/2^2$. Continuing in this fashion, we obtain a sequence $\{z_{k_n}\} \subset B_1$ for which $\|z - \Sigma_{n=1}^{m} z_{k_n}/2^{n-1}\| < 1/2^m$, $m = 1,2,3,\ldots$ . Since $V$ is a Banach space, we conclude that $z = \Sigma_{n=1}^{\infty} z_{k_n}/2^{n-1}$, and hence $z = T_1(x_z)$, where $x_z = \{a_m\}$ is such that $a_m = 1/2^{n-1}$ if $m = k_n$, and $a_m = 0$, $m \neq k_n$, $n = 1,2,3,\ldots$ . If $z \neq 0$ is any element of $V$, then $w = z/\|z\| \in B_1$, so there exists some $x_w$ for which $T_1(x_w) = w$, that is, $T_1(\|z\| x_w) = z$. Hence $T_1$ is surjective.

Consequently $T \in L(\ell_1/W, V)$ is bijective, and from the Open Mapping Theorem (Theorem 7.2.1) or Corollary 7.2.2 it follows that $T^{-1}$ is continuous. Therefore $\ell_1/W$ and $V$ are topologically isomorphic. □

It should be remembered, and is not difficult to prove, that, if $V$ is a separable Banach space and $W \subset V$ is a closed linear subspace, then $V/W$ is separable. In particular, $\ell_1/W$ is a separable Banach space.

7.8. <u>The Category of $L_1([-\pi,\pi],dt/2\pi)^{\wedge}$ in $C_0(\mathbb{Z})$</u>. We recall from Example 3.1.7 that, if $f \in L_1([-\pi,\pi],dt/2\pi)$ and

$$\hat{f}(k) = \frac{1}{2\pi}\int_{-\pi}^{\pi} f(t)e^{-ikt}\,dt \qquad (k \in \mathbb{Z}),$$

then $\hat{f} \in C_0(\mathbb{Z})$, where, as before, $\mathbb{Z}$ denotes the locally compact space of the integers with the discrete topology. It can also be shown that the linear transformation $f \to \hat{f}$ from $L_1([-\pi,\pi],dt/2\pi)$ to $C_0(\mathbb{Z})$ is injective, as will be done in Corollary 13.7.1. Moreover, it is evident that $\|\hat{f}\|_\infty \le \|f\|_1$, $f \in L_1([-\pi,\pi],dt/2\pi)$.

Combining these observations we see that the Fourier transformation $f \to \hat{f}$ is an injective continuous linear transformation from $(L_1([-\pi,\pi],dt/2\pi),\|\cdot\|_1)$ to $(C_0(\mathbb{Z}),\|\cdot\|_\infty)$. However, the Fourier transformation is not surjective; indeed, the range of the Fourier transformation is a set of category I. We shall now prove this.

We set $L_1([-\pi,\pi],dt/2\pi)^{\wedge} = \{\hat{f} \mid f \in L_1([-\pi,\pi],dt/2\pi)\}$.

<u>Theorem 7.8.1</u>. $L_1([-\pi,\pi],dt/2\pi)^{\wedge}$ is a set of category I in $(C_0(\mathbb{Z}),\|\cdot\|_\infty)$.

<u>Proof</u>. Define $T(f) = \hat{f}$, $f \in L_1([-\pi,\pi],dt/2\pi)$. Then $T$ is an injective continuous linear mapping from $(L_1([-\pi,\pi],dt/2\pi),\|\cdot\|_1)$ to $(C_0(\mathbb{Z}),\|\cdot\|_\infty)$. Suppose the range of $T$, that is, $L_1([-\pi,\pi],dt/2\pi)^{\wedge}$ is of category II in $C_0(\mathbb{Z})$. Then from Corollary 7.2.1 we see that $T$ is surjective, and so from Corollary 7.2.2 it follows that $T^{-1} \in L(C_0(\mathbb{Z}),L_1([-\pi,\pi],dt/2\pi))$. Consequently $\|f\|_1 \le \|T^{-1}\|\|\hat{f}\|_\infty$ for each $f \in L_1([-\pi,\pi],dt/2\pi)$.

However, consider the functions $D_n(t) = [\sin(n + 1/2)t]/[\sin t/2]$, $-\pi \le t \le \pi$, $n = 0,1,2,\ldots$, that is, the Dirichlet kernels introduced in Section 6.6. We proved there that $\sup_n\|D_n\|_1 = \infty$.

On the other hand, since we know that for each $n = 0,1,2,\ldots$, $D_n(t) = \Sigma_{k=-n}^{n} e^{ikt}$, $-\pi \le t \le \pi$, some straightforward computations using the well-known facts that

$$\frac{1}{2\pi}\int_{-\pi}^{\pi} e^{ikt}\,dt = 1 \quad \text{for} \quad k = 0,$$

$$\frac{1}{2\pi}\int_{-\pi}^{\pi} e^{ikt}\,dt = 0 \quad \text{for} \quad k \neq 0 \qquad (k \in \mathbb{Z}),$$

reveal that for each $n = 0,1,2,\ldots,$

$$\hat{D}_n(k) = 1 \quad \text{for} \quad -n \leq k \leq n,$$

$$\hat{D}_n(k) = 0 \quad \text{for} \quad |k| > n \qquad (k \in \mathbb{Z}).$$

Thus $\|\hat{D}_n\|_\infty = 1$, $n = 0,1,2,\ldots,$ and so the inequality

$$\|D_n\|_1 \leq \|T^{-1}\|\|\hat{D}_n\|_\infty$$

cannot hold for all $n$.

But this contradicts the previous estimate, and therefore $L_1([-\pi,\pi],dt/2\pi)^\wedge$ must be of category I in $C_o(\mathbb{Z})$. □

It should, nevertheless, be noted that $L_1([-\pi,\pi],dt/2\pi)^\wedge$ is a norm dense subalgebra in $(C_o(\mathbb{Z}),\|\cdot\|_\infty)$. (See, for example, [$\text{Ru}_1$, p.9].)

## 7.9. Problems.

1. (Corollary 7.1.1) Let $(X,\mathcal{T})$ and $(Y,\mathcal{U})$ be topological spaces such that $\mathcal{U}$ is a Hausdorff topology. Prove that, if $f : X \to Y$ is continuous, then $f$ is closed.

2. Let $(V,\|\cdot\|)$ be a normed linear space over $\Phi$. If $S,T \in \mathcal{L}'(V)$ are both closed, does it necessarily follow that $S + T$ is closed?

3. Let $(V,\|\cdot\|)$ be a normed linear space over $\Phi$. If $T \in \mathcal{L}'(V)$ is a closed linear transformation and $E \subset V$ is a closed subset of $V$, is $T(E)$ necessarily a closed subset of $V$?

4. Let $(X_1,\mathcal{T}_1)$, $(X_2,\mathcal{T}_2)$, and $(X_3,\mathcal{T}_3)$ be topological spaces. Suppose $f : X_1 \to X_2$ and $g : X_2 \to X_3$.

(a) If $f$ and $g$ are closed, prove that $g \circ f : X_1 \to X_3$ is not necessarily closed.

(b) If $f$ is continuous and $g$ is closed, prove that $g \circ f : X_1 \to X_3$ is closed.

5. Let $(V,\|\cdot\|)$ be a normed linear space over $\Phi$. If $T \in L'(V)$ and $\lambda \in \Phi$, prove that $T - \lambda I$ is closed whenever $T$ is closed.

6. Let $(V_1,\mathcal{T}_1)$ and $(V_2,\mathcal{T}_2)$ be topological linear spaces over $\Phi$ and let $\mathcal{U}_1$ and $\mathcal{U}_2$ be neighborhood bases at the origins of $V_1$ and $V_2$, respectively. Let $T \in L'(V_1,V_2)$. Prove that the graph of $T$ in $V_1 \times V_2$ is closed if and only if

$$\cap\{T(U_1) + U_2 \mid U_1 \in \mathcal{U}_1,\ U_2 \in \mathcal{U}_2\} = \{0\}.$$

7. (Corollary 7.2.1) Let $(V_1,\{p_k\})$ and $(V_2,\{q_m\})$ be Fréchet spaces over $\Phi$ and suppose $T \in L'(V_1,V_2)$ is a closed mapping. Prove that the following are equivalent:

(a) $R(T)$ is of category II in $V_2$.

(b) $T$ is surjective.

8. (Corollary 7.2.2) Let $(V_1,\{p_k\})$ and $(V_2,\{q_m\})$ be Fréchet spaces over $\Phi$. Prove that if $T \in L'(V_1,V_2)$ is a closed bijective mapping, then $T^{-1} \in L(V_1,V_2)$.

9. Let $(V_1,\|\cdot\|_1)$ and $(V_2,\|\cdot\|_2)$ be normed linear spaces over $\Phi$. Suppose that for some injective $T \in L(V_1,V_2)$, we have $T^{-1} \in L(V_1,V_2)$. Prove that $V_1$ is a Banach space if and only if $V_2$ is a Banach space.

10. Let $(V_1,\|\cdot\|_1)$ be a normed linear space and $(V_2,\|\cdot\|_2)$ be a Banach space over $\Phi$. If $T$ is a linear transformation mapping a linear subspace $D_T$ of $V_1$ into $V_2$, that is, $T \in L'(D_T,V_2)$,

then $T$ is said to be closed if whenever $\{x_n\}$ is a sequence in $D_T$, $x \in V_1$, and $y \in V_2$ are such that $\lim_n \|x_n - x\|_1 = 0$ and $\lim_n \|T(x_n) - y\|_2 = 0$, it is the case that $x \in D_T$ and $T(x) = y$. Prove that, if $T \in L'(D_T, V_2)$ is closed and bounded, that is, there exists some $M > 0$ for which $\|T(x)\|_2 \leq M\|x\|_1$, $x \in D_T$, then $D_T$ is a closed linear subspace of $V_1$ and $T \in L(D_T, V_2)$.

11. Let $(V_1, \|\cdot\|_1)$ and $(V_2, \|\cdot\|_2)$ be Banach spaces over $\Phi$ and let $T \in L(V_1, V_2)$. Prove that $R(T)$ is isomorphic to $V_1/N(T)$ if and only if $R(T)$ is closed.

12. Let $(V_1, \|\cdot\|_1)$ and $(V_2, \|\cdot\|_2)$ be Banach spaces over $\Phi$ and let $T \in L'(V_1, V_2)$ be injective and closed. Prove that $R(T)$ is closed in $V_2$ if and only if there exists some $c > 0$ such that $\|x\|_1 \leq c\|T(x)\|_2$, $x \in V_1$.

13. Let $(X_1, T_1)$ and $(X_2, T_2)$ be topological spaces and let $f : X_1 \to X_2$ be a mapping. Suppose $T_2'$ is another topology for $X_2$ such that $T_2' \subset T_2$ (that is, $T_2'$ is weaker than $T_2$) and such that $(X_2, T_2')$ is Hausdorff. If $f : (X_1, T_1) \to (X_2, T_2')$ is continuous, prove that $f : (X_1, T_1) \to (X_2, T_2)$ is closed.

*14. Let $(V, \|\cdot\|)$ be a Banach space over $\Phi$ and let $W \subset V$ be a closed linear subspace. Let $\varphi : V \to V/W$ be the natural mapping $\varphi(x) = x + W$, $x \in V$. Prove that $\varphi$ maps the open unit ball of $V$ onto the open unit ball of $V/W$.

15. Let $(V, \|\cdot\|)$ be a separable Banach space over $\Phi$ and let $W \subset V$ be a closed linear subspace of $V$. Prove that $V/W$ is separable.

16. Let $(V_1, \|\cdot\|_1)$ and $(V_2, \|\cdot\|_2)$ be Banach spaces over $\Phi$ and let $f : V_1 \to V_2$ be such that $y^* \circ f \in V_1^*$, $y^* \in V_2^*$. Prove that $f \in L(V_1, V_2)$.

17. For $f \in L_1([-\pi,\pi],dt/2\pi)$ define $x_k^*(f) = \hat{f}(k)$, $k \in \mathbb{Z}$. Suppose $|||\cdot|||$ is a norm on $L_1([-\pi,\pi],dt/2\pi)$ such that

(i) $(L_1([-\pi,\pi],dt/2\pi), |||\cdot|||)$ is a Banach space.

(ii) $x_k^*$ is a continuous linear functional on the Banach space $(L_1([-\pi,\pi],dt/2\pi), |||\cdot|||)$ for each $k \in \mathbb{Z}$.

Prove that $\|\cdot\|_1$ and $|||\cdot|||$ are equivalent norms on the space $L_1([-\pi,\pi],dt/2\pi)$, where as usual

$$\|f\|_1 = \frac{1}{2\pi}\int_{-\pi}^{\pi} |f(t)|\, dt.$$

*18. Let $E \subset \mathbb{Z}$ and define

$$C_E = \{f \mid f \in C([-\pi,\pi]),\ \hat{f}(k) = 0,\ k \notin E\}.$$

(a) Prove that $C_E$ is a closed linear subspace of $(C([-\pi,\pi]), \|\cdot\|_\infty)$.

(b) $E \subset \mathbb{Z}$ is said to be a <u>Sidon set</u> if there exists some constant $M > 0$ such that $\Sigma_{k=-\infty}^{\infty}|\hat{f}(k)| \leq M\|f\|_\infty$, $f \in C_E$. Prove that $E \subset \mathbb{Z}$ is a Sidon set if and only if $\Sigma_{k=-\infty}^{\infty}|\hat{f}(k)| < \infty$, $f \in C_E$.

*19. Let $(V, \|\cdot\|)$ be a normed linear space over $\Phi$ and let $B_1 = \{x \mid x \in V,\ \|x\| \leq 1\}$. Suppose $A \subset V$ has the following property: there exists a number $\lambda > 1$ such that to each $x \in B_1$ there corresponds an $a \in A$ such that $\|x - a\| < 1/\lambda$. Prove that to each $x \in B_1$ there corresponds a sequence $\{a_n\}_{n=0}^{\infty}$ of points of $A$ such that $x = \Sigma_{n=0}^{\infty} a_n/\lambda^n$.

20. Use Problem 19 to prove that a closed linear map from a Banach space $V$ to a Banach space $W$ is continuous.

21. Let $\mathcal{T}_1$ and $\mathcal{T}_2$ be topologies for a linear space $V$ such that $(V,\mathcal{T}_1)$ and $(V,\mathcal{T}_2)$ are both Fréchet spaces over $\Phi$. Prove that the following are equivalent:

(a) $\mathcal{T}_1 \nsubseteq \mathcal{T}_2$.

(b) The topology $\mathcal{T}$ with subbase $\mathcal{T}_1 \cup \mathcal{T}_2$ is not complete.

(c) The topology $\mathcal{T}_1 \cap \mathcal{T}_2$ is not Hausdorff.

(d) There is an $x \in V$, $x \neq 0$, such that $x \in U_1 + U_2$ for each $\mathcal{T}_1$-neighborhood $U_1$ of the origin and each $\mathcal{T}_2$-neighborhood $U_2$ of the origin.

(e) There is a sequence $\{x_n\}$ in $V$ such that $\{x_n\}$ converges to zero relative to $\mathcal{T}_1$ and to $y \neq 0$ relative to $\mathcal{T}_2$.

22. Let $(V_n, \mathcal{T}_n)$, $n = 1,2,3,\ldots$, be Fréchet spaces over $\Phi$. Let $W$ be a topological linear space and for each $n$ let $T_n \in L(V_n, W)$. If $W = \cup_{n=1}^{\infty} T_n(V_n)$, prove that every continuous linear transformation $T$ from $W$ onto any Frechet space $F$ is open, that is, $T(U)$ is open in $F$ whenever $U$ is open in $W$.

23. Let $(V_1, \|\cdot\|_1)$, $(V_2, \|\cdot\|_2)$, and $(V_3, \|\cdot\|_3)$ be Banach spaces over $\Phi$. Let $T \in L(V_1, V_3)$ and $S \in L(V_2, V_3)$. Suppose that for every $x \in V_1$ the equation $T(x) = S(y)$ has a unique solution $y$. Prove that the mapping $A(x) = y$ from $V_1$ to $V_2$ so determined is linear and bounded.

24. Let $(V_1, \|\cdot\|_1)$, $(V_2, \|\cdot\|_2)$, and $(V_3, \|\cdot\|_3)$ be Banach spaces over $\Phi$. Suppose $T \in L(V_1, V_3)$ is a composite $T(x) = AB(x)$ for all $x \in V_1$, where $B \in L'(V_1, V_2)$ and $A \in L(V_2, V_3)$. If $A$ is injective, prove that $B$ is continuous.

25. Let $X$ be a locally compact Hausdorff topological space and let $\mu \in M(X)$. If $f : X \to \mathbb{C}$ is a function such that $fg$ is integrable with respect to $\mu$ for every $g \in L_p(X,\mu)$, prove that $f \in L_q(X,\mu)$, where $1/p + 1/q = 1$, $1 < p < \infty$.

26. Suppose $1 \leq p,q \leq \infty$, and let $\{a_{kj}\}_{k,j=1}^{\infty}$ be an infinite matrix of complex numbers such that, if $x = \{x_j\} \in \ell_p$, then the series $y_k = \Sigma_{j=1}^{\infty} a_{kj} x_j$ converges for each $k = 1,2,3,\ldots$, and the sequence $y = \{y_k\} \in \ell_q$. Prove that the transformation $T : \ell_p \to \ell_q$, defined by $T(x) = y$, $x \in \ell_p$, belongs to $L(\ell_p, \ell_q)$.

27. Prove that there exists no sequence $\{c_n\}$ of complex numbers such that an infinite series $\Sigma_{n=1}^{\infty} a_n$ of complex numbers converges absolutely, that is, $\Sigma_{n=1}^{\infty} |a_n| < \infty$, if and only if $\{c_n a_n\}$ is a bounded sequence.

28. Let $E$ denote the subset of $L_2([0,1],dt)$ consisting of all functions of the form

$$f(s) = a + \int_0^s g(t)\,dt \qquad (s \in [0,1]),$$

where $g \in L_2([0,1],dt)$, $a \in \Phi$, and $\int_0^1 g(t)\,dt = 0$. Define $T : E \to L_2([0,1],dt)$ by $T(f) = f'$, where $f'$ is the first derivative of $f$. Prove that $T$ is closed.

*29. Let $K(t,s)$ be a continuous function on $[0,1] \times [0,1]$ and let $m = \|K\|_\infty > 0$. Define $T \in L'(C([0,1]))$ by

$$T(f)(t) = f(t) + \lambda \int_0^1 K(t,s)f(s)\,ds \qquad (t \in [0,1];\ f \in C([0,1])),$$

where $\lambda \in \mathbb{C}$ is a given number such that $m|\lambda| < 1$.

(a) Prove that $T$ is injective, surjective, and continuous.

(b) Use Corollary 7.2.2 to prove that $T^{-1}$ is continuous. If $T(f) = g$, prove that for each $t \in [0,1]$ there exists a measure $\mu_t \in M([0,1])$ such that

$$f(t) = \int_0^1 g(s)\,d\mu_t(s).$$

30. Let $W$ be the linear subspace of $C([0,1])$ consisting of those functions $f$ in $C([0,1])$ that have continuous first and second derivatives and satisfy $f(0) = f(1) = 0$. Let $T$ be the linear differential operator $T : W \to C([0,1])$, defined by

$$T(f)(t) = a_0(t)f''(t) + a_1(t)f'(t) + a_2(t)f(t) \qquad (t \in [0,1]),$$

where the fixed coefficients $a_0, a_1$, and $a_2$ are in $C([0,1])$. Assuming that the differential equation $T(f) = g$ has a unique solution $f \in W$ for each choice of $g \in C([0,1])$, prove that $T^{-1}$ exists and is continuous.

31. A linear space $V$ is a <u>direct sum</u> of two of its linear subspaces $M$ and $N$ if every $x \in V$ has a unique decomposition $x = y + z$, where $y \in M$ and $z \in N$. In this case we write $\mathbf{V} = M \oplus N$. If $V$ is a Banach space, then a <u>projection</u> $P$ is an idempotent transformation in $L(V)$, that is, $P^2 = P$, and a <u>direct summand</u> of $V$ is a closed linear subspace $M$ such that for some other closed linear subspace $N$, $V = M \oplus N$.

(a) If $V$ is a Banach space and $P \in L(V)$ is a projection, prove that $N(P)$ and $R(P)$ are direct summands of $V$ and $V = R(P) \oplus N(P)$.

(b) Let $V$ be a Banach space and let $M$ be a direct summand of $V$. Suppose that $V = M \oplus N$ and define $P : V \to M$ by setting $P(x) = y$ whenever $x = y + z \in V$, $y \in M$, $z \in N$. Prove that $P$ is a projection.

32. Let $(V, \|\cdot\|)$ be an infinite-dimensional separable Banach space over $\Phi$. A <u>Schauder basis</u> in $V$ is a set $\{x_k\} \subset V$, $\|x_k\| = 1$, such that for every $x \in V$ there is a unique series $\Sigma_{k=1}^{\infty} a_k x_k$ such that $\lim_n \|x - \Sigma_{k=1}^{n} a_k x_k\| = 0$. For each $k = 1,2,3,\ldots$, let $a_k^* : V \to \mathbb{C}$ be the mapping that associates with each $x \in V$ the coefficient $a_k$ of $x_k$ in the Schauder expansion of $x$. In other words, $x = \Sigma_{k=1}^{\infty} a_k^*(x) x_k$. Prove that for each $k = 1,2,3,\ldots$, $a_k^* \in V^*$.

*33. (a) Give an example of an unbounded linear transformation $T$ from a normed linear space to a Banach space such that $T$ has a closed graph.

(b) Give an example of an unbounded linear transformation $T$ from a Banach space to a normed linear space such that $T$ has a closed graph.

## CHAPTER 8

## REFLEXIVITY

8.0. Introduction. In this chapter we introduce the notion of a reflexive Banach space -- a Banach space $V$ that can be identified in a natural manner with the space of continuous linear functionals on its dual space $V^*$. This notion is an important and useful one, as will be apparent from the development of this and the succeeding two chapters. After giving a precise definition of a reflexive Banach space and establishing some basic results about such spaces, we shall discuss the concept of uniform convexity and prove Mil'man's Theorem, which asserts that every uniformly convex Banach space is reflexive. A key tool in this development will be the Hahn-Banach Theorem and its consequences. In the last section of the chapter we shall use the previously obtained results to show that $L_p(X,\mathcal{S},\mu)$, $1 < p < \infty$, is a reflexive Banach space and to identify its dual space $L_p(X,\mathcal{S},\mu)^*$ with $L_q(X,\mathcal{S},\mu)$, $1/p + 1/q = 1$.

Many of the concepts and results discussed here have validity in a more general context, but we shall concentrate our attention only on Banach spaces. Expositions of these more general results can be found in several of the references listed in the bibliography.

8.1. Reflexive Spaces. We have already noted (Theorem 3.2.2) that, if $V$ is a normed linear space, then $V^*$ is a Banach space. Clearly, one can then consider the continuous linear functionals on $V^*$, that is, the space $(V^*)^* = V^{**}$. By the preceding remark this is again a Banach space. Similarly, we can consider $V^{***}$, $V^{****}, \ldots$ .

Suppose that $(V,\|\cdot\|)$ is a normed linear space over $\Phi$ and let $x \in V$. We claim that $x$ defines a continuous linear functional on $(V^*, \|\cdot\|)$, that is, $x$ determines an element of $V^{**}$. Indeed, given $x \in V$, define $\tau(x)(x^*) = x^*(x)$, $x^* \in V^*$. It is apparent that $\tau(x)$ is a linear functional on $V^*$, and the estimate

$$|\tau(x)(x^*)| = |x^*(x)| \leq \|x^*\|\|x\| \qquad (x^* \in V^*),$$

shows that $\tau(x) \in V^{**}$ and, moreover, that $\|\tau(x)\| \leq \|x\|$.

In this way we obtain a mapping $\tau$ from $V$ to $V^{**}$. The next theorem gives the basic properties of $\tau$.

<u>Theorem</u> <u>8.1.1</u>. Let $(V, \|\cdot\|)$ be a normed linear space over $\Phi$ and suppose $\tau : V \to V^{**}$ is defined by

$$\tau(x)(x^*) = x^*(x) \qquad (x \in V,\ x^* \in V^*).$$

Then $\tau$ is an isometric isomorphism. Moreover, if $(V, \|\cdot\|)$ is a Banach space over $\Phi$, then $\tau(V)$ is a closed linear subspace of $(V^{**}, \|\cdot\|)$.

<u>Proof</u>. The preceding discussion shows that $\tau$ maps $V$ into $V^{**}$, and a simple argument reveals that $\tau$ is linear. Since $\|\tau(x)\| \leq \|x\|$, $x \in V$, the isometric and injective nature of $\tau$ will be proved on showing that $\|\tau(x)\| \geq \|x\|$, $x \in V$.

If $x = 0$, then clearly $\tau(x) = 0$ and $\|\tau(x)\| = \|x\|$. So suppose $x \neq 0$. Then from Corollary 4.2.5 to the Hahn-Banach Theorem we see that there exists some $x_0^* \in V^*$ such that $\|x_0^*\| = 1$ and $x_0^*(x) = \|x\|$. Thus

$$\begin{aligned} \|x\| = |x_0^*(x)| &= |\tau(x)(x_0^*)| \\ &\leq \sup_{\substack{\|x^*\| = 1 \\ x^* \in V^*}} |\tau(x)(x^*)| \\ &= \|\tau(x)\|, \end{aligned}$$

and so $\tau$ is an isometry.

The last assertion of the theorem is obvious. □

The mapping $\tau$ is called the <u>canonical embedding</u> of $V$ into $V^{**}$. It allows us, albeit with an abuse of terminology, to think of $V$ as a linear subspace of $V^{**}$. At times $V^{**}$ will be referred to as the <u>second conjugate</u>, or <u>bidual</u>, of $V$, and its generic elements will be denoted by $x^{**}$.

We give spaces $V$ a special name when $\tau$ is surjective.

<u>Definition 8.1.1</u>. Let $(V,\|\cdot\|)$ be a normed linear space over $\Phi$. Then $V$ is said to be <u>reflexive</u> if the canonical embedding $\tau : V \to V^{**}$ is surjective.

Thus, if $V$ is reflexive, then every continuous linear functional on $V^*$ is of the form $\tau(x)$ for some $x \in V$.

It is important to note, however, that the property of reflexivity is not identical with the assertion that there exists an isometric isomorphism between $V$ and $V^{**}$, but that this isomorphism must be the canonical one $\tau$. Indeed, it is possible to construct an example of a normed linear space $V$ and a mapping $\varphi : V \to V^{**}$ such that $\varphi$ is a surjective isometric isomorphism but for which $\tau(V)$ is a linear subspace of codimension one in $V^{**}$. One such example is due to R.C.James, and an exposition of it can be found in [Da, p. 72].

If $V$ is a reflexive normed linear space, then $V$ is isometrically isomorphic to the Banach space $V^{**}$, and so $V$ must itself be a Banach space. We include this observation as part of the next result.

<u>Theorem 8.1.2</u>. Let $(V,\|\cdot\|)$ be a normed linear space over $\Phi$. If $V$ is reflexive, then

(i) $(V,\|\cdot\|)$ is a Banach space over $\Phi$.

(ii) $V^*$ is reflexive.

(iii) If $W \subset V$ is a closed linear subspace, then $W$ is a reflexive Banach space.

<u>Proof</u>. Part (i) is proved in the paragraph preceding the theorem. To prove part (ii) we consider the canonical embedding of $V^*$ into $V^{***}$, which we denote by $\tau'$ to distinguish it from $\tau$. We must show that $\tau'$ is surjective; that is, given $x_o^{***} \in V^{***}$, we must find some $x_o^* \in V^*$ such that $\tau'(x_o^*) = x_o^{***}$. Define $x_o^*$ by $x_o^*(x) = x_o^{***}[\tau(x)]$, $x \in V$. It is apparent that $x_o^* \in V^*$, and to see that $\tau'(x_o^*) = x_o^{***}$ we need only show that

$$\tau'(x_o^*)(x^{**}) = x_o^{***}(x^{**}) \qquad (x^{**} \in V^{**}).$$

However, given any $x^{**} \in V^{**}$, let $x \in V$ be such that $\tau(x) = x^{**}$. This is possible because $V$ is reflexive. Thus for this $x$ we would have

$$\begin{aligned}
\tau'(x_o^*)(x^{**}) &= x^{**}(x_o^*) \\
&= \tau(x)(x_o^*) \\
&= x_o^*(x) \\
&= x_o^{***}[\tau(x)] \\
&= x_o^{***}(x^{**}).
\end{aligned}$$

Hence $\tau'(x_o^*) = x_o^{***}$, and $V^*$ is reflexive.

Suppose now that $W \subset V$ is a closed linear subspace. Then $W$ is a Banach space, as $V$ is. Let $\tau_W$ denote the canonical embedding of $W$ into $W^{**}$. We must show that $\tau_W$ is surjective.

To this end let $y_o^{**} \in W^{**}$ and define $x_o^{**} \in V^{**}$ by $x_o^{**}(x^*) = y_o^{**}[\psi(x^*)]$, $x^* \in V^*$, where $\psi$ denotes the natural mapping of $V^*$ onto $W^*$, determined by the fact that $W^*$ is isometrically isomorphic to $V^*/W^{\perp}$. The latter assertion was established in Theorem 4.6.2.

It is readily verified that $x_o^{**}$ is indeed an element of $V^{**}$. Appealing again to the reflexivity of $V$, we choose $x_o \in V$ such that $\tau(x_o) = x_o^{**}$. We claim that $x_o \in W$, because if $x_o \notin W$, then by Corollary 4.2.4 to the Hahn-Banach Theorem there exists some $x_o^* \in V^*$ such that $x_o^*(x_o) \neq 0$ and $x_o^*(y) = 0$, $y \in W$. Thus, in particular, $x_o^* \in W^{\perp}$ and so $\psi(x_o^*) = 0$, from which it follows that $y_o^{**}[\psi(x_o^*)] = 0$. However, by the definition of $x_o^{**}$ we see that

$$\begin{aligned} y_o^{**}[\psi(x_o^*)] &= x_o^{**}(x_o^*) \\ &= \tau(x_o)(x_o^*) \\ &= x_o^*(x_o) \\ &= 0 \end{aligned}$$

by the choice of $x_o^*$. This contradiction implies that $x_o \in W$.

Consequently $\tau_W(x_o) \in W^{**}$. Moreover, from the proof of Theorem 4.6.2, since $x_o \in W$, we have $\psi(x^*)(x_o) = x^*(x_o)$, $x^* \in V^*$. Thus for any $y^* \in W^*$, if $x^* \in V^*$ is such that $\psi(x^*) = y^*$, then

$$\begin{aligned} \tau_W(x_o)(y^*) &= y^*(x_o) \\ &= \psi(x^*)(x_o) \\ &= x^*(x_o) \\ &= \tau(x_o)(x^*) \\ &= x_o^{**}(x^*) \\ &= y_o^{**}[\psi(x^*)] \end{aligned}$$

$$= y_o^{**}(y^*).$$

Therefore $\tau_W(x_o) = y_o^{**}$, and $\tau_W$ is surjective. □

Actually part (ii) of Theorem 8.1.2 is a necessary and sufficient condition for the reflexivity of $V$.

Theorem 8.1.3. Let $(V, \|\cdot\|)$ be a Banach space over $\Phi$. Then the following are equivalent:

(i) $V$ is reflexive.

(ii) $V^*$ is reflexive.

Proof. That part (i) implies part (ii) is a portion of Theorem 8.1.2. Conversely, suppose that $V$ is not reflexive. Then from Theorem 8.1.1 we see that $\tau(V)$ is a proper closed linear subspace of $V^{**}$, and so, by Corollary 4.2.4 again, there exists some $x_o^{***} \in V^{***}$ such that $x_o^{***} \neq 0$ and $x_o^{***}[\tau(x)] = 0$, $x \in V$. We claim that $x_o^{***} \notin \tau'(V^*)$, where $\tau'$ again denotes the canonical embedding of $V^*$ into $V^{***}$.

Indeed, if $x_o^* \in V^*$ were such that $\tau'(x_o^*) = x_o^{***}$, then

$$\begin{aligned} x_o^*(x) &= \tau(x)(x_o^*) \\ &= x_o^{***}[\tau(x)] \\ &= 0 \qquad\qquad (x \in V), \end{aligned}$$

from which it follows at once that $x_o^* = 0$. But then $\tau'(x_o^*) = x_o^{***} = 0$, as $\tau'$ is linear, contrary to the choice of $x_o^{***}$.

Therefore $\tau'(V^*) \neq V^{***}$, and part (ii) implies part (i). □

Let us look at two other questions whose answers contribute to the interest in reflexive spaces.

Suppose $(V, \|\cdot\|)$ is a Banach space and $x^* \in V^*$. Then by definition we have $\|x^*\| = \sup_{\|x\| \leq 1} |x^*(x)|$.

An obvious question to ask is whether there exists some $x_0 \in V$, $\|x_0\| \leq 1$, such that $|x^*(x_0)| = \|x^*\|$; that is, does $x^*$ attain its maximum absolute value at some point of the closed unit ball $B_1 = \{x \mid x \in V, \|x\| \leq 1\}$? To answer this question we cannot make an appeal to the continuity of $|x^*(\cdot)|$ on $B_1$, since, as remarked in the comments following Theorem 1.3.3, $B_1$ is not compact unless $V$ is finite-dimensional. However, if $V$ is reflexive, then we can assert that $x^*$ attains its maximum absolute value on $B_1$.

Theorem 8.1.4. Let $(V, \|\cdot\|)$ be a reflexive Banach space over $\Phi$. If $x^* \in V^*$, then there exists some $x_0 \in V$, $\|x_0\| = 1$, such that $x^*(x_0) = \|x^*\|$.

Proof. If $x^* = 0$, then the assertion is evident, whereas if $x^* \neq 0$, then by Corollary 4.2.5 to the Hahn-Banach Theorem there exists some $x_0^{**} \in V^{**}$ such that $x_0^{**}(x^*) = \|x^*\|$ and $\|x_0^{**}\| = 1$. If $x_0 \in V$ is such that $\tau(x_0) = x_0^{**}$, then it is apparent that $\|x_0\| = 1$ and $x^*(x_0) = \|x^*\|$. □

In general the point $x_0$ need not be unique. However, if $(V, \|\cdot\|)$ is a uniformly convex Banach space, a concept to be discussed in the next section, then $x^* \neq 0$ attains its maximum absolute value at a unique point in the unit ball. We shall return to this stronger result in Section 8.2.

The converse of Theorem 8.1.4 is also valid: if $(V, \|\cdot\|)$ is a Banach space such that, whenever $x^* \in V^*$, there exists some $x_0 \in V$, $\|x_0\| = 1$, for which $x^*(x_0) = \|x^*\|$, then $V$ is reflexive. We do not give the proof but instead refer the reader to the original source (see [J]).

Theorem 8.1.4 can be employed to show that certain spaces are not reflexive.

<u>Corollary 8.1.1</u>. $(c_o, \|\cdot\|_\infty)$ is not a reflexive Banach space.

<u>Proof</u>. In view of Theorem 8.1.4 it suffices to exhibit a continuous linear functional on $c_o$ that does not attain its maximum absolute value on the closed unit ball; that is, to construct some $x^* \in c_o^*$ such that $|x^*(x)| < \|x^*\|$, $x \in c_o$, $\|x\|_\infty \le 1$. We now proceed to do this.

For each $x = \{a_k\} \in c_o$ we define $x^*(x) = \Sigma_{k=1}^{\infty} a_k/k!$. Clearly, $x^*$ is a linear functional on $c_o$. Moreover, since $|a_k| \le \|\{a_k\}\|_\infty$, $k = 1,2,3,\ldots,$ it is apparent that, if $x \in c_o$, $\|x\|_\infty \le 1$, then $|x^*(x)| \le \Sigma_{k=1}^{\infty} 1/k!$. Hence $x^* \in c_o^*$ and $\|x^*\| \le \Sigma_{k=1}^{\infty} 1/k!$. Actually, $\|x^*\| = \Sigma_{k=1}^{\infty} 1/k!$ since for each positive integer $n$, if $x^n = \{a_k^n\} \in c_o$ is such that $a_k^n = 1$, $k = 1,2,\ldots,n$, and $a_k^n = 0$, $k > n$, then $\|x^n\|_\infty = 1$ and $x^*(x^n) = \Sigma_{k=1}^{n} 1/k!$. Consequently

$$\begin{aligned}\|x^*\| &= \sup_{\|x\|_\infty = 1} |x^*(x)| \\ &\ge \sup_n x^*(x^n) \\ &\ge \sup_n \sum_{k=1}^{n} \frac{1}{k!} \\ &= \sum_{k=1}^{\infty} \frac{1}{k!}.\end{aligned}$$

However, there exists no $x \in c_o$, $\|x\|_\infty \le 1$, for which $|x^*(x)| = \|x^*\|$, as can be seen from the following argument: Suppose $x = \{a_k\} \in c_o$, $\|x\|_\infty \le 1$. Since $\lim_k a_k = 0$, there exists some $n$ such that $|a_n| < 1$, and hence

$$\begin{aligned}|x^*(x)| &= \left|\sum_{k=1}^{\infty} \frac{a_k}{k!}\right| \\ &\le \sum_{\substack{k=1 \\ k \ne n}}^{\infty} \frac{|a_k|}{k!} + \frac{|a_n|}{n!}\end{aligned}$$

$$< \sum_{\substack{k=1 \\ k \neq n}}^{\infty} \frac{1}{k!} + \frac{1}{n!}$$

$$= \sum_{k=1}^{\infty} \frac{1}{k!}$$

$$= \|x^*\|.$$

Therefore $c_o$ is not reflexive. □

Another interesting question is the following one: Suppose $K$ is a nonempty closed convex set in a Banach space $V$ and let $d = \inf_{x \in K}\|x\|$. Clearly, if $0 \in K$, then $d = 0$, and there exists some point $x_o \in K$ for which $\|x_o\| = d$, namely, $x_o = 0$. On the other hand, if $0 \notin K$, then $d > 0$, but it is not entirely clear under what conditions there exists an $x_o \in K$ for which $\|x_o\| = d$. Indeed, there may exist no such $x_o$. However, if $V$ is reflexive, then such an $x_o \in K$ with minimal norm always exists. The proof of this requires some knowledge of weak topologies and will be delayed until Chapter 9 (Corollary 9.9.2). In the next section we shall prove that such a unique $x_o \in K$ exists whenever $V$ is uniformly convex.

The notion of reflexivity also plays an important role in the study of weak topologies on Banach spaces, a subject we shall pursue in some detail in Chapters 9 and 10.

## 8.2. Uniform Convexity and Mil'man's Theorem.

As indicated in the preceding section, it is of some interest to develop means for determining whether a given Banach space is reflexive. Such a determination is not always easy. There is, however, one property of a Banach space that is amenable to verification and ensures that the space is reflexive. This property is the uniform convexity of the norm. The main purpose of this section is to establish this result.

Definition 8.2.1. Let $(V,\|\cdot\|)$ be a normed linear space over $\Phi$. Then $(V,\|\cdot\|)$ is said to be uniformly convex if for each $\epsilon > 0$ there exists some $\delta > 0$ such that, whenever $x,y \in V$, $\|x\| \leq 1$, $\|y\| \leq 1$, and $\|x - y\| \geq \epsilon$, then $\|(x + y)/2\| \leq 1 - \delta$.

Strictly speaking it is not a linear space that is uniformly convex, but rather the norm on the space, and if one takes the same linear space with two different norms, then one space may be uniformly convex and the other space may fail to be so.

For instance, consider $V = \mathbb{R}^2$. Then we claim that $(\mathbb{R}^2,\|\cdot\|_2)$ is uniformly convex, whereas $(\mathbb{R}^2,\|\cdot\|_1)$ is not. Indeed, note first that, if $x = (a_1,a_2)$ and $y = (b_1,b_2)$ are in $\mathbb{R}^2$, then

$$\begin{aligned}
\left(\left\|\frac{x + y}{2}\right\|_2\right)^2 + \left(\left\|\frac{x - y}{2}\right\|_2\right)^2 &= \frac{1}{4}\sum_{k=1}^{2}(a_k + b_k)^2 + \frac{1}{4}\sum_{k=1}^{2}(a_k - b_k)^2 \\
&= \frac{1}{4}\sum_{k=1}^{2}(2a_k^2 + 2b_k^2) \\
&= \frac{1}{2}\left(\sum_{k=1}^{2} a_k^2\right) + \frac{1}{2}\left(\sum_{k=1}^{2} b_k^2\right) \\
&= \frac{1}{2}(\|x\|_2)^2 + \frac{1}{2}(\|y\|_2)^2.
\end{aligned}$$

But then, if $\epsilon > 0$ is given, let $\delta > 0$ be such that $(1 - \delta)^2 = 1 - \epsilon^2/4$. Then for any $x,y \in \mathbb{R}^2$, $\|x\|_2 \leq 1$, $\|y\|_2 \leq 1$, and $\|x - y\|_2 \geq \epsilon$ we have

$$\begin{aligned}
\left(\left\|\frac{x + y}{2}\right\|_2\right)^2 &= \frac{1}{2}(\|x\|_2)^2 + \frac{1}{2}(\|y\|_2)^2 - \left(\left\|\frac{x - y}{2}\right\|_2\right)^2 \\
&\leq 1 - \left(\left\|\frac{x - y}{2}\right\|_2\right)^2 \\
&\leq 1 - \frac{\epsilon^2}{4} \\
&= (1 - \delta)^2,
\end{aligned}$$

that is, $\|(x + y)/2\|_2 \leq 1 - \delta$. Hence $(\mathbb{R}^2,\|\cdot\|_2)$ is uniformly convex. Evidently the same argument mutatis mutandis shows that $(\mathbb{R}^n,\|\cdot\|_2)$ and $(\mathbb{C}^n,\|\cdot\|_2)$ are uniformly convex.

On the other hand, let $x = (1,0)$ and $y = (0,1)$. Then $\|x\|_1 = \|y\|_1 = 1$ and $\|x - y\|_1 = 2 = \epsilon > 0$, whereas

$$\left\|\frac{x + y}{2}\right\|_1 = 1 > 1 - \delta \qquad (\delta > 0).$$

Thus $(\mathbb{R}^2, \|\cdot\|_1)$ is not uniformly convex.

Note also that the norms $\|\cdot\|_2$ and $\|\cdot\|_1$ are equivalent norms on $\mathbb{R}^2$.

Another important class of uniformly convex spaces is the inner product spaces introduced in Example 1.2.10. Indeed, suppose $V$ is a linear space over $\Phi$ and $\psi : V \times V \to \Phi$ is an inner product. Then, as indicated previously, $(V, \|\cdot\|)$ is a normed linear space over $\Phi$ if $\|x\| = [\psi(x,x)]^{1/2}$, $x \in V$. The space $(V, \|\cdot\|)$ is uniformly convex. We shall prove this later when we discuss inner product spaces in detail in Chapter 13. In particular, every Hilbert space is uniformly convex.

Geometrically the notion of uniform convexity is an indication of the "roundness" of the closed unit ball about the origin in the space. It says, in effect, that, given $\epsilon > 0$, there exists a $\delta > 0$ such that, if $x,y$ are in the closed unit ball and are at least a distance $\epsilon$ from one another, then the midpoint of the line segment between $x$ and $y$ is a least a distance $\delta$ from the surface of the closed unit ball. Note that uniform convexity thus asserts more than that the midpoint of the line segment between two points in the closed unit ball lies in the interior of the ball. It actually gives some measure of how far in the interior the midpoint lies.

The weaker notion that $\|x\| \leq 1$, $\|y\| \leq 1$ implies $\|(x + y)/2\| < 1$, is generally called <u>strict convexity</u>, or <u>rotundity</u>. We do not pursue this notion further here, but instead refer the interested reader to [Da, pp. 111-115; K, pp. 342-347; $W_1$, pp. 107 and 108].

We now prove that every uniformly convex Banach space is reflexive. The proof is due to Kakutani [Kt] and makes crucial use of the corollary to Helly's Theorem (Theorem 4.10.1).

Theorem 8.2.1 (Mil'man's Theorem). Let $(V,\|\cdot\|)$ be a Banach space over $\Phi$. If $(V,\|\cdot\|)$ is uniformly convex, then $V$ is reflexive.

Proof. Given $x_0^{**} \in V^{**}$, we must show that there is some $x_0 \in V$ such that $\tau(x_0) = x_0^{**}$. Since this can obviously be done if $x_0^{**} = 0$, we may without loss of generality assume that $\|x_0^{**}\| = 1$.

From the definition of $\|x_0^{**}\|$ we see that for each positive integer $k$ there exists some $y_k^* \in V^*$, $\|y_k^*\| = 1$, for which $|x_0^{**}(y_k^*)| > 1 - 1/k$. On setting $x_k^* = \exp[-i \arg x_o^{**}(y_k^*)]y_k^*$, $k = 1,2,3,\ldots,$ we obtain a sequence $\{x_k^*\} \subset V^*$ such that $\|x_k^*\| = 1$ and $x_0^{**}(x_k^*) > 1 - 1/k$, $k = 1,2,3,\ldots$ .

Now for each positive integer $n$ we may apply Corollary 4.10.1 to Helly's Theorem with $\epsilon = 1/n$ to deduce the existence of some $x_n \in V$ such that

$$\text{(i)} \quad \|x_n\| \leq \|x_0^{**}\| + \frac{1}{n} = 1 + \frac{1}{n}.$$

$$\text{(ii)} \quad x_k^*(x_n) = x_0^{**}(x_k^*) \qquad (k = 1,2,\ldots,n).$$

Repeating this for each $n = 1,2,3,\ldots,$ we obtain a sequence $\{x_n\} \subset V$ whose elements satisfy conditions (i) and (ii) and which is such that for each $n$

$$x_n^*(x_m) = x_n^*(x_n) = x_0^{**}(x_n^*) \qquad (m \geq n).$$

We claim that $\{x_n\} \subset V$ is a Cauchy sequence.

Indeed, suppose this is not the case. Then there exist some $\epsilon > 0$ and a strictly increasing sequence $\{n_k\}$ of positive integers such that $\|x_{n_{k+1}} - x_{n_k}\| \geq \epsilon$, $k = 1,2,3,\ldots$ . For this $\epsilon > 0$

there exists, by the uniform convexity of $(V,\|\cdot\|)$, some $\delta > 0$ such that, whenever $\|x\| \leq 1$, $\|y\| \leq 1$, and $\|x - y\| \geq \epsilon/2$, we have $\|(x + y)/2\| \leq 1 - \delta$. Thus, on the one hand, since $\|x_{n_k}\| \leq 1 + 1/n_k$ we see that

$$\left\|\frac{x_{n_k}}{1 + 1/n_k}\right\| \leq 1,$$

$$\left\|\frac{x_{n_{k+1}}}{1 + 1/n_k}\right\| \leq 1,$$

and

$$\left\|\frac{x_{n_{k+1}} - x_{n_k}}{1 + 1/n_k}\right\| \geq \frac{\epsilon}{1 + n_k} \geq \frac{\epsilon}{2},$$

and hence

$$\left\|\frac{x_{n_{k+1}} + x_{n_k}}{2(1 + 1/n_k)}\right\| \leq 1 - \delta \qquad (k = 1,2,3,\ldots).$$

Consequently $\|x_{n_{k+1}} + x_{n_k}\| \leq (1 + 1/n_k)2(1 - \delta)$, from which we conclude that

$$\limsup_k \|x_{n_{k+1}} + x_{n_k}\| \leq 2(1 - \delta) < 2.$$

On the other hand, since $n_k < n_{k+1}$ we see that

$$x^*_{n_k}(x_{n_{k+1}}) = x^*_{n_k}(x_{n_k}) = x_0^{**}(x^*_{n_k}) \quad (k = 1,2,3,\ldots)$$

by the choice of the sequence $\{x_n\}$. Hence, since $\|x^*_{n_k}\| = 1$, we have

$$\begin{aligned}
\|x_{n_{k+1}} + x_{n_k}\| &= \|x^*_{n_k}\|\|x_{n_{k+1}} + x_{n_k}\| \\
&\geq |x^*_{n_k}(x_{n_{k+1}} + x_{n_k})| \\
&= 2x_0^{**}(x^*_{n_k}) \\
&> 2\left(1 - \frac{1}{n_k}\right) \qquad (k = 1,2,3,\ldots).
\end{aligned}$$

Thus

$$\liminf_k \|x_{n_{k+1}} + x_{n_k}\| \geq 2,$$

which clearly contradicts the previously obtained estimate.

This contradiction proves that $\{x_n\}$ is a Cauchy sequence in the Banach space $V$. Let $x_0 \in V$ be such that $\lim_n \|x_n - x_0\| = 0$. Obviously $\lim_n \|x_n\| = \|x_0\|$, and this fact, combined with the estimates

$$1 - \frac{1}{n} < x_0^{**}(x_n^*) = x_n^*(x_n) \leq \|x_n\| \leq 1 + \frac{1}{n} \qquad (n = 1,2,3,\ldots),$$

allows us to deduce that $\|x_0\| = 1$. Moreover, by the choice of the sequence $\{x_n\}$, we have for each positive integer $n$

$$x_0^{**}(x_n^*) = x_n^*(x_n) = x_n^*(x_{n+k}) \qquad (k = 1,2,3,\ldots),$$

and so from the continuity of $x_n^*$ it follows that $x_0^{**}(x_n^*) = x_n^*(x_0)$, $n = 1,2,3,\ldots$ . Thus $\tau(x_0)(x_n^*) = x_0^{**}(x_n^*)$, $n = 1,2,3,\ldots$ . To complete the proof we need only show that the last equation holds for arbitrary $x_0^* \in V^*$.

With this in mind, we note first that there is precisely one $x_0 \in V$ such that $\|x_0\| = 1$ and $x_0^{**}(x_n^*) = x_n^*(x_0)$, $n = 1,2,3,\ldots$, as shown by the following argument: Suppose $y_0 \in V$ satisfies the same conditions and that $\|x_0 - y_0\| = \epsilon > 0$. Then by the uniform convexity of $(V,\|\cdot\|)$ there is some $\delta > 0$ such that, whenever $\|x\| \leq 1$, $\|y\| \leq 1$, and $\|x - y\| \geq \epsilon$, we have $\|(x + y)/2\| \leq 1 - \delta$. Thus, in particular, we would have $\|(x_0 + y_0)/2\| \leq 1 - \delta$, that is, $\|x_0 + y_0\| \leq 2(1 - \delta) < 2$. However, arguing as before, we see that

$$\begin{aligned}\|x_0 + y_0\| &\geq |x_n^*(x_0 + y_0)| \\ &= 2x_0^{**}(x_n^*) \\ &> 2(1 - \tfrac{1}{n}) \qquad (n = 1,2,3,\ldots),\end{aligned}$$

and so $\|x_0 + y_0\| \geq 2$, contradicting the previous estimate. Hence $x_0 = y_0$.

Finally, let $x_0^* \in V^*$. If $x_0^* = 0$, then $\tau(x_0)(x_0^*) = 0$, and we may assume without loss of generality that $\|x_0^*\| = 1$. Let $\{x_k^*\} \subset V^*$ be as before and for each nonnegative integer $n$ apply Corollary 4.10.1 to Helly's Theorem to obtain a sequence $x_0', x_1', x_2', \ldots$ of elements of $V$ such that

(a) $\|x_0'\| \leq 3$.

(b) $\|x_n'\| \leq \|x_0^{**}\| + \frac{1}{n} = 1 + \frac{1}{n}$ $(n = 1,2,3,\ldots)$.

(c) $x_k^*(x_n') = x_0^{**}(x_k^*)$ $(k = 0,1,2,\ldots,n;\ n = 1,2,3,\ldots)$.

Note that we now have applied Corollary 4.10.1 to the sequence $x_0^*, x_1^*, x_2^*, \ldots$, whereas before we applied it to $x_1^*, x_2^*, x_3^*, \ldots$ .

Repeating our previous arguments, we deduce that $\{x_n'\}$ is a Cauchy sequence in $V$, and hence there exists some $y_0 \in V$, $\|y_0\| = 1$, such that $x_0^{**}(x_n^*) = x_n^*(y_0)$, $n = 1,2,3,\ldots$ . From the uniqueness of $x_0$, which we have established, we conclude that $x_0 = y_0$, whereas $x_0^{**}(x_0^*) = x_0^*(y_0)$ shows that $\tau(x_0)(x_0^*) = x_0^{**}(x_0^*)$.

Therefore, since $x_0^* \in V^*$ is arbitrary, we see that $\tau(x_0) = x_0^{**}$, and so $V$ is reflexive. □

The converse of Mil'man's Theorem may fail. Indeed, there exist reflexive Banach spaces $(V, \|\cdot\|)$ such that no norm equivalent to $\|\cdot\|$ is uniformly convex. For the details the reader is referred to [K, p. 361].

We shall obtain some necessary and sufficient conditions for a space to be reflexive in Chapters 9 and 10 (Theorem 9.9.1, Corollary 9.9.1, and Corollary 10.3.1) when we discuss weak topologies.

We note also that all finite-dimensional Banach spaces are reflexive. This follows at once from Mil'man's Theorem, the uniform convexity of $(\mathbb{R}^n, \|\cdot\|_2)$ and $(\mathbb{C}^n, \|\cdot\|_2)$, $n = 1,2,3,\ldots$, and the fact that all such spaces are topologically isomorphic to either $(\mathbb{R}^n, \|\cdot\|_2)$ or $(\mathbb{C}^n, \|\cdot\|_2)$ (Theorem 1.3.1).

One further comment about Mil'man's Theorem is necessary in reference to the proof: the argument used in the proof to show that the sequence $\{x_n\}$, constructed by means of Corollary 4.10.1 to Helly's Theorem, is a Cauchy sequence, can readily be modified to obtain the following proposition. The details are left to the reader. We shall use the proposition in the proof of Theorem 8.2.2.

Proposition 8.2.1. Let $(V, \|\cdot\|)$ be a uniformly convex normed linear space over $\Phi$. If $\{x_n\} \subset V$ is a sequence such that

(i) $\lim_n \|x_n\| = 1$,
(ii) $\lim_{n,k} \|x_n + x_k\| = 2$,

then $\{x_n\}$ is a Cauchy sequence.

Let us now see how the two questions explicitly raised at the end of the preceding section can be answered in uniformly convex spaces. First we address ourselves to the existence of an element of minimal norm in a closed convex set.

Theorem 8.2.2. Let $(V, \|\cdot\|)$ be a uniformly convex Banach space over $\Phi$. If $K \subset V$ is a nonempty closed convex set, then there exists a unique $x_o \in K$ such that $\|x_o\| = \inf_{x \in K} \|x\|$.

Proof. If $0 \in K$, then clearly $x_o = 0$. Suppose $0 \notin K$ and set $d = \inf_{x \in K} \|x\| > 0$. Let $\{x_n\} \subset K$ be a sequence such that $\lim_n \|x_n\| = d$. Clearly, $\{x_n/d\}$ is then a sequence such that $\lim_n \|x_n/d\| = 1$. Moreover, for each $n$ and $k$ the estimates

$$\left\|\frac{x_n + x_k}{d}\right\| \leq \frac{1}{d}(\|x_n\| + \|x_k\|)$$

and

$$\left\|\frac{x_n + x_k}{d}\right\| = \frac{2}{d}\left\|\frac{x_n + x_k}{2}\right\| \geq 2$$

show that $\lim_{n,k}\|(x_n + x_k)/d\| = 2$. The latter estimate is valid because $K$ is convex.

Thus from Proposition 8.2.1 we conclude that $\{x_n/d\}$ is a Cauchy sequence in $V$, and so is $\{x_n\}$. Let $x_o \in V$ be such that $\lim_n\|x_n - x_o\| = 0$. Obviously, by the choice of $\{x_n\}$ and the fact that $K$ is closed, we have $x_o \in K$ and $\|x_o\| = d$.

Suppose that there exist some $y_o \in K$, $\|y_o\| = d$, and $\|x_o - y_o\| = \epsilon > 0$. Then by the uniform convexity of $(V,\|\cdot\|)$ we see that there exists some $\delta > 0$ such that $\|(x_o + y_o)/2\| \leq (1 - \delta)d < d$. However, since $K$ is convex

$$\left\|\frac{x_o + y_o}{2}\right\| \geq \inf_{x \in K} \|x\| = d,$$

which contradicts the previous estimate.

Therefore $x_o = y_o$, and $x_o$ is unique. □

The completeness assumption cannot be dropped.

<u>Corollary 8.2.1</u>. Let $(V,\|\cdot\|)$ be a uniformly convex Banach space over $\Phi$. If $K \subset V$ is a nonempty closed convex set and $y_o \notin K$, then there exists a unique $x_o \in K$ such that

$$\|y_o - x_o\| = \inf_{x \in K} \|y_o - x\|.$$

A second corollary to Theorem 8.2.2 provides us with the promised improvement of Theorem 8.1.4.

<u>Corollary 8.2.2</u>. Let $(V,\|\cdot\|)$ be a uniformly convex Banach space over $\Phi$. If $x^* \in V^*$, $x^* \neq 0$, then there exists a unique $x_o \in V$, $\|x_o\| = 1$, such that $x^*(x_o) = \|x^*\|$.

<u>Proof</u>. Let $L = \{x \mid x \in V, x^*(x) = \|x^*\|\}$. Then by Proposition 5.1.1 $L$ is a closed hyperplane in $V$ that does not contain the origin. In particular, $L$ is a nonempty closed convex set such that $0 \notin L$, and hence by Theorem 8.2.2 there exists a unique $x_o \in L$

such that $\|x_o\| = \inf_{x \in L}\|x\| > 0$. Clearly $x^*(x_o) = \|x^*\|$, from which it follows that $\|x_o\| \geq 1$.

On the other hand, since $(V,\|\cdot\|)$ is uniformly convex and hence reflexive by Mil'man's Theorem, we deduce from Theorem 8.1.4 that there exists some $y_o \in V$, $\|y_o\| = 1$, such that $x^*(y_o) = \|x^*\|$. Thus $y_o \in L$ and $\|y_o\| \leq \|x_o\|$. Consequently $y_o = x_o$, as $x_o$ is the unique element in $L$ of minimal norm.

Therefore $\|x_o\| = 1$, and the proof is complete. □

<u>8.3. Reflexivity of $L_p(X,S,\mu)$, $1 < p < \infty$.</u> We now wish to apply Mil'man's Theorem to prove that the spaces $L_p(X,S,\mu)$, $1 < p < \infty$, introduced in Example 1.2.4 are reflexive Banach spaces. As a corollary we shall see that $L_p(X,S,\mu)^*$ can be identified with $L_q(X,S,\mu)$, $1/p + 1/q = 1$.

First we prove an elementary lemma.

<u>Lemma 8.3.1</u>. If $2 \leq p < \infty$ and $a,b \in \mathbb{R}$, $a \geq 0$ and $b \geq 0$, then

(i) $a^p + b^p \leq (a^2 + b^2)^{p/2}$.

(ii) $(a^2 + b^2)^{p/2} \leq 2^{(p-2)/2}(a^p + b^p)$.

<u>Proof</u>. If either $a$ or $b$ is zero, then the inequalities are trivial, and we may assume without loss of generality that neither $a$ nor $b$ is zero. Obviously, $a^2/(a^2 + b^2) \leq 1$ and $b^2/(a^2 + b^2) \leq 1$, and so

$$\frac{a^p}{(a^2 + b^2)^{p/2}} + \frac{b^p}{(a^2 + b^2)^{p/2}} = \left[\frac{a^2}{a^2 + b^2}\right]^{p/2} + \left[\frac{b^2}{a^2 + b^2}\right]^{p/2}$$

$$\leq \frac{a^2}{a^2 + b^2} + \frac{b^2}{a^2 + b^2}$$

$$= 1,$$

as $p/2 \geq 1$. Thus $a^p + b^p \leq (a^2 + b^2)^{p/2}$.

To prove part (ii) of the lemma we note that the inequality is trivial if $p = 2$. If $p > 2$, then we set $p' = p/2 > 1$, and $q' = p'/(p' - 1) = p/(p - 2)$. Then $1/p' + 1/q' = 1$, and so by Hölder's Inequality for finite sums we see that

$$\begin{aligned} a^2 + b^2 &\leq [(a^2)^{p'} + (b^2)^{p'}]^{1/p'}[1^{q'} + 1^{q'}]^{1/q'} \\ &= [a^p + b^p]^{2/p}[2^{(p-2)/p}]. \end{aligned}$$

It is thus apparent that $(a^2 + b^2)^{p/2} \leq 2^{(p-2)/2}(a^p + b^p)$. □

We shall use this lemma to show that $L_p(X,S,\mu)$ is uniformly convex for $p \geq 2$.

Theorem 8.3.1. Let $(X,S,\mu)$ be a positive measure space. If $2 \leq p < \infty$, then $(L_p(X,S,\mu), \|\cdot\|_p)$ is a uniformly convex Banach space.

Proof. Let $f,g \in L_p(X,S,\mu)$. We then apply Lemma 8.3.1 twice: once with $a = |f(t) + g(t)|$ and $b = |f(t) - g(t)|$; and once with $a = |f(t)|$ and $b = |g(t)|$, to deduce for almost all $t \in X$ that

$$|f(t) + g(t)|^p + |f(t) - g(t)|^p \leq [|f(t) + g(t)|^2 + |f(t) - g(t)|^2]^{p/2}$$

and

$$[|f(t)|^2 + |g(t)|^2]^{p/2} \leq 2^{(p-2)/2}[|f(t)|^p + |g(t)|^p].$$

Furthermore, using the definition of the absolute value of a complex number, we see that for almost all $t \in X$

$$\begin{aligned} |f(t) + g(t)|^p + |f(t) - g(t)|^p &\leq [|f(t) + g(t)|^2 + |f(t) - g(t)|^2]^{p/2} \\ &= [2|f(t)|^2 + 2|g(t)|^2]^{p/2} \\ &\leq 2^{p-1}[|f(t)|^p + |g(t)|^p]. \end{aligned}$$

Consequently, on integrating, we obtain the inequality

$$(\|f + g\|_p)^p + (\|f - g\|_p)^p \leq 2^{p-1}[(\|f\|_p)^p + (\|g\|_p)^p]$$

which is valid for each $f$ and $g$ in $L_p(X,\mathcal{S},\mu)$.

Now given $\epsilon, 0 < \epsilon < 2$, let $\delta > 0$ be such that $1 - \delta = [1 - (\epsilon/2)^p]^{1/p}$. If $f,g \in L_p(X,\mathcal{S},\mu)$ are such that $\|f\|_p \leq 1$, $\|g\|_p \leq 1$, and $\|f - g\|_p \geq \epsilon$, then

$$\begin{aligned}(\|\tfrac{f+g}{2}\|_p)^p &\leq 2^{p-1}[(\|\tfrac{f}{2}\|_p)^p + (\|\tfrac{g}{2}\|_p)^p] - (\|\tfrac{f-g}{2}\|_p)^p \\ &\leq 1 - (\tfrac{\epsilon}{2})^p \\ &= (1 - \delta)^p.\end{aligned}$$

Therefore $L_p(X,\mathcal{S},\mu)$ is uniformly convex. □

<u>Corollary 8.3.1</u>. Let $(X,\mathcal{S},\mu)$ be a positive measure space. If $2 \leq p < \infty$, then $(L_p(X,\mathcal{S},\mu),\|\cdot\|_p)$ is a reflexive Banach space.

<u>Proof</u>. Apply Mil'man's Theorem (Theorem 8.2.1). □

It is, furthermore, true that $L_p(X,\mathcal{S},\mu)$ is reflexive for all $p$, $1 < p < \infty$, and we shall establish this fact in a moment. Before that, however, we wish to give another lemma. In the interest of completeness we make the following definition:

<u>Definition 8.3.1</u>. Let $V$ and $W$ be linear spaces over $\Phi$. Then a mapping $\varphi : V \to W$ is said to be <u>antilinear</u> if

$$\varphi(ax + by) = \bar{a}\varphi(x) + \bar{b}\varphi(y) \qquad (x,y \in V;\ a,b \in \Phi).$$

If $\Phi = \mathbb{R}$, then clearly "linear" and "antilinear" are identical.

<u>Lemma 8.3.2</u>. Let $(X,\mathcal{S},\mu)$ be a positive measure space and let $1 < p < \infty$. If $1/p + 1/q = 1$, then for each $g \in L_q(X,\mathcal{S},\mu)$ the formula

$$x_g^*(f) = \int_X f(t)\overline{g(t)}\, d\mu(t) \qquad (f \in L_p(X,S,\mu))$$

defines an element $x_g^* \in L_p(X,S,\mu)^*$. Moreover, the correspondence $g \to x_g^*$ defines an antilinear isometry from $(L_q(X,S,\mu), \|\cdot\|_q)$ to $(L_p(X,S,\mu)^*, \|\cdot\|)$.

<u>Proof</u>. Obviously $x_g^*$ is linear, and by Hölder's Inequality it is clear that

$$|x_g^*(f)| \le \|\overline{g}\|_q \|f\|_p$$

$$= \|g\|_q \|f\|_p \qquad (f \in L_p(X,S,\mu)).$$

Thus $x_g^* \in L_p(X,S,\mu)^*$ and $\|x_g^*\| \le \|g\|_q$.

Furthermore, let $f$ be defined almost everywhere on $X$ by the expression $f(t) = \{|\overline{g(t)}|^{q-1}\exp[-i \arg \overline{g(t)}]\}/(\|g\|_q)^{q/p}$. Then, since $(q - 1)p = q$, we see that

$$\|f\|_p = [\frac{1}{\|\overline{g}\|_q}]^{q/p} [\int_X |\overline{g(t)}|^{(q-1)p}\, d\mu(t)]^{1/p}$$

$$= \left[\frac{\|\overline{g}\|_q}{\|\overline{g}\|_q}\right]^{q/p}$$

$$= 1,$$

and so $f \in L_p(X,S,\mu)$ and $\|f\|_p = 1$. Moreover,

$$x_g^*(f) = [\frac{1}{\|\overline{g}\|_q}]^{q/p} \int_X |\overline{g(t)}|^q\, d\mu(t)$$

$$= [\int_X |g(t)|^q\, d\mu(t)]^{1 - 1/p}$$

$$= \|g\|_q.$$

Thus $\|x_g^*\| = \|g\|_q$.

The last assertion of the lemma is now apparent. □

Theorem 8.3.2. Let $(X,\mathcal{S},\mu)$ be a positive measure space. If $1 < p < \infty$, then $(L_p(X,\mathcal{S},\mu),\|\cdot\|_p)$ is a reflexive Banach space.

Proof. If $p \geq 2$, the assertion is established by Corollary 8.3.1. Therefore suppose $1 < p < 2$. Lemma 8.3.2 asserts that $L_p(X,\mathcal{S},\mu)$ can be identified via the antilinear isometry $g \to x_g^*$ with a closed linear subspace of $L_q(X,\mathcal{S},\mu)^*$, where $1/p + 1/q = 1$. But $2 < q < \infty$, as $1 < p < 2$, and so $L_q(X,\mathcal{S},\mu)^*$ is reflexive by Theorem 8.1.2(ii), since it is the dual space of the reflexive Banach space $L_q(X,\mathcal{S},\mu)$. Therefore, by Theorem 8.1.2(iii), $L_p(X,\mathcal{S},\mu)$ is reflexive. □

Corollary 8.3.2. Let $(X,\mathcal{S},\mu)$ be a positive measure space and let $1 < p < \infty$. If $1/p + 1/q = 1$, then the mapping $g \to x_g^*$, defined by

$$x_g^*(f) = \int_X f(t)\overline{g(t)}\, d\mu(t) \qquad (f \in L_p(X,\mathcal{S},\mu)),$$

is a surjective antilinear isometry from $(L_q(X,\mathcal{S},\mu),\|\cdot\|_q)$ to $(L_p(X,\mathcal{S},\mu)^*,\|\cdot\|)$.

Proof. We know by Lemma 8.3.2 that $g \to x_g^*$ is an antilinear isometry from $L_q(X,\mathcal{S},\mu)$ to $L_p(X,\mathcal{S},\mu)^*$. Suppose $x^{**} \in L_p(X,\mathcal{S},\mu)^{**}$ is such that $x^{**}(x_g^*) = 0$, $g \in L_q(X,\mathcal{S},\mu)$. Since $L_p(X,\mathcal{S},\mu)$ is reflexive, there exists some $f \in L_p(X,\mathcal{S},\mu)$ such that $\tau(f) = x^{**}$. But then

$$\begin{aligned}\int_X f(t)\overline{g(t)}\, d\mu(t) &= x_g^*(f)\\ &= \tau(f)(x_g^*)\\ &= x^{**}(x_g^*)\\ &= 0 \qquad (g \in L_q(X,\mathcal{S},\mu)),\end{aligned}$$

from which it follows that $f = 0$.

Therefore from Corollary 4.2.8 to the Hahn-Banach Theorem we conclude that the mapping $g \to x_g^*$ is surjective. □

We close this chapter with a few remarks about $L_1(X,S,\mu)$. Except in trivial situations, $(L_1(X,S,\mu),\|\cdot\|_1)$ is not uniformly convex. For suppose $E_1,E_2 \in S$ are two disjoint sets with positive finite measure and let $f = \chi_{E_1}/\mu(E_1)$ and $g = \chi_{E_2}/\mu(E_2)$, where $\chi_{E_k}$ denotes the characteristic function of $E_k$, $k = 1,2$. Then it is easy to verify that $\|f\|_1 = \|g\|_1 = 1$ and $\|f - g\|_1 = 2$, whereas $\|(f + g)/2\|_1 = 1$.

It also happens that $L_1(X,S,\mu)$ fails to be reflexive unless it is finite dimensional (see Section 11.3).

However, it can be shown that $L_1(X,S,\mu)^*$ can be identified with $L_\infty(X,S,\mu)$, at least when $\mu$ is $\sigma$-finite. This was alluded to in Example 3.1.8. We state this result precisely in the next theorem.

<u>Theorem</u> <u>8.3.3</u>. Let $(X,S,\mu)$ be a $\sigma$-finite positive measure space. Then the mapping $g \to x_g^*$, defined by

$$x_g^*(f) = \int_X f(t)\overline{g(t)}\, d\mu(t) \qquad (f \in L_1(X,S,\mu)),$$

is a surjective antilinear isometry from $(L_\infty(X,S,\mu),\|\cdot\|_\infty)$ to $(L_1(X,S,\mu)^*,\|\cdot\|)$.

A proof of this result, which we do not give, can be found in [$DS_1$, pp. 289 and 290; Ry, pp. 246-248].

It is perhaps worthwhile mentioning that the mappings in Corollary 8.3.2 and Theorem 8.3.3 can be replaced by surjective linear isometries if we define $x_g^*$ by

$$x_g^*(f) = \int_X f(t)g(t)\, d\mu(t) \qquad (f \in L_p(X,S,\mu)),$$

where $g \in L_q(X,S,\mu)$, $1 \le p < \infty$, and $1/p + 1/q = 1$.

We have chosen the antilinear isometry so that the description of $L_2(X,S,\mu)^*$ will coincide with the usual characterization of the continuous linear functionals on the Hilbert space $L_2(X,S,\mu)$ in terms of inner products. This latter result will be discussed in Section 13.4.

## 8.4. Problems.

1. Let $(V,\|\cdot\|)$ be a Banach space over $\Phi$ and let $\tau$ denote the canonical embedding of $V$ into $V^{**}$. Prove that, if $V$ is not reflexive, then $\tau(V)$ is of category I in $V^{**}$.

2. Use the canonical embedding $\tau : V \to V^{**}$ to prove that every normed linear space $V$ is isometrically isomorphic to a dense subset of a Banach space. (Compare this with Problem 15 of Chapter 4.)

3. Let $(V,\|\cdot\|)$ be a Banach space over $\Phi$. Prove that, if $V$ is not reflexive, then its successive second duals $V^{**}$, $V^{****}, \ldots,$ are all distinct.

4. Let $(V_1,\|\cdot\|_1)$ and $(V_2,\|\cdot\|_2)$ be Banach spaces over $\Phi$ and let $\tau_1$ and $\tau_2$ denote the canonical embeddings of $V_1$ into $V_1^{**}$ and $V_2$ into $V_2^{**}$, respectively. If $T \in L(V_1,V_2)$, let $T^{**}$ denote the adjoint of the adjoint of $T$. Prove that $T^{**} \circ \tau_1 = \tau_2 \circ T$.

5. Prove that $(c,\|\cdot\|_\infty)$ is not reflexive.

6. Recall that $V'$ denotes the space of all linear functionals on $V$. It is possible to embed $V$ in $(V')' = V''$ by $\tau' : V \to V''$, defined as $\tau'(x)(x') = x'(x), x' \in V'$. Prove that $\tau'$ is surjective if and only if $V$ is finite dimensional. Conclude that $\tau : V \to V^{**}$ is more useful than $\tau'$.

7. (Proposition 8.2.1) Let $(V,\|\cdot\|)$ be a uniformly convex normed linear space over $\Phi$. Prove that, if $\{x_n\} \subset V$ is a sequence such that

(i) $\lim_n \|x_n\| = 1$

(ii) $\lim_{n,k} \|x_n + x_k\| = 2$,

then $\{x_n\}$ is a Cauchy sequence.

8. (Corollary 8.2.1) Let $(V, \|\cdot\|)$ be a uniformly convex Banach space over $\Phi$. Prove that, if $K \subset V$ is a nonempty closed convex set and $y_o \notin K$, then there exists a unique $x_o \in K$ such that $\|y_o - x_o\| = \inf_{x \in K} \|y_o - x\|$.

9. Let $(V, \|\cdot\|)$ be a uniformly convex Banach space over $\Phi$ and let $K \subset V$ be a nonempty closed convex locally compact set. Define a mapping $P_K : V \to K$ as follows: If $y_o \in K$, let $P_K(y_o) = y_o$, while if $y_o \notin K$, let $P_K(y_o) = x_o$, where $x_o$ is the unique point of $K$ such that $\|y_o - x_o\| = \inf_{x \in K} \|y_o - x\|$ ($x_o$ exists and is unique by Corollary 8.2.1). Prove that this "nearest point mapping" $P_K$ is continuous.

10. Let $(V, \|\cdot\|)$ be a uniformly convex Banach space over $\Phi$. By Corollary 8.2.2, if $x^* \in V^*$, $x^* \neq 0$, then there exists a unique $x_o \in V$, $\|x_o\| = 1$, such that $x^*(x_o) = \|x^*\|$. Let $y_o = \|x^*\| x_o$. Then $y_o$ is the unique point of $V$ such that $\|y_o\| = \|x^*\|$ and $x^*(y_o) = \|x^*\| \|y_o\|$. Define a mapping $T : V^* \to V$ by $T(x^*) = y_o$, $x^* \neq 0$, and $T(0) = 0$. Prove that $T$ is norm-preserving and surjective, but need not be injective (and hence need not be linear).

11. Let $(V, \|\cdot\|)$ be a uniformly convex normed linear space over $\Phi$. If $x, y \in V$ are such that $\|x + y\| = \|x\| + \|y\|$, prove that $x = ay$ for some $a \in \Phi$.

*12. Let $(V, \|\cdot\|)$ be a Banach space over $\Phi$. Consider the following convergence property (*): given a sequence $\{x_n\} \subset V$ and $x_o \in V$, if $\lim_n x^*(x_n) = x^*(x_o)$, $x^* \in V^*$, and if $\lim_n \|x_n\| = \|x_o\|$, then $\lim_n \|x_n - x_o\| = 0$.

(a) Prove that, if $(V, \|\cdot\|)$ is uniformly convex, then $V$ has the property (*). (For the case $V = L_p(X, S, \mu)$, $1 < p < \infty$, this is called the Radon-Riesz Theorem.)

(b) Property (*) does not characterize uniform convexity. Prove this by giving an example of a nonuniformly convex Banach space that has the property (*).

13. Let $(V,\|\cdot\|)$ be a uniformly convex Banach space over $\Phi$. If $\epsilon > 0$ is given, let $\delta(\epsilon)$ be such that if $\|x\| \leq 1$, $\|y\| \leq 1$, and $\|x - y\| \geq \epsilon$, then $\|(x + y)/2\| \leq 1 - \delta(\epsilon)$. If $x^* \in V^*$, $\|x^*\| = 1$, let $E = \{x \mid x \in V, |x^*(x) - 1| < \delta(\epsilon)/2\}$. Prove that, if $x,y \in E$ for some $x^* \in V^*$ and if $\|x\| \leq 1$, $\|y\| \leq 1$, then $\|x - y\| < \epsilon$.

14. Let $V$ be the space of all sequences $\{x_n\}$ of complex numbers such that $x_n \neq 0$ for at most finitely many values of $n$ and give $V$ the $\ell_2$ norm, so that $\|\{x_n\}\|_2 = (\Sigma_{n=1}^{\infty}|x_n|^2)^{1/2}$.

(a) Prove that $(V,\|\cdot\|_2)$ is uniformly convex.

(b) Give an example of an $x^* \in V^*$ such that for no $\{x_n\} \in V$, $\|\{x_n\}\|_2 = 1$, is it true that $x^*(\{x_n\}) = \|x^*\|$. Does this contradict Corollary 8.2.2?

15. Let $(V,\|\cdot\|)$ be a normed linear space over $\Phi$. $V$ is said to be <u>rotund</u> if $\|x\| \leq 1$ and $\|y\| \leq 1$, $x \neq y$, imply $\|(x + y)/2\| < 1$.

(a) Prove that every uniformly convex normed linear space is rotund.

(b) Prove that every finite-dimensional rotund space is uniformly convex.

(c) Give an example to show that the assumption of finite dimensionality cannot be dropped in problem (b).

16. Letting $(V,\|\cdot\|)$ be a uniformly convex normed linear space over $\Phi$ and letting $W \subset V$ be a linear subspace, prove each of the following:

(a) $W$ is uniformly convex.

(b) If $W$ is a closed linear subspace, then $V/W$ is uniformly convex.

17. Let $(V,\|\cdot\|)$ be a uniformly convex normed linear space over $\Phi$ and let $1 < p < \infty$. Prove that, given $\epsilon > 0$, there exists some $\delta_p > 0$ such that $\|x\| \leq 1$, $\|y\| \leq 1$, and $\|x - y\| \geq \epsilon$ imply $\|(x + y)/2\|^p \leq (1 - \delta_p)[(\|x\|^p + \|y\|^p)/2]$.

*18. Let $(V_n,\|\cdot\|_n)$, $n = 1,2,3,\ldots$, be Banach spaces over $\Phi$. If $1 \leq p < \infty$, let $\ell_p(\{V_n\})$ denote the space of all sequences $x = \{x_n\}$, $x_n \in V_n$, such that $\Sigma_{n=1}^{\infty}(\|x_n\|_n)^p < \infty$, with componentwise addition and scalar multiplication. Define a norm $\|\cdot\|_p$ on $\ell_p(\{V_n\})$ by

$$\|\{x_n\}\|_p = [\sum_{n=1}^{\infty}(\|x_n\|_n)^p]^{1/p}.$$

(a) Prove that $(\ell_p(\{V_n\}),\|\cdot\|_p)$ is a Banach space.

(b) If $1 < p < \infty$ and $1/p + 1/q = 1$, prove that the dual of $\ell_p(\{V_n\})$ is $\ell_q(\{V_n^*\})$. Conclude that $\ell_p(\{V_n\})$ is reflexive if and only if each $V_n$ is reflexive.

(c) Let $\ell_p(V)$ denote $\ell_p(\{V_n\})$ when $V_n = V$, $n = 1,2,3,\ldots$, $1 < p < \infty$. Prove that $\ell_p(V)$ is uniformly convex if and only if $V$ is uniformly convex. (It can be shown that $\ell_p(\{V_n\})$ is uniformly convex if and only if each $V_n$ is uniformly convex and, given $\epsilon > 0$, there exists a $\delta > 0$ that satisfies the uniform convexity condition for all $V_n$ simultaneously. See, for instance, the article M.M. Day, "Some more uniformly convex spaces", Bulletin of the American Mathematical Society, 47, 504-507 (1941).)

*19. (Theorem 8.3.3) Let $(X,\mathcal{S},\mu)$ be a $\sigma$-finite positive measure space. Prove that the mapping $g \to x_g^*$, defined by

$$x_g^*(f) = \int_X f(t)\overline{g(t)}\,d\mu(t) \qquad (f \in L_1(X,\mathcal{S},\mu)),$$

is a surjective antilinear isometry from $(L_\infty(X,\mathcal{S},\mu),\|\cdot\|_\infty)$ to $(L_1(X,\mathcal{S},\mu)^*,\|\cdot\|)$.

20. Give an example of an $x^* \in L_\infty([0,1])^*$ such that $x^* \notin L_1([0,1])$.

21. Let $(V, \|\cdot\|)$ be a reflexive Banach space over $\Phi$, let $M > 0$, and suppose $\{c_\alpha\}_{\alpha \in \Lambda} \subset \Phi$ and $\{x_\alpha^*\}_{\alpha \in \Lambda} \subset V^*$. Prove that the following are equivalent:

(a) There exists some $x \in V$ such that $\|x\| \leq M$ and $x_\alpha^*(x) = c_\alpha$, $\alpha \in \Lambda$.

(b) $|\Sigma_\alpha a_\alpha c_\alpha| \leq M\|\Sigma_\alpha a_\alpha x_\alpha^*\|$, where the sums are taken over all possible subsets $\{a_\alpha\}_{\alpha \in \Lambda} \subset \Phi$ where at most a finite number of the $a_\alpha$ are not zero.

# CHAPTER 9

# WEAK TOPOLOGIES

9.0. Introduction. In this and the following chapter we shall discuss topologies on linear spaces that are generated by families of linear functionals. This topic is a fundamental and substantial one in functional analysis, and our treatment is neither definitive nor exhaustive. After a discussion of some of the general properties of such topologies, we shall specialize to two particular ones -- the weak and weak* topologies -- and restrict our attention primarily to normed linear spaces. These two topologies, the first on $V$ and the second on the dual space $V^*$, are generated respectively by the continuous linear functionals on $V$, and by the continuous linear functionals on $V^*$ that are determined by the elements of $V$, that is, by $\tau(V)$.

The general properties of the weak and weak* topologies will be developed in Sections 9.2 and 9.3, and in Section 9.4 we shall prove one of the most important theorems of the subject, the Banach-Alaoglu Theorem. This theorem, in the form we shall prove, says that a set in the dual space of a Banach space is compact in the weak* topology if and only if it is closed in that topology and norm bounded. This result provides us with an appropriate replacement in dual spaces of the standard Euclidean characterization of compact sets as the closed and bounded sets. Some consequences of the Banach-Alaoglu Theorem concerning metrizability and sequential compactness in the weak and weak* topologies will also be discussed. Sections 9.5 through 9.8 contain various applications of the Banach-Alaoglu Theorem.

In Section 9.9 we shall give a characterization of reflexive Banach spaces in terms of the compactness of the norm closed unit ball in the weak topology and an application of this result to show that in a reflexive Banach space every nonempty norm closed convex set contains a vector of minimum norm. The final section is devoted to a theorem concerning adjoint transformations.

The general subject will be pursued further in Chapter 10, in which we shall prove the Krein-Šmulian and Eberlein-Šmulian Theorems.

9.1. $\mathcal{F}$-topologies. Our goal in this section is to describe how one can introduce a topology into a linear space $V$ by means of families of linear functionals on $V$ and to examine some of the basic properties of such topologies.

To be more precise, suppose $V$ is a linear space over $\Phi$ and let $\mathcal{F} \subset V'$ be a family of linear functionals on $V$. We wish to define a topology $\mathcal{T}_{\mathcal{F}}$ on $V$ such that $(V,\mathcal{T}_{\mathcal{F}})$ is a locally convex topological linear space for which a net $\{x_\alpha\} \subset V$ converges to $x \in V$ in the topology $\mathcal{T}_{\mathcal{F}}$ if and only if $\lim_\alpha x'(x_\alpha) = x'(x)$, $x' \in \mathcal{F}$, and for which the functionals in $\mathcal{F}$ are continuous linear functionals on $(V,\mathcal{T}_{\mathcal{F}})$. Defining a topology $\mathcal{T}_{\mathcal{F}}$ on $V$ that satisfies the last two requirements is not difficult.

Indeed, recalling our earlier discussion of topologies in seminormed linear spaces (Section 1.5), we define a family $P = \{p_{x'} \mid x' \in \mathcal{F}\}$ of seminorms on $V$ by setting $p_{x'}(x) = |x'(x)|$, $x \in V$ and $x' \in \mathcal{F}$. It is easily seen that each $p_{x'}$ is a seminorm. Then the topology $\mathcal{T}_P$ generated by $P$, that is, the topology whose neighborhood base at a point $x \in V$ consists of sets of the form

$$U(x;\epsilon;p_{x_1'},p_{x_2'},\ldots,p_{x_n'}) = \{y \mid y \in V,\ p_{x_k'}(x - y) < \epsilon,\ k = 1,2,\ldots,n\},$$

where $\epsilon > 0$, $n \in \mathbb{Z}$, $n > 0$, and the choice of $x_1',x_2',\ldots,x_n'$ in $\mathcal{F}$ is arbitrary, is such that $\{x_\alpha\}$ converges to $x$ in $\mathcal{T}_P$ if and only if $\lim_\alpha x'(x_\alpha) = x'(x)$, $x' \in \mathcal{F}$. Moreover, it is apparent that the elements of $\mathcal{F}$ are continuous with respect to $\mathcal{T}_P$.

However, the pair $(V, T_P)$ may not be a seminormed linear space -- equivalently, not a locally convex topological linear space -- since we have made no requirement that $|x'(x)| = p_{x'}(x) = 0$, $x' \in F$, should imply $x = 0$. If this were done, then from Proposition 1.5.1 and Theorems 1.5.1 and 2.3.1 we could immediately deduce that $(V, T_P)$ was a locally convex topological linear space with the desired properties. It is easily seen that $F \subset V'$ has the indicated property if and only if $F$ separates points.

We summarize this discussion in the next theorem, leaving the details of the proof to the reader.

Theorem 9.1.1. Let $V$ be a linear space over $\Phi$ and suppose $F \subset V'$ separates points. If $P = \{p_{x'} \mid x' \in F\}$ where

$$p_{x'}(x) = |x'(x)| \qquad (x \in V,\ x' \in F),$$

then $P$ is a family of seminorms on $V$ such that $(V, P)$ is a seminormed linear space over $\Phi$. Moreover, if $T_F = T_P$ is the topology on $V$ determined by the family $P$, then

(i) $(V, T_F)$ is a locally convex topological linear space over $\Phi$.

(ii) A net $\{x_\alpha\} \subset V$ converges to $x \in V$ in $T_F$ if and only if $\lim_\alpha x'(x_\alpha) = x'(x)$, $x' \in F$.

(iii) If $x' \in F$, then $x'$ is a continuous linear functional on $(V, T_F)$.

(iv) $T_F$ is the weakest topology on $V$ for which the elements of $F$ are continuous.

We shall often refer to the topology $T_F$ constructed in this way as the $F$-*topology*.

Before discussing some basic results about $F$-topologies, we wish to present a useful alternative way of considering such topologies.

Suppose $V$ is a linear space over $\Phi$ and $F \subset V'$ separates points. For each $x' \in F$ we set $\Phi_{x'} = \Phi$ and define the mapping

$T : V \to \Pi_{x' \in \mathcal{F}} \Phi_{x'}$ by setting $T(x) = \{x'(x)\}$, $x \in V$. That is, $T(x)$ is that element of the product space $\Pi_{x' \in \mathcal{F}} \Phi_{x'}$ whose x'th coordinate is $x'(x)$. Let us consider $V$ with the $\mathcal{F}$-topology $\mathcal{T}_{\mathcal{F}}$ and the product space $\Pi_{x' \in \mathcal{F}} \Phi_{x'}$ as a linear space with coordinatewise addition and scalar multiplication, and the usual product topology $\mathcal{P}$. Then $(V, \mathcal{T}_{\mathcal{F}})$ and $(\Pi_{x' \in \mathcal{F}} \Phi_{x'}, \mathcal{P})$ are topological linear spaces, and $T : V \to \Pi_{x' \in \mathcal{F}} \Phi_{x'}$ is linear and continuous. The mapping $T$ is also injective since, if $T(x) = T(y)$, then $x'(x) = x'(y)$, $x' \in \mathcal{F}$, and so $x = y$, as $\mathcal{F}$ separates points. Thus $T : V \to T(V)$ is a bijective continuous linear mapping, and hence $T^{-1} : T(V) \to V$ exists. Considering $T(V)$ as a topological space with the relative topology inherited from $\Pi_{x' \in \mathcal{F}} \Phi_{x'}$, we see at once that $T^{-1}$ is continuous. Hence $T$ is a linear homeomorphism between $(V, \mathcal{T}_{\mathcal{F}})$ and $(T(V), \mathcal{P}_{T(V)})$, where $\mathcal{P}_{T(V)}$ denotes the relative topology on $T(V)$, and thus one can consider the $\mathcal{F}$-topology $\mathcal{T}_{\mathcal{F}}$ on $V$ as just the relative topology on $T(V)$ considered as a subspace of $\Pi_{x' \in \mathcal{F}} \Phi_{x'}$.

The usefulness of this alternative description of the $\mathcal{F}$-topology is exemplified by the following theorem, whose proof is immediate on recalling that a countable product of metric spaces is again a metric space:

<u>Theorem</u> <u>9.1.2</u>. Let $V$ be a linear space over $\Phi$ and suppose $\mathcal{F} \subset V'$ separates points. If $\mathcal{F}$ is countable, then the $\mathcal{F}$-topology $\mathcal{T}_{\mathcal{F}}$ is metrizable.

The next two results are straightforward, and their proofs are left to the reader.

<u>Proposition</u> <u>9.1.1</u>. Let $V$ be a linear space over $\Phi$ and suppose $\mathcal{F}_k \subset V'$ separates points, $k = 1,2$. If $\mathcal{F}_1 \subset \mathcal{F}_2$, then $\mathcal{T}_{\mathcal{F}_1} \subset \mathcal{T}_{\mathcal{F}_2}$.

<u>Proposition</u> <u>9.1.2</u>. Let $(V, \mathcal{T})$ be a locally convex topological linear space over $\Phi$. If $\mathcal{F} \subset V^*$ separates points, then $\mathcal{T}_{\mathcal{F}} \subset \mathcal{T}$.

We noted in Theorem 9.1.1 that each $x' \in F$ is a continuous linear functional on $(V, T_F)$. Actually $F$ is precisely the set of continuous linear functionals on $(V, T_F)$, provided $F$ is a linear space.

Theorem 9.1.3. Let $V$ be a linear space over $\Phi$ and suppose $F \subset V'$ is a linear subspace that separates points. Then the following are equivalent:

(i) $x^* \in F$.

(ii) $x^*$ is a continuous linear functional on $(V, T_F)$.

Proof. As already indicated, part (i) implies part (ii). Conversely, suppose $x^*$ is a continuous linear functional on $(V, T_F)$ and, without loss of generality, assume that $x^* \neq 0$. Since $x^*$ is continuous at $0$, there exists some $\epsilon > 0$ and $x_1', x_2', \ldots, x_n'$ in $F$ such that, if $x \in U(0;\epsilon;p_{x_1'}, p_{x_2'}, \ldots, p_{x_n'})$, then $|x^*(x)| < 1$. We claim that $x^*$ is a linear combination of $x_1', x_2', \ldots, x_n'$.

Indeed, suppose $x \in \bigcap_{k=1}^n N(x_k')$. Then $x_k'(x) = 0$, $k = 1,2,\ldots,n$, implies $x \in U(0;\epsilon;p_{x_1'}, p_{x_2'}, \ldots, p_{x_n'})$, and so $|x^*(x)| < 1$. But each $N(x_k')$ is a linear subspace of $V$, and so $\bigcap_{k=1}^n N(x_k')$ is a linear subspace. Hence, if $x \in \bigcap_{k=1}^n N(x_k')$, then $mx \in \bigcap_{k=1}^n N(x_k')$, $m = 1,2,3,\ldots$, and so $|x^*(mx)| < 1$, $m = 1,2,3,\ldots$ . Consequently $x^*(x) = 0$, $x \in \bigcap_{k=1}^n N(x_k')$; that is, $\bigcap_{k=1}^n N(x_k') \subset N(x^*)$. Thus from Corollary 3.3.1 we see that $x^*$ is linearly dependent on $x_1', x_2', \ldots, x_n'$ and hence can be expressed as a linear combination of these elements of $F$.

Therefore $x^* \in F$ as $F$ is a linear space, and the proof is complete. □

This result, combined with Corollary 5.3.3 to the Hahn-Banach Theorem in its geometric form, immediately yields the following corollary:

Corollary 9.1.1. Let $V$ be a linear space over $\Phi$ and suppose $F \subset V'$ is a linear subspace that separates points. If $W \subset V$ is a linear subspace, then the following are equivalent:

(i) $W$ is a proper closed linear subspace of $(V, T_F)$.

(ii) If $x_o \notin W$, then there exists some $x' \in F$ such that $x'(x_o) = 1$ and $x'(x) = 0$, $x \in W$.

9.2. The Weak and Weak* Topologies. For the remainder of this chapter we concentrate our attention on two particular $F$-topologies, known generally as the weak and weak* topologies. In this section we define these topologies and discuss some elementary results.

Definition 9.2.1. Let $(V, T)$ be a locally convex topological linear space over $\Phi$. Then the $F$-topology on $V$ corresponding to $F = V^* \subset V'$ is called the weak topology on $V$. In this case we set $T_F = T^W$.

By Corollary 4.2.1 to the Hahn-Banach Theorem $V^* \subset V'$ separates points.

In view of the discussion in the preceding section we see that the basic neighborhoods in $T^W$ are sets of the form

$$U(x;\epsilon;x_1^*,x_2^*,\ldots,x_n^*) = \{y \mid y \in V, |x_k^*(x) - x_k^*(y)| < \epsilon, k = 1,2,\ldots,n\}$$

and that a net $\{x_\alpha\} \subset V$ converges to $x \in V$ in $T^W$; that is, $\{x_\alpha\}$ converges weakly to $x$, if and only if $\lim_\alpha x^*(x_\alpha) = x^*(x)$, $x^* \in V^*$.

Furthermore, from Theorems 9.1.1 and 9.1.3 we see that $(V, T^W)$ is a locally convex topological linear space over $\Phi$ and that $x' \in V'$ is a continuous linear functional on $(V, T^W)$ if and only if $x' \in V^*$; that is, a linear functional on $V$ is weakly continuous if and only if it is continuous.

We deduce the next proposition at once from Proposition 9.1.2.

Proposition 9.2.1. Let $(V,\mathcal{T})$ be a locally convex topological linear space over $\Phi$. Then $\mathcal{T}^w \subset \mathcal{T}$.

The weak* topology is an $\mathcal{F}$-topology, not on $V$, but on $V^*$. Before we can define this, we need to make a few preliminary observations.

Suppose $(V,\mathcal{T})$ is a locally convex topological linear space over $\Phi$. We claim, in analogy with our discussion of reflexive normed linear spaces in Section 8.1, that every $x \in V$ defines an element of $(V^*)'$. Indeed, as before, we define for each $x \in V$

$$\tau(x)(x^*) = x^*(x) \qquad (x^* \in V^*).$$

Then it is readily verified that $\tau(x) \in (V^*)'$, $x \in V$, and that $\tau : V \to (V^*)'$ is an injective linear mapping. Hence $V$ can be identified with $\tau(V)$ as a linear subspace of $(V^*)'$. Moreover, $\tau(V)$ separates points, since $\tau(x)(x^*) = x^*(x) = 0$, $x \in V$, implies $x^* = 0$.

We can now define the weak* topology.

Definition 9.2.2. Let $(V,\mathcal{T})$ be a locally convex topological linear space over $\Phi$. Then the $\mathcal{F}$-topology on $V^*$ corresponding to $\mathcal{F} = \tau(V) \subset (V^*)'$ is called the weak* topology on $V^*$. In this case we set $\mathcal{T}_{\mathcal{F}} = \mathcal{T}^{w^*}$.

Again, in view of Section 9.1, we observe the following facts: The basic neighborhoods in $\mathcal{T}^{w^*}$ are of the form

$$\begin{aligned} &U(x^*;\epsilon;\tau(x_1),\tau(x_2),\ldots,\tau(x_n)) \\ &\quad = \{y^* \mid y^* \in V^*, |\tau(x_k)(x^*) - \tau(x_k)(y^*)| < \epsilon, k = 1,2,\ldots,n\} \\ &\quad = \{y^* \mid y^* \in V^*, |x^*(x_k) - y^*(x_k)| < \epsilon, k = 1,2,\ldots,n\} \\ &\quad = U(x^*;\epsilon;x_1,x_2,\ldots,x_n) \end{aligned}$$

and $\{x_\alpha^*\}$ converges to $x^*$ in $\mathcal{T}^{w^*}$; that is, $\{x_\alpha^*\}$ <u>converges weak* to</u> $x^*$, if and only if $\lim_\alpha x_\alpha^*(x) = x^*(x)$, $x \in V$. Furthermore, $(V^*,\mathcal{T}^{w^*})$ is a locally convex topological linear space over $\Phi$, and $x'' \in (V^*)'$ is continuous in the topology $\mathcal{T}^{w^*}$ if and only if $x'' = \tau(x)$ for some $x \in V$. That is, a linear functional on $V^*$ is weak* continuous if and only if it belongs to $\tau(V) \subset (V^*)'$.

Note that, if $V$ is a normed linear space, then $V^*$ is a Banach space, and it can be equipped with both a weak and weak* topology by taking $\mathcal{F} \subset (V^*)'$ to be $V^{**}$ and $\tau(V)$, respectively. Propositions 9.1.1 and 9.1.2 then immediately show that the following proposition is valid:

<u>Proposition 9.2.2</u>. Let $(V,\|\cdot\|)$ be a normed linear space over $\Phi$ and denote the norm topology on $V^*$ by $\mathcal{T}$. Then $\mathcal{T}^{w^*} \subset \mathcal{T}^{w} \subset \mathcal{T}$.

We shall investigate the weak and weak* topologies on Banach spaces in considerable detail in this and the following chapter.

As we have indicated, the notions of continuity and weak continuity for linear functionals on locally convex topological linear spaces are equivalent. The same is true for linear transformations, in general, between Fréchet spaces.

<u>Theorem 9.2.1</u>. Let $(V_1,\mathcal{T}_1)$ and $(V_2,\mathcal{T}_2)$ be Fréchet spaces over $\Phi$ and suppose $T \in \mathcal{L}'(V_1,V_2)$. Then the following are equivalent:

(i) $T \in \mathcal{L}(V_1,V_2)$.
(ii) $T : (V_1,\mathcal{T}_1^w) \to (V_2,\mathcal{T}_2^w)$ is continuous.

<u>Proof</u>. Suppose $T \in \mathcal{L}(V_1,V_2)$ and let $U(0;\epsilon;y_1^*,y_2^*,\ldots,y_n^*)$ be a $\mathcal{T}_2^w$ neighborhood of $0$. Let $x_k^*(x) = y_k^*[T(x)]$, $x \in V$, $k = 1,2,\ldots,n$. Then $x_k^* \in V_1^*$, and if $x \in U(0;\epsilon;x_1^*,x_2^*,\ldots,x_n^*)$, then $|y_k^*[T(x)]| = |x_k^*(x)| < \epsilon$, $k = 1,2,\ldots,n$; that is, $T(x) \in U(0;\epsilon;y_1^*,y_2^*,\ldots,y_n^*)$. Thus $T$ is weakly continuous at $0$,

and hence $T$ is weakly continuous by Proposition 3.2.2. Hence part (i) implies part (ii).

Conversely, suppose $T$ is weakly continuous. We shall show that $T \in L(V_1,V_2)$ by proving that $T$ is a closed mapping and applying the Closed Graph Theorem (Theorem 7.3.1). With this in mind, for each $y^* \in V_2^*$ we again define $x^*(x) = y^*[T(x)]$, $x \in V_1$. Then clearly $x^* \in V_1'$. We claim that $x^* \in V_1^*$. To see this it suffices, by Theorem 9.1.3, to show that $x^*$ is weakly continuous. But this follows at once since $y^* \in V_2^*$ is continuous, and so weakly continuous, and $T\colon (V_1,\mathcal{T}_1^w) \to (V_2,\mathcal{T}_2^w)$ is assumed to be continuous. Hence $x^* \in V_1^*$.

Now suppose $\{x_n\} \subset V_1$, $x \in V_1$, and $y \in V_2$ are such that $\{x_n\}$ converges to $x$ in $\mathcal{T}_1$ and $\{T(x_n)\}$ converges to $y$ in $\mathcal{T}_2$. Then for each $y^* \in V_2^*$ we have, on the one hand,

$$\lim_n y^*[T(x_n)] = y^*(y),$$

and, on the other hand,

$$\begin{aligned}\lim_n y^*[T(x_n)] &= \lim_n x^*(x_n)\\ &= x^*(x)\\ &= y^*[T(x)].\end{aligned}$$

Thus $y^*(y) = y^*[T(x)]$, $y^* \in V_2^*$, and so from Corollary 4.2.1 to the Hahn-Banach Theorem we conclude that $y = T(x)$.

Therefore $T$ is a closed mapping and so continuous by the Closed Graph Theorem. □

Thus a linear transformation between two Fréchet spaces is continuous if and only if it is weakly continuous. Note that the only if portion of this equivalence, that is, part (i) of Theorem 9.2.1 implying part (ii), remains valid for any locally convex topological linear space.

As indicated in Proposition 9.2.1, the weak topology on a locally convex topological linear space $(V,\mathcal{T})$ is weaker than the original topology, so that, in general, a subset $E \subset V$ that is closed in $\mathcal{T}^W$ must be closed in $\mathcal{T}$, but the converse need not always hold. Nevertheless, if we restrict our attention to certain algebraic types of sets in $V$, then the two notions of being closed are equivalent.

Theorem 9.2.2. Let $(V,\mathcal{T})$ be a locally convex topological linear space over $\Phi$ and suppose $K \subset V$ is a nonempty convex set. Then the following are equivalent:

(i) $K$ is closed in $\mathcal{T}^W$.
(ii) $K$ is closed in $\mathcal{T}$.

Proof. Clearly part (i) implies part (ii), as $\mathcal{T}^W \subset \mathcal{T}$, so we assume that $K$ is closed in $\mathcal{T}$. From Corollary 5.4.2 to the geometric form of the Hahn-Banach Theorem we know that $K$ is the intersection of all the closed half-spaces that contain $K$. Since each of these half-spaces is of the form $\{x \mid x \in V, x^*(x) \leq a\}$ for some $a \in \mathbb{R}$ and for some real continuous linear functional on $V$, we see that each such half-space is weakly closed.

Hence $K$ is closed in $\mathcal{T}^W$, and part (ii) of the theorem implies part (i). □

In particular, we note that a linear subspace $W$ of a locally convex topological linear space $V$ is closed if and only if it is weakly closed.

As an easy corollary of the theorem, whose proof is left to the reader, we have the next result:

Corollary 9.2.1. Let $(V,\mathcal{T})$ be a locally convex topological linear space over $\Phi$. If $\{x_\alpha\} \subset V$ is a net that converges in $\mathcal{T}^W$ to $x \in V$, then there exists a net $\{y_\beta\} \subset V$, where each $y_\beta$ is a finite sum of the form $\Sigma_\alpha a_{\beta\alpha} x_\alpha$, $a_{\beta\alpha} \geq 0$, $\Sigma_\alpha a_{\beta\alpha} = 1$, such that $\{y_\beta\}$ converges in $\mathcal{T}$ to $x$.

Thus, if a net $\{x_\alpha\}$ converges weakly to $x$, then there is a net $\{y_\beta\}$ whose elements are convex linear combinations of the elements of $\{x_\alpha\}$ and which converges to $x$.

Concrete examples of sets whose closure in $\mathcal{T}$ and $\mathcal{T}^w$ do not coincide are easy to come by. For example, let $V = \ell_2$ with the usual norm topology given by the norm $\|\{a_k\}\|_2 = (\Sigma_{k=1}^{\infty}|a_k|^2)^{1/2}$. Let, as in Section 7.6, $e^m = \{e_k^m\} \in \ell_2$ denote the sequence defined by $e_k^m = 0$, $k \neq m$, $e_m^m = 1$, $m = 1,2,3,\ldots$ . Then it is easily seen that $\|e^m - e^n\|_2 = \sqrt{2}$, $m \neq n$, from which it follows that the sequence $\{e^m\}$ in $\ell_2$ does not converge in the norm topology. In particular, the origin does not belong to the norm closure of the set consisting of the sequences $e^m$, $m = 1,2,3,\ldots$ .

However, from Corollary 8.3.2 we know that $\ell_2^*$ can be identified with $\ell_2$. Indeed, if $x^* \in \ell_2^*$ and $x^*$ corresponds to $\{b_k\} \in \ell_2$, then $x^*(x) = \Sigma_{k=1}^{\infty} a_k \overline{b}_k$, $x = \{a_k\} \in \ell_2$. But then we see that for each $x^* \in \ell_2^*$

$$\lim_m x^*(e^m) = \lim_m \overline{b}_m = 0,$$

as $\Sigma_{k=1}^{\infty}|b_k|^2 < \infty$. Thus the origin belongs to the closure of the set consisting of the sequences $e^m$, $m = 1,2,3,\ldots$, in $\mathcal{T}^w$.

The analog of Theorem 9.2.2 with $\mathcal{T}^w$ replaced by $\mathcal{T}^{w^*}$ fails to be valid. For instance, it is not difficult to give examples of proper closed linear subspaces that are weak* dense. This will follow easily from the next theorem, which is of independent interest.

<u>Theorem 9.2.3</u>. Let $(V,\|\cdot\|)$ be a normed linear space over $\Phi$ and set

$$B_1 = \{x \mid x \in V,\ \|x\| \leq 1\},$$

$$B_1^{**} = \{x^{**} \mid x^{**} \in V^{**},\ \|x^{**}\| \leq 1\}.$$

If $\tau : V \to V^{**}$ is the canonical embedding of $V$ into $V^{**}$, then $\tau(B_1)$ is dense in $B_1^{**}$ in $(V^{**},\mathcal{T}^{w^*})$.

Proof. Let $x_o^{**} \in B_1^{**}$ and suppose $U(x_o^{**};\epsilon;x_1^*,x_2^*,\ldots,x_n^*)$ is an open neighborhood of $x_o^{**}$ in $\mathcal{T}^{w^*}$. It suffices to show that there exists some $x_o \in V$, $\|x_o\| \le 1$, such that

$$\tau(x_o) \in U(x_o^{**};\epsilon;x_1^*,x_2^*,\ldots,x_n^*).$$

To this end let $\delta = \epsilon/2 \max_{k=1,2,\ldots,n}\|x_k^*\|$ and set $y_o^{**} = (1-\delta)x_o^{**}$.

It is readily seen that $\|y_o^{**}\| < 1$ and

$$y_o^{**} \in U(x_o^{**};\epsilon;x_1^*,x_2^*,\ldots,x_n^*).$$

From Corollary 4.10.1 to Helly's Theorem we deduce the existence of some $x_o \in V$ such that $\|x_o\| \le \|y_o^{**}\| + (1 - \|y_o^{**}\|)/2 < 1$ and $x_k^*(x_o) = y_o^{**}(x_k^*)$, $k = 1,2,\ldots,n$. Evidently $x_o \in B_1$, and

$$\begin{aligned} |x_o^{**}(x_k^*) - \tau(x_o)(x_k^*)| &= |x_o^{**}(x_k^*) - x_k^*(x_o)| \\ &= |x_o^{**}(x_k^*) - y_o^{**}(x_k^*)| \\ &< \epsilon \qquad\qquad (k = 1,2,\ldots,n) \end{aligned}$$

shows that $\tau(x_o) \in U(x_o^{**};\epsilon;x_1^*,x_2^*,\ldots,x_n^*)$.

Therefore $\tau(B_1)$ is weak* dense in $B_1^{**}$. □

Corollary 9.2.2. Let $(V,\|\cdot\|)$ be a normed linear space over $\Phi$. If $\tau : V \to V^{**}$ is the canonical embedding of $V$ into $V^{**}$, then $\tau(V)$ is dense in $(V^{**},\mathcal{T}^{w^*})$.

In particular, if $V$ is a nonreflexive Banach space (for example, $c_o$), then $\tau(V)$ is a proper norm closed linear subspace of $V^{**}$ that is weak* dense in $V^{**}$, so that the norm closure and weak* closure of $\tau(V)$ do not coincide.

9.3. Completeness in the Weak and Weak* Topologies. From this point through Chapter 10 we shall concentrate our attention solely on normed linear spaces. We do this primarily to give the development of the succeeding sections a somewhat more concrete flavor than would be the case if we tried to treat the various matters discussed in their fullest generality. Some partial references to more general treatments will be indicated at appropriate points.

Our concern in this section will be the completeness properties of the weak and weak* topologies. We shall see that $\mathcal{T}^{w^*}$ is always sequentially complete when $V$ is a Banach space, that $\mathcal{T}^{w}$ is sequentially complete provided $V$ is reflexive, but that neither topology is complete if $V$ is infinite-dimensional.

Theorem 9.3.1. Let $(V,\|\cdot\|)$ be a normed linear space over $\Phi$.

(i) If $\{x_k\} \subset V$ is a Cauchy sequence in $(V,\mathcal{T}^{w})$, then there exists some $M > 0$ such that $\sup_k\|x_k\| \leq M$.

(ii) If $(V,\|\cdot\|)$ is a Banach space and $\{x_k^*\} \subset V^*$ is a Cauchy sequence in $(V^*,\mathcal{T}^{w^*})$, then there exists some $M > 0$ such that $\sup_k\|x_k^*\| \leq M$.

(iii) If $(V,\|\cdot\|)$ is a Banach space, then $(V^*,\mathcal{T}^{w^*})$ is sequentially complete.

(iv) If $(V,\|\cdot\|)$ is a reflexive Banach space, then $(V,\mathcal{T}^{w})$ is sequentially complete.

Proof. The proofs are all essentially applications of the Uniform Boundedness Theorem (Theorem 6.2.1). For instance, suppose $\{x_k\} \subset V$ is a Cauchy sequence in $(V,\mathcal{T}^{w})$. Then $\{x^*(x_k)\}$ is a Cauchy sequence for each $x^* \in V^*$, and so there exists some $M_{x^*} > 0$ such that

$$\sup_k |\tau(x_k)(x^*)| = \sup_k |x^*(x_k)| \leq M_{x^*}.$$

Hence by the Uniform Boundedness Theorem and Theorem 8.1.1 there exists some $M > 0$ for which

$$\sup_k \|x_k\| = \sup_k \|\tau(x_k)\| \leq M.$$

This proves part (i) of the theorem.

A similar argument establishes part (ii), and part (iii) is proved through an application of the Banach-Steinhaus Theorem (Theorem 6.2.2). The details are left to the reader.

If $(V,\|\cdot\|)$ is a reflexive Banach space and $\{x_k\}$ is a Cauchy sequence in $(V,\mathcal{T}^w)$, then it is readily verified from the basic definitions that $\{\tau(x_k)\}$ is a Cauchy sequence in $(V^{**},\mathcal{T}^{w^*})$. Since the weak* topology on $V^{**}$ is sequentially complete, it follows at once that $\{\tau(x_k)\}$ converges in $(V^{**},\mathcal{T}^{w^*})$, and so $\{x_k\}$ converges in $(V,\mathcal{T}^w)$, as $\tau(V) = V^{**}$.

Therefore $(V,\mathcal{T}^w)$ is sequentially complete, and part (iv) of the theorem is proved. □

The argument used to prove the last portion of Theorem 9.3.1 is worth noting, as it is a special case of a general principle. Namely, if some property has been shown to hold for the weak* topology, then the same property must be valid for the weak topology on any reflexive Banach space V. This follows immediately from the definition of the weak and weak* topologies in V and $V^{**}$, respectively, and the fact that $\tau(V) = V^{**}$. We shall make repeated use of this observation in the sequel.

Among other things, Theorem 9.3.1 shows that $L_p(X,\mathcal{S},\mu)$, $1 < p < \infty$, is sequentially complete in both the weak and weak* topologies. It is well to note that a nonreflexive Banach space may or may not be weakly sequentially complete, depending on the space in question. For example, $L_1(X,\mathcal{S},\mu)$ is sequentially complete in the weak topology when $\mu$ is $\sigma$-finite, whereas $c_0$ is not.

<u>Theorem 9.3.2</u>. Let $(X,\mathcal{S},\mu)$ be a $\sigma$-finite positive measure space. Then $(L_1(X,\mathcal{S},\mu),\mathcal{T}^w)$ is sequentially complete.

<u>Proof</u>. Suppose that $\{f_k\} \subset L_1(X,\mathcal{S},\mu)$ is a weak Cauchy sequence. Then, since $L_1(X,\mathcal{S},\mu)^* = L_\infty(X,\mathcal{S},\mu)$ (Theorem 8.3.3), we see that $\{\int_E f_k(t)\, d\mu(t)\}$ is a Cauchy sequence of complex numbers for each

$E \in S$, as the characteristic function $\chi_E$ of $E$ belongs to $L_\infty(X,S,\mu)$. In particular, each such sequence converges, and so we can define a set function $\sigma$ on $S$ by

$$\sigma(E) = \lim_k \int_E f_k(t)\, d\mu(t) \qquad (E \in S).$$

It can be shown that $\sigma$ is a bounded complex-valued measure on $X$ that is absolutely continuous with respect to $\mu$. The nontrivial portion of this assertion is the fact that $\sigma$ is a measure -- that is, $\sigma$ is countably additive. We do not give the details of this here but instead refer the reader to the discussion of the Vitali-Hahn-Saks Theorem (see, for example, [$DS_1$, pp. 158-160; Y, pp. 70-72]), of which the preceding assertion is a special case.

Now, by the Lebesgue-Radon-Nikodym Theorem (see, for example, $DS_1$, p. 176]), we deduce the existence of some $f \in L_1(X,S,\mu)$ such that

$$\begin{aligned} \sigma(E) &= \lim_k \int_E f_k(t)\, d\mu(t) \\ &= \int_E f(t)\, d\mu(t) \qquad (E \in S). \end{aligned}$$

Furthermore, if $g$ is any finite linear combination of characteristic functions of sets in $S$, then the preceding identity shows that

$$\lim_k \int_X f_k(t)\overline{g(t)}\, d\mu(t) = \int_X f(t)\overline{g(t)}\, d\mu(t).$$

But such functions are norm dense in $L_\infty(X,S,\mu)$, from which we deduce by a simple triangle inequality argument that

$$\lim_k \int_X f_k(t)\overline{h(t)}\, d\mu(t) = \int_X f(t)\overline{h(t)}\, d\mu(t) \qquad (h \in L_\infty(X,S,\mu)).$$

Therefore $\{f_k\}$ converges weakly to $f$, as $L_1(X,S,\mu)^* = L_\infty(X,S,\mu)$, and so $L_1(X,S,\mu)$ is weakly sequentially complete. □

To see that $c_o$ is not weakly sequentially complete we need only exhibit a weak Cauchy sequence in $c_o$ that does not converge. In order to do this we need to know that $c_o^*$ can be identified with

$\ell_1$, a fact whose proof we leave to the reader. The identification is obtained in the expected manner, namely, $\{a_k\} \in \ell_1$ corresponds to $x^* \in c_o^*$, defined by

$$x^*(y) = \sum_{k=1}^{\infty} b_k a_k \qquad (y = \{b_k\} \in c_o).$$

Now for each positive integer $n$ define $y^n = \{b_k^n\} \in c_o$ by $b_k^n = 1$, $k = 1,2,\ldots,n$, $b_k^n = 0$, $k > n$. Then we see that for any $x = \{a_k\} \in \ell_1$ and $n > m$

$$\begin{aligned} |x^*(y^n) - x^*(y^m)| &= \left|\sum_{k=1}^{n} a_k - \sum_{k=1}^{m} a_k\right| \\ &= \left|\sum_{k=m+1}^{n} a_k\right| \\ &\leq \sum_{k=m+1}^{n} |a_k|, \end{aligned}$$

from which it follows immediately that $\{x^*(y^n)\}$ is a Cauchy sequence, as $\Sigma_{k=1}^{\infty}|a_k| < \infty$. Thus $\{y^n\}$ is a weak Cauchy sequence in $c_o$. But $\{y^n\}$ cannot converge weakly to any element of $c_o$, because if $y = \{b_k\} \in c_o$ were such an element, then we would have

$$\begin{aligned} \sum_{k=1}^{\infty} b_k a_k &= x^*(y) \\ &= \lim_n x^*(y^n) \\ &= \lim_n \sum_{k=1}^{n} a_k \\ &= \sum_{k=1}^{\infty} a_k \qquad (\{a_k\} \in \ell_1), \end{aligned}$$

and so $b_k = 1$, $k = 1,2,3,\ldots$, contradicting the assumption that $\lim_k b_k = 0$.

Hence $c_o$ is not weakly sequentially complete. If we shift our attention from sequential completeness to completeness, then the weak and weak* topologies are complete if and only if $V$ is finite dimensional.

Theorem 9.3.3. Let $(V,\|\cdot\|)$ be a normed linear space over $\Phi$. Then the following are equivalent:

(i) $V$ is finite dimensional.

(ii) $(V,\mathcal{T}^W)$ is a complete locally convex topological linear space over $\Phi$.

(iii) $(V^*,\mathcal{T}^{W^*})$ is a complete locally convex topological linear space over $\Phi$.

Proof. Since all n-dimensional topological linear spaces over $\Phi$ are topologically isomorphic (Corollary 2.2.1), it follows easily that part (i) implies parts (ii) and (iii). Conversely, if $V$ is infinite-dimensional, then we shall construct a net in $V$ that is a Cauchy net in $(V,\mathcal{T}^W)$, but is not convergent in $(V,\mathcal{T}^W)$. This will prove that part (ii) implies part (i). The implication from part (iii) to part (i) is left to the reader.

With this in mind we consider the directed set

$$\Lambda = \{\alpha \mid \alpha \subset V^*,\ \alpha \text{ finite}\},$$

where $\alpha > \beta$ if and only if $\alpha \supset \beta$ and $\alpha,\beta \in \Lambda$. Let $x'' \in (V^*)' \sim V^{**}$. We have not as yet proved the existence of such a discontinuous linear functional on an infinite-dimensional normed linear space, and so before continuing with the proof of the theorem we shall do so.

Let $\{x_k\} \subset V$ be a sequence of linearly independent vectors in $V$, where $V$ is an infinite-dimensional normed linear space, and for each positive integer $k$ let $\epsilon_k > 0$ be such that $\|\epsilon_k x_k\| < 1/k$. Then, as is readily verified, $\{\epsilon_k x_k\}$ is a sequence of linearly independent vectors such that $\lim_k \|\epsilon_k x_k\| = 0$. Let $W$ be the linear subspace of $V$ spanned by the sequence $\{x_k\}$ and define a linear functional $y'$ on $W$ by setting $y'(x_k) = 1/\epsilon_k$, $k = 1,2,3,\ldots$, and extending the definition of $y'$ to all of $W$ by linearity. That is, if $y = \Sigma\, a_k x_k \in W$, where the sum is of course finite, then $y'(y) = \Sigma\, a_k y'(x_k)$. Next, let $B \subset V$ be an algebraic basis for $V$ such that $B \supset \{x_k\}$. Then on defining $x'(x_k) = y'(x_k)$, $k = 1,2,3,\ldots$, $x'(x) = 0$, $x \in B \sim \{x_k\}$, and extending the definition to all of $V$ by linearity, we obtain a linear functional $x'$ defined on all of $V$. Moreover, $x' \in V' \sim V^*$, as $\lim_k \|\epsilon_k x_k\| = 0$ and $\lim_k x'(\epsilon_k x_k) = 1$. Thus we see that, when $V$ is an infinite-dimensional normed linear space, there always exists some $x' \in V' \sim V^*$.

Consider now any such discontinuous linear functional $x''$ on the infinite-dimensional Banach space $V^*$. The dimension of $V^*$ must be infinite, since if it were finite, then $V^{**}$ and hence $\tau(V) \subset V^{**}$, would be finite dimensional. This would imply that $V$ is finite dimensional, contrary to assumption. If $\alpha = \{x_1^*, x_2^*, \ldots, x_n^*\}$ is in $\Lambda$, then we denote by $V_\alpha^*$ the linear subspace of $V^*$ spanned by $\alpha$, and by $x_\alpha''$ the restriction of $x''$ to $V_\alpha^*$. From Proposition 3.2.4 we see that $x_\alpha''$ is a continuous linear functional with norm $\|x_\alpha''\|$ on the finite-dimensional space $V_\alpha^*$, and so for any $a_1, a_2, \ldots, a_n$ in $\Phi$ we have that

$$\begin{aligned} \left|\sum_{k=1}^{n} a_k x''(x_k^*)\right| &= \left|x''\left(\sum_{k=1}^{n} a_k x_k^*\right)\right| \\ &= \left|x_\alpha''\left(\sum_{k=1}^{n} a_k x_k^*\right)\right| \\ &\leq \|x_\alpha''\| \left\|\sum_{k=1}^{n} a_k x_k^*\right\|. \end{aligned}$$

Consequently, by an appeal to Helly's Theorem (Theorem 4.10.1), for each $\alpha \in \Lambda$ we deduce the existence of some $x_\alpha \in V$ such that $x^*(x_\alpha) = x''(x^*)$, $x^* \in \alpha$. Moreover, it is apparent that, if $\alpha > \beta$, then $x^*(x_\alpha) = x^*(x_\beta)$, $x^* \in \beta$, as $\alpha > \beta$ if and only if $\alpha \supset \beta$.

We claim that the net $\{x_\alpha\} \subset V$ as here constructed is a Cauchy net in $(V, \mathcal{T}^w)$. Indeed, if $x^* \in V^*$ and $\gamma \in \Lambda$ is such that $x^* \in \gamma$, then for any $\alpha, \beta \in \Lambda$, where $\alpha > \gamma$ and $\beta > \gamma$, we obtain $x^*(x_\alpha) - x^*(x_\beta) = x''(x^*) - x''(x^*) = 0$, and so $\{x_\alpha\}$ is a weak Cauchy net.

But $\{x_\alpha\}$ cannot converge in $\mathcal{T}^w$ to any element of V, because, if $x \in V$ is such that $\{x_\alpha\}$ converges to x in $(V,\mathcal{T}^w)$, then for each $x^* \in V^*$ we have $\lim_\alpha x^*(x_\alpha) = x^*(x)$. Furthermore, if $x^* \in V^*$ and $\gamma \in \Lambda$ is such that $x^* \in \gamma$, then for any $\alpha > \gamma$ we have $x^*(x_\alpha) = x''(x^*)$. Hence $x''(x^*) = \lim_\alpha x^*(x_\alpha) = x^*(x) = \tau(x)(x^*)$ for each $x^* \in V^*$, from which we conclude that $x'' = \tau(x) \in V^{**}$. This, however, contradicts the fact that $x''$ is discontinuous on $V^*$.

Therefore $\{x_\alpha\}$ does not converge in $(V,\mathcal{T}^w)$, and so $(V,\mathcal{T}^w)$ is not complete.

□

It is easily seen that, if we begin with a nonreflexive normed linear space $(V,\|\cdot\|)$, then the proof that $(V,\mathcal{T}^w)$ is not complete when V is infinite dimensional can be somewhat shortened. Indeed, in this case we need only take $x'' \in V^{**} \sim \tau(V)$ and argue as before. The existence of $x''$ is now assured by the definition of reflexivity. If, however, V is reflexive, then we must use the discontinuous linear functional $x''$ as done in the proof of the theorem.

Next let us give some necessary and sufficient conditions for the convergence of a sequence in the weak or weak* topology. The proofs are straightforward and are left to the reader.

<u>Theorem 9.3.4</u>. Let $(V,\|\cdot\|)$ be a normed linear space over $\Phi$. Then

(i) A sequence $\{x_k\} \subset V$ converges to x in $(V,\mathcal{T}^w)$ if and only if $\sup_k\|x_k\| < \infty$ and $\lim_k x^*(x_k) = x^*(x_k)$ for each $x^*$ in some norm dense subset of $(V^*,\|\cdot\|)$.

(ii) A sequence $\{x_k^*\} \subset V^*$ converges to $x^*$ in $(V^*,\mathcal{T}^{w^*})$ if and only if $\sup_k\|x_k^*\| < \infty$ and $\lim_k x_k^*(x) = x^*(x)$ for each x in some norm dense subset of $(V,\|\cdot\|)$.

Arguing much as we did in proving that $L_1(X,S,\mu)$ is weakly sequentially complete, together with an appeal to Theorem 9.3.1(i), provides us with a proof of the following corollary to Theorem 9.3.4:

Corollary 9.3.1. Let $(X,S,\mu)$ be a $\sigma$-finite positive measure space and suppose $\{f_k\} \subset L_1(X,S,\mu)$. Then the following are equivalent:

(i) There exists some $f \in L_1(X,S,\mu)$ such that the sequence $\{f_k\}$ converges weakly to $f$.

(ii) $\sup_k \|f_k\|_1 < \infty$ and, for each $E \in S$, the sequence $\{\int_E f_k(t)\, d\mu(t)\}$ converges.

While we are on the subject of describing weak convergence let us take a look at $C_o(X)$.

Theorem 9.3.5. Let $X$ be a locally compact Hausdorff topological space and suppose $\{f_k\} \subset C_o(X)$, $f \in C_o(X)$. Then the following are equivalent:

(i) The sequence $\{f_k\}$ converges weakly to $f$.

(ii) $\sup_k \|f_k\|_\infty < \infty$ and $\lim_k f_k(t) = f(t)$, $t \in X$.

Proof. If part (i) holds, then from Theorem 9.3.1(i) we see that $\sup_k \|f_k\|_\infty < \infty$. Moreover, for each $t \in X$ the formula $x_t^*(g) = g(t)$, $g \in C_o(X)$, clearly defines a continuous linear functional $x_t^*$ on $C_o(X)$. Since $\{f_k\}$ converges weakly to $f$, it follows immediately that $\lim_k f_k(t) = f(t)$, $t \in X$. Thus part (i) implies part (ii).

Conversely suppose that part (ii) holds and let $x^* \in C_o(X)^*$. Then by the Riesz Representation Theorem (Theorem 4.8.2) there exists a unique bounded, regular, complex-valued Borel measure $\mu_{x^*}$ such that

$$x^*(g) = \int_X g(t)\, d\mu_{x^*}(t) \qquad (g \in C_o(X)).$$

Let $M = \sup_k \|f_k\|_\infty$. Then the function $g_M(t) = M$, $t \in X$, belongs to $L_1(X,\mu_{x^*})$, as $\mu_{x^*}$ is bounded, and $|f_k(t)| \le M$, $t \in X$, $k = 1,2,3,\ldots$ . Hence, since the sequence $\{f_k\}$ converges pointwise to $f$, we may appeal to the Lebesgue Dominated Convergence Theorem

[Ry, p. 229] to conclude that

$$\begin{aligned}\lim_k x^*(f_k) &= \lim_k \int_X f_k(t)\, d\mu_{x^*}(t) \\ &= \int_X f(t)\, d\mu_{x^*}(t) \\ &= x^*(f).\end{aligned}$$

Therefore, since $x^* \in C_o(X)^*$ is arbitrary, we see that $\{f_k\}$ converges weakly to $f$, and part (ii) of the theorem implies part (i). □

9.4. The Banach-Alaoglu Theorem. We have mentioned several times (for example, after Theorem 1.3.3) that closed bounded sets in infinite-dimensional normed linear spaces need not be compact in the norm topology, and so norm compactness in such spaces is not equivalent to norm closedness and norm boundedness. It is one of the chief virtues of the weak* topology that we can recover an analog of this Euclidean characterization of compactness in the dual space of a Banach space, namely, that a set's being weak* closed and norm bounded is equivalent to its being weak* compact. This is the Banach-Alaoglu Theorem, whose precise statement and proof, which is essentially an application of Tikhonov's Theorem that the topological product of compact spaces is compact and the Uniform Boundedness Theorem, shall be our first concern in this section.

After discussing some general consequences of the Banach-Alaoglu Theorem we shall devote the succeeding four sections to some examples of applications of this important result. We shall then return to general considerations of weak topologies, reflexivity, and compactness.

Theorem 9.4.1 (Banach-Alaoglu Theorem). Let $(V, \|\cdot\|)$ be a Banach space over $\Phi$ and let $E \subset V^*$. Then the following are equivalent:

(i) $E$ is a compact set in $(V^*, \mathcal{T}^{w^*})$.

(ii) $E$ is a closed set in $(V^*, \mathcal{T}^{w^*})$ and a bounded set in $(V^*, \|\cdot\|)$.

Proof. Suppose $E$ is compact in $(V^*, \mathcal{T}^{w^*})$. Since $\mathcal{T}^{w^*}$ is a Hausdorff topology, it follows at once that $E$ is weak* closed. Recall from our discussion of $\mathcal{F}$-topologies following Theorem 9.1.1 that $\mathcal{T}^{w^*}$ can be considered as the relative topology on $T(V^*)$ considered as a subspace of the topological product space $\Pi_{x \in V} \Phi_{\tau(x)}$, where $T(x^*) = \{\tau(x)(x^*)\}$ and each $\Phi_{\tau(x)} = \Phi$, $x \in V$. For each $y \in V$ let $P_y : \Pi_{x \in V} \Phi_{\tau(x)} \to \Phi_{\tau(y)}$ denote the projection of the product space onto $\Phi_{\tau(y)}$. Since each $P_y$ is continuous and $T$ is a homeomorphism, we conclude that $P_y \circ T : V^* \to \Phi_{\tau(y)}$ is continuous for each $y \in V$.

In particular, $P_y \circ T(E) = \{x^*(y) \mid x^* \in E\}$ is a compact subset of $\Phi_{\tau(y)} = \Phi$ for each $y \in V$. Thus there exists some $M_y > 0$ such that $\sup_{x^* \in E} |x^*(y)| \leq M_y$.

Consequently from Corollary 6.2.1 to the Uniform Boundedness Theorem we see that there exists some $M > 0$ such that

$$\sup_{x^* \in E} \|x^*\| \leq M,$$

that is, $E$ is norm bounded. Hence part (i) implies part (ii).

Conversely, suppose $E$ is weak* closed and norm bounded. We claim without loss of generality we may assume that

$$E = B_1^* = \{x^* \mid x^* \in V^*, \|x^*\| \leq 1\}.$$

Obviously $B_1^*$ is norm bounded, and an elementary argument shows that $B_1^*$ is weak* closed. Suppose then that $B_1^*$ is weak* compact and $E$ is any weak* closed and norm bounded set in $V^*$. Clearly, since $E$ is norm bounded, there exists some $a > 0$ such that $E \subset aB_1^*$. But $aB_1^*$ is weak* compact, since scalar multiplication is a homeomorphism in $(V^*, \mathcal{T}^{w^*})$ (Theorem 2.1.2), and so $E$ is weak* compact, as it is a weak* closed subset of a weak* compact set. Thus we may assume that $E = B_1^*$.

Now for each $x \in V$ let $K_{\tau(x)} = \{a \mid a \in \Phi, |a| \leq \|\tau(x)\| = \|x\|\}$.

Clearly each $K_{\tau(x)}$ is a compact subset of $\Phi_{\tau(x)} = \Phi$, and $T(B_1^*) \subset \Pi_{x \in V} K_{\tau(x)}$. The latter assertion is valid since, if $x^* \in B_1^*$, then $T(x^*) = \{x^*(x)\}$ and $|x^*(x)| \leq \|x^*\|\|x\| \leq \|x\|$. We claim that $T(B_1^*)$ is a compact subset of $T(V^*)$ in the relative topology on $T(V^*)$ considered as a subspace of $\Pi_{x \in V} \Phi_{\tau(x)}$. Indeed, suppose $\{x_\alpha^*(x)\}$ is a net in $T(B_1^*)$. Since $T(B_1^*) \subset \Pi_{x \in V} K_{\tau(x)}$ and $\Pi_{x \in V} K_{\tau(x)}$ is compact in the product topology, by Tikhonov's Theorem [Ry, pp. 166 and 167], we see that there exists a subnet $\{x_\beta^*(x)\}$ of $\{x_\alpha^*(x)\}$ and some $\varphi = \{\varphi_x\} \in \Pi_{x \in V} K_{\tau(x)}$ such that $\{x_\beta^*(x)\}$ converges to $\varphi$ in the product topology. Moreover, $\varphi \in T(B_1^*)$.

To see this we first show that $x^*(x) = \varphi_x$, $x \in V$, defines an element $x^*$ of $B_1^*$. With this in mind, let $x,y \in V$ be given and suppose $\epsilon > 0$. Let

$$U_x = \{a \mid a \in \Phi, |a - \varphi_x| < \tfrac{\epsilon}{3}\},$$

$$U_y = \{a \mid a \in \Phi, |a - \varphi_y| < \tfrac{\epsilon}{3}\},$$

$$U_{x+y} = \{a \mid a \in \Phi, |a - \varphi_{x+y}| < \tfrac{\epsilon}{3}\},$$

and let $U = U_x \times U_y \times U_{x+y} \times \Pi_{z \neq x,y,x+y} \Phi_{\tau(z)}$. Clearly $U$ is an open neighborhood of $\varphi$ in $\Pi_{x \in V} \Phi_{\tau(x)}$, and so there exists some $\beta_0$ such that for $\beta > \beta_0$ we have

$$\begin{aligned} |x^*(x+y) - [x^*(x) + x^*(y)]| &\leq |\varphi_{x+y} - x_\beta^*(x+y)| \\ &\quad + |x_\beta^*(x) - \varphi_x| + |x_\beta^*(y) - \varphi_y| \\ &< \frac{\epsilon}{3} + \frac{\epsilon}{3} + \frac{\epsilon}{3} \\ &= \epsilon. \end{aligned}$$

Since $\epsilon > 0$ is arbitrary, we conclude that

$$x^*(x + y) = x^*(x) + x^*(y) \qquad (x,y \in V).$$

Similarly one proves that $x^*(ax) = ax^*(x)$, $a \in \Phi$ and $x \in V$. Thus $x^* \in V'$. But evidently $|x^*(x)| = |\varphi_x| \leq \|x\|$, $x \in V$, as $\varphi$ belongs to $\Pi_{x \in V} K_{\tau(x)}$, and so $x^* \in B_1^*$. Consequently $\varphi = T(x^*) \in T(B_1^*)$, and $T(B_1^*)$ is compact. Since $T$ is a homeomorphism, it follows at once that $B_1^*$ is weak* compact.

Therefore part (ii) of the theorem implies part (i). □

One observation about the proof is in order before we turn to some corollaries. It is not difficult to show that a set of the form $\{x^* \mid x^* \in V^*, \|x^*\| \leq a\}$ is a closed set in $(V^*, \mathcal{T}^{w^*})$; that is, every norm closed ball about the origin in $V^*$ is weak* closed. It is not, however, the case that norm bounded norm closed sets in $V^*$ are necessarily weak* closed. For instance, if $V$ is not reflexive and $B_1 = \{x \mid x \in V, \|x\| \leq 1\}$, then

$$\tau(B_1) \subset B_1^{**} = \{x^{**} \mid x^{**} \in V^{**}, \|x^{**}\| \leq 1\}$$

is norm bounded and norm closed as a subset of $(V^{**}, \|\cdot\|)$ (Theorem 8.1.1), and $\tau(B_1) \neq B_1^{**}$. But $\tau(B_1)$ is not closed in $(V^{**}, \mathcal{T}^{w^*})$, as $\tau(B_1)$ is weak* dense in $B_1^{**}$ (Theorem 9.2.3).

Also it should be noted that $\{x^* \mid x^* \in V^*, \|x^*\| = 1\}$ need not be compact in $(V^*, \mathcal{T}^{w^*})$. We leave it to the reader to supply an appropriate example.

Furthermore, we remark that, if $V$ is infinite-dimensional, then it is never the case that $B_1^* \subset V^*$ has a nonempty interior in $\mathcal{T}^{w^*}$ that contains the origin. Indeed, if $B_1^*$ had this property, then from the Banach-Alaoglu Theorem and Theorem 2.2.2 we would immediately conclude that $V^*$, and hence $V$, was finite-dimensional, contrary to assumption.

An examination of the proof of Theorem 9.4.1 reveals that the assumption that $V$ is a Banach space was used only to prove that part (i) implies part (ii). Hence we have the first corollary.

Corollary 9.4.1. Let $(V,\|\cdot\|)$ be a normed linear space over $\Phi$ and let $E \subset V^*$. If $E$ is weak* closed and norm bounded, then $E$ is weak* compact.

Recalling our remarks following Theorem 9.3.1 concerning reflexivity and weak topologies, we obtain at once the next result.

Corollary 9.4.2. Let $(V,\|\cdot\|)$ be a reflexive Banach space over $\Phi$ and let $E \subset V$. Then the following are equivalent:

(i) $E$ is a compact set in $(V,\mathcal{T}^w)$.
(ii) $E$ is a closed set in $(V,\mathcal{T}^w)$ and a bounded set in $(V,\|\cdot\|)$.

The next corollary is an immediate consequence of Corollary 9.4.1 to the Banach-Alaoglu Theorem, since a set $E$ in a Hausdorff topological space is compact if and only if every net in $E$ has a subnet that converges to a point of $E$.

Corollary 9.4.3. Let $(V,\|\cdot\|)$ be a normed linear space over $\Phi$. If $\{x_\alpha^*\} \subset V^*$ is a net such that $\sup_\alpha \|x_\alpha^*\| \leq M$, then $\{x_\alpha^*\}$ has a subnet $\{x_\beta^*\}$ that converges to some point of

$$\{x^* \mid x^* \in V^*,\ \|x^*\| \leq M\}$$

in $(V^*,\mathcal{T}^{w^*})$.

Note carefully that the last corollary does not assert that a norm bounded net in $V^*$ has a weak* convergent subsequence, not even in the case when the original net is itself a sequence.

For example, consider $\ell_\infty$ and for each positive integer $n$ define $x_n^*(x) = a_n$ for $x = \{a_k\} \in \ell_\infty$. Clearly $x_n^* \in \ell_\infty^*$ and $\|x_n^*\| = 1$, $n = 1,2,3,\ldots$ . However, the sequence $\{x_n^*\} \subset \ell_\infty^*$ has no weak* convergent subsequence, as shown by the following argument: Suppose that $\{x_{n_k}^*\}$ is any subsequence of $\{x_n^*\}$ and define $y = \{b_j\} \in \ell_\infty$ by $b_j = 1$ if $j = n_k$ and $k$ is even, and $b_j = 0$ for all other $j$. Then

$$x^*_{n_k}(y) = b_{n_k} = 1 \quad \text{for} \quad k = 2,4,6,\ldots,$$

$$x^*_{n_k}(y) = b_{n_k} = 0 \quad \text{for} \quad k = 1,3,5,\ldots,$$

and so the sequence $\{x^*_{n_k}(y)\}$ does not converge. Hence $\{x^*_n\}$ has no weak* convergent subsequence.

For the sake of completeness we make the following definition:

Definition 9.4.1. A set $E$ in a Hausdorff topological space $X$ is said to be sequentially compact if every sequence in $E$ has a subsequence that converges to a point of $E$.

The preceding example thus shows that the unit ball in $\ell^*_\infty$ is not sequentially compact in the weak* topology. The unit ball is, of course, weak* compact by the Banach-Alaoglu Theorem. It is, however, the case that weak* closed norm bounded sets in $V^*$ are not only weak* compact but also weak* sequentially compact if $V$ is a separable normed linear space. This will follow from Theorem 9.4.2. First we need a lemma.

Lemma 9.4.1. Let $(V,\|\cdot\|)$ be a normed linear space over $\Phi$ and suppose that $\mathcal{F} \subset V$. If $\mathcal{F}$ is a norm dense subset of $V$, then the restrictions to any norm bounded subset of $V^*$ of the $\mathcal{F}$-topologies $\mathcal{T}_{\tau(\mathcal{F})}$ and $\mathcal{T}^{w^*}$ are equivalent.

Proof. By Proposition 9.1.1 we have $\mathcal{T}_{\tau(\mathcal{F})} \subset \mathcal{T}^{w^*}$. Hence it suffices to show that, if $\{x^*_\alpha\} \subset V^*$ is a net such that $\sup_\alpha \|x^*_\alpha\| \le M$ and $\{x^*_\alpha\}$ converges to $x^*$ in $(V^*,\mathcal{T}_{\tau(\mathcal{F})})$, then $\{x^*_\alpha\}$ converges to $x^*$ in $(V^*,\mathcal{T}^{w^*})$.

However, given any $x \in V$ and $\epsilon > 0$, let $y \in \mathcal{F}$ be such that $\|x - y\| < \epsilon/3 \max(M,\|x^*\|)$. If $\alpha_0$ is such that $|x^*_\alpha(y) - x^*(y)| < \epsilon/3$ whenever $\alpha > \alpha_0$, then we also have

$$|x^*_\alpha(x) - x^*(x)| \le |x^*_\alpha(x) - x^*_\alpha(y)| + |x^*_\alpha(y) - x^*(y)| + |x^*(y) - x^*(x)|$$

$$\le \|x^*_\alpha\|\|x - y\| + |x^*_\alpha(y) - x^*(y)| + \|x^*\|\|y - x\|$$

$$< \epsilon$$

whenever $\alpha > \alpha_0$.

Thus $\{x^*_\alpha\}$ converges to $x^*$ in $(V^*, \mathcal{T}^{w^*})$. □

Note that, if $\mathcal{F} \subset V$ is norm dense in $V$, then $\tau(\mathcal{F}) \subset V^{**}$ separates the points of $V^*$, and $\tau(\mathcal{F})$ does indeed define an $\mathcal{F}$-topology $\mathcal{T}_{\tau(\mathcal{F})}$ on $V^*$.

Theorem 9.4.2. Let $(V, \|\cdot\|)$ be a normed linear space over $\Phi$ and let $B_1^* = \{x^* \mid x^* \in V^*, \|x^*\| \le 1\}$. Then the following are equivalent:

(i) $(V, \|\cdot\|)$ is separable.

(ii) The weak* topology $\mathcal{T}^{w^*}$ on $V^*$ restricted to $B_1^*$ is metrizable.

Proof. If $V$ is separable, let $\mathcal{F} \subset V$ be a countable norm dense subset of $V$. Then by Theorem 9.1.2 the topology $\mathcal{T}_{\tau(\mathcal{F})}$ on $V^*$ is metrizable, and hence by Lemma 9.4.1 the weak* topology restricted to $B_1^*$ is metrizable. Thus part (i) implies part (ii).

Conversely, suppose the weak* topology on $B_1^*$ is metrizable. Then, in particular, there exists a sequence of open neighborhoods $U_n$ of the origin in $\mathcal{T}^{w^*}$ such that $\bigcap_{n=1}^{\infty} U_n = \{0\}$. Without loss of generality we may assume that

$$U_n = U_n(0, \epsilon_n, F_n) = \{x^* \mid x^* \in V^*, |x^*(x)| < \epsilon_n, x \in F_n\},$$

where $F_n \subset V$ is a finite set. Let $F = \bigcup_{n=1}^{\infty} F_n$. Clearly $F$ is countable. Moreover, suppose $x^* \in V^*$ is such that $x^*(x) = 0$, $x \in F$. Then $|x^*(x)| < \epsilon_n$, $x \in F_n$, $n = 1,2,3,\ldots$, which shows that $x^* \in \bigcap_{n=1}^{\infty} U_n = \{0\}$. Hence $x^* = 0$.

If $\mathcal{F}$ denotes the collection of all finite linear combinations of elements of F with rational coefficients, then $\mathcal{F}$ is obviously a countable subset of V that is norm dense in V by the previous observation and Corollary 4.2.8 to the Hahn-Banach Theorem.

Therefore V is separable, and part (ii) of the theorem implies part (i). □

It is evident that $B_1^*$ in the theorem could be replaced by any norm closed bounded ball $\{x^* \mid x^* \in V^*, \|x^*\| \leq a\}$. Similarly, the proof of the first portion of the theorem shows that the weak* topology on any norm bounded subset of V* is metrizable, provided V is separable.

These observations, combined with the fact that compactness and sequential compactness are equivalent in metric spaces, give us the next two corollaries.

<u>Corollary 9.4.4</u>. Let $(V,\|\cdot\|)$ be a separable normed linear space over $\Phi$. If $E \subset V^*$ is weak* closed and norm bounded, then E is weak* sequentially compact.

<u>Corollary 9.4.5 (Helly's Selection Theorem)</u>. Let $(V,\|\cdot\|)$ be a separable normed linear space over $\Phi$. If $\{x_n^*\} \subset V^*$ is a sequence such that $\sup_n \|x_n^*\| \leq M$, then $\{x_n^*\}$ has a subsequence that is weak* convergent to some point in $\{x^* \mid x^* \in V^*, \|x^*\| \leq M\}$.

For reflexive spaces the preceding results combined with Theorem 4.5.1 give us the following corollaries. The details are left to the reader.

<u>Corollary 9.4.6</u>. Let $(V,\|\cdot\|)$ be a reflexive Banach space over $\Phi$ and let $B_1 = \{x \mid x \in V, \|x\| \leq 1\}$. Then the following are equivalent:

(i) $(V,\|\cdot\|)$ is separable.

(ii) The weak topology $\mathcal{T}^w$ restricted to $B_1$ is metrizable.

Corollary 9.4.7. Let $(V,\|\cdot\|)$ be a separable reflexive Banach space over $\Phi$. If $E \subset V$ is weakly closed and norm bounded, then $E$ is weakly sequentially compact.

Furthermore, by either arguing as in the proof of Theorem 9.4.2 *mutatis mutandis*, or by applying that result to the dual and bidual of a normed linear space $V$ and using some of the properties of the canonical embedding of $V$ into $V^{**}$, we can prove the next result. Again the details are left to the reader.

Theorem 9.4.3. Let $(V,\|\cdot\|)$ be a normed linear space over $\Phi$ and let $B_1 = \{x \mid x \in V, \|x\| \leq 1\}$. Then the following are equivalent:

(i) $(V^*,\|\cdot\|)$ is separable.

(ii) The weak topology $T^w$ on $V$ restricted to $B_1$ is metrizable.

As seen by the example of the closed unit ball in $\ell_\infty^*$, a weak* compact set need not be weak* sequentially compact. Corollary 9.4.4 asserts that weak* compactness implies weak* sequential compactness if $V$ is a separable Banach space, while Corollaries 9.4.2 and 9.4.1 together assert that weak compactness implies weak sequential compactness in separable reflexive Banach spaces. It is an extremely profound, and perhaps surprising, fact that weak compactness and weak sequential compactness are equivalent in any Banach space. This is the content of the Eberlein-Šmulian Theorem (Theorem 10.3.1), to be proved in the next chapter. No such result is available for the weak* topology.

The Banach-Alaoglu Theorem has a valid analog in the context of locally convex topological linear spaces. We neither discuss nor use this generalization. The interested reader is referred to [K, pp. 245-249; KeNa, p. 155; $W_1$, pp. 236- 241].

9.5. Banach Spaces as Spaces of Continuous Functions. In this short section we shall see how every complex Banach space can be represented as a space of continuous functions on a compact Hausdorff topological space.

Theorem 9.5.1. Let $(V,\|\cdot\|)$ be a Banach space over $\mathbb{C}$. Then there exists a compact Hausdorff topological space $X$ such that $(V,\|\cdot\|)$ is isometrically isomorphic to a closed linear subspace of $(C(X),\|\cdot\|_\infty)$.

Proof. Let $X = B_1^* = \{x^* \mid x^* \in V^*, \|x^*\| \leq 1\}$. Then, by the Banach-Alaoglu Theorem (Theorem 9.4.1), $X$ is a compact Hausdorff topological space in the relative weak* topology. For each $x \in V$ define $f_x(x^*) = x^*(x)$, $x^* \in X$. Clearly $f_x$ is a complex-valued function on $X$, and since $\{x_\alpha^*\}$ converges to $x^*$ in the weak* topology if and only if $\{x_\alpha^*(x)\}$ converges to $x^*(x)$ for each $x \in V$, we know at once that $f_x \in C(X)$. It is easily seen that the mapping $\psi : V \to C(X)$ defined by $\psi(x) = f_x$, $x \in V$, is linear. Moreover,

$$\begin{aligned}
\|\psi(x)\|_\infty &= \|f_x\|_\infty \\
&= \sup_{x^* \in X} |f_x(x^*)| \\
&= \sup_{x^* \in B_1^*} |\tau(x)(x^*)| \\
&= \|\tau(x)\| \\
&= \|x\|
\end{aligned}$$

shows that $\psi$ is an isometry.

The conclusion of the theorem is now evident. □

The function $f_x$ is, of course, just $\tau(x)$ restricted to $B_1^*$.

Obviously a similar result for Banach spaces over $\mathbb{R}$ and spaces of real-valued continuous functions is also valid. Moreover, there is an analog for arbitrary locally convex topological linear spaces: every such space is topologically isomorphic to a linear subspace of $C'(X)$, where $X$ is a suitable locally compact Hausdorff topological space. Refer to Example 1.2.1 for the definition of $C'(X)$. The details are available, for example, in [K, pp. 250 and 251].

9.6. Banach Limits Revisited. We discussed Banach limits in Section 4.3 as an application of the Hahn-Banach Theorem. Recall that the aim was to extend the notion of the limit of a convergent sequence in a reasonable way to all bounded sequences, "reasonable" meaning that the extension should be linear, invariant under shifts, and in agreement with the usual notion of limit for convergent sequences. We now want to again prove the existence of such Banach limits on $\ell_\infty$, but this time we use the Banach-Alaoglu Theorem.

Actually this time we only consider real sequences -- that is, the Banach space $(\ell_\infty^R, \|\cdot\|_\infty)$ over $\mathbb{R}$. We wish to prove the existence of some $x^* \in (\ell_\infty^R)^*$ such that

(i) If $\{a_k\} \in \ell_\infty^R$ converges, then $x^*(\{a_k\}) = \lim_k a_k$.

(ii) For each $n = 1,2,3,\ldots$, if $\{a_k\} \in \ell_\infty^R$ and $c_k = a_{k+n}$, $k = 1,2,3,\ldots$, then $x^*(\{a_k\}) = x^*(\{c_k\})$.

(iii) If $\{a_k\} \in \ell_\infty^R$ and $a_k \geq 0$, $k = 1,2,3,\ldots$, then $x^*(\{a_k\}) \geq 0$.

Condition (iii) was not imposed on a Banach limit before. The virtue of this condition is elucidated by the following lemma:

Lemma 9.6.1. Suppose that $x^* \in (\ell_\infty^R)^*$ is such that

(i) If $e_k = 1$, $k = 1,2,3,\ldots$, then $x^*(\{e_k\}) = 1$.

(ii) For each $n = 1,2,3,\ldots$, if $\{a_k\} \in \ell_\infty^R$ and $c_k = a_{k+n}$, $k = 1,2,3,\ldots$, then $x^*(\{a_k\}) = x^*(\{c_k\})$.

(iii) If $\{a_k\} \in \ell_\infty^R$, $a_k \geq 0$, $k = 1,2,3,\ldots$, then $x^*(\{a_k\}) \geq 0$.

Then for each $\{a_k\} \in \ell_\infty^R$

$$\liminf_k a_k \le x^*(\{a_k\}) \le \limsup_k a_k.$$

Proof. Let $\{a_k\} \in \ell_\infty^R$ and $\epsilon > 0$ be given. Then there exists some positive integer $N$ such that $a_k < \limsup_m a_m + \epsilon$ for $k \ge N$. Consider a sequence $\{b_k\} \in \ell_\infty^R$ defined by

$$b_k = \min\{a_k, \limsup_m a_m + \epsilon\} \quad (k = 1,2,3,\ldots).$$

Clearly $\{a_k\}$ and $\{b_k\}$ differ in at most a finite number of components, and hence there exists some positive integer $n$ such that $a_{k+n} = b_{k+n}$, $k = 1,2,3,\ldots$ . Let $\{c_k\}$ be the sequence defined by $c_k = a_{k+n} = b_{k+n}$, $k = 1,2,3,\ldots$ . Then $x^*(\{a_k\}) = x^*(\{c_k\}) = x^*(\{b_k\})$ by part (ii) of the hypotheses.

But it is evident that $b_k \le \limsup_m a_m + \epsilon$, $k = 1,2,3,\ldots$, and so from parts (i) and (ii) of the hypotheses we conclude that

$$\begin{aligned} x^*(\{a_k\}) &= x^*(\{b_k\}) \\ &\le (\limsup_m a_m + \epsilon)x^*(\{e_k\}) \\ &= \limsup_m a_m + \epsilon. \end{aligned}$$

Since $\epsilon > 0$ is arbitrary, we see that $x^*(\{a_k\}) \le \limsup_m a_m$. The inequality $\liminf_m a_m \le x^*(\{a_k\})$ is proved in a similar manner. □

In particular, if $x^* \in (\ell_\infty^R)^*$ satisfies all three conditions of the lemma, then $x^*(\{a_k\}) = \lim_k a_k$ whenever $\{a_k\}$ is convergent, and so $x^*$ is a Banach limit.

The proof of the next theorem then reduces to showing the existence of some $x^* \in (\ell_\infty^R)^*$ that satisfies all three conditions of Lemma 9.6.1.

Theorem 9.6.1. There exists some $x^* \in (\ell_\infty^R)^*$ such that

(i) If $\{a_k\} \in \ell_\infty^R$ is convergent, then $x^*(\{a_k\}) = \lim_k a_k$.

(ii) For each $n = 1,2,3,\ldots$, if $\{a_k\} \in \ell_\infty^R$ and $c_k = a_{k+n}$, $k = 1,2,3,\ldots$, then $x^*(\{a_k\}) = x^*(\{c_k\})$.

(iii) If $\{a_k\} \in \ell_\infty^R$, $a_k \geq 0$, $k = 1,2,3,\ldots$, then $x^*(\{a_k\}) \geq 0$.

(iv) If $\{a_k\} \in \ell_\infty^R$, then

$$\liminf_k a_k \leq x^*(\{a_k\}) \leq \limsup_k a_k.$$

Proof. As indicated, we need only prove the existence of some $x^* \in (\ell_\infty^R)^*$ that satisfies the three conditions of Lemma 9.6.1. With this in mind, for each positive integer $m$ we define

$$x_m^*(\{a_k\}) = \frac{1}{m} \sum_{k=1}^{m} a_k \qquad (\{a_k\} \in \ell_\infty^R).$$

It is easily seen that $x_m^* \in (\ell_\infty^R)^*$ and $\|x_m^*\| \leq 1$. But, since $\|\{e_k\}\|_\infty = 1$ and $x_m^*(\{e_k\}) = 1$, we conclude that $\|x_m^*\| = 1$, $m = 1,2,3,\ldots$ .

Let $F_j$ denote the closure of $\{x_m^* \mid m \geq j\}$ in $((\ell_\infty^R)^*, \mathcal{T}^{w^*})$, $j = 1,2,3,\ldots$ . It is apparent that

$$F_j \subset B_1^* = \{x^* \mid x^* \in (\ell_\infty^R)^*, \|x^*\| \leq 1\} \qquad (j = 1,2,3,\ldots),$$

and that the family $\{F_j\}$ has the finite intersection property; that is, any finite number of the sets $F_j$ have a nonempty intersection. Since $B_1^*$ is weak* compact, from the Banach-Alaoglu Theorem (Theorem 9.4.1), we see that $\bigcap_{j=1}^\infty F_j \neq \emptyset$.

Let $x^* \in \bigcap_{j=1}^\infty F_j$. We claim that $x^*$ has the desired properties. Indeed, it is evident that $\|x^*\| \leq 1$. Moreover, given $\epsilon > 0$, consider the open weak* neighborhood $U(x^*,\epsilon,\{e_k\})$ of $x^*$. Since $x^* \in \bigcap_{j=1}^\infty F_j$, we see that there exists a positive integer $m$ such that $x_m^* \in U(x^*,\epsilon,\{e_k\})$. Hence

$$\begin{aligned} |x^*(\{e_k\}) - x_m^*(\{e_k\})| &= |x^*(\{e_k\}) - 1| \\ &< \epsilon, \end{aligned}$$

and so $x^*(\{e_k\}) = 1$ and $\|x^*\| = 1$, as $\epsilon > 0$ is arbitrary.

A similar argument shows that, if $\{a_k\} \in \ell_\infty^R$, $a_k \geq 0$, $k = 1,2,3,\ldots$, then $x^*(\{a_k\}) \geq 0$, since $x_m^*(\{a_k\}) \geq 0$, $m = 1,2,3,\ldots$ .

To prove that $x^*$ satisfies part (ii) of the theorem it clearly suffices to show that, if $\{a_k\} \in \ell_\infty^R$ and $c_k = a_{k+1}$, $k = 1,2,3,\ldots$, then $x^*(\{a_k\}) = x^*(\{c_k\})$. We note first that

$$\begin{aligned} x_m^*(\{a_k\}) - x_m^*(\{c_k\}) &= \frac{1}{m}\Big(\sum_{k=1}^{m} a_k - \sum_{k=2}^{m+1} a_k\Big) \\ &= \frac{1}{m}(a_1 - a_{m+1}) \qquad (m = 1,2,3,\ldots). \end{aligned}$$

Consequently

$$|x_m^*(\{a_k\}) - x_m^*(\{c_k\})| \leq \frac{2\|\{a_k\}\|_\infty}{m} \qquad (m = 1,2,3,\ldots).$$

Now let $\epsilon > 0$ be given and consider the weak* neighborhood $U(x^*,\epsilon/3,\{a_k\},\{c_k\})$. Then, since $x^* \in \bigcap_{j=1}^\infty F_j$, there exists some $m > 6\|\{a_k\}\|_\infty/\epsilon$ such that $x_m^* \in U(x^*,\epsilon/3,\{a_k\},\{c_k\})$, and so

$$\begin{aligned} |x^*(\{a_k\}) - x^*(\{c_k\})| &\leq |x^*(\{a_k\}) - x_m^*(\{a_k\})| \\ &\quad + |x_m^*(\{a_k\}) - x_m^*(\{c_k\})| \\ &\quad + |x_m^*(\{c_k\}) - x^*(\{c_k\})| \\ &< \frac{2\epsilon}{3} + \frac{2\|\{a_k\}\|_\infty}{m} \\ &< \epsilon. \end{aligned}$$

Therefore $x^*(\{a_k\}) = x^*(\{c_k\})$, as $\epsilon > 0$ is arbitrary, and the proof is complete. □

<u>9.7. Fourier Series of Functions in $L_p([-\pi,\pi],dt/2\pi)$, $1<p\leq\infty$.</u> As is easily seen from an application of Hölder's Inequality, the linear spaces $L_p([-\pi,\pi],dt/2\pi)$, $1 < p \leq \infty$, are all linear subspaces

of $L_1([-\pi,\pi],dt/2\pi)$, and hence we can meaningfully discuss the Fourier transform

$$\hat{f}(k) = \frac{1}{2\pi}\int_{-\pi}^{\pi} f(t)e^{-ikt}\,dt \qquad (k \in \mathbb{Z})$$

for $f \in L_p([-\pi,\pi],dt/2\pi)$, $1 \leq p \leq \infty$. Thus associated formally with each $f \in L_p([-\pi,\pi],dt/2\pi)$ we have its Fourier series

$$\sum_{k=-\infty}^{\infty} \hat{f}(k)e^{ikt}.$$

Note we are here making no assertions about the convergence of this series. Our purpose in this section is to give some necessary and sufficient conditions for a formal series $\Sigma_{k=-\infty}^{\infty} c_k e^{ikt}$ to be the Fourier series of some $f \in L_p([-\pi,\pi],dt/2\pi)$; that is, conditions to ensure that there exists some $f \in L_p([-\pi,\pi],dt/2\pi)$ such that $\hat{f}(k) = c_k$, $k \in \mathbb{Z}$.

In Section 6.6 we indicated that the nth symmetric partial sum of the Fourier series of an $f \in L_p([-\pi,\pi],dt/2\pi)$, call it $s_n(f)$, could be written as

$$s_n(f)(s) = \sum_{k=-n}^{n} \hat{f}(k)e^{iks}$$

$$= \frac{1}{2\pi}\int_{-\pi}^{\pi} f(t)D_n(s-t)\,dt \qquad (s \in [-\pi,\pi]),$$

where $D_n(u) = \sin[(n + 1/2)u]/\sin(u/2)$, $n = 0,1,2,\ldots,$ is the Dirichlet kernel. If one considers the arithmetic average of the first $n + 1$ partial sums $s_n(f)$, that is,

$$\sigma_n(f) = \frac{1}{n+1}\sum_{k=0}^{n} s_k(f)$$

then some reasonably straightforward computations reveal that

$$\sigma_n(f)(s) = \sum_{k=-n}^{n} \left(1 - \frac{|k|}{n+1}\right)\hat{f}(k)e^{iks}$$

$$= \frac{1}{2\pi}\int_{-\pi}^{\pi} f(t)F_n(s-t)\,dt \qquad (s \in [-\pi,\pi]),$$

where

$$F_n(u) = \sum_{k=-n}^{n} \left(1 - \frac{|k|}{n+1}\right) e^{iku}$$

$$= \frac{1}{n+1} \left[\frac{\sin[(n+1)u/2]}{\sin(u/2)}\right]^2 \quad (n = 0,1,2,\ldots).$$

The function $\sigma_n(f)$ is, for obvious reasons, called the nth <u>Cesàro mean</u> of the Fourier series of $f$, and $F_n$ is known as the <u>Fejér kernel</u>. A discussion of this can be found in [$E_2$, pp. 78 and 79]. We omit the details.

More generally, if one considers a formal series $\Sigma_{k=-\infty}^{\infty} c_k e^{iks}$, then one can discuss the nth Cesàro mean of this series, call it $\sigma_n(\{c_k\})$, and show that

$$\sigma_n(\{c_k\})(s) = \sum_{k=-n}^{n} \left(1 - \frac{|k|}{n+1}\right) c_k e^{iks} \quad (s \in [-\pi,\pi]).$$

We claim that such a formal series is the Fourier series of some $f \in L_p([-\pi,\pi],dt/2\pi)$, $1 < p \leq \infty$, if and only if

$$\sup_n \|\sigma_n(\{c_k\})\|_p < \infty.$$

It is apparent that each $\sigma_n(\{c_k\}) \in L_p([-\pi,\pi],dt/2\pi)$ for all $p$, $1 < p \leq \infty$, as each such Cesàro sum is a continuous function on $[-\pi,\pi]$.

Before we can prove the indicated result we need to note a preliminary fact concerning the convolution of functions in $L_p([-\pi,\pi],dt/2\pi)$ and in $L_1([-\pi,\pi],dt/2\pi)$. The reader is referred to [$E_2$, p. 56] for the details. We assume tacitly, as usual, that the elements of $L_p([-\pi,\pi],dt/2\pi)$ have been extended periodically to the entire real line.

<u>Proposition 9.7.1</u>. Let $f \in L_1([-\pi,\pi],dt/2\pi)$ and let $g \in L_p([-\pi,\pi],dt/2\pi)$, $1 \leq p \leq \infty$. Then the formal expression

$$f * g(s) = \frac{1}{2\pi} \int_{-\pi}^{\pi} f(s-t)g(t)\, dt,$$

defined for almost all $s \in [-\pi,\pi]$, determines an element $f * g \in L_p([-\pi,\pi],dt/2\pi)$ and $\|f * g\|_p \le \|f\|_1\|g\|_p$.

This should be compared with Proposition 4.7.1.

Theorem 9.7.1. For any doubly infinite sequence $\{c_k\} \subset \mathbb{C}$ and $1 < p \le \infty$, the following are equivalent:

(i) There exists some $f \in L_p([-\pi,\pi],dt/2\pi)$ such that $\hat{f}(k) = c_k$, $k \in \mathbb{Z}$.

(ii) $\sup_n\|\sigma_n(\{c_k\})\|_p < \infty$.

Proof. Suppose there is some $f \in L_p([-\pi,\pi],dt/2\pi)$ such that $\hat{f}(k) = c_k$, $k \in \mathbb{Z}$. Then for each nonnegative integer $n$ we have $\sigma_n(\{c_k\}) = \sigma_n(f) = F_n * f$. Hence by Proposition 9.7.1

$$\|\sigma_n(\{c_k\})\|_p = \|F_n * f\|_p \le \|F_n\|_1\|f\|_p.$$

But, since $F_n(t) \ge 0$, $t \in [-\pi,\pi]$, a simple computation shows that

$$\begin{aligned}\|F_n\|_1 &= \frac{1}{2\pi}\int_{-\pi}^{\pi} |F_n(t)|\ dt \\ &= \frac{1}{2\pi}\int_{-\pi}^{\pi} F_n(t)\ dt \\ &= \frac{1}{2\pi}\sum_{k=-n}^{n}\left(1 - \frac{|k|}{n+1}\right)\int_{-\pi}^{\pi} e^{ikt}\ dt \\ &= 1 \qquad\qquad (n = 0,1,2,\ldots).\end{aligned}$$

Thus $\sup_n\|\sigma_n(\{c_k\})\|_p \le \|f\|_p < \infty$, and part (i) of the theorem implies part (ii).

Conversely, suppose there exists some $M > 0$ such that $\sup_n\|\sigma_n(\{c_k\})\|_p \le M$. Consequently from Corollary 8.3.2 and Theorem 8.3.3 we see that the sequence $\{\sigma_n(\{c_k\})\}$ lies in the norm closed ball of radius $M$ about the origin in $L_q([-\pi,\pi],dt/2\pi)^* = L_p([-\pi,\pi],dt/2\pi)$, $1/p + 1/q = 1$. Since $L_q([-\pi,\pi],dt/2\pi)$ is a separable Banach space, we deduce from Helly's Selection Theorem (Corollary 9.4.5) the existence of some $f \in L_p([-\pi,\pi],dt/2\pi)$ and some subsequence $\{\sigma_{n_m}(\{c_k\})\}$ of $\{\sigma_n(\{c_k\})\}$ such that $\{\sigma_{n_m}(\{c_k\})\}$ converges to $f$ in the weak* topology on $L_q([-\pi,\pi],dt/2\pi)^*$

Since $e^{ij\cdot} \in L_q([-\pi,\pi],dt/2\pi)$, $j \in \mathbb{Z}$, we conclude, given $j \in \mathbb{Z}$, that

$$\lim_m \frac{1}{2\pi}\int_{-\pi}^{\pi} \sigma_{n_m}(\{c_k\})(t)e^{-ijt}\,dt = \frac{1}{2\pi}\int_{-\pi}^{\pi} f(t)e^{-ijt}\,dt$$
$$= \hat{f}(j) \qquad (j \in \mathbb{Z}).$$

However, given $j \in \mathbb{Z}$, we see that if $n_m > j$, then

$$\frac{1}{2\pi}\int_{-\pi}^{\pi} \sigma_{n_m}(\{c_k\})(t)e^{-ijt}\,dt = \sum_{k=-n_m}^{n_m} \left(1 - \frac{|k|}{n_m+1}\right)c_k\left(\frac{1}{2\pi}\int_{-\pi}^{\pi} e^{ikt}e^{-ijt}\,dt\right)$$
$$= \left(1 - \frac{|j|}{n_m+1}\right)c_j,$$

since

$$\frac{1}{2\pi}\int_{-\pi}^{\pi} e^{i(k-j)t}\,dt = 1 \quad \text{for} \quad k = j,$$
$$\frac{1}{2\pi}\int_{-\pi}^{\pi} e^{i(k-j)t}\,dt = 0 \quad \text{for} \quad k \neq j \qquad (k,j \in \mathbb{Z}).$$

The last assertion is easily verified. Hence we find that

$$\hat{f}(j) = \lim_m \frac{1}{2\pi}\int_{-\pi}^{\pi} \sigma_{n_m}(\{c_k\})(t)e^{-ijt}\,dt$$
$$= \lim_m \left(1 - \frac{|j|}{n_m+1}\right)c_j$$
$$= c_j \qquad (j \in \mathbb{Z}).$$

Therefore part (ii) of the theorem implies part (i). □

The second implication in Theorem 9.7.1 could also have been proved by arguing directly from the Banach-Alaoglu Theorem (Theorem 9.4.1) since $\{x^* \mid x^* \in L_q([-\pi,\pi],dt/2\pi)^*, \|x^*\| \leq M\}$ is a weak* closed norm bounded set and hence weak* compact. The details are left to the reader.

In the case $p = 2$ one can show that $\{c_k\}$ is such that $c_k = \hat{f}(k)$, $k \in \mathbb{Z}$, for some $f \in L_2([-\pi,\pi],dt/2\pi)$ if and only if $\Sigma_{k=-\infty}^{\infty}|c_k|^2 < \infty$. We shall return to this when we discuss Hilbert spaces in Chapter 13.

9.8. Multipliers. An argument like the one utilized in the preceding section can be used to characterize those $T$ belonging to $L(L_1(\mathbb{R},dt),L_p(\mathbb{R},dt))$ that commute with convolution. We make the following definition:

Definition 9.8.1. If $1 \leq p \leq \infty$ and $T \in L(L_1(\mathbb{R},dt),L_p(\mathbb{R},dt))$, then $T$ is said to be a multiplier from $L_1(\mathbb{R},dt)$ to $L_p(\mathbb{R},dt)$ if $T(f) * g = T(f * g)$, $f,g \in L_1(\mathbb{R},dt)$.

Before we characterize multipliers, at least in the case $1 < p \leq \infty$, we note the following lemma:

Lemma 9.8.1. There exists a sequence $\{u_k\} \subset L_1(\mathbb{R},dt)$ such that

(i) $\|u_k\|_1 = 1$, $k = 1,2,3,\ldots$ .
(ii) If $f \in L_1(\mathbb{R},dt)$, then $\lim_k\|u_k * f - f\|_1 = 0$.

Proof. Take $u_k = (k/2)\chi_{[-1/k,1/k]}$, $k = 1,2,3,\ldots,$ where $\chi_{[-1/k,1/k]}$ is the characteristic function of the closed interval $[-1/k,1/k]$.

□

The sequence $\{u_k\}$ is called an approximate identity for $L_1(\mathbb{R},dt)$.

Theorem 9.8.1. Let $1 < p \leq \infty$ and $T \in L'(L_1(\mathbb{R},dt),L_p(\mathbb{R},dt))$. Then the following are equivalent:

(i) There exists some $h \in L_p(\mathbb{R},dt)$ such that $T(f) = f * h$, $f \in L_1(\mathbb{R},dt)$.
(ii) $T$ is a multiplier from $L_1(\mathbb{R},dt)$ to $L_p(\mathbb{R},dt)$.

Proof. That part (i) implies part (ii) is immediate from Propositions 4.7.1(i) through (iv). Conversely, suppose that $T$ is a multiplier. If $\{u_n\} \subset L_1(\mathbb{R},dt)$ is a sequence that satisfies the conclusions of Lemma 9.8.1, then for each $f \in L_1(\mathbb{R},dt)$ we have

$$\|T(f) - T(u_n) * f\|_p = \|T(f) - T(u_n * f)\|_p$$
$$\leq \|T\|\|f - u_n * f\|_1,$$

from which it follows that $\lim_n \|T(f) - T(u_n) * f\|_p = 0$. Moreover, $\|T(u_n)\|_p \leq \|T\|\|u_n\|_1 = \|T\|$, $n = 1,2,3,\ldots,$ shows that $\{T(u_n)\}$ is a bounded sequence in $L_p(\mathbb{R},dt) = L_q(\mathbb{R},dt)^*$, $1/p + 1/q = 1$.

Since $L_q(\mathbb{R},dt)$ is separable, as $1 \leq q < \infty$, we can once again apply Helly's Selection Theorem (Corollary 9.4.5) to deduce the existence of some $h \in L_p(\mathbb{R},dt)$ and some subsequence $\{T(u_{n_k})\}$ of $\{T(u_n)\}$ that converges to $h$ in $(L_p(\mathbb{R},dt),\mathcal{T}^{w^*})$. Consequently for any $f \in L_p(\mathbb{R},dt)$ and $g \in L_q(\mathbb{R},dt)$ we obtain

$$\begin{aligned}
\int_{\mathbb{R}} T(f)(t)g(-t)\,dt &= \lim_k \int_{\mathbb{R}} T(u_{n_k}) * f(t)g(-t)\,dt \\
&= \lim_k [T(u_{n_k}) * f] * g(0) \\
&= \lim_k T(u_{n_k}) * (f * g)(0) \\
&= \lim_k \int_{\mathbb{R}} T(u_{n_k})(t)f * g(-t)\,dt \\
&= \int_{\mathbb{R}} h(t)f * g(-t)\,dt \\
&= h * (f * g)(0) \\
&= (h * f) * g(0) \\
&= \int_{\mathbb{R}} h * f(t)g(-t)\,dt,
\end{aligned}$$

where we have used the properties of convolution as presented in Proposition 4.7.1 several times. If $1 < p < \infty$, then a direct appeal to Corollary 4.2.6 to the Hahn-Banach Theorem allows us to conclude that $T(f) = h * f$, $f \in L_1(\mathbb{R},dt)$, whereas, if $p = \infty$, then the weak* denseness of $L_1(\mathbb{R},dt)$ in $L_\infty(\mathbb{R},dt)^*$ (Corollary 9.2.2) combined with the preceding result shows that $x^*[T(f)] = x^*(f * h)$, $x^* \in L_\infty(\mathbb{R},dt)^*$, and hence $T(f) = f * h$, $f \in L_1(\mathbb{R},dt)$.

Therefore part (ii) of the theorem implies part (i), and the equivalence is established. □

If $p = 1$, then the multipliers from $L_1(\mathbb{R},dt)$ to $L_1(\mathbb{R},dt)$ can also be characterized as convolution operators, but this time with bounded, regular, complex-valued Borel measures on $\mathbb{R}$. One can give a proof that, as might be expected, utilizes the Banach-Alaoglu Theorem (Theorem 9.4.1) and the Riesz Representation Theorem (Theorem 4.8.2) for $C_o(\mathbb{R})$. We do not go into the details here, but give the appropriate definitions and statement of the results.

<u>Lemma 9.8.2</u>. If $f \in L_1(\mathbb{R},dt)$ and $\mu \in M(\mathbb{R})$, then the formal expression

$$f * \mu(t) = \int_{\mathbb{R}} f(t - s)\, d\mu(s),$$

defined for almost all $t \in \mathbb{R}$, determines an element $f * \mu \in L_1(\mathbb{R},dt)$. Moreover, $\|f * \mu\|_1 \leq \|\mu\|\|f\|_1$.

The reader is referred to Example 1.2.6 for a discussion of $M(\mathbb{R})$. A proof of the lemma can be found in $[HR_1$, p. 292].

<u>Theorem 9.8.2</u>. If $T \in L'(L_1(\mathbb{R},dt))$, then the following are equivalent:

(i) There exists some $\mu \in M(\mathbb{R})$ such that $T(f) = f * \mu$, $f \in L_1(\mathbb{R},dt)$.

(ii) $T$ is a multiplier from $L_1(\mathbb{R},dt)$ to $L_1(\mathbb{R},dt)$.

One can also show (see, for example [L, pp. 67-69]) that the multipliers from $L_1(\mathbb{R},dt)$ to $L_p(\mathbb{R},dt)$, $1 \leq p \leq \infty$, are precisely those $T \in L(L_1(\mathbb{R},dt), L_p(\mathbb{R},dt))$ for which $T\tau_s = \tau_s T$, $s \in \mathbb{R}$, where $\tau_s f(t) = f(t - s)$.

9.9. Weak Compactness and Reflexivity. We now wish to return to the discussion of some further general results about the weak and weak* topologies. Specifically, we shall show that a Banach space is reflexive if and only if the norm closed unit ball about the origin is weakly compact. As a corollary to this result we shall prove that every nonempty norm closed convex set in a reflexive Banach space contains an element of minimal norm.

Theorem 9.9.1. Let $(V,\|\cdot\|)$ be a Banach space over $\Phi$ and let $B_1 = \{x \mid x \in V, \|x\| \leq 1\}$. Then the following are equivalent:

(i) $(V,\|\cdot\|)$ is reflexive.
(ii) $B_1$ is compact in $(V,\mathcal{T}^W)$.

Proof. That part (i) implies part (ii) follows at once from Corollary 9.4.1 to the Banach-Alaoglu Theorem.

Conversely, suppose $B_1$ is weakly compact. We claim that the canonical embedding $\tau$ of $V$ into $V^{**}$ is a homeomorphism from $(V,\mathcal{T}^W)$ to $(\tau(V),\mathcal{T}^{W^*})$. By the latter pairing we mean, of course, the linear subspace $\tau(V)$ of $V^{**}$ with the relative topology induced by $\mathcal{T}^{W^*}$. It is apparent that $\tau$ is bijective, and since a net $\{x_\alpha\} \subset V$ converges weakly to $x \in V$ if and only if $\lim_\alpha x^*(x_\alpha) = x^*(x)$, $x^* \in V^*$, if and only if $\lim_\alpha \tau(x_\alpha)(x^*) = \tau(x)(x^*)$, $x^* \in V^*$, if and only if $\{\tau(x_\alpha)\}$ converges weak* to $\tau(x)$, we conclude that $\tau$ is indeed a homeomorphism from $(V,\mathcal{T}^W)$ to $(\tau(V),\mathcal{T}^{W^*})$.

Hence $\tau(B_1)$ is compact in $(\tau(V),\mathcal{T}^{W^*})$, as $B_1$ is weakly compact, and so $\tau(B_1)$ is compact as a subset of $(V^{**},\mathcal{T}^{W^*})$. Consequently $\tau(B_1)$ is a closed subset of $(V^{**},\mathcal{T}^{W^*})$, as $\mathcal{T}^{W^*}$ is a Hausdorff topology.

However, from Theorem 9.2.3 we know that $\tau(B_1)$ is dense in $B_1^{**} = \{x^{**} \mid x^{**} \in V^{**}, \|x^{**}\| \leq 1\}$ in the weak* topology on $V^{**}$, and so $\tau(B_1) = B_1^{**}$.

It follows easily from this that $\tau(V) = V^{**}$, and so $V$ is reflexive.

□

Some other necessary and sufficient conditions for reflexivity can be easily deduced.

<u>Corollary 9.9.1</u>. Let $(V,\|\cdot\|)$ be a Banach space over $\Phi$. Then the following are equivalent:

(i) $(V,\|\cdot\|)$ is reflexive.

(ii) If $E \subset V$ is closed in $(V,\mathcal{T}^w)$ and bounded in $(V,\|\cdot\|)$, then $E$ is compact in $(V,\mathcal{T}^w)$.

(iii) $(V^*,\mathcal{T}^{w^*})$ and $(V^*,\mathcal{T}^w)$ are topologically isomorphic.

<u>Proof</u>. The fact that part (i) implies part (ii) is contained in Corollary 9.4.2, and if part (ii) holds, then, in particular, $B_1$ is weakly compact since it is a weakly closed norm bounded set. Thus part (ii) implies part (i) by Theorem 9.9.1.

On the other hand, if $V$ is reflexive, then $\tau(V) = V^{**}$, and it follows immediately that the identity mapping between $(V^*,\mathcal{T}^{w^*})$ and $(V^*,\mathcal{T}^w)$ is a linear homeomorphism. Thus part (i) implies part (iii).

Finally, if $(V^*,\mathcal{T}^{w^*})$ and $(V^*,\mathcal{T}^w)$ are topologically isomorphic, then $B_1^* = \{x^* \mid x^* \in V^*, \|x^*\| \leq 1\}$ is compact in $(V^*,\mathcal{T}^w)$ since it is compact in $(V^*,\mathcal{T}^{w^*})$ by the Banach-Alaoglu Theorem (Theorem 9.4.1). Thus from Theorem 9.9.1 we see that $(V^*,\|\cdot\|)$ is reflexive, and so $(V,\|\cdot\|)$ is also reflexive, by Theorem 8.1.3.

Therefore part (iii) implies part (i), and the proof is complete. □

We can also now prove the result, indicated in Section 8.1, that in a reflexive Banach space every nonempty closed convex set contains an element of minimal norm.

Corollary 9.9.2. Let $(V,\|\cdot\|)$ be a reflexive Banach space over $\Phi$. If $K \subset V$ is a nonempty norm closed convex set, then there exists some $x_o \in K$ such that $\|x_o\| = \inf_{x \in K}\|x\|$.

Proof. Since $K$ is nonempty, there exists some $a > 0$ such that $aB_1 = \{x \mid x \in V, \|x\| \leq a\}$ intersects $K$. Clearly $aB_1 \cap K$ is then a nonempty norm closed convex set in $V$, and so from Theorem 9.2.2 we see that $aB_1 \cap K$ is weakly closed and norm bounded. Thus by Corollary 9.9.1 the set $aB_1 \cap K$ is weakly compact, $V$ being reflexive.

Now consider the family of sets

$$\mathcal{K} = \{bB_1 \cap K \mid 0 \leq b \leq a,\ bB_1 \cap K \neq \phi\}.$$

By the argument of the preceding paragraph we see that each set $bB_1 \cap K \in \mathcal{K}$ is a weakly closed subset of $aB_1 \cap K$, and it is also apparent that $\mathcal{K}$ has the finite intersection property. Hence $\bigcap_{\mathcal{K}}(bB_1 \cap K) \neq \phi$, as $aB_1 \cap K$ is weakly compact. Moreover, if $b_o = \inf\{b \mid 0 \leq b \leq a,\ bB_1 \cap K \neq \phi\}$, then it follows at once that $b_oB_1 \cap K = \bigcap_{\mathcal{K}}(bB_1 \cap K) \neq \phi$.

Evidently, if $x_o \in b_oB_1 \cap K$, then $\|x_o\| = \inf_{x \in K}\|x\|$. □

9.10. A Theorem Concerning the Adjoint Transformation. In this section we wish to employ the Open Mapping Theorem, the Hahn-Banach Theorem, the Banach-Alaoglu Theorem, as well as the description of $c_o^*$ as $\ell_1$, to prove a useful result concerning the adjoint of a continuous linear transformation. We shall apply the result in Section 11.4 to another topic from harmonic analysis. The reader should refer to Section 4.4 for the definition of the adjoint of a continuous linear transformation.

Theorem 9.10.1. Let $(V_1,\|\cdot\|_1)$ and $(V_2,\|\cdot\|_2)$ be Banach spaces over $\Phi$ and let $T \in \mathcal{L}(V_1,V_2)$. If $T$ is injective and $T(V_1)$ is norm dense in $V_2$, then the following are equivalent:

(i) $T^*$ is surjective.

(ii) $T$ is surjective.

(iii) There exists some $m > 0$ such that $\|T^*(y^*)\| \geq m\|y^*\|$, $y^* \in V_2^*$.

Proof. If $T^*$ is surjective, then an easy argument using the Open Mapping Theorem (Theorem 7.2.1) demonstrates the existence of some $\delta > 0$ such that $\{x^* \mid x^* \in V_1^*, \|x^*\| \leq \delta\} \subset T^*(B_1^*)$, where $B_1^* = \{y^* \mid y^* \in V_2^*, \|y^*\| \leq 1\}$. Thus for each $x \in V_1$ we have, by Corollary 4.2.5 to the Hahn-Banach Theorem,

$$\begin{aligned}
\|T(x)\|_2 &= \sup_{y^* \in B_1^*} |y^*[T(x)]| \\
&= \sup_{y^* \in B_1^*} |T^*(y^*)(x)| \\
&\geq \sup_{\substack{x^* \in V_1^* \\ \|x^*\| \leq \delta}} |x^*(x)| \\
&= \sup_{\substack{x^* \in V_1^* \\ \|x^*\| \leq 1}} |\delta x^*(x)| \\
&= \delta\|x\|_1.
\end{aligned}$$

It readily follows from this estimate that $T(V_1)$ is a norm closed linear subspace of $V_2$, whence, since $T(V_1)$ is norm dense in $V_2$, we deduce that $T(V_1) = V_2$, that is, $T$ is surjective.

Thus part (i) implies part (ii).

Similarly, if $T$ is surjective, then by the Open Mapping Theorem (Theorem 7.2.1) there exists some $m > 0$ such that $\{y \mid y \in V_2, \|y\|_2 \leq m\} \subset T(B_1)$, where $B_1 = \{x \mid x \in V_1, \|x\|_1 \leq 1\}$. Consequently we deduce that for each $y^* \in V_2^*$

$$\|T^*(y^*)\| = \sup_{x \in B_1} |T^*(y^*)(x)|$$

$$
\begin{aligned}
&= \sup_{x \in B_1} |y^*[T(x)]| \\
&\geq \sup_{\substack{y \in V_2 \\ \|y\|_2 \leq m}} |y^*(y)| \\
&= \sup_{\substack{y \in V_2 \\ \|y\|_2 \leq 1}} |y^*(my)| \\
&= m\|y^*\|,
\end{aligned}
$$

from which we conclude that part (ii) of the theorem implies part (iii).

Finally, suppose there exists some $m > 0$ such that $\|T^*(y^*)\| \geq m\|y^*\|$, $y^* \in V_2^*$, and let $W = T^*(V_2^*)$. The hypothesis on $T^*$ entails that $W$ is a norm closed linear subspace of $V_1^*$ and that, if $x^* \in W$ and $y^* \in V_2^*$ are such that $T^*(y^*) = x^*$, then $\|y^*\| \leq \|T^*(y^*)\|/m = \|x^*\|/m$. Assume now that $W \neq V_1^*$, meaning that $T^*$ is not surjective. It follows at once that there exists some $x_o^* \in V_1^*$ for which $\|x_o^* - x^*\| > 1$, $x^* \in W$. For each positive integer $n$ we define $W_n = \{x^* \mid x^* \in W, \|x^* - x_o^*\| \leq n\}$. Evidently $W_1 = \phi$ and $W_n = (nB_1^* + x_o^*) \cap W$, $n = 2,3,4,\ldots,$ where

$$B_1^* = \{x^* \mid x^* \in V_1^*, \|x^*\| \leq 1\}.$$

We claim that $W_n$ is compact in $(V_1^*, \mathcal{T}^{W^*})$.

Indeed, suppose $\{x_\alpha^*\} \subset W_n$ is a net and let $y_\alpha^* \in V_2^*$ be such that $T^*(y_\alpha^*) = x_\alpha^*$. From the previous observation we see that for each $\alpha$

$$\|y_\alpha^*\| \leq \frac{\|x_\alpha^*\|}{m} \leq \frac{n + \|x_o^*\|}{m},$$

and thus from the Banach-Alaoglu Theorem (Theorem 9.4.1) we deduce the existence of some $y^* \in V_2^*$ and a subnet $\{y_\beta^*\}$ of $\{y_\alpha^*\}$ that converges to $y^*$ in $(V_2^*, \mathcal{T}^{W^*})$. Let $x^* = T^*(y^*)$. Then $x^* \in W$ and

$$\begin{aligned}\lim_\beta x^*_\beta(x) &= \lim_\beta T^*(y^*_\beta)(x)\\ &= \lim_\beta y^*_\beta[T(x)]\\ &= y^*[T(x)]\\ &= x^*(x) \qquad (x \in V_1).\end{aligned}$$

Thus the subnet $\{x^*_\beta\}$ of $\{x^*_\alpha\}$ converges to $x^*$ in $(V^*_1, \mathcal{T}^{w^*})$. But $nB^*_1 + x^*_o$ is weak* closed, and so $x^* \in nB^*_1 + x^*_o$. Hence $W_n = (nB^*_1 + x^*_o) \cap W$ is weak* compact.

For any finite set $F \subset V_1$ we set

$$F^o = \{x^* \mid x^* \in V^*_1, |x^*(x) - x^*_o(x)| \leq 1, x \in F\}.$$

Clearly each such set $F^o$ is closed in $(V^*_1, \mathcal{T}^{w^*})$. We claim that there exist finite sets $F_n \subset V_1$, $n = 0,1,2,\ldots$, with $F_n \subset (1/n)B_1$, $n = 1,2,3,\ldots$, such that

$$F^o_0 \cap F^o_1 \cap \cdots \cap F^o_{n-1} \cap W_n = \emptyset \qquad (n = 1,2,3,\ldots).$$

Since $W_1 = \emptyset$, we may take any finite subset of $V_1$ for $F_0$. Suppose the finite sets $F_0, F_1, \ldots, F_{n-1}$ in $V_1$ have been so chosen that $F_k \subset (1/k)B_1$, $k = 1,2,\ldots,n-1$, and

$$F^o_0 \cap F^o_1 \cap \cdots \cap F^o_{k-1} \cap W_k = \emptyset \qquad (k = 1,2,\ldots,n).$$

In particular, if $E_n = F^o_0 \cap F^o_1 \cap \cdots \cap F^o_{n-1}$, then $E_n \cap W_n = \emptyset$. Assume, for every finite subset $F \subset (1/n)B_1$, that $F^o \cap E_n \cap W_{n+1} \neq \emptyset$. It is apparent that for any finite set $F \subset (1/n)B_1$ the subset $F^o \cap E_n \cap W_{n+1}$ is a weak* closed subset of the weak* compact set $E_n \cap W_{n+1}$ and that the collection of all such sets $F^o \cap E_n \cap W_{n+1}$ has the finite intersection property. Thus we see that

$$\bigcap_{\substack{F \subset (1/n)B_1 \\ F \text{ finite}}} (F^o \cap E_n \cap W_{n+1}) \neq \emptyset.$$

Suppose $x^*$ belongs to this intersection. Then $|x^*(x) - x_o^*(x)| \leq 1$ for all $x \in (1/n)B_1$, from which we deduce that

$$\begin{aligned} \|x^* - x_o^*\| &= \sup_{x \in B_1} |x^*(x) - x_o^*(x)| \\ &= \sup_{x \in (1/n)B_1} |x^*(nx) - x_o^*(nx)| \\ &= n[\sup_{x \in (1/n)B_1} |x^*(x) - x_o^*(x)|] \\ &\leq n. \end{aligned}$$

Hence $x^* \in E_n \cap W_n$, contradicting the fact that $E_n \cap W_n = \emptyset$.

Consequently there exist finite sets $F_n \subset V_1$ such that $F_n \subset (1/n)B_1$ and $F_0^o \cap F_1^o \cap \cdots \cap F_{n-1}^o \cap W_n = \emptyset$, $n = 1,2,3,\ldots$ . Suppose such a sequence of finite sets $\{F_n\}$ is given and let $\cup_{n=1}^{\infty} F_n = \{x_k\}$. Obviously $\lim_k \|x_k\|_1 = 0$. Furthermore, we claim that $\sup_k |x^*(x_k) - x_o^*(x_k)| > 1$, $x^* \in W$, as if $x^* \in W$, then there exists some $n$ for which $x^* \in W_n$, and so $x^* \notin F_0^o \cap F_1^o \cap \cdots \cap F_{n-1}^o$; hence there exists some $j$, $0 \leq j \leq n - 1$, such that $x^* \notin F_j^o$, that is, $|x^*(x) - x_o^*(x)| > 1$, $x \in F_j$. The indicated estimate is now apparent.

Now consider the mapping $S : V_1^* \to c_o$ defined by $S(x^*) = \{x^*(x_k)\}$. Clearly $S \in L(V_1^*, c_o)$ and $\|S\| \leq \sup_k \|x_k\|_1$, as $\lim_k \|x_k\|_1 = 0$. Moreover,

$$\begin{aligned} \inf_{x^* \in W^*} \|S(x^*) - S(x_o^*)\|_\infty &= \inf_{x^* \in W^*} [\sup_k |x^*(x_k) - x_o^*(x_k)|] \\ &\geq 1. \end{aligned}$$

Thus from Corollary 4.2.4 to the Hahn-Banach Theorem and the fact that $c_o^*$ can be identified with $\ell_1$ (Example 3.1.6) we deduce the existence of some sequence $\{a_k\} \in \ell_1$ such that

(a) $\Sigma_{k=1}^{\infty} a_k x_o^*(x_k) = 1$,

(b) $\Sigma_{k=1}^{\infty} a_k x^*(x_k) = 0$ $(x^* \in W)$.

But, since $x_o = \Sigma_{k=1}^{\infty} a_k x_k$ is obviously an element of $V_1$, the preceding equations say precisely that $x_o^*(x_o) = 1$ and $x^*(x_o) = 0$, $x^* \in W^*$. However, since $W = T^*(V_2^*)$, then $x^*(x_o) = 0$, $x^* \in W$, means that $y^*[T(x_o)] = 0$, $y^* \in V_2^*$, and so from Corollary 4.2.6 to the Hahn-Banach Theorem we conclude that $T(x_o) = 0$, and hence that $x_o = 0$, since $T$ is injective. This conclusion, however, contradicts the fact that $x_o^*(x_o) = 1$. Consequently $W = V_1^*$, and part (iii) of the theorem implies part (i).

Therefore all three parts of the theorem are equivalent. □

A construction similar to that used in the last portion of the proof will be used again during the investigation of neighborhood bases for the bounded weak* topology to be discussed in Section 10.1.

## 9.11. Problems.

1. (Theorem 9.1.1) Let $V$ be a linear space over $\Phi$ and suppose $F \subset V'$ separates points. Prove that, if

$$P = \{p_{x'} \mid x' \in F\},$$

where $p_{x'}(x) = |x'(x)|$, $x \in V$ and $x' \in F$, then $P$ is a family of seminorms on $V$ such that $(V,P)$ is a seminormed linear space. Furthermore, let $T_F = T_P$ be the topology on $V$ determined by the family $P$ and prove the following:

(a) $(V,T_F)$ is a locally convex topological linear space over $\Phi$.

(b) A net $\{x_\alpha\} \subset V$ converges to $x \in V$ in $T_F$ if and only if $\lim_\alpha x'(x_\alpha) = x'(x)$ for each $x' \in F$.

(c) If $x' \in F$, then $x'$ is a continuous linear functional on $(V,T_F)$.

(d) $T_F$ is the weakest topology on $V$ for which the elements of $F$ are continuous.

2. (Theorem 9.1.2) Let $V$ be a linear space over $\Phi$ and suppose $F \subset V'$ separates points. Prove that, if $F$ is countable, then the $F$-topology $T_F$ is metrizable.

3. (Proposition 9.1.1) Let $V$ be a linear space over $\Phi$ and suppose $F_k \subset V'$, $k = 1,2$, separates points. Prove that, if $F_1 \subset F_2$, then $T_{F_1} \subset T_{F_2}$.

4. (Proposition 9.1.2) Let $(V,T)$ be a locally convex topological linear space over $\Phi$. Prove that, if $F \subset V^*$ separates points, then $T_F \subset T$.

5. (Corollary 9.1.1) Let $V$ be a linear space over $\Phi$ and suppose $F \subset V'$ is a linear subspace that separates points. Prove that, if $W \subset V$ is a linear subspace, then the following are equivalent:

(i) $W$ is a proper closed linear subspace of $(V,T_F)$.

(ii) If $x_o \notin W$, then there exists some $x' \in F$ for which $x'(x_o) = 1$ and $x'(x) = 0$, $x \in W$.

6. Let $V$ be a linear space over $\Phi$ and suppose both $F_1$ and $F_2$ are linear subspaces of $V'$ that separate points. Prove that, if $T_{F_1} = T_{F_2}$, then $F_1 = F_2$.

7. (Proposition 9.2.2) Let $(V,\|\cdot\|)$ be a normed linear space over $\Phi$ and denote the norm topology on $V^*$ by $T$. Prove that $T^{w^*} \subset T^w \subset T$.

8. (Corollary 9.2.1) Let $(V,T)$ be a locally convex topological linear space over $\Phi$. Prove that, if $\{x_\alpha\} \subset V$ is a net that converges in $T^w$ to $x \in V$, then there exists a net $\{y_\beta\} \subset V$, each $y_\beta$ being a finite sum of the form $\Sigma_\alpha a_{\beta\alpha} x_\alpha$, $a_{\beta\alpha} \geq 0$, $\Sigma_\alpha a_{\beta\alpha} = 1$, such that $\{y_\beta\}$ converges in $T$ to $x$.

9. (Corollary 9.2.2) Let $(V,\|\cdot\|)$ be a normed linear space over $\Phi$. Prove that, if $\tau : V \to V^{**}$ is the canonical embedding

of $V$ into $V^{**}$, then $\tau(V)$ is dense in $(V^{**}, \mathcal{T}^{w^*})$.

10. (Theorem 9.3.1) Let $(V, \|\cdot\|)$ be a Banach space over $\Phi$ and prove each of the following:

(a) If $\{x_k^*\} \subset V^*$ is a Cauchy sequence in $(V^*, \mathcal{T}^{w^*})$, then there exists some $M > 0$ such that $\sup_k \|x_k^*\| \leq M$.

(b) $(V^*, \mathcal{T}^{w^*})$ is sequentially complete.

*11. (Theorem 9.3.3) Let $(V, \|\cdot\|)$ be a normed linear space over $\Phi$ and suppose $(V^*, \mathcal{T}^{w^*})$ is a complete locally convex topological linear space over $\Phi$. Prove that $V$ is finite-dimensional.

12. (Theorem 9.3.4) Let $(V, \|\cdot\|)$ be a normed linear space over $\Phi$ and prove the following:

(a) A sequence $\{x_k\} \subset V$ converges to $x$ in $(V, \mathcal{T}^w)$ if and only if $\sup_k \|x_k\| < \infty$ and $\lim_k x^*(x_k) = x^*(x)$ for each $x^*$ is some norm dense subset of $(V^*, \|\cdot\|)$.

(b) A sequence $\{x_k^*\} \subset V^*$ converges to $x^*$ in $(V^*, \mathcal{T}^{w^*})$ if and only if $\sup_k \|x_k^*\| < \infty$ and $\lim_k x_k^*(x) = x^*(x)$ for each $x$ in some norm dense subset of $(V, \|\cdot\|)$.

13. (Corollary 9.3.1) Let $(X, \mathcal{S}, \mu)$ be a $\sigma$-finite positive measure space and suppose $\{f_k\} \subset L_1(X, \mathcal{S}, \mu)$. Prove that the following are equivalent:

(i) There exists some $f \in L_1(X, \mathcal{S}, \mu)$ such that the sequence $\{f_k\}$ converges weakly to $f$.

(ii) $\sup_k \|f_k\|_1 < \infty$, and for each $E \in \mathcal{S}$ the sequence $\{\int_E f_k(t)\, d\mu(t)\}$ converges.

*14. Let $E$ be a subset of $\ell_p$, $1 < p < \infty$, consisting of the vectors $\{x_{mn} \mid 1 \leq m < n < \infty\}$, where the $m$th coordinate of $x_{mn}$ is 1, the $n$th coordinate is $m$, and all other coordinates are 0. Prove that the origin of $\ell_p$ is in the weak closure of $E$, but that no sequence of elements of $E$ converges weakly to zero. (This example is due to von Neumann.)

15. Prove that $C_o(\mathbb{R})$ is weak* dense in $L_\infty(\mathbb{R})$.

16. Let $(V,\|\cdot\|)$ be an infinite-dimensional normed linear space over $\Phi$. Prove that $V$ is the union of countably many sets that are closed and nowhere dense in $(V,\mathcal{T}^W)$.

17. Recall that both $c^*$ and $c_o^*$ can be identified with $\ell_1$. Thus $\ell_1$ has two weak* topologies $\mathcal{T}_{F_1}$ and $\mathcal{T}_{F_2}$, where $F_1 = \tau_1(c)$ for $\tau_1 : c \to c^{**}$ and $F_2 = \tau_2(c_o)$ for $\tau_2 : c_o \to c_o^{**}$. Prove that the topologies $\mathcal{T}_{F_1}$ and $\mathcal{T}_{F_2}$ on $\ell_1$ are incomparable -- that is, we have neither $\mathcal{T}_{F_1} \subset \mathcal{T}_{F_2}$ nor $\mathcal{T}_{F_2} \subset \mathcal{T}_{F_1}$.

18. Let $(V,\|\cdot\|)$ be a Banach space over $\Phi$. Prove that $V^*$ is weak* sequentially closed in $V'$. (Sequentially closed means that, if $\{x_k^*\}$ is a sequence in $V^*$ and $\lim_k x_k^* = x^*$, then $x^* \in V^*$.)

19. Let $(V,\|\cdot\|)$ be a Banach space over $\Phi$ and for $a \geq 0$ let $B_a = \{x^* \mid x^* \in V^*, \|x^*\| \leq a\}$. Prove that $B_a$ is closed in $(V^*,\mathcal{T}^{W^*})$.

*20. Give an example of a Banach space $V$ for which $\{x^* \mid x^* \in V^*, \|x^*\| = 1\}$ is not compact in $(V^*,\mathcal{T}^{W^*})$.

21. (Corollary 9.4.1) Let $(V,\|\cdot\|)$ be a normed linear space over $\Phi$ and let $E \subset V^*$. Prove that, if $E$ is weak* closed and norm bounded, then $E$ is weak* compact.

22. (Corollary 9.4.2) Let $(V,\|\cdot\|)$ be a reflexive Banach space over $\Phi$ and let $E \subset V$. Prove that the following are equivalent:

(i) $E$ is a compact set in $(V,\mathcal{T}^W)$.
(ii) $E$ is a closed set in $(V,\mathcal{T}^W)$ and a bounded set in $(V,\|\cdot\|)$.

23. (Corollary 9.4.3) Let $(V,\|\cdot\|)$ be a normed linear space over $\Phi$. Prove that, if $\{x_\alpha^*\} \subset V^*$ is a net such that $\sup_\alpha \|x_\alpha^*\| \leq M$,

then $\{x^*_\alpha\}$ has a subnet $\{x^*_\beta\}$ that converges to some point of $\{x^* \mid x^* \in V^*, \|x^*\| \leq M\}$ in $(V^*, T^{w^*})$.

24. (Corollary 9.4.4) Let $(V, \|\cdot\|)$ be a separable normed linear space over $\Phi$. Prove that, if $E \subset V^*$ is weak* closed and norm bounded, then $E$ is weak* sequentially compact.

25. (Corollary 9.4.5) Let $(V, \|\cdot\|)$ be a separable normed linear space over $\Phi$. Prove that, if $\{x^*_n\} \subset V^*$ is a sequence such that $\sup_n \|x^*_n\| \leq M$, then $\{x^*_n\}$ has a subsequence $\{x^*_{n_k}\}$ that converges to some point of $\{x^* \mid x^* \in V^*, \|x^*\| \leq M\}$ in $(V^*, T^{w^*})$.

26. (Corollary 9.4.6) Let $(V, \|\cdot\|)$ be a reflexive Banach space over $\Phi$ and let $B_1 = \{x \mid x \in V, \|x\| \leq 1\}$. Prove that the following are equivalent:

(i) $(V, \|\cdot\|)$ is separable.
(ii) The weak topology $T^w$ restricted to $B_1$ is metrizable.

27. (Corollary 9.4.7) Let $(V, \|\cdot\|)$ be a separable reflexive Banach space over $\Phi$. Prove that, if $E \subset V$ is weakly closed and norm bounded, then $E$ is weakly sequentially compact.

28. (Theorem 9.4.3) Let $(V, \|\cdot\|)$ be a normed linear space over $\Phi$ and let $B_1 = \{x \mid x \in V, \|x\| \leq 1\}$. Prove that the following are equivalent:

(i) $(V^*, \|\cdot\|)$ is separable.
(ii) The weak topology $T^w$ on $V$ restricted to $B_1$ is metrizable.

29. Let $(V, \|\cdot\|)$ be a separable Banach space over $\Phi$ and let $E \subset V^*$ be a convex subset of $V^*$. Prove that $E$ is weak* closed in $V^*$ if and only if $\{x^*_n\} \subset E$ and $\lim_n x^*_n(x) = x^*(x)$, $x \in V$, imply that $x^* \in E$.

30. Let $(V, \|\cdot\|)$ be a Banach space over $\Phi$ and let

$B_1 = \{x \mid x \in V, \|x\| \leq 1\}$. Prove that, if the weak topology $\mathcal{T}^w$ on $V$ restricted to $B_1$ is metrizable, then $V^*$ contains a countable family $F$ that separates points.

31. Let $(V,\mathcal{T})$ be a locally convex topological linear space over $\Phi$ and let $F \subset V^*$ be a linear subspace of $V^*$. Prove that $F$ is weak* dense in $V^*$ if and only if $F$ separates points.

32. Let $(V_1,\mathcal{T}_1)$ and $(V_2,\mathcal{T}_2)$ be topological linear spaces over $\Phi$ and let $T$ be any mapping from $V_1$ to $V_2$. Prove that $T$ is continuous relative to the weak topology on $V_2$; that is, $T : (V_1,\mathcal{T}_1) \to (V_2,\mathcal{T}_2^w)$ is continuous, if and only if

$$x^* \circ T : (V_1,\mathcal{T}_1) \to \mathbb{C}$$

is continuous for every $x^* \in V_2^*$.

33. Let $(V,\|\cdot\|)$ be a normed linear space over $\Phi$ and let $\{x_n^*\} \subset V^*$, $x^* \in V^*$. Prove that, if $\{x_n^*\}$ converges to $x^*$ in $(V^*,\mathcal{T}^{w^*})$, then $\|x^*\| \leq \liminf_n \|x_n^*\|$.

34. Let $(V,\|\cdot\|)$ be a normed linear space over $\Phi$. Defining $p : V^* \to \mathbb{R}$ by $p(x^*) = \|x^*\|$, $x^* \in V^*$, prove that $p$ is lower semicontinuous on $(V^*,\mathcal{T}^{w^*})$.

35. Let $(V,\|\cdot\|)$ be a normed linear space over $\Phi$. A linear subspace $W \subset V^*$ is said to be <u>saturated</u> if for each $x^* \in V^* \sim W$ there exists some $x \in W_\perp$ such that $x^*(x) \neq 0$. Prove that a subspace $W \subset V^*$ is saturated if and only if $W$ is weak* closed.

36. Let $(V_1,\|\cdot\|_1)$ and $(V_2,\|\cdot\|_2)$ be normed linear spaces over $\Phi$ and let $T \in L(V_1,V_2)$. If $\{x_n\}$ converges weakly to $x$ in $V_1$, prove that $\{T(x_n)\}$ converges weakly to $T(x)$ in $V_2$.

37. Let $(V,\|\cdot\|)$ be a reflexive Banach space over $\Phi$. If $W \subset V$ is a closed linear subspace of $V$, prove that $V/W$ is reflexive.

38. We have seen that it is possible for a subset $E$ of $\ell_2$ to have $x \in \ell_2$ as a weak closure point and yet have no sequence of elements of $E$ that converges weakly to $x$ (see Problem 14). Prove that, if $x \in \ell_2$ is in the weak closure of a bounded set $E \subset \ell_2$, then $x$ is the weak limit of a sequence of elements of $E$.

39. (Proposition 9.7.1) Let $f \in L_1([-\pi,\pi],dt/2\pi)$ and $g \in L_p([-\pi,\pi],dt/2\pi)$, $1 \leq p \leq \infty$. Prove that the expression

$$f * g(t) = \frac{1}{2\pi} \int_{-\pi}^{\pi} f(t - s)g(s)\, ds$$

defined for almost all $t \in [-\pi,\pi]$, determines an element $f * g$ of $L_p([-\pi,\pi],dt/2\pi)$ and $\|f * g\|_p \leq \|f\|_1 \|g\|_p$.

*40. (a) If $1 \leq p \leq \infty$ and $T \in L(L_1(\mathbb{R},dt),L_p(\mathbb{R},dt))$, prove that $T$ is a multiplier if and only if $T\tau_s = \tau_s T$ for each $s \in \mathbb{R}$, where $\tau_s f(t) = f(t - s)$, $t \in \mathbb{R}$. Thus $T \in L(L_1(\mathbb{R},dt),L_p(\mathbb{R},dt))$, $1 \leq p \leq \infty$, is a multiplier if and only if $T$ commutes with translation.

(b) If $1 \leq p < \infty$ and $T \in L(L_p(\mathbb{R},dt),L_\infty(\mathbb{R},dt))$, prove that the following are equivalent:

(i) $T\tau_s = \tau_s T$, $s \in \mathbb{R}$.

(ii) There exists some $h \in L_q(\mathbb{R},dt)$, $1/p + 1/q = 1$, such that $T(f) = f * h$, $f \in L_p(\mathbb{R},dt)$.

(c) Prove by means of an example that there exist $T$ in $L(L_\infty(\mathbb{R},dt),L_\infty(\mathbb{R},dt))$ for which $T\tau_s = \tau_s T$, $s \in \mathbb{R}$, but such that $T(f) = f * h$, $f \in L_\infty(\mathbb{R},dt)$, holds for no $h \in L_1(\mathbb{R},dt)$. (Hint: Let $W$ be the linear subspace of $L_\infty(\mathbb{R},dt)$ consisting of those $f \in C(\mathbb{R})$ for which $\lim_{|t|\to\infty} f(t)$ exists. If $y^*$ on $W$ is defined by $y^*(f) = \lim_{|t|\to\infty} f(t)$, $f \in W$, then $y^*$ is a continuous linear functional on $W$ that commutes with translation. By Theorem 4.3.1 this functional has an extension $x^* \in L_\infty(\mathbb{R},dt)^*$ that also commutes with translation. Define $T \in L(L_\infty(\mathbb{R},dt),L_\infty(\mathbb{R},dt))$ by

$$T(f)(t) = x^*[\tau_t(f)] \qquad (t \in \mathbb{R},\ f \in L_\infty(\mathbb{R},dt)).)$$

41. (Lemma 9.8.2) Prove that, if $f \in L_1(\mathbb{R},dt)$ and $\mu \in M(\mathbb{R})$, then the expression

$$f * \mu(t) = \int_{\mathbb{R}} f(t - s)\, d\mu(s),$$

defined for almost all $t \in \mathbb{R}$, determines an element $f * \mu$ in $L_1(\mathbb{R},dt)$ and $\|f * \mu\|_1 \leq \|\mu\|\|f\|_1$.

*42. (Theorem 9.8.2) Prove that, if $T \in L'(L_1(\mathbb{R},dt))$, then the following are equivalent:

(i) There exists some $\mu \in M(\mathbb{R})$ such that $T(f) = f * \mu$, $f \in L_1(\mathbb{R},dt)$.

(ii) $T$ is a multiplier from $L_1(\mathbb{R},dt)$ to $L_1(\mathbb{R},dt)$.

43. Let $(V,\|\cdot\|)$ be a Banach space over $\Phi$. Prove that $V$ is reflexive if and only if every norm closed linear subspace of $V^*$ is weak* closed.

44. Let $(V,\|\cdot\|)$ be a Banach space over $\Phi$ and suppose that every linear subspace $F \subset V^*$ that separates points is dense in $(V^*,\|\cdot\|)$. Prove that $V$ is reflexive.

45. Let $(V,\|\cdot\|)$ be a reflexive Banach space over $\Phi$ and suppose there is a countable set $F \subset V^*$ that separates points. Prove that $V^*$ is separable.

46. Let $(V_1,\|\cdot\|_1)$ and $(V_2,\|\cdot\|_2)$ be Fréchet spaces over $\Phi$ and let $T \in L'(V_1,V_2)$. If $T^*$ denotes the adjoint of $T$, prove that $T$ is weakly continuous if and only if $T^*$ is weakly continuous.

## CHAPTER 10

## THE KREIN-ŠMULIAN AND EBERLEIN-ŠMULIAN THEOREMS

10.0. Introduction. The development in the preceding chapter revealed, among other things, that the unit ball in the dual of a Banach space, although always compact in the weak* topology, could fail to be weak* sequentially compact. On the other hand, we saw that in a separable reflexive Banach space the unit ball is both compact and sequentially compact in the weak topology. The main purpose of this chapter is to show that this latter situation holds for all Banach spaces; that is, a weakly closed set in a Banach space is weakly compact if and only if it is weakly sequentially compact. This is the content of the Eberlein-Šmulian Theorem, which will be proved in Section 10.3.

The Eberlein-Šmulian Theorem is a profound result, and its proof, which is quite involved, will utilize much of the machinery we have developed in the preceding nine chapters, as well as some new ideas and results to be discussed in the next two sections. These new results include the introduction of a new topology on the dual space, called the bounded weak* topology, and the Krein-Šmulian Theorem, which says that a convex set in the dual space of a Banach space is weak* closed if and only if its intersection with every norm closed ball about the origin is weak* closed.

10.1. The Bounded Weak* Topology. Our purpose here is to introduce a new topology on the dual space $V^*$ of a Banach space $V$ that is stronger than the weak* topology but is such that the continuous linear functionals on the topological linear space so obtained are precisely the weak* continuous linear functionals on $V^*$. The majority of the development in this section is actually devoted to describing various neighborhood bases for this new topology.

Definition 10.1.1. Let $(V,\|\cdot\|)$ be a normed linear space over $\Phi$ and let $B_1^* = \{x^* \mid x^* \in V^*, \|x^*\| \leq 1\}$. A set $U \subset V^*$ is said to belong to $\mathcal{T}^{bw^*}$ if for each $a \geq 0$ the set $U \cap aB_1^*$ is an open subset of $aB_1^*$, the latter being considered a topological space in the relative weak* topology.

In other words, $U \in \mathcal{T}^{bw^*}$ if and only if the intersection of $U$ with every norm closed ball about the origin in $V^*$ is open in the relative weak* topology on that ball.

It is not difficult to verify that the following proposition is valid. The details are left to the reader.

Proposition 10.1.1. Let $(V,\|\cdot\|)$ be a normed linear space over $\Phi$.

(i) $U \in \mathcal{T}^{bw^*}$ if and only if for each $a \geq 0$ and $x^* \in U \cap aB_1^*$ there exist some $\epsilon > 0$ and $x_1, x_2, \ldots, x_n$ in $V$ such that

$$U(x^*;\epsilon;x_1,x_2,\ldots,x_n) \cap aB_1^*$$

$$= \{y^* \mid y^* \in V^*, \|y^*\| \leq a, |x^*(x_k) - y^*(x_k)| < \epsilon, k = 1,2,\ldots,n\}$$

is contained in $U$.

(ii) $\mathcal{T}^{bw^*}$ is a topology on $V^*$ such that $(V^*,\mathcal{T}^{bw^*})$ is a locally convex topological linear space over $\Phi$.

(iii) $\mathcal{T}^{w^*} \subset \mathcal{T}^{bw^*}$.

We call $\mathcal{T}^{bw^*}$ the bounded weak* topology on $V^*$. It is apparent that a set $E \subset V^*$ is closed in $(V^*,\mathcal{T}^{bw^*})$ if and only if $E \cap aB_1^*$ is closed in $(V^*,\mathcal{T}^{w^*})$ for each $a \geq 0$. This follows at once since each $aB_1^*$ is weak* closed.

Next we wish to give three descriptions of a neighborhood base at the origin for the bounded weak* topology. The first description involves a fairly long proof, but the other two follow easily from the first.

Theorem 10.1.1. Let $(V,\|\cdot\|)$ be a normed linear space over $\Phi$. Then the sets

$$U(0,1,\{x_k\}) = \{x^* \mid x^* \in V^*, |x^*(x_k)| < 1, k = 1,2,3,\ldots\},$$

where $\{x_k\}$ is any sequence in $V$ such that $\lim_k\|x_k\| = 0$, form a neighborhood base at the origin for $\mathcal{T}^{bw^*}$.

Proof. First we must show that each such set belongs to $\mathcal{T}^{bw^*}$. So let $\{x_k\} \subset V$ be such that $\lim_k \|x_k\| = 0$ and let $a > 0$. Choose a positive integer $n$ such that $\|x_k\| < 1/a$, $k > n$. We claim that

$$U(0,1,\{x_k\}) \cap aB_1^* = U(0;1;x_1,x_2,\ldots,x_n) \cap aB_1^*.$$

Clearly

$$U(0,1,\{x_k\}) \cap aB_1^* \subset U(0;1;x_1,x_2,\ldots,x_n) \cap aB_1^*.$$

Conversely, suppose $x^* \in U(0;1;x_1,x_2,\ldots,x_n) \cap aB_1^*$. Then $|x^*(x_k)| < 1$, $k = 1,2,\ldots,n$, whereas if $k > n$, then

$$|x^*(x_k)| \le \|x^*\|\|x_k\| < a(\tfrac{1}{a}) = 1,$$

and we see that $x^* \in U(0,1,\{x_k\}) \cap aB_1^*$. Thus

$$U(0,1,\{x_k\}) \cap aB_1^* = U(0;1;x_1,x_2,\ldots,x_n) \cap aB_1^*,$$

and the latter set is open in the relative weak* topology on $aB_1^*$. Since the result holds trivially when $a = 0$, we conclude that $U(0,1,\{x_k\}) \in \mathcal{T}^{bw^*}$.

It is evident that the intersection of any two sets of the form $U(0,1,\{x_k\})$ contains a third, so to prove that these sets are a neighborhood base at the origin it remains to show only that every $U \in \mathcal{T}^{bw^*}$ such the $0 \in U$ contains one of these sets; that is, given $U \in \mathcal{T}^{bw^*}$ such that $0 \in U$, it suffices to prove the existence of a sequence $\{x_k\} \subset V$ such that $\lim_k\|x_k\| = 0$ and $U(0,1,\{x_k\}) \subset U$. The proof of this will require a bit of labor.

In general, if $F \subset V$ is a finite set, then we write

$$F^o = \{x^* \mid x^* \in V^*, |x^*(x)| \leq 1, x \in F\}.$$

Note that each such $F^o$ is a weak* closed subset of $V^*$. Now, since $U$ is open in $\mathcal{T}^{bw*}$, there exist, by Proposition 10.1.1(i), some $\epsilon > 0$ and $x_1, x_2, \ldots, x_n$ in $V$ such that

$$U(0;1;x_1,x_2,\ldots,x_n) \cap B_1^* \subset U.$$

Obviously, on setting $y_k = 2x_k/\epsilon$, $k = 1,2,\ldots,n$, we see that $F_1^o \cap B_1^* \subset U$, where $F_1 = \{y_1,y_2,\ldots,y_n\}$. Using a similar argument, we wish to define a particular sequence $\{F_n\}$ of finite subsets of $V$. These sets will be obtained as follows:

Beginning with $F_1$, choose a finite set $E_1 \subset V$ such that $\|y\| \leq 1$, $y \in E_1$, and $(F_1 \cup E_1)^o \cap 2B_1^* \subset U$. Set $F_2 = F_1 \cup E_1$. Having defined $F_1, F_2, \ldots, F_n$, choose a finite set $E_n \subset V$ such that $\|y\| \leq 1/n$, $y \in E_n$, and $(F_n \cup E_n)^o \cap (n+1)B_1^* \subset U$. Set $F_{n+1} = F_n \cup E_n$. Of course, it is not entirely clear that such choices of the sets $E_n$ can be made. To see that this can be done it clearly suffices to show that, if $F_1, F_2, \ldots, F_n$ are given, then $E_n$ with the appropriate properties can be found.

Suppose this is not the case. Then for any finite set $E \subset V$ such that $\|y\| \leq 1/n$, $y \in E$, we have

$$(F_n \cup E)^o \cap (n+1)B_1^* \cap U^c \neq \emptyset,$$

where $U^c$ denotes the complement of $U$ in $V^*$, whereas $F_n^o \cap nB_1^* \subset U$. Consider the family of sets

$$\mathcal{S} = \{(F_n \cup E)^o \cap (n+1)B_1^* \cap U^c \mid E \subset V,\ E \text{ finite},\ \|y\| \leq 1/n,\ y \in E\}.$$

Clearly all the sets in $\mathcal{S}$ are weak* closed subsets of $(n+1)B_1^*$, which, by the Banach-Alaoglu Theorem (Theorem 9.4.1), is weak* compact. Moreover, $\mathcal{S}$ has the finite intersection property since, if $E_1, E_2, \ldots, E_m$ are finite subsets of $V$ such that $\|y\| \leq 1/n$, $y \in E_k$,

$k = 1,2,\ldots,m,$ then

$$\bigcap_{k=1}^{m} [(F_n \cup E_k)^o \cap (n + 1)B_1^* \cap U^c] = [F_n \cup (\bigcup_{k=1}^{m} E_k)]^o \cap (n + 1)B_1^* \cap U^c \neq \emptyset.$$

Consequently we see that $\bigcap_S [(F_n \cup E)^o \cap (n + 1)B_1^* \cap U^c] \neq \emptyset$.

Suppose, however, that $x^* \in \bigcap_S [(F_n \cup E)^o \cap (n + 1)B_1^* \cap U^c]$. Then, in particular, we see that $|x^*(x)| \leq 1$ for each $x \in V$ such that $\|x\| \leq 1/n$, from which we deduce at once that $\|x^*\| \leq n$. Thus

$$\bigcap_S [(F_n \cup E)^o \cap (n + 1)B_1^* \cap U^c] \subset [F_n^o \cap (n + 1)B_1^* \cap U^c] \cap nB_1^*$$
$$= F_n^o \cap nB_1^* \cap U^c,$$

and so $F_n^o \cap nB_1^* \cap U^c \neq \emptyset$. This, however, contradicts the hypothesis that $F_n^o \cap nB_1^* \subset U$. Therefore there exists some finite set $E_n \subset V$ such that $\|y\| \leq 1/n$, $y \in E_n$, and $(F_n \cup E_n)^o \cap (n + 1)B_1^* \subset U$.

We have now shown that there exists a sequence $\{F_n\}$ of finite subsets of $V$ such that $F_n^o \cap nB_1^* \subset U$. Let $F = \bigcup_{n=1}^{\infty} F_n$. Clearly $F$ is a countable subset of $V$, and so we may enumerate it as a sequence, say $\{x_k\}$. Moreover, $\lim_k \|x_k\| = 0$. This is apparent on noting that for each positive integer $n$ all but a finite number of the members of $F$ have norm less than or equal to $1/n$.

We claim, furthermore, that $U(0,1,\{x_k\}) \subset U$. Indeed, suppose $x^* \in U(0,1,\{x_k\})$ and let $n \geq \|x^*\|$. Then $|x^*(x_k)| < 1$, $k = 1,2,3,\ldots,$ shows at once that $x^* \in F_n^o \cap nB_1^* \subset U$. Thus $U(0,1,\{x_k\}) \subset U$, and the proof is complete.

□

A couple of observations are in order. First, given $U \in \mathcal{T}^{bw^*}$ for which $0 \in U$, it is clear that, if $\{x_k\} \subset V$ is the sequence constructed in the proof such that $U(0,1,\{x_k\}) \subset U$, then it also happens that $\{x^* \mid |x^*(x_k)| \leq 1, k = 1,2,3,\ldots\} \subset U$. This is apparent from the definition of $F^o$. Moreover, it is also evident that for any $\epsilon$, $0 < \epsilon < 1$, we have $U(0,\epsilon,\{x_k\}) \subset U$, where, of course, $U(0,\epsilon,\{x_k\}) = \{x^* \mid x^* \in V^*, |x^*(x_k)| < \epsilon, k = 1,2,3,\ldots\}$. The latter observation immediately yields the following corollary:

Corollary 10.1.1. Let $(V,\|\cdot\|)$ be a normed linear space over $\Phi$. Then for each $\epsilon$, $0 < \epsilon \leq 1$, the sets

$$U(0,\epsilon,\{x_k\}) = \{x^* \mid x^* \in V^*, |x^*(x_k)| < \epsilon, k = 1,2,3,\ldots\},$$

where $\{x_k\}$ is any sequence in $V$ such that $\lim_k \|x_k\| = 0$, form a neighborhood base at the origin for $\mathcal{T}^{bw^*}$.

The final description of a neighborhood base for $\mathcal{T}^{bw^*}$ is contained in the next corollary.

Corollary 10.1.2. Let $(V,\|\cdot\|)$ be a normed linear space over $\Phi$. Then the sets

$$K^o = \{x^* \mid x^* \in V^*, |x^*(x)| \leq 1, x \in K\},$$

where $K$ is any compact subset of $(V,\|\cdot\|)$, form a neighborhood base at the origin for $\mathcal{T}^{bw^*}$.

Proof. Suppose $K \subset V$ is compact in the norm topology and $a > 0$. Then, since $(V,\|\cdot\|)$ is a metric space, we see that $K$ is totally bounded, and so there exist $x_1,x_2,\ldots,x_n$ in $K$ such that $K \subset \bigcup_{k=1}^n \{x \mid x \in V, \|x - x_k\| \leq 1/2a\}$. Let $x^*$ belong to the set $U(0;1/2;x_1,x_2,\ldots,x_n) \cap aB_1^*$. Then, if $x \in K$, there exists some $x_k$, $k = 1,2,\ldots,n$, such that $\|x - x_k\| \leq 1/2a$, and so

$$\begin{aligned} |x^*(x)| &\leq |x^*(x - x_k)| + |x^*(x_k)| \\ &\leq \|x^*\|\|x - x_k\| + \frac{1}{2} \\ &\leq a\left(\frac{1}{2a}\right) + \frac{1}{2} \\ &= 1. \end{aligned}$$

Thus $U(0;1/2;x_1,x_2,\ldots,x_n) \cap aB_1^* \subset K^o$, and so from Proposition 10.1.1(i) we conclude that $K^o \in \mathcal{T}^{bw*}$. Clearly $0 \in K^o$, and if $K_1, K_2 \subset V$ are compact, then $K_1 \cup K_2$ is compact and

$$(K_1 \cup K_2)^o \subset K_1^o \cap K_2^o.$$

Consequently it remains only to show that, if $U \in \mathcal{T}^{bw*}$ is such that $0 \in U$, then there exists some norm compact $K \subset V$ for which $K^o \subset U$. But from Theorem 10.1.1 we know that there exists a sequence $\{x_k\} \subset V$ such that $\lim_k \|x_k\| = 0$ and $U(0,1,\{x_k\}) \subset U$. Moreover, as remarked after Theorem 10.1.1, we may even assume that $\{x^* \mid x^* \in V^*, |x^*(x_k)| \leq 1,\ k = 1,2,3,\ldots\} \subset U$.

Clearly then $K = \{x_k\} \cup \{0\}$ is a norm compact subset of $V$ and $K^o \subset U$.

□

We saw in Proposition 10.1.1(iii) that the bounded weak* topology is stronger than the weak* topology. We now wish to show that the continuous linear functionals on $(V^*, \mathcal{T}^{bw*})$ are precisely those on $(V^*, \mathcal{T}^{w*})$, provided $V$ is a Banach space.

<u>Theorem</u> <u>10.1.2</u>. Let $(V, \|\cdot\|)$ be a Banach space over $\Phi$ and suppose $x'' \in (V^*)'$. Then the following are equivalent:

(i) $x'' : (V^*, \mathcal{T}^{w*}) \to \Phi$ is continuous.
(ii) $x'' : (V^*, \mathcal{T}^{bw*}) \to \Phi$ is continuous.

<u>Proof</u>. Since $\mathcal{T}^{w*} \subset \mathcal{T}^{bw*}$, it is evident that part (i) implies part (ii). On the other hand, suppose $x'' : (V^*, \mathcal{T}^{bw*}) \to \Phi$ is continuous. Then from Theorem 10.1.1 we see that there exists some sequence $\{x_k\} \subset V$ for which $\lim_k \|x_k\| = 0$ and $|x''(x^*)| < 1$ whenever $x^* \in U(0,1,\{x_k\})$. Moreover, for each $\epsilon$, $0 < \epsilon \leq 1$, if $x^* \in U(0,\epsilon,\{x_k\})$, then $|(x^*/\epsilon)(x_k)| < 1$, $k = 1,2,3,\ldots$, and so $|x''(x^*/\epsilon)| < 1$. Thus $|x''(x^*)| < \epsilon$ whenever $x^* \in U(0,\epsilon,\{x_k\})$.

Now define $T : V^* \to c_o$ by setting $T(x^*) = \{x^*(x_k)\}$, $x^* \in V^*$. Then $T(V^*)$ is a linear subspace of $c_o$, and we define a linear functional $y^*$ on $T(V^*)$ by setting $y^*[T(x^*)] = x''(x^*)$, $x^* \in V^*$. We need to note first that $y^*$ is well-defined; that is, if $T(x_1^*) = T(x_2^*)$, then $x''(x_1^*) = x''(x_2^*)$. But if $T(x_1^*) = T(x_2^*)$, then $x_1^*(x_k) = x_2^*(x_k)$, $k = 1,2,3,\ldots$, and so $x_1^* - x_2^* \in U(0,\epsilon,\{x_k\})$ for all $\epsilon$, $0 < \epsilon \leq 1$. Hence $|x''(x_1^* - x_2^*)| < \epsilon$ for all $\epsilon$, $0 < \epsilon \leq 1$, and so $x''(x_1^*) = x''(x_2^*)$. Thus $y^*$ is well-defined and clearly linear.

Furthermore, suppose $\{x_n^*\} \subset V^*$ is a sequence such that $\lim_n \|\{x_n^*(x_k)\}\|_\infty = 0$; that is, $\{\{x_n^*(x_k)\}\} = \{T(x_n^*)\}$ converges to the zero sequence in $c_o$. Then, given $\epsilon > 0$, there exists some positive integer $N$ such that, if $n \geq N$, then $|x_n^*(x_k)| < \epsilon$, $k = 1,2,3,\ldots$ . Hence $|y^*[T(x_n^*)]| = |x''(x_n^*)| < \epsilon$ for $n \geq N$. Consequently $y^*$ is continuous at the origin, and so $y^*$ is a continuous linear functional on $T(V^*)$, by Proposition 3.2.2.

From Theorem 4.2.2 we see that we can extend $y^*$ to a continuous linear functional on $c_o$, which we again denote by $y^*$. Since, as noted in Example 3.1.6, $c_o^* = \ell_1$, we deduce that there exists a sequence $\{a_k\} \in \ell_1$ such that $y^*[T(x^*)] = \Sigma_{k=1}^\infty x^*(x_k)a_k$, $x^* \in V^*$. But then

$$\begin{aligned} x''(x^*) &= y^*[T(x^*)] \\ &= \sum_{k=1}^{\infty} x^*(x_k)a_k \\ &= \lim_n \sum_{k=1}^{n} x^*(x_k)a_k \\ &= \lim_n x^*\Big(\sum_{k=1}^{n} a_k x_k\Big) \\ &= x^*\Big(\sum_{k=1}^{\infty} a_k x_k\Big) \\ &= \tau\Big(\sum_{k=1}^{\infty} a_k x_k\Big)(x^*) \qquad (x^* \in V^*). \end{aligned}$$

The penultimate equality is valid since $x^* \in V^*$ and $\Sigma_{k=1}^{\infty} a_k x_k$ is obviously an element of $V$, as $\{a_k\} \in \ell_1$ and $\lim_k \|x_k\| = 0$.

Therefore $x'' = \tau(\Sigma_{k=1}^{\infty} a_k x_k)$, and so, by Theorem 9.1.3, $x''$ is a weak* continuous linear functional on $V^*$. □

<u>10.2. The Krein-Šmulian Theorem.</u> We are now almost in a position to prove the Krein-Šmulian Theorem. However, one additional preliminary result is necessary.

<u>Lemma 10.2.1.</u> Let $(V,\mathcal{T}_1)$ and $(V,\mathcal{T}_2)$ be locally convex topological linear spaces over $\Phi$ and suppose that a linear functional on $V$ is continuous on $(V,\mathcal{T}_1)$ if and only if it is continuous on $(V,\mathcal{T}_2)$. If $K \subset V$ is convex, then the following are equivalent:

(i) $K$ is closed in $(V,\mathcal{T}_1)$.
(ii) $K$ is closed in $(V,\mathcal{T}_2)$.

<u>Proof.</u> If $K$ is empty or all of $V$, then the result is trivial. Therefore suppose $K$ is closed in $(V,\mathcal{T}_1)$, $K \neq V$, and $K$ is nonempty. If $x_0 \notin K$, then by Corollary 5.4.1 and Proposition 5.1.1 there exists some closed real hyperplane $L$ that strictly separates $x_0$ and $K$; that is, there exists some continuous real linear functional $x^*$ on $(V,\mathcal{T}_1)$ and some $a \in \mathbb{R}$ such that $x^*(x_0) < a < x^*(x)$, $x \in K$. Let $\epsilon = a - x^*(x_0) > 0$ and set

$$U = \{y \mid y \in V, |x^*(x_0) - x^*(y)| < \epsilon\}.$$

Since $(V,\mathcal{T}_1)$ and $(V,\mathcal{T}_2)$ have the same continuous linear functionals, it is apparent that $U$ is an open set in $(V,\mathcal{T}_2)$, and clearly $x_0 \in U$. Moreover, $U \cap K = \emptyset$, because if $z \in U \cap K$, then $|x^*(x_0) - x^*(z)| < \epsilon = a - x^*(x_0)$ shows that $x^*(z) < a$, which contradicts the fact that $x^*(z) > a$, as $z \in K$. Thus $K$ is closed in $(V,\mathcal{T}_2)$.

Therefore part (i) of the lemma implies part (ii), and the same argument *mutatis mutandis* establishes the reverse implication. □

*Theorem 10.2.1* (*Krein-Šmulian Theorem*). Let $(V,\|\cdot\|)$ be a Banach space over $\Phi$ and suppose $K \subset V^*$ is convex. Then the following are equivalent:

(i) $K$ is closed in $(V^*,\mathcal{T}^{w^*})$.

(ii) For each $a \geq 0$, the set $K \cap aB_1^*$ is closed in $(V^*,\mathcal{T}^{w^*})$, where $B_1^* = \{x^* \mid x^* \in V^*, \|x^*\| \leq 1\}$.

*Proof*. From Proposition 10.1.1(ii) and Theorem 10.1.2 we see that $(V^*,\mathcal{T}^{w^*})$ and $(V^*,\mathcal{T}^{bw^*})$ are locally convex topological linear spaces over $\Phi$ with the same continuous linear functionals. By Lemma 10.2.1, therefore, $K$ is closed in $(V^*,\mathcal{T}^{w^*})$ if and only if $K$ is closed in $(V^*,\mathcal{T}^{bw^*})$. However, from the remark following Proposition 10.1.1 we know that $K$ is closed in $(V^*,\mathcal{T}^{bw^*})$ precisely when part (ii) of the present theorem holds, which completes the proof. □

Before utilizing this result in the proof of the Eberlein-Šmulian Theorem we examine a few corollaries. The first two have simple proofs, which are left to the reader.

*Corollary 10.2.1*. Let $(V,\|\cdot\|)$ be a reflexive Banach space over $\Phi$ and suppose $K \subset V$ is convex. Then the following are equivalent:

(i) $K$ is closed in $(V,\mathcal{T}^{w})$.

(ii) For each $a \geq 0$, the set $K \cap aB_1$ is closed in $(V,\mathcal{T}^{w})$, where $B_1 = \{x \mid x \in V, \|x\| \leq 1\}$.

*Corollary 10.2.2*. Let $(V, \|\cdot\|)$ be a Banach space over $\Phi$ and suppose $W \subset V^*$ is a linear subspace. Then the following are equivalent:

(i) $W$ is closed in $(V^*, \mathcal{T}^{w^*})$.

(ii) $W \cap B_1^*$ is closed in $(V^*, \mathcal{T}^{w^*})$, where

$$B_1^* = \{x^* \mid x^* \in V^*, \|x^*\| \leq 1\}.$$

Corollary 10.2.3. Let $(V, \|\cdot\|)$ be a Banach space over $\Phi$ and suppose $W \subset V^*$ is a linear subspace. Then the following are equivalent:

(i) $W$ is closed in $(V^*, \mathcal{T}^{w^*})$.

(ii) There exist a norm bounded set $E \subset W$ that is closed in $(V^*, \mathcal{T}^{w^*})$, some $x_o^* \in W$, and some $\epsilon > 0$ such that

$$E \supset \{x^* \mid x^* \in W, \|x^* - x_o^*\| < \epsilon\}.$$

Proof. If $W$ is closed in $(V^*, \mathcal{T}^{w^*})$, then set $E = W \cap B_1^*$, where $B_1^* = \{x^* \mid x^* \in V^*, \|x^*\| \leq 1\}$. Clearly $E$ is weak* closed and norm bounded, and $E \supset \{x^* \mid x^* \in W, \|x^*\| \leq 1\}$. Thus part (i) implies part (ii) on taking $x_o^* = 0$ and $\epsilon = 1$.

Conversely, suppose part (ii) holds. Since scalar multiplication and translation are homeomorphisms on $(V^*, \mathcal{T}^{w^*})$ (Theorem 2.1.2), it is immediately obvious that for each $a > 0$ we have $(2a/\epsilon)(E - x_o^*)$ is closed in $(V^*, \mathcal{T}^{w^*})$, while the assertion is trivially true if $a = 0$. Obviously part (ii) entails that $W \cap (\epsilon/2)B_1^* \subset E - x_o^*$, and so for each $a \geq 0$

$$\begin{aligned} W \cap aB_1^* &= W \cap \left(\frac{a\epsilon}{\epsilon}\right) B_1^* \\ &= \frac{2a}{\epsilon}\left[W \cap \left(\frac{\epsilon}{2}\right) B_1^*\right] \\ &\subset \frac{2a}{\epsilon}(E - x_o^*), \end{aligned}$$

as $W$ is a linear subspace. Hence, for each $a \geq 0$, we deduce that $W \cap aB_1^* = (2a/\epsilon)(E - x_o^*) \cap aB_1^*$, and the latter set is closed in $(V^*, \mathcal{T}^{w^*})$.

Therefore, by the Krein-Šmulian Theorem, we see that $W$ is closed in $(V^*, \mathcal{T}^{w^*})$, and part (ii) of the corollary implies part (i). $\square$

In other words, Corollary 10.2.3 asserts that, if $(V,\|\cdot\|)$ is a Banach space, then a linear subspace $W \subset V^*$ is weak* closed if and only if it contains some weak* closed norm bounded set $E$ that in turn contains some open ball in the normed linear space $(W,\|\cdot\|)$.

Finally, we wish to apply Corollary 10.2.3 to derive a result for the weak* topology analogous to Theorem 9.2.2, which asserts that a nonempty convex set is closed if and only if it is weakly closed. A lemma is necessary, which we state in a form suitable to our needs. It is valid, however, in a more general context, as can be seen in [$DS_1$, pp. 415 and 416]. The proof is left to the reader.

<u>Definition 10.2.1</u>. Let $V$ be a linear space over $\Phi$ and suppose $E \subset V$. Then the <u>convex hull</u> of $E$ is defined as

$$\mathrm{co}(E) = \{\sum_{k=1}^{n} a_k x_k \mid a_k \geq 0,\ \sum_{k=1}^{n} a_k = 1,\ x_k \in E,\ n = 1,2,3,\ldots\}.$$

If $(V,\mathcal{T})$ is a topological linear space over $\Phi$, then the <u>closed convex hull</u> of $E$ is defined as the closure in $(V,\mathcal{T})$ of $\mathrm{co}(E)$. It will be denoted by $\overline{\mathrm{co}}(E)$.

It is not difficult to see that $\mathrm{co}(E)$ and $\overline{\mathrm{co}}(E)$ are, respectively, the smallest convex and the smallest closed convex sets that contain $E$. Indeed, it can be shown that

$$\mathrm{co}(E) = \bigcap\{K \mid K \supset E,\ K \text{ convex}\}$$

and

$$\overline{\mathrm{co}}(E) = \bigcap\{K \mid K \supset E,\ K \text{ closed and convex}\}.$$

The details are left to the reader.

We shall discuss an important theorem about convex hulls in the next chapter. Now, however, we need only the next result.

<u>Lemma 10.2.2</u>. Let $(V,\|\cdot\|)$ be a Banach space over $\Phi$ and suppose $K_1, K_2 \subset V^*$ are convex sets that are compact in $(V^*,\mathcal{T}^{w^*})$. Then $\overline{\mathrm{co}}(K_1 \cup K_2) = \mathrm{co}(K_1 \cup K_2)$.

Theorem 10.2.2. Let $(V,\|\cdot\|)$ be a Banach space over $\Phi$ and suppose $K \subset V^*$ is a convex set that is closed in $(V^*,\mathcal{T}^{w^*})$. If $W$ denotes the linear subspace of $V^*$ spanned by $K$, then the following are equivalent:

(i) $W$ is closed in $(V^*,\mathcal{T}^{w^*})$.
(ii) $W$ is closed in $(V^*,\|\cdot\|)$.

Proof. For simplicity we assume that $\Phi = \mathbb{R}$. The case where $\Phi = \mathbb{C}$ is left to the reader. Since $\mathcal{T}^{w^*}$ is weaker than the norm topology on $V^*$ it follows at once that part (i) implies part (ii).

Conversely, suppose $W$ is norm closed and for each positive integer $n$ set $K_n = K \cap nB_1^*$, where $B_1^* = \{x^* \mid x^* \in V^*, \|x^*\| \leq 1\}$. From the Banach-Alaoglu Theorem (Theorem 9.4.1) it is apparent that each $K_n$ is a weak* compact convex set, and so from Lemma 10.2.2 we see that $K_n' = \text{co}(K_n \cup -K_n)$ is weak* closed. However, by Proposition 9.2.2 the weak* topology on $V^*$ is weaker than the weak topology on $V^*$, and hence each $K_n'$ is closed in $(V^*,\mathcal{T}^w)$. Since each $K_n'$ is convex, it follows from Theorem 9.2.2 that each $K_n'$ is closed in $(V^*,\|\cdot\|)$.

Now set $K' = \text{co}(K \cup -K)$. It is evident that each $x \in W$ can be written as $x = \Sigma_{k=1}^m a_k x_k - \Sigma_{k=m+1}^n a_k x_k$, where $x_k \in K$ and $a_k > 0$, $k = 1,2,\ldots,n$. Thus, on setting $a = \Sigma_{k=1}^n a_k > 0$, we see that $x \in aK'$, as $\Sigma_{k=1}^m (a_k/a)x_k - \Sigma_{k=m+1}^n (a_k/a)x_k \in K'$. Moreover, $0 \in K'$, from which it follows at once that $aK' \subset bK'$, $0 \leq a \leq b$. Consequently, given $x \in W$, there exists some positive integer $n_x$ such that $x \in nK'$, $n \geq n_x$. We claim this implies that $W = \bigcup_{n=1}^\infty K_n'$.

Indeed, it is evident that $\bigcup_{n=1}^\infty K_n' \subset W$. On the other hand, if $x \in W$, then $x \in nK'$, $n \geq n_x$. Clearly $K' = \bigcup_{j=1}^\infty K_j'$, and so there exists some $j_x$ for which $x \in n_x K_{j_x}'$. If $1 \leq j_x \leq n_x$, then $x = n_x y$ for some $y \in K_{j_x}'$. Consequently, since

$$x = \left[\frac{j_x n_x}{j_x}\right] y = j_x\left(\left[\frac{n_x}{j_x}\right] y + \left[1 - \frac{n_x}{j_x}\right] 0\right),$$

we conclude that $x \in j_x K'_{j_x}$ since $y \in K'_{j_x}$ and $0 \in K'_{j_x}$. Hence $x \in \bigcup_{n=1}^{\infty} nK'_n$, and so $W = \bigcup_{n=1}^{\infty} K'_n$.

But $W$, being a norm closed linear subspace of $V^*$, is a Banach space over $\Phi$, and so, by the Baire Category Theorem (Theorem 6.1.1), we conclude that there exists some $n_o$ such that the norm closed set $K'_{n_o}$ has a nonempty interior in $W$. However, $K'_{n_o}$ is also weak* closed and norm bounded. Thus, appealing to Corollary 10.2.3, we see that $W$ must be closed in $(V^*, \mathcal{T}^{w^*})$.

Therefore part (ii) of the theorem implies part (i). □

<u>10.3</u>. <u>The Eberlein-Šmulian Theorem</u>. We have now reached the point where we can prove the Eberlein-Šmulian Theorem, which says that a weakly closed set in a Banach space is weakly compact if and only if it is weakly sequentially compact. The proof is quite long and involved. The key results used in proving the "only if" portion of the theorem are the Hahn-Banach, Uniform Boundedness, and Banach-Alaoglu Theorems, together with the fact that the weak and norm closures of a convex set in a Banach space coincide; the proof of the "if" portion of the theorem employs the Uniform Boundedness and Banach-Alaoglu Theorems, the coincidence of the weak and norm closures of a convex set in a Banach space, and the Krein-Šmulian Theorem.

<u>Theorem 10.3.1</u> (<u>Eberlein-Šmulian Theorem</u>). Let $(V, \|\cdot\|)$ be a Banach space over $\Phi$ and suppose $E \subset V$ is closed in $(V, \mathcal{T}^w)$. Then the following are equivalent:

(i) $E$ is compact in $(V, \mathcal{T}^w)$.
(ii) $E$ is sequentially compact in $(V, \mathcal{T}^w)$.

<u>Proof</u>. Suppose that $E$ is weakly compact and let $\{x_k\} \subset E$ be a sequence. We must show that $\{x_k\}$ has a weakly convergent subsequence.

Let $W$ denote the norm closure of the linear space spanned by

$\{x_k\}$. Then $(W, \|\cdot\|)$ is a separable Banach space over $\Phi$. Set $B_1^* = \{x^* \mid x^* \in W^*, \|x^*\| \leq 1\}$. From Theorem 9.4.2 we see that the weak* topology on $W^*$ restricted to $B_1^*$ is metrizable, $W$ being separable, and from the Banach-Alaoglu Theorem (Theorem 9.4.1) we see that $B_1^*$ is weak* compact. Thus $B_1^*$ is a compact metric space in the weak* topology and hence a separable metric space. A similar argument reveals that for each positive integer $n$, the space $nB_1^*$ is a separable metric space in the weak* topology on $W^*$, and so the weak* topology on $W^* = \bigcup_{n=1}^{\infty} nB_1^*$ must also be separable. Let $F_o \subset W^*$ be a countable set that is weak* dense in $W^*$.

First we note that $F_o$ is a family of linear functionals on $W$ that separates the points of $W$, as if $x \in W$ and $x^*(x) = 0$, $x^* \in F_o$, then, since $F_o$ is weak* dense in $W^*$, it follows at once that $x^*(x) = 0$, $x^* \in W^*$. Hence, by Corollary 4.2.6 to the Hahn-Banach Theorem, we conclude that $x = 0$, and so $F_o$ separates points. Furthermore, extending each $x^* \in F_o$ by Theorem 4.2.1 to some continuous linear functional on $V$, we obtain a countable set $F \subset V^*$. Let $F = \{x_n^*\}$.

Since $E$ is weakly compact, it follows that $E$ is norm bounded. Indeed, consider $T : V \to \Pi_{x^* \in V^*} \Phi_{x^*}$, where $T(x) = \{x^*(x)\}$ and $\Phi_{x^*} = \Phi$. Then from the discussion of $F$-topologies following Theorem 9.1.1 we know that $T^W$ can be considered as the relative topology on $T(V)$ viewed as a subspace of the topological product space $\Pi_{x^* \in V^*} \Phi_{x^*}$. Furthermore, for each fixed $y^* \in V^*$ it is easily seen that, if $P_{y^*}$ denotes the projection of $\Pi_{x^* \in V^*} \Phi_{x^*}$ onto $\Phi_{y^*}$, then $P_{y^*} \circ T(E) = \{y^*(x) \mid x \in E\}$ is a compact subset of $\Phi_{y^*} = \Phi$, and so there exists some $M_{y^*} > 0$ such that

$$\sup_{x \in E} |T(x)(y^*)| = \sup_{x \in E} |y^*(x)| \leq M_{y^*}.$$

Consequently Corollary 6.2.1 to the Uniform Boundedness Theorem leads us to conclude that there exists some $M > 0$ such that

$$\sup_{x \in E} \|\tau(x)\| = \sup_{x \in E} \|x\| \le M,$$

that is, $E$ is norm bounded.

Hence the sequence $\{x_1^*(x_k)\}$ is a bounded sequence of numbers and thus has a convergent subsequence. Consequently there exists a subsequence $\{x_{k(1)}\}$ of $\{x_k\}$ such that $\{x_1^*(x_{k(1)})\}$ converges. The same argument applied to the sequence $\{x_2^*(x_{k(2)})\}$ assures us of the existence of a subsequence $\{x_{k(2)}\}$ of $\{x_{k(1)}\}$, and hence of $\{x_k\}$, such that $\{x_2^*(x_{k(2)})\}$ converges. Clearly $\{x_1^*(x_{k(2)})\}$ is also convergent. Continuing in this fashion, we obtain the sequences $\{x_{k(j)}\}$, $j = 1,2,3,\ldots,$ each of which is a subsequence of $\{x_k\}$ and is such that $\{x_n^*(x_{k(j)})\}$ converges, $n = 1,2,3,\ldots,j$. Now take $\{y_k\}$ to be the diagonal sequence; that is, set $y_k = x_{k(k)}$, $k = 1,2,3,\ldots$ . Then evidently $\{y_k\}$ is a subsequence of $\{x_k\}$, and $\{x_n^*(y_k)\}$ converges for each $n = 1,2,3,\ldots$ .

Now $\{y_k\} \subset E$, and $E$ is weakly compact. Hence there exists some $y_o \in E$ such that every open neighborhood of $y_o$ in $\mathcal{T}^w$ contains infinitely many terms of the sequence $\{y_k\}$. Moreover, since $\{y_k\} \subset W$ and $W$ is a norm closed linear subspace of $V$, it follows from Theorem 9.2.2 that $W$ is weakly closed, and so $y_o \in W$.

Furthermore, we claim that $\lim_k x_n^*(y_k) = x_n^*(y_o)$, $n = 1,2,3,\ldots$ . Indeed, suppose $n$ is given and let $\lim_k x_n^*(y_k) = a$. Given $\epsilon > 0$, there exists some positive integer $N$ such that $|x_n^*(y_k) - a| < \epsilon/2$, $k \ge N$. Since each weak neighborhood of $y_o$ contains infinitely many terms of $\{y_k\}$, we see that there exists some $k_o \ge N$ for which $y_{k_o} \in U(y_o, \epsilon/2, x_n^*)$. Thus

$$\begin{aligned} |x_n^*(y_o) - a| &\le |x_n^*(y_o) - x_n^*(y_{k_o})| + |x_n^*(y_{k_o}) - a| \\ &< \frac{\epsilon}{2} + \frac{\epsilon}{2} \\ &= \epsilon, \end{aligned}$$

from which we conclude that $x_n^*(y_o) = a$, since $\epsilon$ is arbitrary.

To prove that $\{y_k\}$ actually converges weakly to $y_o$ it remains only to show that $\lim_k x^*(y_k) = x^*(y_o)$ for each $x^* \in V^*$. Suppose this were not the case. Then there would exist some $x_o^* \in V^*$, some $\epsilon > 0$, and a subsequence $\{y_{k_j}\}$ of $\{y_k\}$ such that

$$|x_o^*(y_{k_j} - y_o)| \geq \epsilon \qquad (j = 1,2,3,\ldots).$$

However, since $\{y_{k_j}\} \subset E$ and $E$ is weakly compact, there exists some $z_o \in E$ such that every open neighborhood of $z_o$ in $\mathcal{T}^w$ contains infinitely many terms of $\{y_{k_j}\}$. As before, we deduce that $z_o \in W$ and $\lim_j x_n^*(y_{k_j}) = x_n^*(z_o)$, $n = 1,2,3,\ldots$ . Moreover, since $\{y_{k_j}\}$ is a subsequence of $\{y_k\}$, we must also have $\lim_j x_n^*(y_{k_j}) = x_n^*(y_o)$, $n = 1,2,3,\ldots$, and so $x_n^*(y_o - z_o) = 0$, $n = 1,2,3,\ldots$ . But the sequence $\{x_n^*\}$ consists of extensions to all of $V$ of the family of linear functionals $F_o$ on $W$ that separates the points of $W$, and so we conclude that $y_o = z_o$.

Thus, on the one hand, we see that every open neighborhood of $y_o$ in $\mathcal{T}^w$ contains infinitely many terms of $\{y_{k_j}\}$, while on the other hand, the inequalities $|x_o^*(y_{k_j} - y_o)| \geq \epsilon$, $j = 1,2,3,\ldots$, show that $y_{k_j} \notin U(y_o,\epsilon,x_o^*)$, $j = 1,2,3,\ldots$ . This contradiction shows that $\{y_k\}$ converges to $y_o$ in $(V,\mathcal{T}^w)$.

Therefore $E$ is weakly sequentially compact, and part (i) of the theorem implies part (ii).

Conversely, suppose $E$ is weakly sequentially compact. Now, if $\tau$ denotes the canonical embedding of $V$ into $V^{**}$, then, as noted previously (e.g., in the proof of Theorem 9.9.1), $\tau : (V,\mathcal{T}^w) \to (\tau(V),\mathcal{T}^{w^*})$ is a homeomorhpism. Thus $E$ will be compact in $(V,\mathcal{T}^w)$ if and only if $\tau(E)$ is compact in $(\tau(V),\mathcal{T}^{w^*})$, and the latter will clearly be the case (Theorem 9.4.1) if we can show that $\tau(E)$ is norm bounded and closed in $(V^{**},\mathcal{T}^{w^*})$. Since compactness and sequential compactness are equivalent in $\Phi$, it follows, by essentially the same argument as that given in the first half of the proof, that $\tau(E)$ is norm bounded. It remains then only to show that $\tau(E)$ is weak* closed -- that is, that the closure of $\tau(E)$ in $(V^{**}, \mathcal{T}^{w^*})$ is contained in $\tau(E)$. To accomplish this if suffices to prove that the closure of $\tau(E)$ in $(V^{**}, \mathcal{T}^{w^*})$ lies in $\tau(V)$, since E is weakly closed. Thus, if $x_0^{**}$ belongs to the closure of $\tau(E)$ in $(V^{**}, \mathcal{T}^{w^*})$, we must prove that $x_0^{**} \in \tau(V)$. This will require considerable work.

First, suppose $x_1^*, x_2^*, \ldots, x_n^*$ are in $V^*$. We claim that there exists some $y \in E$ such that $x_0^{**}(x_k^*) = x_k^*(y)$, $k = 1,2,\ldots,n$. Corollary 4.10.1 to Helly's Theorem, of course, shows that such a $y \in V$ always exists. We require, however, that $y \in E$. But since $x_0^{**}$ is in the weak* closure of $\tau(E)$, for each positive integer k there exists some $y_k \in E$ such that $\tau(y_k) \in U(x_0^{**}; 1/k; x_1^*, x_2^*, \ldots, x_n^*)$, and the weak sequential compactness of E then allows us to deduce the existence of a subsequence $\{y_{k_j}\}$ of $\{y_k\}$ and a $y \in E$ such that $\{y_{k_j}\}$ converges weakly to y. The element y belongs to E, as E is weakly closed. In particular then

$$
\begin{aligned}
|x_0^{**}(x_m^*) - x_m^*(y)| &\leq |x_0^{**}(x_m^*) - \tau(y_{k_j})(x_m^*)| \\
&\quad + |x_m^*(y_{k_j}) - x_m^*(y)| \\
&< \frac{1}{k_j} + |x_m^*(y_{k_j}) - x_m^*(y)| \qquad (m = 1,2,\ldots,n),
\end{aligned}
$$

from which we conclude that $x_0^{**}(x_m^*) = x_m^*(y)$, $m = 1,2,\ldots,n$.

With this preliminary observation made, we now note that to prove $x_0^{**} \in \tau(V)$ it suffices, by Theorem 9.1.3, to show that $x_0^{**}$ is a weak* continuous linear functional on $V^*$, and to do this we need only show, in view of Theorem 3.3.2, that

$$N(x_0^{**}) = \{x^* \mid x^* \in V^*, \ x_0^{**}(x^*) = 0\}$$

is closed in $(V^*, \mathcal{T}^{w^*})$. Finally, to establish this we need only prove, by Corollary 10.2.2 to the Krein-Šmulian Theorem, that $N(x_0^{**}) \cap B_1^*$ is closed in $(V^*, \mathcal{T}^{w^*})$, where, as usual

$$B_1^* = \{x^* \mid x^* \in V^*, \|x^*\| \leq 1\}.$$

To this end, let $x_0^*$ belong to the weak* closure of $N(x_0^{**}) \cap B_1^*$ in $V^*$ and let $\epsilon > 0$ be given. By our previous observation, there exists some $y_1 \in E$ such that $x_0^{**}(x_0^*) = x_0^*(y_1)$. The fact that $x_0^*$ is in the weak* closure of $N(x_0^{**}) \cap B_1^*$ implies the existence of some $x_1^* \in N(x_0^{**}) \cap B_1^*$ such that $x_1^* \in U(x_0^*,\epsilon/2,y_1)$. Suppose now we have chosen $y_1, y_2, \ldots, y_n$ in $E$ and $x_1^*, x_2^*, \ldots, x_n^*$ in $N(x_0^{**}) \cap B_1^*$ such that

$$x_0^{**}(x_m^*) = x_m^*(y_k) \qquad (m = 0,1,2,\ldots,k - 1;\ k = 1,2,3,\ldots,n)$$

and $x_m^* \in U(x_0^*;\epsilon/2;y_1,y_2,\ldots,y_m)$, $m = 1,2,3,\ldots,n$. Then there exists some $y_{n+1} \in E$ such that $x_0^{**}(x_m^*) = x_m^*(y_{n+1})$, $m = 1,2,\ldots,n$, by our previous observation. Choose $x_{n+1}^* \in N(x_0^{**}) \cap B_1^*$ so that $x_{n+1}^* \in U(x_0^*;\epsilon/2;y_1,y_2,\ldots,y_{n+1})$. In this way we obtain sequences $\{y_k\} \subset E$ and $\{x_m^*\} \subset V^*$ such that

(a) $x_m^* \in N(x_0^{**}) \cap B_1^*$ $\qquad (m = 1,2,3,\ldots)$.

(b) $x_0^{**}(x_m^*) = x_m^*(y_k)$ $\qquad (m = 0,1,\ldots,k - 1;\ k = 1,2,3,\ldots)$.

(c) $x_m^* \in U(x_0^*;\epsilon/2;y_1,y_2,\ldots,y_m)$ $\qquad (m = 1,2,3,\ldots)$.

Moreover, since $x_m^* \in N(x_0^{**}) \cap B_1^*$, $m = 1,2,3,\ldots,$ we see that

(d) $x_0^{**}(x_m^*) = x_m^*(y_k) = 0$ $\qquad (m = 1,2,\ldots,k - 1;\ k = 1,2,3,\ldots)$.

(e) $\|x_m^*\| \leq 1$ $\qquad (m = 1,2,3,\ldots)$.

Furthermore, statements (b) and (c) show that

(f) $|x_0^{**}(x_0^*) - x_m^*(y_k)| < \frac{\epsilon}{2}$ $\qquad (k = 1,2,\ldots,m;\ m = 1,2,3,\ldots)$.

Now, since $\{y_k\} \subset E$ and $E$ is weakly sequentially compact and weakly closed, there exists a subsequence of $\{y_k\}$ that converges weakly to some $y_o \in E$. In order to avoid some notational awkwardness, we may assume without loss of generality that $\{y_k\}$ itself converges weakly to $y_o$. Then $x_m^*(y_o) = \lim_k x_m^*(y_k) = 0$, $m = 1,2,3,\ldots,$ since by statement (d) $x_m^*(y_k) = 0$, $k \geq m - 1$.

Finally, since $\{y_k\}$ converges weakly to $y_0$, there exist $a_1, a_2, \ldots, a_n$ in $\Phi$, $a_k \geq 0$, $\Sigma_{k=1}^n a_k = 1$ such that, if $z = \Sigma_{k=1}^n a_k y_k$, then $\|z - y_0\| < \epsilon/2$. This follows at once from Theorem 9.2.2 and Corollary 9.2.1. Thus, for $m = n$, we see by statement (f) that

$$\begin{aligned} |x_0^{**}(x_0^*) - x_n^*(z)| &= |x_0^{**}(\sum_{k=1}^{n} a_k x_0^*) - x_n^*(\sum_{k=1}^{n} a_k y_k)| \\ &\leq \sum_{k=1}^{n} a_k |x_0^{**}(x_0^*) - x_n^*(y_k)| \\ &< \frac{\epsilon}{2}. \end{aligned}$$

Consequently

$$\begin{aligned} |x_0^{**}(x_0^*)| &\leq |x_0^{**}(x_0^*) - x_n^*(z)| + |x_n^*(z) - x_n^*(y_0)| + |x_n^*(y_0)| \\ &< \frac{\epsilon}{2} + \|x_n^*\|\|z - y_0\| + |x_n^*(y_0)| \\ &< \epsilon \end{aligned}$$

by statement (e) and the final observation of the preceding paragraph.

Hence, since $\epsilon > 0$ is arbitrary, we see that $x_0^{**}(x_0^*) = 0$; that is, $x_0^* \in N(x_0^{**})$. Moreover, since $x_0^*$ belongs to the weak* closure of $N(x_0^{**}) \cap B_1^*$ and $B_1^*$ is weak* closed, we conclude that $x_0^* \in N(x_0^{**}) \cap B_1^*$. Thus $N(x_0^{**}) \cap B_1^*$, and hence $N(x_0^{**})$ is closed in $(V^*, T^{w^*})$.

Therefore $x_0^{**} \in \tau(V)$ and $E$ is compact in $(V, T^w)$. This completes the proof that part (ii) implies part (i), and of the theorem as a whole. □

It is important to note that the assumption that $E$ is closed in $(V, T^w)$ was only used explicitly to prove the implication from part (ii) to part (i). If $E$ is weakly compact, it is of course automatically weakly closed and hence weakly sequentially compact.

It is apparent, moreover, that we could just as well have phrased the preceding theorem as follows: the weak closure of $E$ is weakly compact if and only if it is weakly sequentially compact. We prefer, however, the previous formulation.

An immediate corollary of the Eberlein-Šmulian Theorem is the following result:

Corollary 10.3.1. Let $(V,\|\cdot\|)$ be a Banach space over $\Phi$ and $B_1 = \{x \mid x \in V, \|x\| \leq 1\}$. Then the following are equivalent:

(i) $(V,\|\cdot\|)$ is reflexive.
(ii) $B_1$ is compact in $(V,\mathcal{T}^W)$.
(iii) $B_1$ is sequentially compact in $(V,\mathcal{T}^W)$.

A portion of the argument used in the proof of the first implication of the Eberlein-Šmulian Theorem can be also used to strengthen a previous result. In Corollary 9.4.6 we showed that, if $V$ is a separable reflexive Banach space, then the weak topology restricted to the closed unit ball is metrizable. The assumption of reflexivity here is actually unnecessary, as shown by the following theorem:

Theorem 10.3.2. Let $(V,\|\cdot\|)$ be a separable Banach space over $\Phi$. If $E \subset V$ is compact in $(V,\mathcal{T}^W)$, then the weak topology restricted to $E$ is metrizable.

Proof. We shall only sketch the proof, leaving the details to the reader. Let $\{x_k\}$ be a norm dense countable subset of $V$ and construct, using $\{x_k\}$, as in the proof of the Eberlein-Šmulian Theorem, a countable family $\mathcal{F} \subset V^*$ that separates points. Let $\mathcal{F} = \{x_k^*\}$. Then define a metric $\rho$ on $V$ by setting

$$\rho(x,y) = \sum_{k=1}^{\infty} \frac{|x_k^*(x - y)|}{2^k[1 + |x_k^*(x - y)|]} \qquad (x,y \in V).$$

Then the topology $\mathcal{T}_\rho$ on $V$ determined by this metric is weaker than $\mathcal{T}^W$. Consequently the identity mapping from $(V,\mathcal{T}^W)$ to $(V,\mathcal{T}_\rho)$ is a continuous bijection, and its restriction to any weakly compact set $E \subset V$ is a homeomorphism. □

We can combine the Eberlein-Šmulian Theorem and Theorem 9.3.5 to give a description of weakly compact sets in $C_o(X)$.

<u>Theorem</u> <u>10.3.3</u>. Let $X$ be a locally compact Hausdorff topological space and suppose $E \subset C_o(X)$. Then the following are equivalent:

(i) $E$ is compact in $(C_o(X),\mathcal{T}^W)$.

(ii) $E$ is closed in $(C_o(X),\mathcal{T}^W)$ and bounded in $(C_o(X),\|\cdot\|_\infty)$; moreover, if $\{f_k\}$ is a sequence in $E$, then there exists a subsequence $\{f_{k_j}\}$ of $\{f_k\}$ and some $f \in E$ such that

$$\lim_j f_{k_j}(t) = f(t) \qquad (t \in X).$$

<u>Proof</u>. If $E$ is weakly compact, then it is weakly closed and, as seen in the proof of the Eberlein-Šmulian Theorem, norm bounded. If $\{f_k\}$ is a sequence in $E$, then, since $E$ is weakly sequentially compact, there exists a subsequence $\{f_{k_j}\}$ of $\{f_k\}$ and some $f \in E$ such that $\{f_{k_j}\}$ converges weakly to $f$. By Theorem 9.3.5 we conclude that $\lim_j f_{k_j}(t) = f(t)$, $t \in X$, and so part (i) implies part (ii).

Conversely, suppose part (ii) holds. Then, given any sequence $\{f_k\} \subset E$, there exists some subsequence $\{f_{k_j}\}$ of $\{f_k\}$ and some $f \in E$ such that $\lim_j f_{k_j}(t) = f(t)$, $t \in X$. Since the sequence $\{f_{k_j}\}$ is norm bounded, we see from Theorem 9.3.5 that $\{f_{k_j}\}$ converges weakly to $f$. Hence $E$ is weakly closed and weakly sequentially compact, and so, by the Eberlein-Šmulian Theorem, $E$ is weakly compact.

Therefore part (ii) implies part (i). □

This completes our discussion of weak topologies <u>per</u> <u>se</u>. The reader should, however, be aware that we have in no sense examined the subject in its full generality. For further reading we refer the reader to [$DS_1$, pp. 418-435; $E_1$, pp. 500-615; K, pp. 233-274 and 310-330; KeNa, pp. 137-223; $W_1$, pp. 233-249; Y, pp. 141-145].

## 10.4. Problems.

1. (Proposition 10.1.1) Let $(V,\|\cdot\|)$ be a normed linear space over $\Phi$. Prove each of the following:

(a) $U \in \mathcal{T}^{bw*}$ if and only if for each $a \geq 0$ and $x^* \in U \cap aB_1^*$ there exist some $\epsilon > 0$ and $x_1,x_2,\ldots,x_n$ in $V$ such that

$$U(x^*;\epsilon;x_1,x_2,\ldots,x_n) \cap aB_1^*$$
$$= \{y^* \mid y^* \in V^*,\ \|y^*\| \leq a,\ |x^*(x_k) - y^*(x_k)| < \epsilon,\ k = 1,2,\ldots,n\}$$

is contained in $U$, where $B_1^* = \{x^* \mid x^* \in V^*,\ \|x^*\| \leq 1\}$.

(b) $\mathcal{T}^{bw*}$ is a topology on $V^*$ such that $(V^*,\mathcal{T}^{bw*})$ is a locally convex topological linear space over $\Phi$.

(c) $\mathcal{T}^{w*} \subset \mathcal{T}^{bw*}$.

2. (Corollary 10.1.1) Let $(V,\|\cdot\|)$ be a normed linear space over $\Phi$. Prove that for each $\epsilon$, $0 < \epsilon \leq 1$, the sets

$$U(0,\epsilon,\{x_k\}) = \{x^* \mid x^* \in V^*,\ |x^*(x_k)| < \epsilon,\ k = 1,2,3,\ldots\},$$

where $\{x_k\}$ is any sequence in $V$ such that $\lim_k \|x_k\| = 0$, form a neighborhood base at the origin for $\mathcal{T}^{bw*}$.

3. Let $(V,\|\cdot\|)$ be a Banach space over $\Phi$.

(a) If $x^{**} \in V^{**} \sim \tau(V)$ and $c \geq 0$, define L by

$$L = \{x^* \mid x^* \in V^*,\ x^{**}(x^*) = c\}.$$

Prove that the bounded weak* closure of $L$ is $V^*$.

(b) Prove that $x^{**} \in \tau(V)$ if and only if $N(x^{**})$ is closed in $(V^*,\mathcal{T}^{bw*})$.

*4. Give an example of a normed linear space $V$ and a set $E \subset V^*$ such that $E$ is closed in $(V^*, \mathcal{T}^{bw^*})$, but not in $(V^*, \mathcal{T}^{w^*})$.

5. Let $(V, \|\cdot\|)$ be a normed linear space over $\Phi$. As in the proof of Theorem 10.1.1, if $F \subset V$, let

$$F^o = \{x^* \mid x^* \in V^*, |x^*(x)| \leq 1, x \in F\}.$$

$F^o$ is called the <u>polar</u> in $V^*$ of $F$. Prove each of the following:

(a) $F^o$ is convex, balanced, and weak* closed.
(b) If $F_1 \subset F_2 \subset V$, then $F_1^o \supset F_2^o$.
(c) If $a \in \Phi$, $a \neq 0$, then $(aF)^o = (1/a)F^o$.
(d) If $F_n \subset V$, $n = 1,2,3,\ldots$, then $(\bigcup_{n=1}^{\infty} F_n)^o = \bigcap_{n=1}^{\infty} F_n^o$.

6. Let $(V, \|\cdot\|)$ be a Banach space over $\Phi$ and let $U \subset V$ be such that $0 \in \text{int}(U)$. Prove that $U^o$, the polar of $U$ in $V^*$, is compact in $(V^*, \mathcal{T}^{w^*})$.

7. (Corollary 10.2.1) Let $(V, \|\cdot\|)$ be a reflexive Banach space over $\Phi$ and suppose $K \subset V$ is convex. Prove that the following are equivalent:

(i) $K$ is closed in $(V, \mathcal{T}^w)$.
(ii) For each $a \geq 0$, the set $K \cap aB_1$ is closed in $(V, \mathcal{T}^w)$, where $B_1 = \{x \mid x \in V, \|x\| \leq 1\}$.

8. (Corollary 10.2.2) Let $(V, \|\cdot\|)$ be a Banach space over $\Phi$ and suppose $W \subset V^*$ is a linear subspace. Prove that the following are equivalent:

(i) $W$ is closed in $(V^*, \mathcal{T}^{w^*})$.
(ii) $W \cap B_1^*$ is closed in $(V^*, \mathcal{T}^{w^*})$, where

$$B_1^* = \{x^* \mid x^* \in V^*, \|x^*\| \leq 1\}.$$

9. (Lemma 10.2.2) Let $(V,\|\cdot\|)$ be a Banach space over $\Phi$ and suppose $K_1, K_2 \subset V^*$ are convex sets that are compact in $(V^*, \mathcal{T}^{w^*})$. Prove that $\overline{co}(K_1 \cup K_2) = co(K_1 \cup K_2)$.

10. Let $(V,\|\cdot\|)$ be a reflexive Banach space over $\Phi$ and let $\{x_n\}$ be a bounded sequence in $V$. For each $n = 1,2,3,\ldots$, let $K_n = \overline{co}\{x_n, x_{n+1}, \ldots\}$. Prove that the sequence $\{x_n\}$ converges weakly to $x_o$ if and only if $\{x_o\} = \cap_{n=1}^{\infty} K_n$.

11. (Corollary 10.3.1) Let $(V,\|\cdot\|)$ be a Banach space over $\Phi$ and let $B_1 = \{x \mid x \in V, \|x\| \le 1\}$. Prove that the following are equivalent:

(i) $(V,\|\cdot\|)$ is reflexive.
(ii) $B_1$ is compact in $(V, \mathcal{T}^w)$.
(iii) $B_1$ is sequentially compact in $(V, \mathcal{T}^w)$.

*12. Let $(V,\|\cdot\|)$ be a Banach space over $\Phi$ and let $K \subset V$ be convex. Prove that $K$ is weakly compact if and only if every decreasing sequence of nonempty closed convex subsets of $K$ has a nonempty intersection. (Hint: prove first that $K$ is bounded if the condition holds and then prove $K$ is weakly sequentially compact and use the Eberlein-Šmulian Theorem.)

*13. Let $(V,\|\cdot\|)$ be a Banach space over $\Phi$. Recall that a family $\mathcal{F} \subset V^*$ is said to be <u>equicontinuous</u> if for every $\epsilon > 0$ there exists a neighborhood $U$ of the origin in $V$ such that $|x^*(x)| < \epsilon$ for all $x \in U$ and $x^* \in \mathcal{F}$. Let $E \subset V$ be any subset of $V$. Prove that the following are equivalent:

(i) $E$ is weakly compact.
(ii) $E$ is weakly closed and norm bounded, and for every sequence $\{x_m\} \subset E$ and every equicontinuous sequence $\{x_n^*\} \subset V^*$ such that the double limits $\lim_m \lim_n x_n^*(x_m)$ and $\lim_n \lim_m x_n^*(x_m)$ exist, it is the case that

$$\lim_m \lim_n x_n^*(x_m) = \lim_n \lim_m x_n^*(x_m).$$

14. Let $(V,\|\cdot\|)$ be a Banach space over $\Phi$ and let $E \subset V$. Suppose $V^*$ is the union of countably many weakly compact subsets. Prove that every weak closure point $x_o$ of $E$ is the weak closure point of some countable subset of $E$.

15. Let $(V,\|\cdot\|)$ be a Banach space over $\Phi$ and let $K \subset V$ be convex. Prove that $K$ is weakly closed if and only if its intersection with every norm bounded and weakly closed set is weakly closed.

*16. Let $(V,\|\cdot\|)$ be a Banach space over $\Phi$ and let $E \subset V^*$ be any subset of $V^*$. Prove that, if $E$ is weak* sequentially compact, then the weak* closure of $E$ is weak* compact. Give an example to prove that the converse is not necessarily true.

*17. Let $(V,\|\cdot\|)$ be a Banach space over $\Phi$ and let $E \subset V$ be any subset of $V$. Prove that, if every separable linear subspace of $V$ meets $E$ in a weakly compact set, then the weak closure of $E$ is weakly compact.

# CHAPTER 11

# THE KREIN-MIL'MAN THEOREM

11.0. Introduction. Suppose that $x_1, x_2$ and $x_3$ are three noncollinear points in the Euclidean plane $\mathbb{R}^2$. Then, as is easily seen, every point that lies interior to or on the triangle determined by these three points can be expressed as a convex linear combination of $x_1, x_2$, and $x_3$, that is, a convex linear combination of the vertices of the triangle. The main result of this chapter, the Krein-Mil'man Theorem, will be an appropriate analog in locally convex topological linear spaces of this elementary Euclidean fact. In particular, given a locally convex topological linear space $(V,\mathcal{T})$ and a nonempty compact convex subset $K$ of $V$, we shall see that $K$ coincides with the closure in $(V,\mathcal{T})$ of the set of all finite convex linear combinations of a certain distinguished subset of $K$, called the extreme points of $K$. The fact that points in $K$ may only be approximated by, rather than expressed as, convex linear combinations of the extreme points is necessitated by the possible infinite dimensionality of locally convex topological linear spaces.

The Krein-Mil'man Theorem has many applications, but we have space to discuss only three of them. Perhaps the most interesting of these is a proof of the Stone-Weierstrass Theorem. The applications will be presented in the final sections of this chapter, preceded by some preliminary definitions and results in Section 11.1 and the Krein-Mil'man Theorem itself in Section 11.2. In addition in Section 11.5 we shall discuss another topic from harmonic analysis that utilizes the Stone-Weierstrass Theorem and the theorem concerning adjoint transformations established in Section 9.10.

11.1. Extreme Points and Extremal Sets. Our main concern in this section is to make precise the notion of an extreme point of a convex set and the related notion of an extremal set.

Definition 11.1.1. Let $V$ be a linear space over $\Phi$ and suppose $K \subset V$ is a convex set. A point $x \in K$ is said to be an extreme point of $K$ if, whenever $x = ax_1 + (1 - a)x_2$, where $x_1, x_2 \in K$ and $0 < a < 1$, it happens that $x = x_1 = x_2$.

In other words, an extreme point of a convex set $K$ is any point in $K$ that is not an interior point of any line segment in $K$.

Given a convex set $K$, we shall denote its set of extreme points by $\text{ext}(K)$.

It is easily seen that $\text{ext}(K)$ may be empty. For example, if $V = \mathbb{R}^2$ and $K = \{(a_1,a_2) \mid a_1^2 + a_2^2 < 1\}$, then $\text{ext}(K) = \emptyset$. Less trivially, and with more relevance to our concerns, this may also happen even if $K$ is a nonempty closed convex subset of a topological linear space. This assertion follows immediately from the next result.

Proposition 11.1.1. For each $p$, $1 \leq p < \infty$, let $B_1^p \subset L_p(\mathbb{R},dt)$ be defined as $B_1^p = \{f \mid f \in L_p(\mathbb{R},dt), \|f\|_p \leq 1\}$, where as usual $dt$ denotes Lebesgue measure on $\mathbb{R}$. Then $\text{ext}(B_1^1) = \emptyset$ and $\text{ext}(B_1^p) = \{f \mid f \in L_p(\mathbb{R},dt), \|f\|_p = 1\}$, $1 < p < \infty$.

Proof. Clearly the sets $B_1^p$, $1 \leq p < \infty$, are norm closed convex sets. If $f \in B_1^p$ is such that $0 < \|f\|_p < 1$, then the equation

$$f = \tfrac{1}{2}[(2 - \|f\|_p)f] + \tfrac{1}{2}[\|f\|_p f]$$

shows that $f$ is not an extreme point of $B_1^p$. Even more trivially, it is apparent that $f = 0$ is not an extreme point. Thus

$$\text{ext}(B_1^p) \subset \{f \mid f \in L_p(\mathbb{R},dt), \|f\|_p = 1\}.$$

Now suppose $p = 1$ and $\|f\|_1 = 1$. Let $E \subset \mathbb{R}$ be the set of points where any representative of the equivalence class of $f$ does not vanish. Since $\|f\|_1 = 1$, we see that $E$ is measurable, and the Lebesgue measure of $E$ is positive. Let $E_1$ and $E_2$ be any disjoint subsets of $E$ with positive Lebesgue measure such that $E = E_1 \cup E_2$. Define $f_1 = f\chi_{E_1}/\|f\chi_{E_1}\|_1$ and $f_2 = f\chi_{E_2}/\|f\chi_{E_2}\|_1$, where, as usual, $\chi_{E_k}$ denotes the characteristic function of $E_k$, $k = 1,2$. It is obvious that $f \neq f_k$, $\|f_k\|_1 = 1$, $k = 1,2$, and

$$\begin{aligned}\|f\chi_{E_1}\|_1 f_1 + \|f\chi_{E_2}\|_1 f_2 &= f\chi_{E_1} + f\chi_{E_2}\\ &= f\chi_E\\ &= f,\end{aligned}$$

since $E = E_1 \cup E_2$. Moreover,

$$\begin{aligned}\|f\chi_{E_1}\|_1 + \|f\chi_{E_2}\|_1 &= \int_{E_1}|f(t)|\,dt + \int_{E_2}|f(t)|\,dt\\ &= \int_E|f(t)|\,dt\\ &= 1,\end{aligned}$$

as $E_1 \cap E_2 = \emptyset$.

Hence $f$ is not an extreme point, and so $\operatorname{ext}(B_1^1) = \emptyset$.

On the other hand, suppose $1 < p < \infty$, $\|f\|_p = 1$, and there exists $f_1, f_2 \in B_1^p$, $0 < a < 1$, such that $f = af_1 + (1 - a)f_2$. Since $\|f\|_p = 1$, it follows at once that $\|f_k\|_p = 1$, $k = 1,2$, and hence

$$\begin{aligned}1 = \|f\|_p &= \|af_1 + (1 - a)f_2\|_p\\ &\le \|af_1\|_p + \|(1 - a)f_2\|_p\\ &= a\|f_1\|_p + (1 - a)\|f_2\|_p\\ &= 1,\end{aligned}$$

that is, $\|af_1 + (1 - a)f_2\|_p = \|af_1\|_p + \|(1 - a)f_2\|_p$. However, recall that, if $g \in L_p(\mathbb{R},dt)$, $1 < p < \infty$, and $h \in L_q(\mathbb{R},dt)$, $1/p + 1/q = 1$, then by Hölder's Inequality [Ry, p. 113] we have

$$\int_{\mathbb{R}} |g(t)h(t)|\ dt = \|g\|_p\|h\|_q$$

if and only if $(\|h\|_q)^q|g(t)|^p = (\|g\|_p)^p|h(t)|^q$ for almost all $t$. Thus, as is easily seen from the proof of the triangle inequality in $L_p(\mathbb{R},dt)$, we conclude that

$$\|af_1 + (1 - a)f_2\|_p = \|af_1\|_p + \|(1 - a)f_2\|_p$$

if and only if

$$(\|af_1 + (1 - a)f_2\|_p)^{p/q}|af_1(t)|^p = (\|af_1\|_p)^p|af_1(t) + (1 - a)f_2(t)|^p$$

for almost all $t$. Since $\|f\|_p = \|f_1\|_p = 1$, we have

$$|af_1(t)|^p = a^p|f(t)|^p$$

for almost all $t$, and so $|f_1(t)| = |f(t)|$ for almost all $t$. Similarly $|f_2(t)| = |f(t)|$ for almost all $t$.

But $f = af_1 + (1 - a)f_2$ implies that for almost all $t$ we have

$$\begin{aligned} |f(t)| &= |af_1(t) + (1 - a)f_2(t)| \\ &\leq |af_1(t)| + |(1 - a)f_2(t)| \\ &= a|f_1(t)| + (1 - a)|f_2(t)| \\ &= |f(t)|, \end{aligned}$$

whence $|af_1(t) + (1 - a)f_2(t)| = |af_1(t)| + |(1 - a)f_2(t)|$ for almost all $t$. Consequently from the properties of the triangle inequality for complex numbers we conclude that $f_1(t)\overline{f_2(t)} \geq 0$ for almost all $t$. But then we must have, for almost all $t$,

$$\begin{aligned}|f_1(t) - f_2(t)|^2 &= [f_1(t) - f_2(t)][\overline{f_1(t)} - \overline{f_2(t)}] \\ &= |f_1(t)|^2 - f_2(t)\overline{f_1(t)} - f_1(t)\overline{f_2(t)} + |f_2(t)|^2 \\ &= 2|f(t)|^2 - 2|f_1(t)\overline{f_2(t)}| \\ &= 2|f(t)|^2 - 2|f(t)|^2 \\ &= 0,\end{aligned}$$

since $\overline{f_1(t)}f_2(t) = f_1(t)\overline{f_2(t)} \geq 0$. Thus $f_1(t) = f_2(t)$ for almost all $t$, and so $f = f_1 = f_2$.

Therefore $f \in \operatorname{ext}(B_1^p)$, and hence

$$\operatorname{ext}(B_1^p) = \{f \mid f \in L_p(\mathbb{R},dt),\ \|f\|_p = 1\}. \qquad \square$$

The reason that the closed unit balls in $L_p(\mathbb{R},dt)$, $1 < p < \infty$, have extreme points but the closed unit ball in $L_1(\mathbb{R},dt)$ has none is to be found in the fact that $B_1^p$, $1 < p < \infty$, is weakly compact, whereas $B_1^1$ is not -- equivalently, because $L_p(\mathbb{R},dt)$, $1 < p < \infty$, is reflexive, whereas $L_1(\mathbb{R},dt)$ is not. This will be apparent when we come to the Krein-Mil'man Theorem (Theorem 11.2.1). Indeed, a portion of this theorem says precisely that a nonempty compact convex set in a locally convex topological linear space always has extreme points.

In proving the Krein-Mil'man Theorem we shall need the notion of an extremal set, which we now define.

Definition 11.1.2. Let $(V,\mathcal{T})$ be a topological linear space over $\Phi$ and suppose $K \subset V$ is a closed convex set. A set $E \subset K$ is said to be an extremal subset of $K$ if $E$ is a nonempty closed convex set and whenever $x,y \in K$ and $0 < a < 1$ are such that $ax + (1 - a)y \in E$, it is the case that $x,y \in E$.

Note that the last portion of the definition of an extremal set

$E$ says that, whenever $E$ contains an interior point of a line segment in $K$, the whole line segment lies in $E$.

As some simple examples of extremal sets, suppose $(V,\|\cdot\|) = (\mathbb{R}^2,\|\cdot\|_2)$ and let $K$ be the square

$$\{(a_1,a_2) \mid 0 \le a_1 \le 1,\ 0 \le a_2 \le 1\}.$$

Then $K$ is clearly a closed convex set in $\mathbb{R}^2$, and the corners, edges, and the square itself are all extremal subsets of $K$.

More generally we have the following proposition:

Proposition 11.1.2. Let $(V,\mathcal{T})$ be a topological linear space over $\Phi$ and suppose $K \subset V$ is a closed convex set.

(i) If $x \in K$, then $\{x\}$ is an extremal subset of $K$ if and only if $x \in \text{ext}(K)$.

(ii) If $E$ is an extremal subset of $K$ and $F$ is an extremal subset of $E$, then $F$ is an extremal subset of $K$.

(iii) If $x^*$ is a continuous real linear functional on $(V,\mathcal{T})$ and $b = \sup_{x \in K} x^*(x)$ is finite, then $E_{x^*} = \{x \mid x \in K, x^*(x) = b\}$ is either empty or an extremal subset of $K$.

Proof. The proofs of parts (i) and (ii) are routine and are left to the reader. Suppose $x^*$ is a real continuous linear functional on $(V,\mathcal{T})$ and $E_{x^*}$ is as here defined. Clearly $E_{x^*}$ is closed and convex. If $E_{x^*} \neq \emptyset$ and $x,y \in K$ and $0 < a < 1$ are such that $ax + (1 - a)y \in E_{x^*}$, then

$$\begin{aligned} b &= x^*[ax + (1 - a)y] \\ &= ax^*(x) + (1 - a)x^*(y) \\ &\le ab + (1 - a)b \\ &= b, \end{aligned}$$

from which we deduce that $b = x^*(x) = x^*(y)$; that is, $x,y \in E_{x^*}$.

Therefore either $E_{x^*} = \emptyset$ or $E_{x^*}$ is extremal. □

In particular, we note that, if $K$ is a nonempty closed convex set, then $K$ is an extremal subset of itself.

11.2. The Krein-Mil'man Theorem. We are now in a position to state and prove the Krein-Mil'man Theorem. The proof is an application of Zorn's Lemma [$DS_1$, p. 6] and the Hahn-Banach Theorem. It is desirable at this point to recall the definition of a convex hull made in Section 10.2.

Theorem 11.2.1 (Krein-Mil'man Theorem). Let $(V,T)$ be a locally convex topological linear space over $\Phi$. If $K \subset V$ is a nonempty compact convex set, then $\text{ext}(K) \neq \emptyset$ and $K = \overline{\text{co}}[\text{ext}(K)]$.

Proof. Let $E$ denote the collection of all extremal subsets of $K$. Clearly $E \neq \emptyset$, as $K \in E$. We introduce a partial ordering into $E$ by defining $E_1 > E_2$ whenever $E_1 \subset E_2$, $E_1, E_2 \in E$. If $\{E_\alpha\}_{\alpha \in \Lambda}$ is a linearly ordered subset of $E$, then we claim that $E_o = \cap_{\alpha \in \Lambda} E_\alpha$ is an upper bound for $\{E_\alpha\}_{\alpha \in \Lambda}$.

Clearly $E_o$ is a closed convex subset of $K$. Since $\{E_\alpha\}_{\alpha \in \Lambda}$ evidently has the finite intersection property and $K$ is compact, it follows that $E_o \neq \emptyset$. Moreover, if $x,y \in K$ and $0 < a < 1$ are such that $ax + (1 - a)y \in E_o$, then $ax + (1 - a)y \in E_\alpha$, $\alpha \in \Lambda$. Since each $E_\alpha$ is an extremal subset of $K$, we conclude that $x,y \in E_\alpha$, $\alpha \in \Lambda$, and so $x,y \in E_o$. Thus $E_o$ is an extremal subset of $K$, that is, $E_o \in E$. Obviously $E_o$ is then an upper bound for $\{E_\alpha\}_{\alpha \in \Lambda}$, as $E_o \subset E_\alpha$, $\alpha \in \Lambda$.

Hence, by Zorn's Lemma, we deduce the existence of a maximal element $E \in E$. We claim that $E$ consists of a single point, as can be seen by the following argument: Suppose $x,y \in E$, $x \neq y$. Then from Theorem 5.3.1 and Proposition 3.3.1 we see that there exists a continuous real linear functional $x^* \in V^*$ such that $x^*(x) < x^*(y)$. Let $b = \sup_{z \in E} x^*(z)$ and set

$$E_{x^*} = \{z \mid z \in E,\ x^*(z) = b\}.$$

We know that $E_{x^*} \neq \emptyset$ since $x^*$ is continuous on the compact set $E$, which is compact because it is a closed subset of the compact set $K$. Thus from Propositions 11.1.2(ii) and (iii) we know that $E_{x^*}$ is an extremal subset of $E$ and hence of $K$. Consequently $E_{x^*} \in \mathcal{E}$ and $E_{x^*} > E$. Moreover, $E_{x^*} \neq E$ since $x \notin E_{x^*}$, as $x^*(x) < x^*(y) \leq b$. But this contradicts the maximality of $E$. Hence $E = \{x\}$, and so $x \in \text{ext}(K)$ by Proposition 11.1.2(i).

Therefore $\text{ext}(K) \neq \emptyset$.

Next we wish to show that $\overline{\text{co}}[\text{ext}(K)] = K$. Clearly $\overline{\text{co}}[\text{ext}(K)] \subset K$. Suppose there exists some $x \in K \sim \overline{\text{co}}[\text{ext}(K)]$. By Corollary 5.4.1 there exists a closed real hyperplane that strictly separates $x$ and $\overline{\text{co}}[\text{ext}(K)]$, since the latter is a nonempty closed convex set. Thus there exists some continuous real linear functional $x^*$ on $V$ such that $\sup_{y \in \overline{\text{co}}[\text{ext}(K)]} x^*(y) < x^*(x)$. Let $b = \sup_{y \in K} x^*(y)$ and $E_{x^*} = \{y \mid y \in K,\ x^*(y) = b\}$. Then we conclude as before that, since $K$ is compact, $E_{x^*} \neq \emptyset$, and so it is an extremal subset of $K$ by Proposition 11.1.2(iii). Moreover, since $E_{x^*}$ is a closed subset of the compact set $K$, we see that $E_{x^*}$ is a nonempty compact convex set. Hence by the first portion of the proof we have $\text{ext}(E_{x^*}) \neq \emptyset$. By Propositions 11.1.2(i) and (ii), if $z \in \text{ext}(E_{x^*})$, then $\{z\}$ is an extremal subset of $E_{x^*}$ and hence of $K$. Thus $z \in \text{ext}(K) \subset \overline{\text{co}}[\text{ext}(K)]$. However, since $z \in E_{x^*}$, we also have

$$\begin{aligned} x^*(z) = b &\geq x^*(x) \\ &> \sup_{y \in \overline{\text{co}}[\text{ext}(K)]} x^*(y) \\ &\geq x^*(z), \end{aligned}$$

which is an absurdity.

Therefore $K = \overline{\text{co}}[\text{ext}(K)]$, and the proof is complete. □

Some easy corollaries of the Krein-Mil'man Theorem are the following results:

<u>Corollary 11.2.1</u>. Let $(V,\|\cdot\|)$ be a normed linear space over $\Phi$ and $B_1^* = \{x^* \mid x^* \in V^*, \|x^*\| \leq 1\}$. Then $\text{ext}(B_1^*) \neq \emptyset$ and $B_1^*$ is the closure in $(V^*,\mathcal{T}^{w^*})$ of $\text{co}[\text{ext}(B_1^*)]$.

<u>Proof</u>. Corollary 9.4.1 to the Banach-Alaoglu Theorem shows that $B_1^*$ is a nonempty compact convex set in $(V^*,\mathcal{T}^{w^*})$. Now apply the Krein-Mil'man Theorem. □

<u>Corollary 11.2.2</u>. Let $(V,\|\cdot\|)$ be a reflexive Banach space over $\Phi$ and $B_1 = \{x \mid x \in V, \|x\| \leq 1\}$. Then $\text{ext}(B_1) \neq \emptyset$ and $B_1$ is the closure in $(V,\mathcal{T}^{w})$ of $\text{co}[\text{ext}(B_1)]$.

<u>Proof</u>. Corollary 9.4.2 to the Banach-Alaoglu Theorem or Theorem 9.9.1 and the Krein-Mil'man Theorem. □

<u>Corollary 11.2.3</u>. Let $(V,\mathcal{T})$ be a locally convex topological linear space over $\Phi$ and suppose $K \subset V$ is a nonempty compact convex set. If $x^* \in V^*$, then there exists some $x \in \text{ext}(K)$ such that $\text{Re}(x^*)(x) = \sup_{y \in K}\text{Re}(x^*)(y)$.

<u>Proof</u>. Since $K$ is compact and $\text{Re}(x^*)$ is continuous by Theorem 3.3.1, we see, as in the proof of the Krein-Mil'man Theorem, that $E_{\text{Re}(x^*)} = \{x \mid x \in K, \text{Re}(x^*)(x) = \sup_{y \in K}\text{Re}(x^*)(y)\}$, is a nonempty compact convex set. Thus $\text{ext}(E_{\text{Re}(x^*)}) \neq \emptyset$. Consequently, by Propositions 11.1.2(i) and (ii), if $x \in \text{ext}(E_{\text{Re}(x^*)})$, then $x \in \text{ext}(K)$ and $\text{Re}(x^*)(x) = \sup_{y \in K}\text{Re}(x^*)(y)$. □

One further consequence of the Krein-Mil'man Theorem will be needed in the last section of this chapter. To prove it we need the full generality of Lemma 10.2.2. We again leave the details to the reader.

Lemma 11.2.1. Let $(V,\mathcal{T})$ be a topological linear space over $\Phi$ and let $E_1, E_2, \ldots, E_n$ be any subsets of $V$. If $\overline{co}(E_k)$, $k = 1,2,\ldots,n$, is compact, then

$$\overline{co}\left(\bigcup_{k=1}^{n} E_k\right) = co\left(\bigcup_{k=1}^{n} \overline{co}(E_k)\right).$$

Theorem 11.2.2. Let $(V,\mathcal{T})$ be a locally convex topological linear space over $\Phi$ and suppose $E \subset V$ is closed. If $\overline{co}(E)$ is compact, then $ext[\overline{co}(E)] \subset E$.

Proof. By the Krein-Mil'man Theorem, $ext[\overline{co}(E)] \neq \emptyset$. To show that $ext[\overline{co}(E)] \subset E$ we shall prove that, if $x \in ext[\overline{co}(E)]$, then $x \in cl(E) = E$, and to do this it suffices to show that, if $U$ is any convex balanced closed neighborhood in $V$ that contains the origin in its interior, then $x \in E + U$, since in this case $(x + U) \cap E \neq \emptyset$.

Let $U$ be any such neighborhood. Then $E \subset \bigcup_{y \in E}(y + U)$. But $E$ is compact, being a closed subset of the compact set $\overline{co}(E)$, and so there exists $x_1, x_2, \ldots, x_n$ in $E$ such that $E \subset \bigcup_{k=1}^{n}(x_k + U)$. Set $E_k = E \cap (x_k + U)$, $k = 1,2,\ldots,n$. Then each $E_k$ is compact, as $x_k + U$ is closed, and $E = \bigcup_{k=1}^{n} E_k$. Let $A_k = \overline{co}(E_k)$. Then, from Lemma 11.2.1 we see that

$$\begin{aligned} co\left(\bigcup_{k=1}^{n} A_k\right) &= co\left(\bigcup_{k=1}^{n} \overline{co}(E_k)\right) \\ &= \overline{co}\left(\bigcup_{k=1}^{n} E_k\right) \\ &= \overline{co}(E), \end{aligned}$$

as $\overline{co}(E_k)$ is compact.

Thus $x \in ext[\overline{co}(E)]$ must be of the form $x = \Sigma_{j=1}^{m} a_j y_j$, where $y_j \in \bigcup_{k=1}^{n} A_k$, $a_j \geq 0$, $j = 1,2,\ldots,m$, and $\Sigma_{j=1}^{m} a_j = 1$. However, since $x$ is an extreme point, it follows at once that $x = y_j$ for some $j$, $1 \leq j \leq m$; that is, $x \in A_j = \overline{co}(E_j)$ for some $j$, $1 \leq j \leq m$. But $\overline{co}(E_j) = \overline{co}[E \cap (x_j + U)] \subset x_j + U \subset E + U$, since $U$, and hence $x_j + U$, is convex.

Since this is what we set out to prove, we conclude that $\text{ext}[\overline{\text{co}}(E)] \subset \text{cl}(E) = E$. □

11.3. $L_1(\mathbb{R}, dt)$ Is Not a Dual Space. An easy application of the Krein-Mil'man Theorem (Theorem 11.2.1) allows us to prove that $L_1(\mathbb{R}, dt)$ is not isometrically isomorphic to the dual space of any normed linear space.

Theorem 11.3.1. There exists no normed linear space $(V, \|\cdot\|)$ over $\mathbb{C}$ such that $(L_1(\mathbb{R}, dt), \|\cdot\|_1)$ is isometrically isomorphic to $(V^*, \|\cdot\|)$.

Proof. Suppose there were such a space $(V, \|\cdot\|)$. Then, by the Banach-Alaoglu Theorem (Theorem 9.4.1), we would have that

$$B_1 = \{f \mid f \in L_1(\mathbb{R}, dt),\ \|f\|_1 \leq 1\}$$

is a nonempty weak* compact convex set. Thus, by the Krein-Mil'man Theorem (Theorem 11.2.1), $\text{ext}(B_1) \neq \emptyset$, contradicting Proposition 11.1.1. Therefore no such space $V$ exists. □

Actually this result, as well as Proposition 11.1.1, remains valid for $L_1(X, \mathcal{S}, \mu)$ whenever $(X, \mathcal{S}, \mu)$ is a positive measure space without atoms. The details are left to the reader.

Since $L_1(\mathbb{R}, dt)$ is not a dual space, it, in particular, cannot be reflexive. We give an independent proof of this corollary, utilizing the techniques used in proving Theorem 11.3.1.

Corollary 11.3.1. $(L_1(\mathbb{R}, dt), \|\cdot\|_1)$ is not a reflexive Banach space.

Proof. If $L_1(\mathbb{R}, dt)$ were reflexive, then, by Theorem 9.9.1, $B_1 = \{f \mid f \in L_1(\mathbb{R}, dt),\ \|f\|_1 \leq 1\}$ would be weakly compact. Thus, by the Krein-Mil'man Theorem, $\operatorname{ext}(B_1) \neq \emptyset$, contradicting Proposition 11.1.1. □

Similar arguments involving the Krein-Mil'man Theorem can be used, for instance, to show that $c_0$ and $C([0,1])$ are not dual spaces. The details are left to the reader.

11.4. The Stone-Weierstrass Theorem. The Stone-Weierstrass Theorem, which gives sufficient conditions for a subalgebra of $C_0(X)$, where $X$ is a locally compact Hausdorff topological space, to be uniformly dense in $C_0(X)$, is certainly one of the most fundamental theorems in analysis. In this section we give a proof of this result based on an application of the Krein-Mil'man Theorem. The approach is due originally to L. deBranges [de]. In order to make explicit the role of the Krein-Mil'man Theorem in the following development and to shorten the proof of the Stone-Weierstrass Theorem itself, we begin by recalling some useful facts and establishing some preliminary lemmas.

As indicated in Example 1.2.1, $(C_0(X), \|\cdot\|_\infty)$ is a Banach space over $\mathbb{C}$, and by the Riesz Representation Theorem (Theorem 4.8.2) its dual space $(C_0(X)^*, \|\cdot\|)$ is isometrically isomorphic with $(M(X), \|\cdot\|)$, the Banach space of bounded, regular, complex-valued Borel measures on $X$ with the total variation norm $\|\mu\| = |\mu|(X)$, $\mu \in M(X)$. The isomorphism is given by the formula

$$x^*(f) = \int_X f(t)\, d\mu(t),$$

where $f \in C_0(X)$, $x^* \in C_0(X)^*$, and $\mu \in M(X)$.

In view of this identification of $C_0(X)^*$ with $M(X)$, it is evident that, if $E \subset C_0(X)$, then

$$E^\perp = \{\mu \mid \mu \in M(X), \int_X f(t)\, d\mu(t) = 0,\ f \in E\}$$

is a norm closed linear subspace of $M(X)$. Moreover, the definition of the weak* topology shows at once that $E^{\perp}$ is even closed in $(M(X),\mathcal{T}^{w^*})$. Further, since $B_1^* = \{\mu \mid \mu \in M(X), \|\mu\| \leq 1\}$ is compact in $(M(X),\mathcal{T}^{w^*})$, from the Banach-Alaoglu Theorem (Theorem 9.4.1), we see that $K = E^{\perp} \cap B_1^*$ is a nonempty compact convex subset of $(M(X),\mathcal{T}^{w^*})$. Thus we can apply the Krein-Mil'man Theorem to the set $K$.

Lemma 11.4.1. Let $X$ be a locally compact Hausdorff topological space and suppose $E \subset C_0(X)$ and $K = E^{\perp} \cap B_1^*$. If $\sigma \in \text{ext}(K)$ and $\sigma \neq 0$, then $\|\sigma\| = 1$.

Proof. Since $K$ is a nonempty compact convex set in $(M(X),\mathcal{T}^{w^*})$, we know from the Krein-Mil'man Theorem (Theorem 11.2.1), that $\text{ext}(K) \neq \emptyset$. Suppose $\sigma \in \text{ext}(K)$ and $\sigma \neq 0$. If $\|\sigma\| < 1$, then, since $\text{ext}(K) \subset K$, we see that $\sigma/\|\sigma\| \in E^{\perp} \cap B_1^* = K$ and

$$\sigma = \|\sigma\|\left(\frac{\sigma}{\|\sigma\|}\right) + (1 - \|\sigma\|)0.$$

However, $0 \in K$, so that $\sigma = \sigma/\|\sigma\| = 0$, as $\sigma \in \text{ext}(K)$, contrary to the assumption that $\sigma \neq 0$.

Therefore $\|\sigma\| = 1$. □

Note that, if $K = E^{\perp} \cap B_1^* \neq \{0\}$, then $\text{ext}(K) \neq \{0\}$, as $K = \overline{\text{co}}[\text{ext}(K)]$.

Lemma 11.4.2. Let $X$ be a locally compact Hausdorff topological space, $E \subset C_0(X)$, $K = E^{\perp} \cap B_1^*$, and suppose $\sigma \in \text{ext}(K)$. If $g \in C_0(X)$ is such that $g(t) \geq 0$, $t \in X$, and

$$\int_X f(t)g(t)\, d\sigma(t) = 0 \qquad (f \in E),$$

then there exists some $b_g \geq 0$ such that $g\, d\sigma = b_g\, d\sigma$.

Proof. If $\sigma = 0$, the assertion is trivially true with $b_g = 0$. Suppose $\sigma \neq 0$. Then, by Lemma 11.4.1, $\|\sigma\| = 1$. Obviously we may assume without loss of generality that $0 \leq g(t) \leq 1$, $t \in X$.

Define the measure $\mu \in M(X)$ by setting $d\mu = g\, d\sigma$. By assumption $\mu \in E^{\perp}$. If $\mu = 0$, then $g\,d\sigma = b_g\,d\sigma$ with $b_g = 0$. Moreover, if $\|\mu\| = 1$, then

$$1 = \|\mu\| = \int_X d|\mu|(t)$$

$$= \int_X g(t)\; d|\sigma|(t),$$

as $0 \leq g(t) \leq 1$, $t \in X$, and so

$$\int_X [1 - g(t)]\; d|\sigma|(t) = 0$$

since $\|\sigma\| = 1$. But $1 - g(t) \geq 0$, $t \in X$, and $g$ is continuous, whence we see that $g(t) = 1$, $t \in X$. Thus, if $X$ is compact and $\|\mu\| = 1$, then $g\,d\sigma = b_g\,d\sigma$ with $b_g = 1$; if $X$ is not compact, then $\|\mu\| < 1$, since $g \in C_o(X)$ and so $g$ cannot be identically one.

Finally, suppose $0 < \|\mu\| < 1$ and define $\sigma_1 = \mu/\|\mu\|$ and $\sigma_2 = (\sigma - \mu)/(1 - \|\mu\|)$. Clearly $\sigma_1 \in K$, as $\mu \in E^{\perp}$ and $\|\sigma_1\| = 1$. Also $\sigma_2 \in E^{\perp}$, as $\sigma, \mu \in E^{\perp}$ and $E^{\perp}$ is a linear space. Moreover, since $0 \leq g(t) \leq 1$, $t \in X$,

$$\|\sigma_2\| = \int_X d|\sigma_2|(t)$$

$$= \left(\frac{1}{1 - \|\mu\|}\right) \int_X [1 - g(t)]\; d|\sigma|(t)$$

$$= \frac{\|\sigma\| - \|\mu\|}{1 - \|\mu\|}$$

$$= 1.$$

Hence $\sigma_2 \in K$. But $\|\mu\|\sigma_1 + (1 - \|\mu\|)\sigma_2 = \mu + \sigma - \mu = \sigma$, and $\sigma \in \mathrm{ext}(K)$, from which we conclude that $\sigma = \sigma_1 = \mu/\|\mu\|$.

Therefore $g\,d\sigma = b_g\,d\sigma$, where $b_g = \|\mu\|$. □

One final lemma is necessary, but before giving it we need to make some further definitions.

Definition 11.4.1. Let $X$ be a locally compact Hausdorff topological space. A measure $\mu$ in $M(X)$ is said to vanish on the open set $U \subset X$ if

$$\int_X f(t)\, d\mu(t) = 0$$

for every $f \in C_o(X)$ that has compact support contained in $U$. The support of $\mu$ is the complement of the largest open set on which $\mu$ vanishes.

The existence of the support of $\mu \in M(X)$ can be established by a routine application of Zorn's Lemma [$DS_1$, p. 6].

If $\mu \in M(X)$ and $\mu \neq 0$, then it is evident that the support of $\mu$ is a nonempty closed subset of $X$.

Definition 11.4.2. Let $X$ be a locally compact Hausdorff topological space and $E \subset C_o(X)$. The set $E$ is said to separate points if, whenever $s,t \in X$, $s \neq t$, there exists some $f \in E$ such that $f(s) \neq f(t)$. Furthermore, $E$ is said to be closed under complex conjugation if $f \in E$ implies $\overline{f} \in E$, where $\overline{f}(t) = \overline{f(t)}$, $t \in X$.

Lemma 11.4.3. Let $X$ be a locally compact Hausdorff topological space and suppose $W \subset C_o(X)$ is a subalgebra that separates points and is closed under complex conjugation. If $\sigma \in M(X)$, $\sigma \neq 0$, is such that for each $g \in W$, $g(t) \geq 0$, $t \in X$, there exists some $b_g \geq 0$ for which $g\, d\sigma = b_g\, d\sigma$, then the support of $\sigma$ is a single point.

Proof. Suppose $s$ and $t$ belong to the support of $\sigma$ and $s \neq t$. Since $W$ is a linear subspace that separates points, we see that there exists some $h \in W$ such that $h(t) = 1$ and $|h(s)| = \delta_1 < 1$. Moreover, since $W$ is a subalgebra that is closed under complex conjugation, $g = h\overline{h} \in W$, $g(t) = 1$, $g(s) = |h(s)|^2 = \delta_1^2 = \delta < 1$, and $g(u) \geq 0$, $u \in X$. By the continuity of $g$ we may choose open neighborhoods $U_1$ and $U_2$ of $t$ and $s$, respectively, such that

$$\text{(i)}\quad g(u) > 1 - \frac{1-\delta}{3} = \frac{2}{3} + \frac{\delta}{3} \qquad (u \in U_1).$$

$$\text{(ii)}\quad g(u) < \delta + \frac{1-\delta}{3} = \frac{2\delta}{3} + \frac{1}{3} \qquad (u \in U_2).$$

$$\text{(iii)}\quad U_1 \cap U_2 = \emptyset.$$

$$\text{(iv)}\quad |\sigma|(U_k) > 0 \qquad (k = 1,2).$$

The last condition on the sets $U_1$ and $U_2$ is possible because $s$ and $t$ belong to the support of $\sigma$.

But then

$$\begin{aligned} b_g|\sigma|(U_1) &= \int_{U_1} b_g \, d|\sigma|(u) \\ &= \int_{U_1} g(u) \, d|\sigma|(u) \\ &> \left(\frac{2+\delta}{3}\right)|\sigma|(U_1), \end{aligned}$$

and

$$\begin{aligned} b_g|\sigma|(U_2) &= \int_{U_2} g(u) \, d|\sigma|(u) \\ &< \left(\frac{2\delta+1}{3}\right)|\sigma|(U_2). \end{aligned}$$

This, however, leads to a contradiction since the first estimate implies that $b_g > 2/3 + \delta/3$, while the second implies that $b_g < 2\delta/3 + 1/3 < \delta/3 + 2/3$, as $\delta < 1$.

Therefore the support of $\sigma$ is a single point. □

Utilizing these three lemmas we can now give a short proof of the Stone-Weierstrass Theorem. We need one final definition, however, in order to make the statement of the theorem clear.

Definition 11.4.3. Let $X$ be a locally compact Hausdorff topological space and suppose $E \subset C_o(X)$. Then

$$Z(E) = \{t \mid t \in X,\ f(t) = 0,\ f \in E\}.$$

Theorem 11.4.1 (Stone-Weierstrass Theorem). Let $X$ be a locally compact Hausdorff topological space. If $W \subset C_o(X)$ is a subalgebra that separates points, is closed under complex conjugation, and is such that $Z(W) = \emptyset$, then $W$ is dense in $(C_o(X), \|\cdot\|_\infty)$.

Proof. From Corollary 4.2.8 to the Hahn-Banach Theorem and the fact that $W$ is a linear subspace, we see that $W$ is dense in $(C_o(X), \|\cdot\|_\infty)$ if and only if $W^\perp \cap B_1^* = K = \{0\}$, where, as before, $B_1^* = \{\mu \mid \mu \in M(X),\ \|\mu\| \leq 1\}$. Suppose this is not the case. Then, since $K$ is a nonempty compact convex set in $(M(X), \mathcal{T}^{W^*})$, we know, by the Krein-Mil'man Theorem (Theorem 11.2.1), that $\operatorname{ext}(K) \neq \emptyset$. From the remark following Lemma 11.4.1 we see that $\operatorname{ext}(K) \neq \{0\}$, and so, by Lemma 11.4.1, there exists some $\sigma \in \operatorname{ext}(K)$ such that $\|\sigma\| = 1$.

Moreover, for each $g \in W$ we have that $fg \in W$, $f \in W$, as $W$ is a subalgebra. Hence, in particular,

$$\int_X f(t)g(t)\, d\sigma(t) = 0 \qquad (f \in W)$$

for each $g \in W$, $g(t) \geq 0$, $t \in X$, since $\sigma \in W^\perp$. Thus by Lemma 11.4.2 for each $g \in W$, $g(t) \geq 0$, $t \in X$, there exists some $b_g \geq 0$ such that $g\, d\sigma = b_g\, d\sigma$. However, since $W$ is a subalgebra that separates points and is closed under complex conjugation, we conclude from Lemma 11.4.3 that the support of $\sigma$ is a single point. Consequently there exist some $t_o \in X$ and some $a \in \mathbb{C}$, $|a| = 1$, such that $\sigma = a\delta_{t_o}$, where $\delta_{t_o}$ denotes the measure in $M(X)$ defined by

$$\delta_{t_o}(S) = 1 \quad \text{for} \quad t_o \in S,$$

$$\delta_{t_o}(S) = 0 \quad \text{for} \quad t_o \notin S$$

for any Borel set $S \subset X$.

Thus, since $\sigma \in W^{\perp}$, we see that

$$\begin{aligned} 0 &= \int_X f(t)\, d\sigma(t) \\ &= a\int_X f(t)\, d\delta_{t_o}(t) \\ &= af(t_o) \qquad\qquad (f \in W), \end{aligned}$$

that is, $t_o \in Z(W)$. This, however, contradicts the assumption that $Z(W) = \phi$.

Therefore $K = \{0\}$, and $W$ is dense in $(C_o(X), \|\cdot\|_\infty)$. □

The usual form of the Stone-Weierstrass Theorem for compact spaces is an easy corollary of Theorem 11.4.1.

Corollary 11.4.1. Let $X$ be a compact Hausdorff topological space. If $W \subset C(X)$ is a subalgebra that separates points, is closed under complex conjugation, and contains the constant functions, then $W$ is dense in $(C(X), \|\cdot\|_\infty)$.

Proof. If $W$ contains the constant functions, then $Z(W) = \phi$. □

11.5. Helson Sets. Suppose $a = \{a_k\} \in L_1(\mathbb{Z})$; that is, $\{a_k\}$ is a doubly infinite sequence of complex numbers such that $\Sigma_{k=-\infty}^{\infty} |a_k| < \infty$. It is evident that the formula

$$\check{a}(e^{it}) = \sum_{k=-\infty}^{\infty} a_k e^{ikt} \qquad (t \in (-\pi, \pi])$$

defines a function $\check{a} \in C(\Gamma)$, where $\Gamma = \{e^{it} \mid t \in (-\pi,\pi]\}$. Letting $E \subset \Gamma$ be a closed set, we wish to determine necessary and sufficient conditions under which for each $f \in C(E)$ there exists some $a = \{a_k\} \in L_1(\mathbb{Z})$ for which the restriction of $\check{a}$ to $E$ is equal to $f$, that is, for which $\check{a}|_E = f$. The theorem we shall establish, with the help of the Stone-Weierstrass Theorem and the theorem on adjoint transformations given in Section 9.10, provides only a functional analytic characterization of such sets $E$. A description in terms of set theoretic, algebraic, or topological properties is unknown. To be complete we first make the following definition:

Definition 11.5.1. A closed set $E \subset \Gamma$ is said to be a Helson set if for each $f \in C(E)$ there exists some $a \in L_1(\mathbb{Z})$ such that $\check{a}|_E = f$.

We now wish to begin a general discussion of the mapping $a \to \check{a}$ from $L_1(\mathbb{Z})$ to $C(\Gamma)$ that will culminate in the indicated theorem. We note first that the mapping $a \to \check{a}$ is linear and continuous, as

$$\|\check{a}\|_\infty \leq \sum_{k=-\infty}^{\infty} |a_k| = \|a\|_1 \qquad (a \in L_1(\mathbb{Z})).$$

Now let $E \subset \Gamma$ be closed and define $I(E) = \{a \mid a \in L_1(\mathbb{Z}),\ \check{a}|_E = 0\}$. It is apparent from the preceding observations that $I(E)$ is a closed linear subspace of $L_1(\mathbb{Z})$, and hence Theorem 1.1.1(iii) shows that $L_1(\mathbb{Z})/I(E)$ is a Banach space with the usual quotient norm

$$|||a + I(E)||| = \inf_{b \in I(E)} \|a + b\|_1 \qquad (a \in L_1(\mathbb{Z})).$$

We define $T : L_1(\mathbb{Z})/I(E) \to C(E)$ by $T[a + I(E)] = \check{a}|_E$, $a \in L_1(\mathbb{Z})$. Clearly $T$ is linear and injective, while the estimates

$$\begin{aligned} \|\check{a}|_E\|_\infty &= \|(\check{a} + \check{b})|_E\|_\infty \\ &\leq \|\check{a} + \check{b}\|_\infty \\ &\leq \|a + b\|_1 \qquad (b \in I(E)) \end{aligned}$$

entail that

$$\begin{aligned} \|\check{a}|_E\|_\infty &\leq \inf_{b \in I(E)} \|a + b\|_1 \\ &= |||a + I(E)||| \qquad (a \in L_1(\mathbb{Z})), \end{aligned}$$

whence $T \in L(L_1(\mathbb{Z})/I(E), C(E))$.

The definition of a Helson set says precisely that $E$ is a Helson set if and only if $T : L_1(\mathbb{Z})/I(E) \to C(E)$ is surjective.

The idea is now to apply Theorem 9.10.1 to obtain conditions equivalent to the surjectivity of $T$. To do this we must first verify that $T$ satisfies all the hypotheses of this theorem; in particular, we need to show that $T[L_1(\mathbb{Z})/I(E)]$ is norm dense in $(C(E),\|\cdot\|_\infty)$.

Without going into all the details, we note that $T[L_1(\mathbb{Z})/I(E)]$ is a subalgebra of $C(E)$. This follows easily on observing that, if $a = \{a_k\} \in L_1(\mathbb{Z})$ and $b = \{b_k\} \in L_1(\mathbb{Z})$, then $c = \{c_k\} \in L_1(\mathbb{Z})$, where $c_k = \Sigma_{j=-\infty}^{\infty} a_{k-j} b_j$, $k \in \mathbb{Z}$, that is, $c = a * b \in L_1(\mathbb{Z})$, and furthermore,

$$\begin{aligned}
\check{c}(e^{it}) &= \sum_{k=-\infty}^{\infty} \Big(\sum_{j=-\infty}^{\infty} a_{k-j} b_j\Big) e^{ikt} \\
&= \sum_{j=-\infty}^{\infty} b_j \Big(\sum_{k=-\infty}^{\infty} a_{k-j} e^{ikt}\Big) \\
&= \sum_{j=-\infty}^{\infty} b_j \Big(\sum_{k=-\infty}^{\infty} a_k e^{i(k+j)t}\Big) \\
&= \Big(\sum_{j=-\infty}^{\infty} b_j e^{ijt}\Big)\Big(\sum_{k=-\infty}^{\infty} a_k e^{ikt}\Big) \\
&= \check{a}(e^{it})\check{b}(e^{it}) \qquad (t \in (-\pi,\pi]).
\end{aligned}$$

The verification of this is left to the reader.

Moreover, $T[L_1(\mathbb{Z})/I(E)]$ is closed under complex conjugation since for any $a = \{a_k\} \in L_1(\mathbb{Z})$ we have

$$\begin{aligned}
\overline{\check{a}}(e^{it}) &= \sum_{k=-\infty}^{\infty} \overline{a}_k e^{-ikt} \\
&= \sum_{k=-\infty}^{\infty} b_k e^{ikt} \\
&= \check{b}(e^{it})
\end{aligned}$$

provided $b_k = \overline{a}_{-k}$, $k \in \mathbb{Z}$. Since $W = \{e^{ikt} \mid k \in \mathbb{Z},\ t \in (-\pi,\pi]\}$ separates the points of $\Gamma$ and $Z(W) = \emptyset$, it follows at once that $T[L_1(\mathbb{Z})/I(E)]$ separates the points of $E$ and $Z(T[L_1(\mathbb{Z})/I(E)]) = \emptyset$. Consequently we may apply the Stone-Weierstrass Theorem (Theorem 11.4.1) to $T[L_1(\mathbb{Z})/I(E)]$ and conclude that $T[L_1(\mathbb{Z})/I(E)]$ is norm dense in $C(E)$.

Thus the hypotheses of Theorem 9.10.1 are fulfilled, and we see that $E \subset \Gamma$ is a Helson set if and only if either $T^*$ is surjective or there exists some $m > 0$ such that $\|T^*(y^*)\| \geq m\|y^*\|$, $y^* \in C(E)^*$. Our next task is to describe $T^*$ more precisely.

From Theorems 4.6.2 and 8.3.3 we see that $(L_1(\mathbb{Z})/I(E))^*$ can be identified with $I(E)^\perp \subset L_1(\mathbb{Z})^* = L_\infty(\mathbb{Z})$, while the Riesz Representation Theorem (Theorem 4.8.2) shows that $C(E)^*$ can be identified with $M(E)$, the space of bounded, regular, complex-valued Borel measures on $E$. Since $E \subset \Gamma$ is closed, it is easily seen that the elements of $M(E)$ can be identified with those measures in $M(\Gamma)$ that have support contained in $E$. Thus for each $\mu \in M(E)$ the following definition of $\check{\mu}$ is meaningful:

$$\check{\mu}(k) = \overline{\int_E e^{ikt}\, d\mu(e^{it})} \qquad (k \in \mathbb{Z}).$$

Obviously $\|\check{\mu}\|_\infty = \sup_{k \in \mathbb{Z}}|\check{\mu}(k)| \leq \|\mu\|$, $\mu \in M(E)$, and so $\check{\mu} \in L_\infty(\mathbb{Z})$. The complex conjugation in the definition of $\check{\mu}$ is necessitated by the manner in which we have identified $L_1(\mathbb{Z})^*$ with $L_\infty(\mathbb{Z})$. You will recall (Theorem 8.3.3) that this identification involved an antilinear, rather than a linear, mapping.

Given $\mu \in M(E)$, we see that $T^*(\mu) = \{b_k\} \in I(E)^\perp \subset L_\infty(\mathbb{Z})$ and

$$\begin{aligned}
\sum_{k=-\infty}^{\infty} a_k\overline{b}_k &= T^*(\mu)(a) \\
&= \mu[T(a)] \\
&= \int_E \check{a}(e^{it})\, d\mu(e^{it}) \\
&= \sum_{k=-\infty}^{\infty} a_k \int_E e^{ikt}\, d\mu(e^{it})
\end{aligned}$$

$$= \sum_{k=-\infty}^{\infty} a_k \overline{\check{\mu}(k)} \qquad (a = \{a_k\} \in L_1(\mathbb{Z})).$$

The interchange of integration and summation is valid since the series expression for $\check{a}$ converges uniformly. We then see at once that $T^*(\mu) = \check{\mu}$, $\mu \in M(E)$.

An application of Theorem 9.10.1 immediately yields the following theorem:

Theorem 11.5.1. Let $E \subset \Gamma = \{e^{it} \mid t \in (-\pi,\pi]\}$ be a closed set. Then the following are equivalent:

(i) $E$ is a Helson set.

(ii) If $\{b_k\} \in L_\infty(\mathbb{Z})$ is such that $\Sigma_{k=-\infty}^{\infty} a_k \overline{b}_k = 0$ for each $\{a_k\} \in L_1(\mathbb{Z})$ for which $\Sigma_{k=-\infty}^{\infty} a_k e^{ikt} = 0$, $e^{it} \in E$, then there exists some $\mu \in M(E)$ such that $\check{\mu}(k) = b_k$, $k \in \mathbb{Z}$.

(iii) There exists an $M > 0$ such that

$$\|\mu\| \leq M\|\check{\mu}\|_\infty = M \sup_{k \in \mathbb{Z}} |\check{\mu}(k)| \qquad (\mu \in M(E)).$$

(iv) $|||\mu||| = \|\check{\mu}\|_\infty$, $\mu \in M(E)$, defines a norm on $M(E)$ that is equivalent to $\|\mu\| = |\mu|(E)$, $\mu \in M(E)$.

Analogs of this result that are valid for arbitrary locally compact Abelian topological groups, as well as some specific examples of Helson sets, can be found in [$E_3$, pp. 233-237; $HR_2$, pp. 563-565; $Ru_1$, pp. 114-120].

## 11.6. The Banach-Stone Theorem.

Suppose that $X_1$ and $X_2$ are compact Hausdorff topological spaces that are homeomorphic. If $\tau : X_1 \to X_2$ is a homeomorphism, then for each $f \in C(X_2)$ define $T(f)(t_1) = f[\tau(t_1)]$, $t_1 \in X_1$. It is easily verified that $T(f) \in C(X_1)$ and that $T : C(X_2) \to C(X_1)$ is an isometric isomorphism. Moreover, $T$ is surjective, because if $g \in C(X_1)$ and we define $f \in C(X_2)$ by setting $f(t_2) = g[\tau^{-1}(t_2)]$, $t_2 \in X_2$, then clearly $T(f) = g$. Thus we see that a homeomorphism between $X_1$ and $X_2$ gives rise to an isometric isomorphism of $C(X_2)$ onto $C(X_1)$. The content of the Banach-Stone Theorem is that the converse assertion is valid.

In order to prove this result we need to know the extreme points of the closed unit ball in $M(X)$.

Lemma 11.6.1. Let $X$ be a compact Hausdorff topological space and let $B_1^* = \{\mu \mid \mu \in M(X), \|\mu\| \leq 1\}$. Then

$$\mathrm{ext}(B_1^*) = \{a\delta_t \mid a \in \mathbb{C}, |a| = 1, t \in X\},$$

where $\delta_t \in M(X)$ is defined by

$$\delta_t(E) = 1 \quad \text{for} \quad t \in E,$$

$$\delta_t(E) = 0 \quad \text{for} \quad t \notin E,$$

for any Borel set $E \subset X$.

Proof. Let $B = \{a\delta_t \mid a \in \mathbb{C}, |a| = 1, t \in X\}$. Clearly $B \subset B_1^*$, and the weak* closure $\overline{\mathrm{co}}(B)$ of $\mathrm{co}(B)$ is contained in $B_1^*$ since $B_1^*$ is weak* closed and convex. Suppose $\mu_o \in M(X)$ and $\mu_o \notin \overline{\mathrm{co}}(B)$. By Corollary 5.4.1 there exists a closed real hyperplane that strictly separates $\mu_o$ and $\overline{\mathrm{co}}(B)$, and so by Theorem 9.1.3 and Proposition 5.1.1 there exists some $f \in C(X)$ such that

$$\begin{aligned} b - \epsilon &= \sup_{\mu \in \overline{\mathrm{co}}(B)} \mathrm{Re}\left[\int_X f(s)\, d\mu(s)\right] \\ &< \mathrm{Re}\left[\int_X f(s)\, d\mu_o(s)\right] \\ &= b \end{aligned}$$

for some suitable $\epsilon > 0$. In particular,

$$\begin{aligned} \mathrm{Re}\left[a\int_X f(s)\, d\delta_t(s)\right] &= \mathrm{Re}[af(t)] \\ &\leq b - \epsilon \qquad (a \in \mathbb{C};\ |a| = 1;\ t \in X), \end{aligned}$$

from which it follows that $\|f\|_\infty \leq b - \epsilon$. Hence

$$b = \mathrm{Re}\left[\int_X f(s)\, d\mu_o(s)\right]$$
$$\leq \|f\|_\infty \|\mu_o\|$$
$$\leq (b - \epsilon)\|\mu_o\|,$$

and so $\|\mu_o\| \geq b/(b - \epsilon) > 1$.

Consequently $B_1^* = \overline{\mathrm{co}}(B)$, and so from Theorems 11.2.2 and 9.4.1 we conclude that $\mathrm{ext}(B_1^*) \subset B$.

On the other hand, let $t \in X$, $a \in \mathbb{C}$, $|a| = 1$, and suppose $\mu_1, \mu_2 \in B_1^*$ are such that $a\delta_t = b\mu_1 + (1 - b)\mu_2$, $0 < b < 1$. Obviously $\|\mu_k\| = 1$, $k = 1,2$. We wish to show that $a\delta_t = \mu_1 = \mu_2$. To accomplish this we shall show that, whenever $f \in C(X)$ is such that

$$f(t) = \int_X f(s)\, d\delta_t(s) = 0,$$

it happens that

$$\int_X f(t)\, d\mu_k(t) = 0 \qquad (k = 1,2)$$

and then shall apply the Riesz Representation Theorem (Theorem 4.8.2) and Corollary 3.3.1 to conclude that there exists some $a_k \in \mathbb{C}$ for which $\mu_k = a_k\delta_t$, $k = 1,2$. The desired conclusion will then be an easy consequence of this fact. However, in order to prove the first assertion we shall have to work a little.

Suppose $U$ is any open neighborhood of $t$, and $h_o \in C(X)$ is such that $\|h_o\| \leq 1$ and $h_o(s) = 0$, $s \in U$. Since $X$ is a normal topological space, being compact and Hausdorff, there exists, by Urysohn's Lemma [$W_2$, p. 55], some $h \in C(X)$ such that $\|h\|_\infty \leq 1$, $h(t) = 1$, and $h(s) = 0$, $s \notin U$. Then

$$b\int_X h(s)\, d\mu_1(s) + (1 - b)\int_X h(s)\, d\mu_2(s) = a\int h(s)\, d\delta_t(s)$$
$$= ah(t)$$
$$= a.$$

But $a \in \mathbb{C}$, $|a| = 1$, implies that $a$ is an extreme point of the closed unit ball about the origin in $\mathbb{C}$, and since

$$\left|\int_X h(s)\, d\mu_k(s)\right| \le \|h\|_\infty \|\mu_k\| \le 1 \qquad (k = 1,2),$$

we deduce that

$$a = \int_X h(s)\, d\mu_k(s) \qquad (k = 1,2).$$

Similarly, since $\|h + h_o\|_\infty \le 1$, we see that

$$b\int_X [h(s) + h_o(s)]\, d\mu_1(s) + (1 - b)\int_X [h(s) + h_o(s)]\, d\mu_2(s)$$

$$= a[h(t) + h_o(t)]$$

$$= a,$$

and so

$$\int_X [h(s) + h_o(s)]\, d\mu_k(s) = a \qquad (k = 1,2),$$

from which it follows that

$$\int_X h_o(s)\, d\mu_k(s) = 0 \qquad (k = 1,2).$$

Thus we have shown that, if $h_o \in C(X)$ is any function that vanishes on an open neighborhood of $t$, then

$$\int_X h_o(s)\, d\mu_k(s) = 0 \qquad (k = 1,2).$$

Now suppose $f \in C(X)$ and $f(t) = 0$. We wish to show that

$$\int_X f(s)\, d\mu_k(s) = 0 \qquad (k = 1,2).$$

Clearly we may assume without loss of generality that $\|f\|_\infty \le 1$. For each positive integer $n$ let $U_n$ be an open neighborhood of $t$ such that $|f(s)| < 1/n$, $s \in U_n$, and let $W_n$ be an open neighborhood of $t$ such that $\mathrm{cl}(W_n) \subset U_n$. The latter is possible because $X$ is normal. Appealing to Urysohn's Lemma, let $h_n \in C(X)$ be such that $h_n(s) = 1$, $s \in \mathrm{cl}(W_n)$; $h_n(s) = 0$, $s \notin U_n$; and $\|h_n\|_\infty \le 1$. Set

$g_n = h_n f$, $n = 1,2,3,\ldots$ . Then $\{g_n\} \subset C(X)$; $g_n(s) = f(s)$, $s \in W_n$; $g_n(s) = 0$, $s \notin U_n$; and $\|g_n\|_\infty < 1/n$, $n = 1,2,3,\ldots$ . Moreover, it is evident that $\|f - g_n\|_\infty \leq 1$ and $(f - g_n)(s) = 0$ for $s \in W_n$, $n = 1,2,3,\ldots$, whence by our previous observations we see that

$$\int_X [f(s) - g_n(s)]\, d\mu_k(s) = 0 \qquad (k = 1,2;\ n = 1,2,3,\ldots),$$

as each $W_n$ is an open neighborhood of $t$. But $\lim_n \|f - g_n - f\|_\infty = 0$, and so

$$\begin{aligned} \int_X f(s)\, d\mu_k(s) &= \lim_n \int_X [f(s) - g_n(s)]\, d\mu_k(s) \\ &= 0 \qquad (k = 1,2), \end{aligned}$$

which is what we set out to prove.

Hence, considering $\delta_t, \mu_1$, and $\mu_2$ as elements of $C(X)^*$, we see that $N(\delta_t) \subset N(\mu_k)$, $k = 1,2$, and so by Corollary 3.3.1 there exist $a_k \in \mathbb{C}$ for which $\mu_k = a_k \delta_t$, where obviously $|a_k| = 1$, $k = 1,2$. Consequently

$$\begin{aligned} a\delta_t &= b\mu_1 + (1 - b)\mu_2 \\ &= [ba_1 + (1 - b)a_2]\delta_t, \end{aligned}$$

and so $a = ba_1 + (1 - b)a_2$. Since $a$ is an extreme point of the closed unit ball in $\mathbb{C}$, we conclude at once that $a = a_1 = a_2$, whence $a\delta_t = \mu_1 = \mu_2$.

Therefore each measure in $B$ is an extreme point of $B_1^*$, and so $B = \text{ext}(B_1^*)$.

□

A second lemma will also be useful.

<u>Lemma</u> <u>11.6.2</u>. Let $X$ be a compact Hausdorff topological space and let $B_1^* = \{\mu \mid \mu \in M(X), \|\mu\| \leq 1\}$. If $\varphi : X \to \text{ext}(B_1^*)$ is defined by $\varphi(t) = \delta_t$, $t \in X$, then $\varphi$ is a homeomorphism from $X$ onto $\varphi(X)$ when the latter set is taken with the relative weak* topology.

Proof. Clearly $\varphi$ is an injective mapping of $X$ into $\mathrm{ext}(B_1^*)$ by Lemma 11.6.1. Consider the weak* neighborhood

$$U(\delta_t;\epsilon;f_1,f_2,\ldots,f_n)\cap\varphi(X)$$

of $\delta_t$ in $\varphi(X)$. Then for each $k = 1,2,\ldots,n$,

$$W_k = \{s \mid |f_k(s) - f_k(t)| < \epsilon\}$$

is an open neighborhood of $t$ in $X$, and hence $W = \bigcap_{k=1}^n W_k$ is also such a neighborhood. But, if $s \in W$, then

$$\left|\int_X f_k(u)\, d\delta_s(u) - \int_X f_k(u)\, d\delta_t(u)\right| = |f_k(s) - f_k(t)|$$

$$< \epsilon \qquad (k = 1,2,\ldots,n),$$

that is, $\varphi(W) \subset U(\delta_t;\epsilon;f_1,f_2,\ldots,f_n)\cap\varphi(X)$.

Thus $\varphi$ is continuous and hence a homeomorphism, as $X$ is compact. □

Theorem 11.6.1 (Banach-Stone Theorem). Let $X_1$ and $X_2$ be compact Hausdorff topological spaces. Then the following are equivalent:

(i) $X_1$ and $X_2$ are homeomorphic.

(ii) $(C(X_2),\|\cdot\|_\infty)$ and $(C(X_1),\|\cdot\|_\infty)$ are isometrically and linearly isomorphic.

Proof. The proof that part (i) implies part (ii) is contained in the introductory remarks to this section. Conversely, suppose $T : C(X_2) \to C(X_1)$ is an isometric surjective linear isomorphism. Then the adjoint mapping $T^* : M(X_1) \to M(X_2)$, defined by

$$\begin{aligned} T^*(\mu_1)(f) &= \int_{X_2} f(t_2)\, dT^*(\mu_1)(t_2) \\ &= \int_{X_1} T(f)(t_1)\, d\mu_1(t_1) \\ &= \mu_1[T(f)] \qquad\qquad (f \in C(X_2)), \end{aligned}$$

is an isometric linear isomorphism from $M(X_1)$ to $M(X_2)$ (Theorem 4.4.1). Moreover, $T^*$ is surjective since if $\mu_2 \in M(X_2)$ and $\mu_1 \in M(X_1)$ is defined by

$$\int_{X_1} f(t_1)\, d\mu_1(t_1) = \int_{X_2} T^{-1}(f)(t_2)\, d\mu_2(t_2) \qquad (f \in C(X_1)),$$

then evidently $T^*(\mu_1) = \mu_2$. Alternatively, one could just apply Theorem 9.10.1. Let $B^*_{1k} = \{\mu \mid \mu \in M(X_k), \|\mu\| \le 1\}$, $k = 1,2$. Since $T^*$ is linear and isometric, a simple argument shows that

$$T^*[\text{ext}(B^*_{11})] = \text{ext}(B^*_{12}).$$

Moreover, it is apparent from the definition of the weak* topology that $T^* : (M(X_1), \mathcal{T}^{w^*}) \to (M(X_2), \mathcal{T}^{w^*})$ is continuous. Hence $T^*$ is a homeomorphism from $B^*_{11}$ onto $B^*_{12}$, since $B^*_{11}$ is weak* compact.

Now let $\varphi_k : X_k \to \text{ext}(B^*_{1k})$, $k = 1,2$, be defined by $\varphi_k(t_k) = \delta_{t_k}$, $t_k \in X_k$, $k = 1,2$. By Lemma 11.6.2 each $\varphi_k$ is a homeomorphism onto its range with the relative weak* topology. If $t_1 \in X_1$, then $T^* \circ \varphi_1(t_1) = \alpha(t_1)\delta_{\tau(t_1)}$, for some $\tau(t_1) \in X_2$ and some $\alpha(t_1) \in \mathbb{C}$ for which $|\alpha(t_1)| = 1$, by Lemma 11.6.1 and the fact that $T^*$ maps $\text{ext}(B^*_{11})$ onto $\text{ext}(B^*_{12})$. Clearly $T^* \circ \varphi_1 : X_1 \to T^* \circ \varphi_1(X_1)$ is a homeomorphism. We claim, moreover, that the mapping $\alpha : X_1 \to X_2$, determined by $T^* \circ \varphi_1(t_1) = \alpha(t_1)\delta_{\tau(t_1)}$, $t_1 \in X_1$, is continuous. Indeed, if $f_o \in C(X_2)$ is such that $f_o(s_2) = 1$, $s_2 \in X_2$, then

$$\begin{aligned} \alpha(t_1) &= \alpha(t_1) \int_{X_2} f_o(s_2)\, d\delta_{\tau(t_1)}(s_2) \\ &= \int_{X_1} T(f_o)(s_1)\, d\delta_{t_1}(s_1) \\ &= T(f_o)(t_1) \qquad (t_1 \in X_1), \end{aligned}$$

from which the continuity of $\alpha$ is evident.

Furthermore, we claim that $\psi : X_1 \to \varphi_2(X_2)$, defined by $\psi(t_1) = \delta_{\tau(t_1)}$, $t_1 \in X_1$, is a bijective continuous mapping. The bijectivity is easily established, and its proof is left to the reader. To establish the continuity of $\psi$, let $\delta_{t_2} = \psi(t_1)$ and consider the open neighborhood $U$ of $\delta_{t_2}$ in the relative weak* topology given by

$$\begin{aligned} U &= U(\delta_{t_2};\epsilon;f_1,f_2,\ldots,f_n) \cap \varphi_2(X_2) \\ &= \{\delta_{s_2} \mid s_2 \in X_2,\ |f_k(s_2) - f_k(t_2)| < \epsilon,\ k = 1,2,\ldots,n\}, \end{aligned}$$

where, of course, $f_1,f_2,\ldots,f_n$ belong to $C(X_2)$. Since $\alpha$ is continuous, there exists some open neighborhood $W_1 \subset X_1$ of $t_1$ such that

$$|\alpha(t_1) - \alpha(s_1)| < \frac{\epsilon}{[2 \sup\limits_{k=1,2,\ldots,n} |T(f_k)(t_1)|]} \qquad (s_1 \in W_1),$$

and from the continuity of $T(f_k)$, $k = 1,2,\ldots,n$, we deduce the existence of an open neighborhood $W_2 \subset X_1$ of $t_1$ such that

$$|T(f_k)(t_1) - T(f_k)(s_1)| < \frac{\epsilon}{2} \qquad (s_1 \in W_2;\ k = 1,2,\ldots,n).$$

Now, if $W = W_1 \cap W_2$, then for each $s_1 \in W$ we have

$$\begin{aligned} |f_k(t_2) - f_k[\tau(s_1)]| &= \left|\int_{X_2} f_k(u_2)\, d\delta_{t_2}(u_2) - \int_{X_2} f_k(u_2)\, d\delta_{\tau(s_1)}(u_2)\right| \\ &= \left|\int_{X_1} T(f_k)(u_1)\overline{\alpha(t_1)}\, d\delta_{t_1}(u_1) - \int_{X_1} T(f_k)(u_1)\overline{\alpha(s_1)}\, d\delta_{s_1}(u_1)\right| \\ &= |\overline{\alpha(t_1)}T(f_k)(t_1) - \overline{\alpha(s_1)}T(f_k)(s_1)| \\ &\le |\overline{\alpha(t_1)} - \overline{\alpha(s_1)}|\,|T(f_k)(t_1)| + |\overline{\alpha(s_1)}|\,|T(f_k)(t_1) - T(f_k)(s_1)| \\ &< \frac{\epsilon}{2} + \frac{\epsilon}{2} = \epsilon \qquad (k = 1,2,\ldots,n). \end{aligned}$$

The equality $T^*[\overline{\alpha(s_1)}\delta_{s_1}] = \delta_{\tau(s_1)}$ follows at once from the linearity of $T^*$ and the facts that $T^*(\delta_{s_1}) = \alpha(s_1)\delta_{\tau(s_1)}$ and $|\alpha(s_1)| = 1$ for $s_1 \in X_1$. The preceding estimates show that $\psi(W) \subset U$, and hence $\psi$ is continuous.

Finally, it is apparent that $\tau = \varphi_2^{-1} \circ \psi$, and so $\tau : X_1 \to X_2$ is a homeomorphism. Therefore part (ii) of the theorem implies part (i). $\square$

An examination of the proof shows at once that, if $T : C(X_2) \to C(X_1)$ is an isometric linear isomorphism between the indicated spaces, then $T$ must have the form $T(f)(t_1) = \alpha(t_1)f[\tau(t_1)]$ for $t_1 \in X$ and $f \in C(X_2)$, where $\alpha$ is a continuous complex-valued function on $X_1$ such that $|\alpha(t_1)| = 1$, $t_1 \in X_1$, and $\tau : X_1 \to X_2$ is a homeomorphism of $X_1$ onto $X_2$. It is also immediate from this that if $T$ is an algebraic isomorphism [that is, $T(fg) = T(f)T(g)$, $f,g \in C(X_2)$], then $\alpha(t_1) = 1$, $t_1 \in X_1$.

The Krein-Mil'man Theorem has many other important applications. For some examples in harmonic analysis, representation theory, Choquet theory, probability, and ergodic theory we refer the reader to [$E_1$, pp. 715-732; Ga, pp. 52-62; $HR_1$, pp. 311-346; Ph].

## 11.7. Problems.

1. Letting $(V,\mathcal{T})$ be a topological linear space over $\Phi$ and letting $E$ and $F$ be subsets of $V$, prove each of the following:

(a) $\operatorname{co}(aE) = a \operatorname{co}(E)$ $(a \in \Phi)$.

(b) $\operatorname{co}(E + F) = \operatorname{co}(E) + \operatorname{co}(F)$.

(c) $\overline{\operatorname{co}}(aE) = a \overline{\operatorname{co}}(E)$ $(a \in \Phi)$.

(d) If $\overline{\operatorname{co}}(E)$ is compact, then $\overline{\operatorname{co}}(E + F) = \overline{\operatorname{co}}(E) + \overline{\operatorname{co}}(F)$.

2. (a) Give an example of a topological linear space $V$ and two closed convex subsets $E$ and $F$ such that $\operatorname{co}(E \cup F) \neq \overline{\operatorname{co}}(E \cup F)$.

(b) Give an example of a topological linear space $V$ and two subsets $E$ and $F$ such that $\overline{\operatorname{co}}(E + F) \neq \overline{\operatorname{co}}(E) + \overline{\operatorname{co}}(F)$.

3. (Proposition 11.1.2) Let $(V,\mathcal{T})$ be a topological linear space over $\Phi$ and suppose $K \subset V$ is a closed convex set. Prove each of the following:

(a) If $x \in K$, then $\{x\}$ is an extremal subset of $K$ if and only if $x \in \operatorname{ext}(K)$.

(b) If $E$ is an extremal subset of $K$ and $F$ is an extremal subset of $E$, then $F$ is an extremal subset of $K$.

*4. (Lemma 11.2.1) Let $(V,\mathcal{T})$ be a topological linear space over $\Phi$ and let $E_1,E_2,\ldots,E_n$ be any subsets of $V$. Prove that, if $\overline{co}(E_k)$ is compact, $k = 1,2,\ldots,n$, then

$$\overline{co}\left(\bigcup_{k=1}^{n} E_k\right) = co\left[\bigcup_{k=1}^{n} \overline{co}(E_k)\right].$$

*5. Let $(V,\|\cdot\|)$ be a Banach space over $\Phi$. If $E \subset V$ is compact, prove that $\overline{co}(E)$ is compact.

*6. Give an example of a locally convex topological linear space $V$ and a compact subset $E \subset V$ such that $\overline{co}(E)$ is not compact.

7. (a) If $V$ is a finite-dimensional topological linear space and $K \subset V$ is compact, prove the $co(K)$ is compact and hence closed.

(b) Give an example of a locally convex topological linear space $V$ and a compact subset $K \subset V$ such that $co(K)$ is not closed.

8. Let $C^R([0,1])$ denote the Banach space of real-valued continuous functions on $[0,1]$ with the usual supremum norm (see Example 1.2.1).

(a) Determine the extreme points of the closed unit ball $B_1 = \{f \mid f \in C^R([0,1]), \|f\|_\infty \le 1\}$.

(b) Prove the $C^R([0,1])$ is not a dual space -- that is, there exists no normed linear space $V$ such that $C^R([0,1])$ is isometrically isomorphic to $V^*$.

9. Prove that the extreme points of the closed unit ball $B_1 = \{f \mid f \in L_\infty([0,1]), \|f\|_\infty \le 1\}$ in $L_\infty([0,1])$ are precisely those $f \in L_\infty([0,1])$ such that $|f(t)| = 1$ for almost all $t \in [0,1]$.

10. (a) Determine the extreme points of the closed unit ball $B_1 = \{x \mid x \in \ell_p, \|x\|_p \le 1\}$ for $1 \le p \le \infty$.

(b) In $\ell_2$, prove that $B_1$ is equal to the weak closure of $\text{ext}(B_1)$.

*11. Consider the space $\ell_2$ with $\Phi = \mathbb{R}$ and let $K$ denote the set of all sequences $x = \{x_k\}$ such that $\Sigma_{k=1}^{\infty}(2^k x_k)^2 \leq 1$. Prove that $K$ is compact and convex, and that $K$ is equal to the norm closure of $\text{ext}(K)$.

12. Let $(V,\|\cdot\|)$ be an infinite-dimensional Banach space over $\Phi$ and let $B_1 = \{x \mid x \in V, \|x\| \leq 1\}$. If $\text{ext}(B_1)$ is a finite set of points, prove that $V$ is not isometrically isomorphic to $W^*$ for any Banach space $W$.

13. Let $(V_1,\mathcal{T}_1)$ and $(V_2,\mathcal{T}_2)$ be locally convex topological linear spaces over $\Phi$ and let $T \in L(V_1,V_2)$. If $K \subset V_1$ is a nonempty compact convex subset of $V_1$, prove that

$$\text{ext}[T(K)] = T[\text{ext}(K)],$$

that is, every extreme point of $T(K)$ is the image under $T$ of an extreme point of $K$.

14. Let $(V_1,\mathcal{T}_1)$ and $(V_2,\mathcal{T}_2)$ be topological linear spaces over $\Phi$ and let $T \in L(V_1,V_2)$. Prove that, if $K \subset V_1$ is a convex set, and $E \subset V_2$ is an extremal subset of $T(K)$, then $T^{-1}(E)$ is an extremal subset of $K$.

15. Let $X$ be a locally compact Hausdorff topological space and let $\mu \in M(X)$ be such that $\|\mu\| = |\mu|(X) \leq 1$. Prove that for each $f \in C_o(X)$ and $\epsilon > 0$ there exist $t_1,t_2,\ldots,t_n$ in $X$ and $z_1,z_2,\ldots,z_n$ in $\mathbb{C}$ such that $\Sigma_{k=1}^{n}|z_k| = 1$ and

$$\left|\int_X f(t)\, d\mu(t) - \sum_{k=1}^{n} z_k f(t_k)\right| < \epsilon.$$

16. Let $(V,\mathcal{T})$ be a locally convex topological linear space over $\Phi$ and let $K \subset V$ be a compact convex set. If $E \subset K$ is any subset of $K$, prove that the following are equivalent:

(i) $\overline{\text{co}}(E) = K$.

(ii) $\text{ext}(K) \subset \text{cl}(E)$.

(iii) For every real continuous linear functional $x^*$ on $V$ it is the case that $\sup_{x \in E} x^*(x) = \sup_{x \in K} x^*(x)$.

17. Let $(V,\mathcal{T})$ be a locally convex topological linear space over $\Phi$ and let $K \subset V$ be a compact convex set. Prove that every real continuous linear functional on $K$ assumes its maximum at an extreme point of $K$.

*18. In Theorem 11.2.2 we saw that if $E$ is a closed subset of a locally convex topological linear space and $\overline{\text{co}}(E)$ is compact, then $\text{ext}[\overline{\text{co}}(E)] \subset E$. Give an example to show that the assumption of compactness for $\overline{\text{co}}(E)$ may not be dropped -- that is, give an example of a locally convex topological linear space $V$ and a closed subset $E \subset V$ such that some extreme point of $\overline{\text{co}}(E)$ is not in $E$.

19. Give an example of a locally convex topological linear space $V$ and a compact convex subset $K \subset V$ such that $\text{ext}(K)$ is not closed. (Hint: an example can be found in $\mathbb{R}^3$.)

20. (a) If $V$ is a finite-dimensional topological linear space over $\Phi$ and $K \subset V$ is a compact convex set, prove that $K = \text{co}[\text{ext}(K)]$.

(b) Give an example of a Banach space $V$ and a compact convex subset $K \subset V$ such that $K \not\subset \text{co}[\text{ext}(K)]$.

21. Let $B_1^p = \{f \mid f \in L_p(\mathbb{R},dt),\ \|f\|_p \leq 1\}$, $1 < p < \infty$. Using the uniform convexity of the norm $\|\cdot\|_p$, prove that

$$\text{ext}(B_1^p) = \{f \mid f \in L_p(\mathbb{R},dt),\ \|f\|_p = 1\}.$$

(This provides another proof of a portion of Proposition 11.1.1.)

## CHAPTER 12

## FIXED POINT THEOREMS

12.0. Introduction. Fixed point theorems come in various forms, of which we shall discuss two. One form says that all continuous mappings $\varphi$ from certain kinds of sets $K$ into themselves have fixed points -- that is, there exists some $t_o \in K$ such that $\varphi(t_o) = t_o$. The point in $K$ that is fixed will, in general, depend on $\varphi$. We shall discuss, without proofs, some theorems of this form in Section 12.1.

The second form of fixed point theorem we shall consider is one which says that certain kinds of continuous mappings from certain kinds of sets into themselves have fixed points. We shall prove two such theorems in Sections 12.2 and 12.3, one dealing with contraction mappings on complete metric spaces, and one, the Markov-Kakutani Fixed Point Theorem, dealing with commuting families of continuous affine mappings on compact convex sets.

In the final two sections of the chapter we shall give applications of these results to the proofs of the existence and uniqueness of solutions to certain ordinary differential equations, and to the existence of a Haar measure -- that is, a nonzero, positive, translation invariant, regular Borel measure -- on a compact Abelian topological group.

12.1. The Fixed Point Property. Before discussing contractions and the Markov-Kakutani Fixed Point Theorem in the next two sections we wish to make some general remarks about fixed points and various fixed point theorems.

Definition 12.1.1. Let $X$ be a topological space. If $\varphi : X \to X$ is continuous, then $t_o \in X$ is said to be a fixed point for $\varphi$ if $\varphi(t_o) = t_o$. The topological space $X$ is said to have the fixed point property if every continuous mapping $\varphi : X \to X$ has a fixed point.

The topological space $X = [-1,1]$, with the usual topology, has the fixed point property, as we can see from the following argument: Suppose $\varphi : [-1,1] \to [-1,1]$ is continuous. Then the mapping $\psi(t) = [\varphi(t) - t]/2$ where $t \in [-1,1]$, is also a continuous mapping from $[-1,1]$ to itself; moreover, $\psi(-1) \geq 0$ and $\psi(1) \leq 0$. If $\psi(-1) = 0$ or $\psi(1) = 0$, then clearly -1 or 1 is a fixed point for $\varphi$. If this is not the case, then $\psi(1) < 0 < \psi(-1)$, and so by the intermediate value theorem of calculus we deduce the existence of some $t_o$, $-1 < t_o < 1$, such that $\psi(t_o) = 0$, that is, $\varphi(t_o) = t_o$. Thus, in any case, $\varphi$ has a fixed point, and thereby $[-1,1]$ has the fixed point property. The analogous result is valid in any space $\mathbb{R}^n$.

Theorem 12.1.1 (Brouwer Fixed Point Theorem). The sets

$$B_1^n = \{x \mid x \in \mathbb{R}^n, \|x\|_2 \leq 1\} \qquad (n = 1,2,3,\ldots)$$

have the fixed point property.

We do not prove this result, since it would involve a rather lengthy digression from our main concerns. A proof that involves a minimum of topological machinery is available in [$DS_1$, pp. 468-470], and a more topological proof can be found in [Du, pp. 340 and 341].

The space $\mathbb{R}^n$ itself does not have the fixed point property. For example, if $x_o \in \mathbb{R}^n$, $x_o \neq 0$, then the mapping $\varphi : \mathbb{R}^n \to \mathbb{R}^n$, defined by $\varphi(x) = x + x_o$ for $x \in \mathbb{R}^n$, is clearly continuous, but has no fixed point.

Furthermore, even though the sets $B_1^n$ in $\mathbb{R}^n$ are closed bounded convex sets that have the fixed point property, it is not the case that such sets in an arbitrary Banach space have the fixed point property. For instance, consider the Banach space $(C^R(\mathbb{R}), \|\cdot\|_\infty)$ over $\mathbb{R}$ and let

$$K = \{f \mid f \in C^R(\mathbb{R}),\ \|f\|_\infty \leq 1,\ \lim_{t \to +\infty} f(t) = 1,\ \lim_{t \to -\infty} f(t) = -1\}.$$

It is easily seen that $K$ is a closed bounded convex subset of $C^R(\mathbb{R})$, but we claim that $K$ does not have the fixed point property.

To see this, consider the mapping $T : C^R(\mathbb{R}) \to C^R(\mathbb{R})$, defined by $T(f)(t) = f(t - 1)$, $t \in \mathbb{R}$. Clearly $T$ is continuous, and if $f \in K$, then

$$\begin{aligned} \|T(f)\|_\infty &= \sup_{t \in \mathbb{R}} |f(t - 1)| \\ &= \sup_{t \in \mathbb{R}} |f(t)| \\ &= \|f\|_\infty \end{aligned}$$

and

$$\lim_{t \to +\infty} T(f)(t) = \lim_{t \to +\infty} f(t - 1) = 1,$$

$$\lim_{t \to -\infty} T(f)(t) = \lim_{t \to -\infty} f(t - 1) = -1.$$

Thus $T$ restricted to $K$ is a continuous mapping from $K$ to $K$. However, $T$ has no fixed point, as can be seen from the following argument: Suppose $f_o \in K$ were such that $T(f_o) = f_o$; that is, $f_o(t - 1) = f_o(t)$, $t \in \mathbb{R}$. Then, in particular, we would have $f_o(0) = f_o(k)$, $k = \pm 1, \pm 2, \pm 3, \ldots$, which contradicts the facts that $\lim_{k \to +\infty} f_o(k) = 1$ and $\lim_{k \to -\infty} f_o(k) = -1$, as $f_o \in K$. Consequently $K$ does not have the fixed point property. Note also that $T$ is even linear.

In the preceding example $K$ fails to have the fixed point property because it is not a compact set, as compact convex sets in

Banach spaces all have the fixed point property. This is an immediate consequence of the next theorem.

Theorem 12.1.2 (Schauder-Tikhonov Theorem). Let $(V,\mathcal{T})$ be a locally convex topological linear space over $\Phi$. If $K \subset V$ is a compact convex set, then $K$ has the fixed point property.

The proof of this result is also omitted, since it depends on the Brouwer Fixed Point Theorem. Proofs are available in [$DS_1$, pp. 453-456; $E_1$, pp. 161-163].

Our approach to fixed points will be rather the reverse of that indicated in Theorems 12.1.1 and 12.1.2. Instead of looking for conditions on sets $K$ ensuring that every continuous mapping $\varphi$ of $K$ into itself has a fixed point -- that is, ensuring that $K$ has the fixed point property -- we shall seek conditions on the mappings $\varphi$ that will ensure their having fixed points.

## 12.2. Contraction Mappings.

The theorem to be proved in this section is really a topological result and does not use any of the techniques of functional analysis. However, its applications (e.g. to proofs of the existence and uniqueness of solutions to differential equations) are so typical of fixed point theorems that inclusion of the theorem seems desirable. The Schauder-Tikhonov Theorem also has important applications in the theory of differential equations, but, as indicated, its proof is not so accessible (see, for example, [$E_1$, pp. 164-166]).

Definition 12.2.1. Let $(V,\rho)$ be a metric space and suppose $T : V \to V$. Then $T$ is said to be a contraction if there exists some $b$, $0 \leq b < 1$, such that $\rho(T(x),T(y)) \leq b\rho(x,y)$ for $x,y \in V$.

For example, if $(V,\rho) = (\ell_2,\|\cdot\|_2)$, where $\rho(x,y) = \|x - y\|_2$, $x,y \in \ell_2$, then $T(x) = T(\{a_k\}) = \{a_k/2\}$, where $x = \{a_k\} \in \ell_2$, is clearly a contraction.

<u>Theorem</u> <u>12.2.1</u>. Let $(V,\rho)$ be a complete metric space. If $T : V \to V$ is a contraction, then $T$ has a unique fixed point.

<u>Proof</u>. Let $b$, $0 \leq b < 1$, be such that $\rho(T(x),T(y)) \leq b\rho(x,y)$, $x,y \in V$. If $b = 0$, then the theorem is trivial. So suppose $0 < b < 1$. Then we see at once that for each positive integer $k$

$$\rho(T^k(x),T^k(y)) \leq b\rho(T^{k-1}(x),T^{k-1}(y)) \leq \dots \leq b^k\rho(x,y)$$

for $x,y \in V$. If $x \in V$, we define $x_k = T^k(x)$, $k = 1,2,3,\dots$ . We claim that $\{x_k\}$ is a Cauchy sequence in $(V,\rho)$. Indeed, we see that for any positive integers n and m

$$\begin{aligned}\rho(x_{n+m},x_n) &\leq \rho(x_{n+m},x_{n+m-1}) + \dots + \rho(x_{n+1},x_n)\\ &= \rho(T^{n+m}(x),T^{n+m-1}(x)) + \dots + \rho(T^{n+1}(x),T^n(x))\\ &= \rho(T^{n+m-1}[T(x)],T^{n+m-1}(x)) + \dots + \rho(T^n[T(x)],T^n(x))\\ &\leq b^{n+m-1}\rho(T(x),x) + \dots + b^n\rho(T(x),x)\\ &= b^n[b^{m-1} + \dots + b + 1]\rho(T(x),x)\\ &\leq \left(\frac{b^n}{1-b}\right)\rho(T(x),x).\end{aligned}$$

It is then apparent that $\{x_k\}$ is a Cauchy sequence, as $\lim_n b^n = 0$ since $0 < b < 1$.

Let $x_o \in V$ be such that $\lim_k \rho(x_k,x_o) = 0$. Clearly $\lim_k \rho(T(x_k),T(x_o)) = 0$, as $\rho(T(x_k),T(x_o)) \leq b\rho(x_k,x_o)$. Moreover, since $T(x_k) = x_{k+1}$, $k = 1,2,3,\dots$ , we see that $\{T(x_k)\}$ is a subsequence of the convergent sequence $\{x_k\}$, and so

$$\lim_k \rho(T(x_k),x_o) = 0.$$

These observations and the estimates

$$\rho(T(x_o),x_o) \leq \rho(T(x_o),T(x_k)) + \rho(T(x_k),x_o) \qquad (k = 1,2,3,\dots)$$

reveal that $T(x_o) = x_o$, and so $T$ has a fixed point.

Furthermore, suppose $y_o \in V$ is such that $T(y_o) = y_o$. Then $\rho(T(x_o),T(y_o)) \leq b\rho(x_o,y_o)$ shows immediately that $\rho(x_o,y_o) = 0$, as $0 < b < 1$.

Therefore the fixed point $x_o$ is unique, and the proof is complete.

□

Note that the proof also indicates how one can actually compute the fixed point of a contraction mapping: one finds the limit of the sequence of iterates $\{T^k(x)\}$ for any $x \in V$. This can be quite useful in practice.

## 12.3. The Markov-Kakutani Fixed Point Theorem.

One consequence of the Markov-Kakutani Fixed Point Theorem is that every continuous linear transformation $T$ on a topological linear space to itself has a fixed point in any compact convex set $K$ such that $T(K) \subset K$. The assumption of compactness cannot be removed, as is seen by the example discussed in Section 12.1. The theorem, however, says more than the previous assertion, as we shall see presently.

First, we need a definition and a lemma.

Definition 12.3.1. Let $V$ be a linear space over $\Phi$ and suppose $T : V \to V$. Then $T$ is said to be *affine* if

$$T[ax + (1 - a)y] = aT(x) + (1 - a)T(y) \qquad (x,y \in V;\ 0 \leq a \leq 1).$$

Clearly any $T \in L'(V)$ is an affine mapping.

Recalling the definition of a bounded set in a topological linear space (Definition 2.4.1), it is not difficult to establish the following lemma:

Lemma 12.3.1. Let $(V,\mathcal{T})$ be a topological linear space. If $K \subset V$ is compact, then $K$ is bounded.

Proof. Let $U$ be an open neighborhood of the origin in $V$. Since scalar multiplication is a continuous mapping from $\Phi \times V$ to $V$, we see that there exist some $\delta > 0$ and some open neighborhood $W$ of the origin in $V$ such that $bW \subset U$, $|b| \leq \delta$. Moreover, for each $x \in V$ there exists some positive integer $k$ such that $x/k \in W$, again by the continuity of scalar multiplication. Consequently $V \subset \bigcup_{k=1}^{\infty} kW$. Then, in particular, $K \subset \bigcup_{k=1}^{\infty} kW$, and so there exists some $m$ such that $K \subset \bigcup_{k=1}^{m} kW$, as $K$ is compact.

Let $\epsilon = \delta/m$. If $n$ is any positive integer such that $1/n < \epsilon$, then $k/n < \epsilon k = \delta k/m \leq \delta$, $k = 1,2,3,\ldots,m$, and so

$$(\frac{1}{n})K \subset \bigcup_{k=1}^{m} (\frac{k}{n})W \subset U.$$

Thus $K \subset nU$, and $K$ is bounded. □

Theorem 12.3.1 (Markov-Kakutani Fixed Point Theorem). Let $(V,\mathcal{T})$ be a topological linear space over $\Phi$ and let $K \subset V$ be a nonempty compact convex set. Suppose $G$ is a family of continuous affine mappings from $V$ to $V$ such that

(i) $TS = ST$ $\qquad (T,S \in G)$.

(ii) $T(K) \subset K$ $\qquad (T \in G)$.

Then there exists some $x_0 \in K$ for which $T(x_0) = x_0$, $T \in G$.

Proof. For each positive integer $n$ and each $T \in G$ we define

$$T_n = \frac{1}{n} \sum_{k=0}^{n-1} T^k$$

where $T^0 = I$, the identity mapping from $V$ to $V$. From the convexity of $K$ and the assumption that $T(K) \subset K$ it follows at once that $T_n(K) \subset K$, $n = 1,2,3,\ldots$, $T \in G$. Moreover, since each $T \in G$ is continuous, it follows that $T_n(K)$ is a compact, and hence closed, subset of $K$. Furthermore, if $n$ and $m$ are positive integers and $T,S \in G$, then $T_nS_m(x) = T_n[S_m(x)] \in T_n(K)$, as $S_m(x) \in K$, and since $TS = ST$, we also have $T_nS_m(x) = S_mT_n(x) \in S_m(K)$, and so $T_nS_m(K) \subset T_n(K) \cap S_m(K)$. This last observation allows us to conclude that the family of sets $E = \{T_n(K) \mid n = 1,2,3,\ldots, T \in G\}$ is a family of closed subsets of the compact set $K$ that has the finite intersection property. Hence $\cap_{T_n(K) \in E} T_n(K) \neq \phi$.

Suppose $x_o \in \cap_{T_n(K) \in E} T_n(K)$. We claim $x_o$ is a fixed point for each $T \in G$ by the following argument: Suppose this is not the case. Then there exists some $T \in G$ such that $T(x_o) \neq x_o$. Since the topology $T$ is Hausdorff there exists an open neighborhood $U$ of the origin in $V$ such that $T(x_o) - x_o \notin U$. Now $x_o$ belongs to all the sets in $E$, and so, in particular, $x_o \in T_n(K)$, $n = 1,2,3,\ldots$ . Thus for each positive integer $n$ there exists some $y_n \in K$ such that

$$x_o = \frac{1}{n} \sum_{k=0}^{n-1} T^k(y_n)$$

and hence, $T$ being affine

$$T(x_o) - x_o = T\left[\frac{1}{n} \sum_{k=0}^{n-1} T^k(y_n)\right] - \frac{1}{n} \sum_{k=0}^{n-1} T^k(y_n)$$

$$= \frac{1}{n} \sum_{k=1}^{n} T^k(y_n) - \frac{1}{n} \sum_{k=0}^{n-1} T^k(y_n)$$

$$= \frac{T^n(y_n) - y_n}{n} .$$

Consequently $T(x_o) - x_o \in (\frac{1}{n})(K - K)$, as $T^n(y_n) \in K$ and $y_n \in K$, $n = 1,2,3,\ldots$ .

This, however, implies that $K - K \not\subset nU$, $n = 1,2,3,\ldots$, because if there were some positive integer $n$ for which $K - K \subset nU$, then

$(\frac{1}{n})(K - K) \subset U$, and so $T(x_o) - x_o \in U$, contrary to the choice of $U$.

But $K$ is compact, and so by the continuity of addition in $(V,\mathcal{T})$ the set $K - K$ must also be compact. Thus by Lemma 12.3.1 there exists some positive integer $n$ such that $K - K \subset nU$. This contradiction of the previous conclusion allows us to conclude that $x_o$ is indeed a fixed point for each $T \in \mathcal{G}$. □

The result indicated at the beginning of the section is now an immediate corollary.

Corollary 12.3.1. Let $(V,\mathcal{T})$ be a topological linear space over $\Phi$ and let $K \subset V$ be a nonempty compact convex set. If $T : V \to V$ is a continuous affine mapping such that $T(K) \subset K$, then there exists some $x_o \in K$ for which $T(x_o) = x_o$.

## 12.4. The Picard Existence Theorem for Ordinary Differential Equations.

We now wish to give a typical application of fixed point theorems to the question of the existence and uniqueness of solutions to ordinary differential equations. The fixed point theorem we use is the one concerning contraction mappings on complete metric spaces.

Consider then the initial value problem

$$y'(t) = f[t,y(t)]$$

$$y(t_o) = y_o,$$

where $y'$ denotes the derivative of $y$ with respect to $t$. By a solution to the initial value problem we mean some function $x = x(t)$ that is defined on some closed interval $[t_o - \delta, t_o + \delta]$, $\delta > 0$, for which $x'(t) = f[t,x(t)]$, $t_o - \delta \leq t \leq t_o + \delta$, and $x(t_o) = y_o$. The idea behind the use of the fixed point theorem to prove the existence of a solution to the initial value problem is to convert the latter into an equivalent problem whose solution is a fixed point of some contraction mapping. The means for doing this is provided by the next lemma, whose elementary proof is left to the reader.

<u>Lemma 12.4.1</u>. Suppose $f$ is a continuous real-valued function defined on a suitable domain in $\mathbb{R}^2$ and $\delta > 0$. Then the following are equivalent:

(i) $y$ is a differentiable function defined on $[t_o - \delta, t_o + \delta]$ such that

$$y'(t) = f[t,y(t)] \qquad (t_o - \delta \le t \le t_o + \delta)$$
$$y(t_o) = y_o.$$

(ii) $y$ is a continuous function defined on $[t_o - \delta, t_o + \delta]$ such that

$$y(t) = y_o + \int_{t_o}^{t} f[s,y(s)]\, ds \qquad (t_o - \delta \le t \le t_o + \delta).$$

The precise domain of definition of $f$ has purposely been left vague. This defect will be remedied in the next theorem.

The lemma tells us that $x$ is a solution to the initial value problem if and only if $T(x) = x$, where

$$T(y)(t) = y_o + \int_{t_o}^{t} f[s,y(s)]\, ds.$$

Thus to apply our theorem about contraction mappings we must be able to introduce an appropriate complete metric space such that the $T$ indicated here is a contraction on that space. This necessitates some further restrictions on $f$.

<u>Theorem 12.4.1</u> (<u>Picard Existence Theorem</u>). Let $a > 0, b > 0$, and $(t_o,y_o) \in \mathbb{R}^2$. Set $D = \{(t,s) \mid |t - t_o| \le a, |s - y_o| \le b\}$ and suppose $f : D \to \mathbb{R}$ is such that

(i) $f$ is continuous.

(ii) There exists some $M > 0$ such that

$$|f(t,s_1) - f(t,s_2)| \le M|s_1 - s_2|$$

whenever $|t - t_o| \le a$ and $|s_k - y_o| \le b, k = 1,2.$

If $M_o = \sup_{(t,s) \in D} |f(t,s)|$ and $\delta = \min(a, b/M_o, 1/2M)$, then there exists a unique differentiable function $x$ defined on $[t_o - \delta, t_o + \delta]$ such that

$$x'(t) = f[t,x(t)] \qquad (t_o - \delta \leq t \leq t_o + \delta)$$
$$x(t_o) = y_o.$$

Proof. Consider the subset $V$ of the metrice space $(C^R([t_o - \delta, t_o + \delta]), \|\cdot\|_\infty)$ defined by:

$$V = \{y \mid y \in C^R([t_o - \delta, t_o + \delta]),\ |y(t) - y_o| \leq M_o\delta,\ t_o - \delta \leq t \leq t_o + \delta\}.$$

It is easily seen that $V$ is a closed subspace of the complete metric space $(C^R([t_o - \delta, t_o + \delta]), \|\cdot\|_\infty)$, and so $(V, \|\cdot\|_\infty)$ is itself a complete metric space. For each $y \in V$ we define

$$T(y)(t) = y_o + \int_{t_o}^{t} f[s,y(s)]\, ds \qquad (t_o - \delta \leq t \leq t_o + \delta).$$

We claim that $T$ is a contraction mapping from $V$ to $V$.

First, if $y \in V$, then $|y(s) - y_o| \leq b$, $s \in [t_o - \delta, t_o + \delta]$, as $|y(s) - y_o| \leq M_o\delta \leq b$. Hence $(s,y(s)) \in D$, $s \in [t_o - \delta, t_o + \delta]$, $y \in V$, and so $T(y) \in C^R([t_o - \delta, t_o + \delta])$. Moreover,

$$\begin{aligned} |T(y)(t) - y_o| &= \left|\int_{t_o}^{t} f[s,y(s)]\, ds\right| \\ &\leq M_o|t - t_o| \\ &\leq M_o\delta \qquad (t \in [t_o - \delta, t_o + \delta];\ y \in V), \end{aligned}$$

and so $T : V \to V$.

To see that $T$ is a contraction, suppose $y,z \in V$. Then

$$\begin{aligned} \rho(T(y),T(z)) &= \|T(y) - T(z)\|_\infty \\ &= \sup_{t_o - \delta \leq t \leq t_o + \delta} \left|\int_{t_o}^{t} f[s,y(s)]\, ds - \int_{t_o}^{t} f[s,z(s)]\, ds\right| \end{aligned}$$

$$\leq \sup_{t_o - \delta \leq t \leq t_o + \delta} \left| \int_{t_o}^{t} |f[s,y(s)] - f[s,z(s)]| \, ds \right|$$

$$\leq \sup_{t_o - \delta \leq t \leq t_o + \delta} \left| \int_{t_o}^{t} M|y(s) - z(s)| \, ds \right|$$

$$\leq \sup_{t_o - \delta \leq t \leq t_o + \delta} M\left[ \sup_{t_o - \delta \leq s \leq t_o + \delta} |y(s) - z(s)| \, |t - t_o| \right]$$

$$\leq M\delta \|y - z\|_\infty$$

$$= M\delta\rho(y,z).$$

However, by the choice of $\delta$ we have $M\delta < 1$, and so $T$ is a contraction.

Therefore, appealing to Theorem 12.2.1, we deduce the existence of a unique $x \in V$ such that $T(x) = x$; that is, in view of Lemma 12.4.1, a unique $x \in V$ such that $x'(t) = f[t,x(t)]$, $t_o - \delta \leq t \leq t_o + \delta$, and $x(t_o) = y_o$. □

Condition (ii) imposed on $f$ in the hypotheses of the theorem is known as a <u>Lipschitz condition</u>. It is necessary in order to ensure that the solution to the initial value problem is unique. It is, however, not necessary for the existence of a solution, which can be established under the assumption that $f$ is continuous. This result can also be obtained by means of a fixed point theorem, this time the Schauder-Tikhonov Theorem (Theorem 12.1.2), as is done in [$E_1$, pp.164 and 165]. A proof with a more classical flavor can be found, for example, in [Hu, pp.10-12].

As indicated after Theorem 12.2.1, the solution to the initial value problem, when $f$ is a continuous function that satisfies a Lipschitz condition, can be approximated by the sequence of iterates $\{T^k(x_o)\}$, where $x_o$ is any element of $V$. It is an instructive exercise to carry this out for some simple initial value problem. The interested reader might do this, for example, with

$$y'(t) = t + y(t)$$
$$y(0) = 0,$$

where $x_o(t) = 0$ for all t.

12.5. Haar Measure on Compact Abelian Topological Groups. To begin this section we introduce the notion of a topological group.

Definition 12.5.1. A topological group G is an algebraic group G equipped with a Hausdorff topology $\mathcal{T}$ such that the following mappings are continuous:

(i) The mapping from $G \times G$, with the product topology, to G, defined by $(s,t) \to s + t$, $s,t \in G$.

(ii) The mapping from G to G, defined by $s \to -s$, $s \in G$.

A topological group G is said to be Abelian if G is an Abelian algebraic group; it is said to be (locally) compact if the topological space $(G,\mathcal{T})$ is (locally) compact.

We have used additive notation for the group operation since we shall ultimately restrict our attention to Abelian groups. It should also be noted that not all authors require that the topology on a topological group be Hausdorff, and indeed much of the theory of topological groups remains valid without this assumption (see, for example, [$HR_1$]).

Some examples of topological groups are, $\mathbb{R}^1$, with the usual Euclidean topology; $\mathbb{Z}$, with the discrete topology; $\Gamma = \{z \mid z \in \mathbb{C}, |z| = 1\}$, with the relative topology inherited from the complex plane $\mathbb{C}$ and multiplication of complex numbers as the group operation; the collection $GL(n,\mathbb{C})$ of all $n \times n$ invertible complex matrices and the collection $U(n,\mathbb{C})$ of all $n \times n$ unitary complex matrices, with matrix multiplication as the group operation and the relative topology on $GL(n,\mathbb{C})$ and $U(n,\mathbb{C})$ considered as subsets of $\mathbb{C}^{n^2}$. All of these are locally compact topological groups. The groups $\mathbb{R}^1$, $\mathbb{Z}$, and $\Gamma$ are Abelian, but $GL(n,\mathbb{C})$ and $U(n,\mathbb{C})$ are not; $\mathbb{R}^1$, $\mathbb{Z}$, and $GL(n,\mathbb{C})$ are noncompact, whereas $\Gamma$ and $U(n,\mathbb{C})$ are compact.

The existence of regular Borel measures $\mu$ on a locally compact topological group $G$ that are invariant under right translation -- that is, such that $\mu(E + s) = \mu(E)$ for each $s \in G$ and each Borel set $E$ -- is a result of fundamental importance. Abstract harmonic analysis and the theory of representations for topological groups, for example, would be unthinkable without such measures. Although we are not in a position to establish the existence of such a measure on an arbitrary locally compact topological group, we can use the Markov-Kakutani Fixed Point Theorem to give a rather simple existence proof for the case of compact Abelian groups.

Theorem 12.5.1. Let $G$ be a compact Abelian topological group. Then there exists a unique $\mu \in M(G)$ such that

(i) $\mu$ is a positive measure.
(ii) $\mu(G) = 1$.
(iii) If $E \subset G$ is a Borel set and $s \in G$, then $\mu(E + s) = \mu(E)$.

Proof. Consider the Banach space $(C(G), \|\cdot\|_\infty)$ and let $K$ denote the subset of $C(G)^*$ consisting of all those $x^* \in C(G)^*$ such that

(a) $x^*(f) \geq 0$ whenever $f \in C(G)$ and $f(t) \geq 0$, $t \in G$.
(b) $x^*(f_o) = 1$, where $f_o(t) = 1$, $t \in G$.
(c) $\|x^*\| = 1$.

Evidently $K \neq \emptyset$, as the continuous linear functionals $x_t^*(f) = f(t)$, $f \in C(G)$, belong to $K$ for each $t \in G$. Moreover, it is easily seen that $K$ is norm bounded, weak* closed, and convex. Consequently by the Banach-Alaoglu Theorem (Theorem 9.4.1) $K$ is a nonempty compact convex subset of $(C(G)^*, T^{w^*})$.

For each $s \in G$ define $\tau_s f(t) = f(t - s)$, $t \in G$ and $f \in C(G)$. It is then evident that $\tau_s \in L(C(G))$, $s \in G$. Let $\tau_s^* : C(G)^* \to C(G)^*$ denote the adjoint transformation of $\tau_s$:

$$\tau_s^*(x^*)(f) = x^*(\tau_s f) \qquad (f \in C(G)).$$

Thus $\tau_s^*$ is linear, and from the definition of the weak* topology, it is immediately apparent that each $\tau_s^*$, $s \in G$, is weak* continuous. Moreover, since $\tau_s\tau_t(f) = \tau_{s+t}(f) = \tau_{t+s}(f) = \tau_t\tau_s(f)$ for $f \in C(G)$ and $s,t \in G$, the group $G$ being Abelian, we conclude that $\tau_s^*\tau_t^* = \tau_t^*\tau_s^*$, $s,t \in G$. Finally, we claim that $\tau_s^*(K) \subset K$, $s \in G$.

Indeed, if $x^* \in K$ and $s \in G$, then for $f \in C(G)$ and $f(t) \geq 0$, $t \in G$, we have $\tau_s^*(x^*)(f) = x^*(\tau_s f) \geq 0$, as $\tau_s f(t) = f(t - s) \geq 0$, $t \in G$. Furthermore,

$$\tau_s^*(x^*)(f_o) = x^*(\tau_s f_o) = x^*(f_o) = 1$$

and

$$\begin{aligned} |\tau_s^*(x^*)(f)| &= |x^*(\tau_s f)| \\ &\leq \|x^*\|\|\tau_s f\|_\infty \\ &= \|f\|_\infty \qquad (f \in C(G)), \end{aligned}$$

which shows that $\|\tau_s^*(x^*)\| = 1$. Hence $\tau_s^*(x^*) \in K$ whenever $x^* \in K$.

Consequently $\mathcal{G} = \{\tau_s^* \mid s \in G\}$ satisfies all the hypotheses of the Markov-Kakutani Fixed Point Theorem (Theorem 12.3.1), and so there exists some $x_o^* \in K$ such that $\tau_s^*(x_o^*) = x_o^*$, $s \in G$.

From the Riesz Representation Theorem (Theorem 4.8.2) we then deduce the existence of a unique $\mu \in M(G)$ such that

$$x_o^*(f) = \int_G f(t)\, d\mu(t) \qquad (f \in C(G)).$$

Combining this last observation with a standard approximation argument we obtain for each Borel set $E \subset G$ and each $s \in G$

$$\begin{aligned}
\mu(E + s) &= \int_G \chi_{E+s}(t)\, d\mu(t) \\
&= \int_G \tau_s \chi_E(t)\, d\mu(t) \\
&= \int_G \chi_E(t)\, d\mu(t) \\
&= \mu(E).
\end{aligned}$$

Thus $\mu \in M(G)$ satisfies properties (i) through (iii).

To see that $\mu$ is unique, suppose that $\sigma \in M(G)$ also satisfies properties (i) through (iii). Then by Fubini's Theorem [Ry, p.269] we find that for each $f \in C(G)$,

$$\begin{aligned}
\int_G\int_G f(s + t)\, d\mu(s) d\sigma(t) &= \int_G [\int_G \tau_{-t} f(s)\, d\mu(s)] d\sigma(t) \\
&= \int_G [\int_G f(s)\, d\mu(s)] d\sigma(t) \\
&= \int_G f(s)\, d\mu(s) \int_G d\sigma(t) \\
&= \int_G f(s)\, d\mu(s)
\end{aligned}$$

and similarly that

$$\begin{aligned}
\int_G\int_G f(s + t)\, d\mu(s) d\sigma(t) &= \int_G\int_G f(s + t)\, d\sigma(t) d\mu(s) \\
&= \int_G f(t)\, d\sigma(t).
\end{aligned}$$

Hence

$$\int_G f(t)\, d(\mu - \sigma)(t) = 0 \qquad (f \in C(G)),$$

and so by the Riesz Representation Theorem (Theorem 4.8.2) we conclude that $\mu = \sigma$.

Therefore $\mu$ is the unique measure that satisfies properties (i) through (iii). □

The measure $\mu$, whose existence is assured by Theorem 12.5.1, is called the Haar measure on the compact Abelian group G. The uniqueness of the Haar measure, as defined here, is due to the requirement that $\mu(G) = 1$. Without this restriction the measure $\mu$ is only unique up to a nonnegative multiplicative constant. General discussions of Haar measure on locally compact topological groups are available, for example, in [Ba, pp.125-187; $E_1$, pp.247-254; $HR_1$, pp.184-230; Lo, pp.112-120; Nb, pp.49-119; We, pp.33-42].

The Markov-Kakutani Fixed Point Theorem can also be used to reprove some of our previous results (for example, Theorem 4.3.1 and the existence of Banach limits). The details are left to the reader. The latter application is also an example of the use of the Markov-Kakutani Fixed Point Theorem in the study of so called invariant means on topological groups, a subject that we do not pursue here. The interested reader is referred to [$E_1$, pp.159-161; Gn, pp.230-261].

## 12.6. Problems.

*1. In $(\ell_2, \|\cdot\|_2)$ let H be the set of all sequences $\{x_n\}$ for which $|x_n| \leq 1/n$, $n = 1,2,3,\ldots$ . H is called the Hilbert cube.

(a) Prove that H is compact and convex.

(b) Use the Brouwer Fixed Point Theorem (Theorem 12.1.1) to prove that H has the fixed point property.

2. Let $(V,\rho) = ([0,1/3], |\cdot|)$ and let $T : V \to V$ be the mapping $T(x) = x^2$. Prove that T is a contraction on V.

3. Let $(X,\mathcal{T})$ be a topological space and $Y \subset X$ a subspace of $X$. $Y$ is said to be a <u>retract</u> of $X$ if there exists a continuous map $f : X \to Y$ such that $f(y) = y$, $y \in Y$. Prove that, if $X$ has the fixed point property and $Y$ is a retract of $X$, then $Y$ has the fixed point property.

4. For fixed $n$, define

$$\rho(x,y) = \max_{1 \le k \le n} |x_k - y_k|$$

for $x = (x_1, x_2, \ldots, x_n)$ and $y = (y_1, y_2, \ldots, y_n)$ in $\mathbb{R}^n$. Let $T : \mathbb{R}^n \to \mathbb{R}^n$ be the mapping defined by $T(x) = y$ where

$$y_k = \sum_{j=1}^{n} a_{kj} x_j + b_k \qquad (1 \le k \le n).$$

Prove that $T$ is a contraction on $(\mathbb{R}^n, \rho)$ if and only if

$$\max_{1 \le k \le n} \Big[ \sum_{j=1}^{n} |a_{kj}| \Big] < 1.$$

*5. Let $(a_{kj})$, $k,j = 1,2,3,\ldots$, be an infinite matrix such that $\sup_k [\sum_{j=1}^{\infty} |a_{kj}|] < 1$. If $\{b_k\} \in \ell_\infty$, prove that the system of equations

$$x_k = \sum_{j=1}^{\infty} a_{kj} x_j + b_k \qquad (k = 1,2,3,\ldots)$$

has a unique solution $\{x_k\} \in \ell_\infty$.

6. Let $(V,\rho)$ be a complete metric space and let $T : V \to V$. If for some positive integer $n \ge 1$, $T^n$ is a contraction, prove that $T$ has a unique fixed point.

7. Let $(V,\rho) = (C([a,b]), \|\cdot\|_\infty)$, where $0 < b - a < \infty$, and let $T : V \to V$ be defined by

$$T(f)(t) = \int_a^t f(s)\, ds \qquad (a \le t \le b).$$

(a) Prove that $T$ is not a contraction if $b - a > 1$.

(b) Use Problem 6 to prove that $T$ has a unique fixed point. What is the fixed point?

8. (Corollary 12.3.1) Let $(V,\mathcal{T})$ be a topological linear space over $\Phi$ and let $K \subset V$ be a nonempty compact convex set. Prove that, if $T : V \to V$ is a continuous affine mapping such that $T(K) \subset K$, then there exists some $x_0 \in K$ such that $T(x_0) = x_0$.

9. Let $(\ell_\infty^R, \|\cdot\|_\infty)$ be the Banach space of bounded real sequences. Use the Markov-Kakutani Fixed Point Theorem to prove the existence of Banach limits for $\ell_\infty^R$; that is, prove the existence of a continuous linear functional $x^* \in (\ell_\infty^R)^*$ such that

(i) $x^*(\{a_n\}) \geq 0$ whenever $a_n \geq 0$, $n = 1,2,3,\ldots$ .
(ii) $x^*(\{1\}) = 1$, where $\{1\} = \{1,1,1,\ldots\}$.
(iii) $\|x^*\| \leq 1$.
(iv) $x^*(\{b_n\}) = x^*(\{a_n\})$, where $b_n = a_{n+1}$, $n = 1,2,3,\ldots$ .

10. (Lemma 12.4.1) Suppose $f$ is a continuous real-valued function defined on a suitable domain in $\mathbb{R}^2$ and $\delta > 0$. Prove that the following are equivalent:

(i) $y$ is a differentiable function defined on $[t_0 - \delta, t_0 + \delta]$ such that

$$y'(t) = f[t,y(t)] \qquad (t_0 - \delta \leq t \leq t_0 + \delta)$$
$$y(t_0) = y_0.$$

(ii) $y$ is a continuous function defined on $[t_0 - \delta, t_0 + \delta]$ such that

$$y(t) = y_0 + \int_{t_0}^{t} f[s,y(s)]\, ds \qquad (t_0 - \delta \leq t \leq t_0 + \delta).$$

11. Prove that the Lipschitz condition on $f$ in the hypotheses of the Picard Existence Theorem (Theorem 12.4.1) is necessary to ensure the uniqueness of the solution to the initial value problem. (Hint: Consider the initial value problem

$$y'(t) = y^{2/3}(t)$$
$$y(0) = 0.$$

Exhibit two solutions and show that $f(x,y) = y^{2/3}$ does not satisfy a Lipschitz condition in any rectangle containing $(0,0)$.)

12. Solve the following initial value problems:

(a) $y'(t) = t + y(t)$
$y(0) = 0.$

(b) $y'(t) = t - y(t)$
$y(0) = 1.$

13. Let $(V,\rho)$ be a complete metric space. A mapping $T : V \to V$ is called <u>contractive</u> if $\rho(T(x),T(y)) < \rho(x,y)$ for $x,y \in V$. Clearly every contraction is contractive.

(a) Let $V = \{x \mid x \in \mathbb{R}, x \geq 1\}$, with the usual Euclidean metric. Let $T : V \to V$ be defined by $T(x) = x + 1/x$. Prove that $T$ is contractive, but $T$ is not a contraction.

(b) Prove that $T$ does not have a fixed point. Since $V$ is a complete metric space, conclude that Theorem 12.2.1 does not generalize to contractive mappings.

*14. Let $(V,\rho)$ be a complete metric space and let $T : V \to V$ be a contractive mapping, that is, $\rho(T(x),T(y)) < \rho(x,y)$ for $x,y \in V$. Suppose there is a point $x_1 \in V$ such that the sequence $\{T^n(x_1)\}$ contains a subsequence converging to a point $x_0 \in V$. Prove that $x_0$ is the unique fixed point of $T$. Thus we have an analog to Theorem 12.2.1 that will guarantee a unique fixed point for a contractive mapping.

15. Let $(V,\rho)$ be a complete metric space and let $T : V \to V$ be a contractive mapping, that is, $\rho(T(x),T(y)) < \rho(x,y)$ for $x,y \in V$. Suppose $T(V)$ is contained in a compact subset of $V$.

Prove that $T$ has a unique fixed point $x_0$ and that $x_0 = \lim_n T^n(x)$ for every $x \in V$.

*16. Let $(V,\|\cdot\|)$ be a reflexive Banach space over $\Phi$ and let $T : V \to V$ be a mapping. Let $I : V \to V$ denote the identity mapping $I(x) = x$, $x \in V$ and prove the following:

(a) If $T$ is linear and $\|T\| < 1$, then $I + T$ maps $V$ onto itself.

(b) Suppose that $T$ is not necessarily linear, but that $T$ is weakly continuous and that

$$\limsup_{\|x\| \to \infty} \frac{\|T(x)\|}{\|x\|} < 1.$$

Then prove that $I + T$ maps $V$ onto itself.

*17. Let $(V,\|\cdot\|)$ be a Banach space over $\Phi$. Let $z_0 \in V$ be fixed and let $T \in L(V)$ be such that $1 + \Sigma_{n=1}^{\infty}\|T^n\| < \infty$. Let $S$ be the transformation defined by

$$S(x) = z_0 + T(x) \qquad (x \in V).$$

Prove that $S$ has a unique fixed point $x_0$ which is given by

$$x_0 = z_0 + \sum_{n=1}^{\infty} S^n(z_0).$$

(This is called the Volterra Fixed Point Theorem. Note that $x_0$ is the solution to the functional equation $x = z_0 + T(x)$, or formally $x = [I - T]^{-1}(z_0)$, so that $I - T$ has an inverse when $T$ satisfies the conditions above.)

18. Let $(V,\|\cdot\|) = (C([a,b]),\|\cdot\|_\infty)$ and let $T : V \to V$ be defined by

$$T(f)(t) = \int_a^t f(s)\, ds \qquad (a \leq t \leq b).$$

Let $S$ be the transformation

$$S(f)(t) = g(t) + T(f)(t) \qquad (a \leq t \leq b)$$

for a fixed $g \in V$. Use Problem 17 to prove that $S$ has a unique fixed point $f_o$ where

$$f_o(t) = g(t) + \int_a^t e^{t-s} g(s)\, ds \qquad (a \leq t \leq b).$$

19. Let $G$ be a compact Abelian topological group and let $\mu \in M(G)$ be the Haar measure on $G$ whose existence is proved in Theorem 12.5.1. Prove that for every Borel set $E \subset G$, $\mu(-E) = \mu(E)$.

20. Let $G$ be a compact Abelian topological group, let $\mu \in M(G)$ be a Haar measure on $G$ and prove the following:

(a) If $U \subset G$ is open, then $\mu(U) > 0$.

(b) The group $G$ cannot contain an uncountable number of pairwise disjoint open subsets.

(c) If $U \subset G$ is open and $\{t_k\}$ is a countable set of distinct points in $G$, then there exists some $j$ and $k$, $j \neq k$, such that $(t_j + U) \cap (t_k + U) \neq \phi$.

# CHAPTER 13

# HILBERT SPACES

13.0. Introduction. Our concern in this last chapter is Hilbert spaces. The theory of Hilbert spaces is an extremely rich and well-developed area of mathematics, and we shall only be able to give a very modest introduction to the subject, but nevertheless an introduction that we hope will provide the reader with a knowledge of some of the fundamental results of the theory and convince him of the utility and attractiveness of further study.

We shall begin with several sections devoted to some basic facts about Hilbert spaces, part of which can be deduced immediately from the development of the preceding chapters. Included in these sections are such results as the Cauchy-Schwarz Inequality and the Parallelogram and Polarization Identities. Next we shall turn our attention to several of the theorems that make Hilbert spaces rather unique: the Orthogonal Decomposition Theorem, which says that every closed linear subspace of a Hilbert space is an orthogonal direct summand; the Riesz Representation Theorem for the continuous linear functionals on a Hilbert space, which says that all such functionals are obtained from elements of the original space by means of the inner product; and the existence of orthonormal bases in a Hilbert space.

These results, together with some elementary facts about self-adjoint and unitary continuous linear transformations on Hilbert spaces, will then be used to study Fourier analysis in $L_2([-\pi,\pi],dt/2\pi)$, to obtain some results from the theory of probability and random series, to prove some theorems in ergodic theory, and to generalize a classic theorem in the theory of analytic functions to a certain subspace of $L_2([-\pi,\pi],dt/2\pi)$.

The final three sections of the chapter deal with the spectral theory of continuous linear transformations on Hilbert spaces, culminating in a spectral decomposition theorem for compact self-adjoint transformations that generalizes a standard result of linear algebra concerning the diagonalization of matrices. We shall develop only enough general spectral theory to enable us to establish the spectral decomposition theorem.

## 13.1. Basic Definitions and Results.

We begin with the definition of a bilinear form on a linear space.

Definition 13.1.1. Let $V$ be a linear space over $\Phi$. A mapping $\psi : V \times V \to \Phi$ is said to be a bilinear form on $V$ if

(i) $\psi(x,y+z) = \psi(x,y) + \psi(x,z)$,

(ii) $\psi(x+y,z) = \psi(x,z) + \psi(y,z)$,

(iii) $\psi(ax,y) = a\psi(x,y)$,

(iv) $\psi(x,ay) = \bar{a}\psi(x,y)$ $\qquad (x,y,z \in V;\ a \in \Phi)$.

A bilinear form $\psi$ on $V$ is said to be nonnegative if $\psi(x,x) \geq 0$, $x \in V$; to be positive definite if it is nonnegative and $\psi(x,x) = 0$ implies $x = 0$; and to be symmetric if $\psi(x,y) = \overline{\psi(y,x)}$, $x,y \in V$.

A bilinear form $\psi$ is thus a mapping from $V \times V$ to $\Phi$ that is linear in its first argument, linear in its second argument if $\Phi = \mathbb{R}$, and antilinear in its second argument if $\Phi = \mathbb{C}$. Also we note that for any bilinear form $\psi$ we have $\psi(x,x) = 0$ when $x = 0$.

We can now give a definition of an inner product space that is shorter than that given in Example 1.2.10.

Definition 13.1.2. Let $V$ be a linear space over $\Phi$. If $\psi$ is a symmetric, positive definite, bilinear form on $V$, then $\psi$ is said to be an inner product on $V$, and the pair $(V,\psi)$ is said to be an inner product space.

In general, given an inner product space $(V,\psi)$, we write $\psi(x,y) = \langle x,y\rangle$, $x,y \in V$, and denote the space itself by the pair $(V,\langle\cdot,\cdot\rangle)$.

Besides the examples of inner product spaces cited in Example 1.2.10, we mention here the spaces $V = \ell_2$ with the inner product

$$\langle\{a_k\},\{b_k\}\rangle = \sum_{k=1}^{\infty} a_k\overline{b}_k \qquad (\{a_k\},\{b_k\} \in \ell_2)$$

and $V = L_2(X,S,\mu)$ with the inner product

$$\langle f,g\rangle = \int_X f(t)\overline{g(t)}\,d\mu(t) \qquad (f,g \in L_2(X,S,\mu)).$$

The verifications are left to the reader.

As indicated previously, an inner product space $(V,\langle\cdot,\cdot\rangle)$ can be made into a normed linear space on setting $\|x\| = \langle x,x\rangle^{1/2}$, $x \in V$. In order to prove this, however, we need a preliminary result of considerable usefulness.

Theorem 13.1.1 (Cauchy-Schwarz Inequality). Let $(V,\langle\cdot,\cdot\rangle)$ be an inner product space over $\Phi$. Then

$$|\langle x,y\rangle| \leq \langle x,x\rangle^{1/2}\langle y,y\rangle^{1/2} \qquad (x,y \in V).$$

Moreover, given $x,y \in V$,

$$|\langle x,y\rangle| = \langle x,x\rangle^{1/2}\langle y,y\rangle^{1/2}$$

if and only if $x$ and $y$ are linearly dependent.

Proof. Given $x,y \in V$, we claim that there exists some $a \in \Phi$ such that

$$|\langle x,y\rangle|^2 = \langle x,x\rangle\langle y,y\rangle - \langle y - ax,y - ax\rangle\langle x,x\rangle.$$

Indeed, if $x = 0$, then the equation is easily seen to hold with $a = 0$. If $x \neq 0$, then $\langle x,x\rangle \neq 0$ and we set $a = \langle y,x\rangle/\langle x,x\rangle$. Then

$$\begin{aligned}\langle y - ax, y - ax\rangle &= \langle y,y\rangle - a\langle x,y\rangle - \bar{a}\langle y,x\rangle + a\bar{a}\langle x,x\rangle \\ &= \langle y,y\rangle - \frac{|\langle x,y\rangle|^2}{\langle x,x\rangle} - \frac{|\langle x,y\rangle|^2}{\langle x,x\rangle} + \frac{|\langle x,y\rangle|^2}{\langle x,x\rangle} \\ &= \langle y,y\rangle - \frac{|\langle x,y\rangle|^2}{\langle x,x\rangle},\end{aligned}$$

which is equivalent to the desired equation.

The conclusions of the theorem follow immediately from the validity of the preceding equation. □

It should be noted that the inequality portion of the theorem remains valid, although with a different proof, if we assume only that $\langle\cdot,\cdot\rangle$ is a symmetric nonnegative bilinear form. The positive definiteness of the inner product is, however, necessary for the second assertion of the theorem. We shall make use of this observation in Section 13.14.

As a corollary we obtain the fact that inner product spaces can be normed.

Corollary 13.1.1. Let $(V,\langle\cdot,\cdot\rangle)$ be an inner product space over $\Phi$ and set $\|x\| = \langle x,x\rangle^{1/2}$, $x \in V$. Then $(V,\|\cdot\|)$ is a normed linear space over $\Phi$.

Proof. It is evident that $\|ax\| = |a|\,\|x\|$, $x \in V$, $a \in \Phi$, and $\|x\| = 0$ implies $x = 0$. Moreover, if $x,y \in V$, then, appealing to the Cauchy-Schwarz Inequality, we see that

$$\begin{aligned}\|x + y\|^2 &= \langle x + y, x + y\rangle \\ &= \langle x,x\rangle + \langle y,x\rangle + \langle x,y\rangle + \langle y,y\rangle\end{aligned}$$

$$= \|x\|^2 + 2\mathrm{Re}\langle x,y\rangle + \|y\|^2$$

$$\leq \|x\|^2 + 2|\langle x,y\rangle| + \|y\|^2$$

$$\leq \|x\|^2 + 2\|x\|\|y\| + \|y\|^2$$

$$= (\|x\| + \|y\|)^2.$$

Thus $\|x + y\| \leq \|x\| + \|y\|$, $x,y \in V$, and so $\|\cdot\|$ is a norm. □

We repeat the next definition from Section 1.2 for the sake of completeness.

Definition 13.1.3. Let $(V,\langle\cdot,\cdot\rangle)$ be an inner product space over $\Phi$. If $(V,\|\cdot\|)$ is a Banach space over $\Phi$, then $V$ is said to be a Hilbert space.

Of course, the norm here, and in all analogous situations, is $\|x\| = \langle x,x\rangle^{1/2}$, $x \in V$. Among the examples of inner product spaces we have mentioned previously, $\mathbb{R}^n, \mathbb{C}^n, \ell_2$, and $L_2(X,\mathcal{S},\mu)$ are all Hilbert spaces.

One further corollary of the Cauchy-Schwarz Inequality will be useful.

Corollary 13.1.2. Let $(V,\langle\cdot,\cdot\rangle)$ be an inner product space over $\Phi$. Then the inner product is a continuous $\Phi$-valued function on $V \times V$, where $V$ is taken with its norm topology determined by the inner product.

Proof. If $x_o,y_o \in V$, then

$$|\langle x,y\rangle - \langle x_o,y_o\rangle| \leq |\langle x,y\rangle - \langle x,y_o\rangle| + |\langle x,y_o\rangle - \langle x_o,y_o\rangle|$$

$$= |\langle x,y - y_o\rangle| + |\langle x - x_o,y_o\rangle|$$

$$\leq \|x\|\|y - y_o\| + \|x - x_o\|\|y_o\|.$$

□

13.2. The Parallelogram and Polarization Identities. In this section we examine two fundamental identities in inner product spaces: the Parallelogram and Polarization Identities. We shall use the former identity to give a characterization of those normed linear spaces that are inner product spaces.

Geometrically the Parallelogram Identity says that the sum of the squares of the lengths of a parallelogram's diagonals is equal to the sum of the squares of the lengths of its sides. Precisely, we have the following theorem:

Theorem 13.2.1 (Parallelogram Identity). Let $(V,\langle\cdot,\cdot\rangle)$ be an inner product space over $\Phi$. Then

$$\|x + y\|^2 + \|x - y\|^2 = 2\|x\|^2 + 2\|y\|^2 \qquad (x,y \in V).$$

Proof. For any $x,y \in V$ we have

$$\begin{aligned}\|x + y\|^2 + \|x - y\|^2 &= \langle x + y,x + y\rangle + \langle x - y,x - y\rangle \\ &= \langle x,x\rangle + \langle y,x\rangle + \langle x,y\rangle + \langle y,y\rangle \\ &\quad + \langle x,x\rangle - \langle y,x\rangle - \langle x,y\rangle + \langle y,y\rangle \\ &= 2\|x\|^2 + 2\|y\|^2.\end{aligned}$$

□

A similar direct computation also establishes the Polarization Identity, which allows one to express the inner product in terms of the norm.

Theorem 13.2.2 (Polarization Identity). Let $(V,\langle\cdot,\cdot\rangle)$ be an inner product space over $\Phi$. If $\Phi = \mathbb{R}$, then

$$\langle x,y\rangle = \left\|\frac{x + y}{2}\right\|^2 - \left\|\frac{x - y}{2}\right\|^2 \qquad (x,y \in V),$$

whereas if $\Phi = \mathbb{C}$, then

$$\langle x,y\rangle = \left\|\frac{x + y}{2}\right\|^2 - \left\|\frac{x - y}{2}\right\|^2 + i\left\|\frac{x + iy}{2}\right\|^2 - i\left\|\frac{x - iy}{2}\right\|^2 \quad (x,y \in V).$$

The form of the Polarization Identity suggests a means of introducing an inner product into a normed linear space. The next result asserts that such an attempt will succeed provided the norm satisfies the Parallelogram Identity.

Theorem 13.2.3. Let $(V,\|\cdot\|)$ be a normed linear space over $\Phi$. If

$$\|x + y\|^2 + \|x - y\|^2 = 2\|x\|^2 + 2\|y\|^2 \qquad (x,y \in V),$$

then there exists an inner product $\langle\cdot,\cdot\rangle$ on $V$ such that $\|x\| = \langle x,x\rangle^{1/2}$, $x \in V$.

Proof. We shall consider the cases $\Phi = \mathbb{R}$ and $\Phi = \mathbb{C}$ separately. Suppose first that $\Phi = \mathbb{R}$ and define $\langle\cdot,\cdot\rangle$ on $V \times V$ by

$$\langle x,y\rangle = \left\|\frac{x + y}{2}\right\|^2 - \left\|\frac{x - y}{2}\right\|^2 \qquad (x,y \in V).$$

We claim that $\langle\cdot,\cdot\rangle$ is an inner product on $V$ such that $\|x\| = \langle x,x\rangle^{1/2}$.

The latter assertion is evident, since $\langle x,x\rangle = \|(x + x)/2\|^2$, $x \in V$. It thus follows that $\langle x,x\rangle \geq 0$, $x \in V$, and $\langle x,x\rangle = 0$ if and only if $x = 0$. Hence $\langle\cdot,\cdot\rangle$ is positive definite. Furthermore, given $x,y \in V$, we see that

$$\begin{aligned}\langle y,x\rangle &= \left\|\frac{y + x}{2}\right\|^2 - \left\|\frac{y - x}{2}\right\|^2 \\ &= \left\|\frac{x + y}{2}\right\|^2 - \left\|\frac{x - y}{2}\right\|^2 \\ &= \langle x,y\rangle,\end{aligned}$$

and so $\langle\cdot,\cdot\rangle$ is symmetric. The proof of the linearity of $\langle\cdot,\cdot\rangle$ requires somewhat more work.

Consider any $x,y,z \in V$. Then by the validity of the Parallelogram Identity in $V$ we have

$$2\|x + y\|^2 + 2\|z + y\|^2 = \|x + z + 2y\|^2 + \|x - z\|^2$$

$$2\|x - y\|^2 + 2\|z - y\|^2 = \|x + z - 2y\|^2 + \|x - z\|^2,$$

whence, on subtracting the lower expression from the upper, we obtain

$$2(\|x + y\|^2 - \|x - y\|^2) + 2(\|z + y\|^2 - \|z - y\|^2)$$
$$= \|x + z + 2y\|^2 - \|x + z - 2y\|^2.$$

Recalling the definition of $\langle\cdot,\cdot\rangle$, we deduce at once that

$$\langle x,y\rangle + \langle z,y\rangle = \tfrac{1}{2}\langle x + z, 2y\rangle \qquad (x,y,z \in V).$$

In particular, if $z = 0$, then

$$\langle x,y\rangle = \tfrac{1}{2}\langle x,2y\rangle \qquad (x,y \in V).$$

Hence

$$\langle u + v,y\rangle = \tfrac{1}{2}\langle u + v,2y\rangle$$
$$= \langle u,y\rangle + \langle v,y\rangle \qquad (u,v,y \in V),$$

and so $\langle\cdot,\cdot\rangle$ is additive in its first argument. Since $\langle x,y\rangle = \langle y,x\rangle$, $x,y \in V$, it follows that $\langle\cdot,\cdot\rangle$ is additive in both arguments.

Finally, we must show that $\langle ax,y\rangle = a\langle x,y\rangle$ for $x,y \in V$ and $a \in \mathbb{R}$. Since $\langle\cdot,\cdot\rangle$ is additive, it is obvious that $\langle nx,y\rangle = n\langle x,y\rangle$, $x,y \in V$, for any integer $n$, and hence, if $n \neq 0$,

$$n\langle \tfrac{x}{n},y\rangle = \langle \tfrac{nx}{n},y\rangle = \langle x,y\rangle \qquad (x,y \in V),$$

that is,

$$\langle \tfrac{x}{n},y\rangle = \tfrac{1}{n}\langle x,y\rangle \qquad (x,y \in V).$$

From these facts we deduce at once that for any rational number $r$

$$\langle rx,y\rangle = r\langle x,y\rangle \qquad (x,y \in V).$$

Now suppose $a \in \mathbb{R}$ and let $\{r_k\}$ be a sequence of rationals such that $\lim_k r_k = a$. Then, since $\langle\cdot,\cdot\rangle$ is continuous, being defined in terms of the norm on $V$, we see that

$$\begin{aligned}\langle ax,y\rangle &= \lim_k \langle r_k x,y\rangle \\ &= \lim_k r_k\langle x,y\rangle \\ &= a\langle x,y\rangle \qquad (x,y \in V)\end{aligned}$$

because $\lim_k \|r_k x - ax\| = 0$. The identity $\langle x,ay\rangle = a\langle x,y\rangle$, $x,y \in V$ and $a \in \mathbb{R}$, again follows, as $\langle x,y\rangle = \langle y,x\rangle$. Thus $\langle\cdot,\cdot\rangle$ is a bilinear form on $V$.

Consequently $\langle\cdot,\cdot\rangle$ is a symmetric positive definite bilinear form, that is, an inner product, on $V$, and the theorem is proved when $\Phi = \mathbb{R}$.

Now suppose $\Phi = \mathbb{C}$. In this case we define $\langle\cdot,\cdot\rangle$ as

$$\langle x,y\rangle = \langle x,y\rangle_R + i\langle x,iy\rangle_R \qquad (x,y \in V),$$

where

$$\langle x,y\rangle_R = \left\|\frac{x+y}{2}\right\|^2 - \left\|\frac{x-y}{2}\right\|^2 \qquad (x,y \in V).$$

As is easily seen, $\langle\cdot,\cdot\rangle$ is so defined because

$$\langle x,y\rangle = \left\|\frac{x+y}{2}\right\|^2 - \left\|\frac{x-y}{2}\right\|^2 + i\left\|\frac{x+iy}{2}\right\|^2 - i\left\|\frac{x-iy}{2}\right\|^2 \quad (x,y \in V),$$

that is, the Polarization Identity is valid.

We must still, however, show that $\langle\cdot,\cdot\rangle$ is an inner product. In the following arguments it is well to keep in mind that $\langle\cdot,\cdot\rangle_R$ is an inner product over $\mathbb{R}$, as established in the first part of the proof. Now some elementary computations reveal that

$$\langle x, ix \rangle_R = 0 \qquad (x \in V)$$

and

$$\langle ix, iy \rangle_R = \langle x, y \rangle_R \qquad (x, y \in V).$$

The first of these two identities shows at once that

$$\begin{aligned} \langle x, x \rangle &= \langle x, x \rangle_R + i\langle x, ix \rangle_R \\ &= \langle x, x \rangle_R \\ &= \|x\|^2 \qquad (x \in V) \end{aligned}$$

and that $\langle x, x \rangle = 0$ if and only if $x = 0$. Thus $\langle \cdot, \cdot \rangle$ is positive definite.

Next we claim that $\langle \cdot, \cdot \rangle$ is symmetric. Indeed, utilizing the second identity, we obtain

$$\begin{aligned} \langle y, x \rangle &= \langle y, x \rangle_R + i\langle y, ix \rangle_R \\ &= \langle y, x \rangle_R + i\langle iy, -x \rangle_R \\ &= \langle x, y \rangle_R - i\langle x, iy \rangle_R \\ &= \overline{\langle x, y \rangle} \qquad (x, y \in V). \end{aligned}$$

The fact that $\langle \cdot, \cdot \rangle$ is additive in both its arguments follows at once from the additivity of $\langle \cdot, \cdot \rangle_R$. Hence to show that $\langle \cdot, \cdot \rangle$ is bilinear we need only prove, in view of the symmetry of $\langle \cdot, \cdot \rangle$, that $\langle ax, y \rangle = a\langle x, y \rangle$, $x, y \in V$ and $a \in \mathbb{C}$. First we note that

$$\begin{aligned} \langle ix, y \rangle &= \langle ix, y \rangle_R + i\langle ix, iy \rangle_R \\ &= \langle ix, y \rangle_R + i\langle x, y \rangle_R \\ &= i[\langle x, y \rangle_R - i\langle ix, y \rangle_R] \end{aligned}$$

$$= i[\langle x,y\rangle_R - i\langle -x,iy\rangle_R]$$

$$= i[\langle x,y\rangle_R + i\langle x,iy\rangle_R]$$

$$= i\langle x,y\rangle \qquad (x,y \in V).$$

Consequently, if $a = c + id \in \mathbb{C}$, we have

$$\begin{aligned}\langle ax,y\rangle &= \langle (c + id)x,y\rangle \\ &= \langle cx + idx,y\rangle \\ &= \langle cx,y\rangle + \langle idx,y\rangle \\ &= \langle cx,y\rangle + i\langle dx,y\rangle \\ &= c\langle x,y\rangle + id\langle x,y\rangle \\ &= a\langle x,y\rangle \qquad (x,y \in V),\end{aligned}$$

where the penultimate equality is valid because $\langle\cdot,\cdot\rangle_R$ is a real bilinear form.

Thus we have shown that $\langle\cdot,\cdot\rangle$ is an inner product on $V$ for which $\langle x,x\rangle = \|x\|^2$, $x \in V$, thereby completing the proof of the theorem. □

Theorem 13.2.3 thus asserts that a normed linear space is an inner product space precisely when the Parallelogram Identity is valid for all elements of the space. This fact also provides us at once with the next corollary.

Corollary 13.2.1. Let $(V,\|\cdot\|)$ be a normed linear space over $\Phi$. If every two-dimensional linear subspace of $V$ is an inner product space over $\Phi$, then $V$ is an inner product space over $\Phi$.

Proof. From the hypothesis of the corollary and Theorem 13.2.1 we see that for each $x,y \in V$

$$\|x + y\|^2 + \|x - y\|^2 = 2\|x\|^2 + 2\|y\|^2. \qquad \square$$

<u>13.3. Some Other General Properties of Hilbert Spaces.</u> In this section we wish to note some properties of Hilbert spaces that follow immediately from the general theory we have developed in the preceding chapters. A preliminary and necessary step, however, is to prove that inner product spaces are uniformly convex (Definition 8.2.1).

<u>Theorem 13.3.1.</u> Let $(V,\langle\cdot,\cdot\rangle)$ be an inner product space over $\Phi$. If $\|x\| = \langle x,x\rangle^{1/2}$, $x \in V$, then $(V,\|\cdot\|)$ is a uniformly convex normed linear space.

<u>Proof.</u> Suppose $0 < \varepsilon < 1$ and let $\delta > 0$ be such that $1 - \delta = (1 - \varepsilon^2/4)^{1/2}$. From the Parallelogram Identity (Theorem 13.2.1) we see that, if $x,y \in V$, $\|x\| \leq 1$, $\|y\| \leq 1$, and $\|x - y\| \geq \varepsilon$, then

$$\begin{aligned}\left\|\frac{x + y}{2}\right\|^2 &= 2\left\|\frac{x}{2}\right\|^2 + 2\left\|\frac{y}{2}\right\|^2 - \left\|\frac{x - y}{2}\right\|^2 \\ &\leq \frac{1}{2} + \frac{1}{2} - \frac{\varepsilon^2}{4} \\ &= (1 - \delta)^2.\end{aligned}$$

Thus $(V,\|\cdot\|)$ is uniformly convex. □

Mil'man's Theorem (Theorem 8.2.1) now provides us with the following corollary:

<u>Corollary 13.3.1.</u> Let $(V,\langle\cdot,\cdot\rangle)$ be a Hilbert space over $\Phi$. Then $(V,\|\cdot\|)$ is a reflexive Banach space where $\|x\| = \langle x,x\rangle^{1/2}$, $x \in V$.

We now can list a sampling of results about Hilbert spaces that are immediate consequences of the preceding theorem and corollary, as well as the material of the preceding chapters.

Let $(V,\langle\cdot,\cdot\rangle)$ be a Hilbert space over $\Phi$. Then the following are valid:

(1) If $K \subset V$ is a nonempty closed convex set, then there

exists a unique $x_o \in K$ such that

$$\|x_o\| = \inf_{x \in K} \|x\|.$$

If $y_o \notin K$, then there exists a unique $x_o \in K$ such that

$$\|y_o - x_o\| = \inf_{x \in K} \|y_o - x\|.$$

(Theorem 8.2.2 and Corollary 8.2.1).

(2) If $x^* \in V^*$ and $x^* \neq 0$, then there exists a unique $x_o \in V$, $\|x_o\| = 1$, such that $x^*(x_o) = \|x^*\|$ (Corollary 8.2.2).

(3) $(V, T^W)$ is sequentially complete (Theorem 9.3.1(iv)).

(4) $E \subset V$ is compact in $(V, T^W)$ if and only if $E$ is a closed set in $(V, T^W)$ and a bounded set in $(V, \|\cdot\|)$. In particular, $B_1 = \{x \mid x \in V, \|x\| \leq 1\}$ is weakly compact and weakly sequentially compact (Corollary 9.4.2 and Theorem 10.3.1).

(5) If $(V, \|\cdot\|)$ is separable, then the weak topology $T^W$ restricted to $B_1$ is metrizable (Corollary 9.4.6).

(6) $(V, \|\cdot\|)$ is separable if and only if $(V^*, \|\cdot\|)$ is separable (Theorem 4.5.1).

(7) If $K \subset V$ is convex, then $K$ is closed in $(V, T^W)$ if and only if $K \cap bB_1$ is closed in $(V, T^W)$ for each $b \geq 0$ (Corollary 10.2.1).

(8) If $K \subset V$ is a nonempty convex set that is compact in $(V, T^W)$, then $\text{ext}(K) \neq \emptyset$ and $K = \overline{\text{co}}[\text{ext}(K)]$, where the closure can be taken in either the weak topology $T^W$ or in the norm topology. In particular, $\text{ext}(B_1) \neq \emptyset$ and $B_1 = \overline{\text{co}}[\text{ext}(B_1)]$ (Theorems 11.2.1 and 9.2.2).

In the next section we shall discuss some results that are more peculiar to Hilbert spaces.

<u>13.4. The Orthogonal Decomposition Theorem and the Riesz Representation Theorem.</u> We have two aims in this section: to show that every closed linear subspace of a Hilbert space $V$ is an orthogonal direct summand and to use this result to prove that $V$ can be identified not only with its second dual space $V^{**}$ but also with $V^*$. These results are generalizations to Hilbert spaces of well-known theorems of linear algebra.

We begin with some definitions.

<u>Definition 13.4.1.</u> Let $V$ be a linear space over $\Phi$ and suppose $W_1$ and $W_2$ are linear subspaces of $V$. Then $V$ is said to be the <u>direct sum</u> of $W_1$ and $W_2$ if

(i) For each $x \in V$ there exist $y \in W_1$ and $z \in W_2$ such that $x = y + z$.

(ii) $W_1 \cap W_2 = \{0\}$.

In this case we write $V = W_1 \oplus W_2$.

It follows easily from the definition that, if $V = W_1 \oplus W_2$, then the decomposition $x = y + z$ is unique. One could also use this observation as the definition of the direct sum.

<u>Definition 13.4.2.</u> Let $(V, \langle\cdot,\cdot\rangle)$ be an inner product space over $\Phi$. Then $x, y \in V$ are said to be <u>orthogonal</u>, denoted by $x \perp y$, if $\langle x,y\rangle = 0$. Two subsets $E_1, E_2 \subset V$ are said to be <u>orthogonal</u>, denoted by $E_1 \perp E_2$, if $\langle x,y\rangle = 0$ whenever $x \in E_1$ and $y \in E_2$.

It is evident that, if $E_1$ and $E_2$ are orthogonal, then $E_1 \cap E_2 \subset \{0\}$.

A criterion for orthogonality is given in the following proposition.

Proposition 13.4.1. Let $(V,\langle\cdot,\cdot\rangle)$ be an inner product space over $\Phi$ and let $x,y \in V$.

(i) If $\Phi = \mathbb{R}$, then $x \perp y$ if and only if

$$\|x + y\|^2 = \|x\|^2 + \|y\|^2.$$

(ii) If $\Phi = \mathbb{C}$, then $x \perp y$ if and only if

$$\|x + y\|^2 = \|x\|^2 + \|y\|^2$$

and

$$\|x + iy\|^2 = \|x\|^2 + \|y\|^2.$$

Proof. The necessity portion of each of the two equivalences is easily established and is left to the reader. Conversely, if $\Phi = \mathbb{R}$ and $\|x + y\|^2 = \|x\|^2 + \|y\|^2$, then routine computation shows that

$$\langle y,x\rangle + \langle x,y\rangle = 2\langle x,y\rangle = 0,$$

whereas if $\Phi = \mathbb{C}$, $\|x + y\|^2 = \|x\|^2 + \|y\|^2$, and $\|x + iy\|^2 = \|x\|^2 + \|y\|^2$, then similar calculations reveal that

$$\begin{aligned}\langle y,x\rangle + \langle x,y\rangle &= \overline{\langle x,y\rangle} + \langle x,y\rangle \\ &= 2\mathrm{Re}\langle x,y\rangle \\ &= 0\end{aligned}$$

and

$$\begin{aligned}i\langle y,x\rangle - i\langle x,y\rangle &= -i[\langle x,y\rangle - \overline{\langle x,y\rangle}] \\ &= -i[2i\mathrm{Im}\langle x,y\rangle] \\ &= 2\mathrm{Im}\langle x,y\rangle \\ &= 0.\end{aligned}$$

Thus $\mathrm{Re}\langle x,y\rangle = \mathrm{Im}\langle x,y\rangle = 0$, and so in either case $x \perp y$. □

Definition 13.4.3. Let $(V,\langle\cdot,\cdot\rangle)$ be an inner product space over $\Phi$. If $E \subset V$, then the orthogonal complement $E^{\perp}$ of $E$ is defined by

$$E^{\perp} = \{x \mid x \in V,\ x \perp E\}.$$

The correspondence of the notation for orthogonal complements and for annihilators of subsets of topological linear spaces (Definition 4.6.1) is not accidental. Indeed, an immediate consequence of the Riesz Representation Theorem (Theorem 13.4.2) is that the two notions are identical in a Hilbert space. In view of this observation the next proposition should come as no surprise. The proof is left to the reader.

Proposition 13.4.2. Let $(V,\langle\cdot,\cdot\rangle)$ be an inner product space over $\Phi$ and let $E \subset V$. Then

(i) $E^{\perp}$ is a closed linear subspace of $V$.
(ii) $E^{\perp\perp} = (E^{\perp})^{\perp} \supset E$.
(iii) If $E$ is a linear subspace, then $E \cap E^{\perp} = \{0\}$.

We turn next to the Orthogonal Decomposition Theorem.

Theorem 13.4.1 (Orthogonal Decomposition Theorem). Let $(V,\langle\cdot,\cdot\rangle)$ be a Hilbert space over $\Phi$. If $W \subset V$ is a closed linear subspace, then $V = W \oplus W^{\perp}$.

Proof. Since $W \cap W^{\perp} = \{0\}$, it suffices to show that each $x_o \in V$ can be expressed as $x_o = y_o + z_o$, where $y_o \in W$ and $z_o \in W^{\perp}$. But, given $x_o \in V$, it is evident that $x_o - W$ is a nonempty closed convex subset of $V$, and hence the results of the preceding section (or Theorem 8.2.2) reveal the existence of a unique $z_o \in x_o - W$ such that

$$\|z_o\| = \inf_{x \in x_o - W} \|x\|.$$

Setting $y_o = x_o - z_o$, we see at once that $y_o \in W$ and $x_o = y_o + z_o$. It only remains to show that $z_o \in W^{\perp}$.

However, recalling the proof of the Cauchy-Schwarz Inequality (Theorem 13.1.1), we know that for each $x \in W$ there exists some $a \in \Phi$ such that

$$|\langle x,z_o\rangle|^2 = \|x\|^2\|z_o\|^2 - \|z_o - ax\|^2\|x\|^2.$$

Moreover, $z_o - ax = x_o - y_o - ax \in x_o - W$ and so $\|z_o - ax\| \geq \|z_o\|$ by the choice of $z_o$. Thus we deduce that

$$\begin{aligned}|\langle x,z_o\rangle|^2 &\leq \|x\|^2\|z_o\|^2 - \|z_o\|^2\|x\|^2\\ &= 0,\end{aligned}$$

and hence $\langle x,z_o\rangle = 0$.

Consequently $z_o \in W^{\perp}$, and $V = W \oplus W^{\perp}$. □

It should be observed that the hypotheses of the preceding theorem cannot be weakened; that is, the theorem may fail if either $W$ is not closed or $V$ is not a complete normed linear space. For example, suppose $V = \ell_2$ with the usual inner product

$$\langle\{a_k\},\{b_k\}\rangle = \sum_{k=1}^{\infty} a_k\overline{b}_k \qquad (\{a_k\},\{b_k\} \in \ell_2).$$

Then $\ell_2$ is a Hilbert space. Let $W$ be the linear subspace of $\ell_2$ consisting of such sequences of complex numbers $\{a_k\}$ that $a_k \neq 0$ for only finitely many $k$. Clearly $W$ is not closed.

Furthermore, we claim that $W^{\perp} = \{0\}$. Indeed, suppose $\{b_k\} \in W^{\perp}$ and, given a positive integer $n$, let $\{a_k^n\}$ be such that $a_k^n = 0$, $k \neq n$, and $a_n^n = 1$. Then $\{a_k^n\} \in W$, and

$$0 = \langle\{a_k^n\},\{b_k\}\rangle = \overline{b}_n.$$

Consequently $\{b_k\} = 0$ and hence $W \oplus W^{\perp} \neq \ell_2$.

For an example demonstrating that $V$ must be complete to ensure the validity of the Orthogonal Decomposition Theorem we take $V$ equal to $W$ of the preceding example. That is, $V$ is the inner product space consisting of all sequences with only a finite number of nonzero terms. As already mentioned, $V$ is not a Hilbert space. For $W$ we take the set of all $\{a_k\} \in V$ such that $\Sigma_{k=1}^{\infty} a_k/k = 0$. It is easily verified that $W$ is a linear subspace of $V$ such that $W \neq \{0\}$ and $W \neq V$. Moreover, $W$ is closed, as we see from the following: Suppose $\{\{a_k^{(n)}\}\} \subset W$ is a sequence that converges to $\{a_k\} \in V$ and let $K$ be a positive integer for which $a_k = 0, k > K$. Then for any $M > K$ and any $n$ we have

$$\begin{aligned}
\left|\sum_{k=1}^{\infty} a_k/k\right| &= \left|\sum_{k=1}^{M} a_k/k\right| \\
&\leq \left|\sum_{k=1}^{M} (a_k - a_k^{(n)})/k\right| + \left|\sum_{k=1}^{M} a_k^{(n)}/k\right| \\
&\leq \left[\sum_{k=1}^{M} 1/k^2\right]^{1/2}\left[\sum_{k=1}^{M} |a_k - a_k^{(n)}|^2\right]^{1/2} + \left|\sum_{k=1}^{M} a_k^{(n)}/k\right| \\
&\leq \left(\frac{\pi^2}{6}\right)^{1/2}\|\{a_k\} - \{a_k^{(n)}\}\|_2 + \left|\sum_{k=1}^{M} a_k^{(n)}/k\right|.
\end{aligned}$$

Thus, if $\epsilon > 0$ is given, let $N_1$ be a positive integer such that, for $n \geq N_1$,

$$\|\{a_k\} - \{a_k^{(n)}\}\|_2 < \frac{\epsilon}{2}\left(\frac{6}{\pi^2}\right)^{1/2}$$

and then choose such an $N_2$ that, if $M > N_2$, then

$$\left|\sum_{k=1}^{M} a_k^{(N_1)}/k\right| < \frac{\epsilon}{2}.$$

The last estimate is possible because $\{a_k^{(N_1)}\} \in W$. Combining these three estimates, we conclude that for any $M > \max(N_2, K)$

$$\left|\sum_{k=1}^{\infty} a_k/k\right| = \left|\sum_{k=1}^{M} a_k/k\right|$$

$$< \frac{\epsilon}{2} + \frac{\epsilon}{2}$$

$$= \epsilon .$$

Since $\epsilon$ is arbitrary, it follows that $\Sigma_{k=1}^{\infty} a_k/k = 0$; that is, $\{a_k\} \in W$. Hence $W$ is a closed linear subspace of $V$.

We claim, however, that $W^{\perp} = \{0\}$. Indeed, suppose $\{b_k\} \in W^{\perp}$. then, given a positive integer $n$, let $\{c_k^n\}$ denote the sequence defined by $c_k^n = 0, k \neq n$ and $k \neq n + 1, c_n^n = n, c_{n+1}^n = -(n + 1)$. Clearly $\{c_k^n\} \in W$, and

$$0 = \langle \{c_k^n\}, \{b_k\} \rangle = \sum_{k=1}^{\infty} c_k^n \overline{b}_k$$

$$= n\overline{b}_n - (n + 1)\overline{b}_{n+1}.$$

Thus we see that $b_{n+1} = [n/(n + 1)]b_n, n = 1,2,3,\ldots,$ and so, if $b_1 \neq 0$, then $b_n \neq 0, n = 1,2,3,\ldots$ . This, however, contradicts the fact that $\{b_k\} \in W^{\perp} \subset V$.

Consequently $W^{\perp} = \{0\}$, and $W \oplus W^{\perp} = W \neq V$.

It is readily apparent from the Orthogonal Decomposition Theorem that, given a closed linear subspace $W$ of a Hilbert space $V$, there exists precisely one linear subspace $X$ of $V$ such that $V = W \oplus X$ and $W \perp X$, namely, $X = W^{\perp}$, and that this subspace $X$ is itself closed (Proposition 13.4.2). If, however, we drop the orthogonality requirement, then there may exist many linear subspaces $X$ for which $V = W \oplus X$. For instance, if $V = \mathbb{R}^2$ and $W = \{(x,0) \mid x \in \mathbb{R}\}$, then any straight line $X$ that passes through the origin and is distinct from the real axis is a linear subspace of $\mathbb{R}^2$ such that $\mathbb{R}^2 = W \oplus X$. In this case $W^{\perp}$ is, of course, the set $\{(0,y) \mid y \in \mathbb{R}\}$.

Let us now see how the Orthogonal Decomposition Theorem provides us with a description of the continuous linear functionals on a

Hilbert space. First, suppose that $(V, \langle\cdot,\cdot\rangle)$ is an inner product space and let $x \in V$. Then define $x^*$ on $V$ by

$$x^*(y) = \langle y,x\rangle \qquad (y \in V).$$

Since the inner product is linear in its first argument, we see at once that $x^* \in V'$, and the Cauchy-Schwarz Inequality (Theorem 13.1.1) asserts that

$$|x^*(y)| = |\langle y,x\rangle| \le \|y\|\|x\| \qquad (y \in V).$$

Hence $x^* \in V^*$ and $\|x^*\| \le \|x\|$. Actually $\|x^*\| = \|x\|$ since $x^*(x) = \langle x,x\rangle = \|x\|^2$, $x \in V$. In this way we obtain an isometric mapping from $V$ into $V^*$. The mapping, however, is not linear since

$$\begin{aligned} ax^*(y) &= a\langle y,x\rangle \\ &= \langle y,\bar{a}x\rangle \\ &= (\bar{a}x)^*(y) \qquad (y \in V;\ a \in \Phi). \end{aligned}$$

The mapping is easily seen to be antilinear.

The preceding development shows that the correspondence $x \to x^*$ is an antilinear isometry from $V$ into $V^*$ when $V$ is an inner product space. The important point is that the mapping is surjective if $V$ is a Hilbert space.

Theorem 13.4.2 (Riesz Representation Theorem). Let $(V, \langle\cdot,\cdot\rangle)$ be a Hilbert space over $\Phi$. Then the following are equivalent:

(i) $x^* \in V^*$.

(ii) There exists a unique $x \in V$ such that $x^*(y) = \langle y,x\rangle$, $y \in V$.

Moreover, the mapping $x \to x^*$ from $V$ to $V^*$, defined by the equation $x^*(y) = \langle y,x\rangle$, $y \in V$, is a surjective antilinear isometry from $V$ to $V^*$.

<u>Proof</u>. Evidently the theorem will be proved if we can show that part(i) implies part(ii), as the remaining assertions are apparent from the previous discussion. So let $x^* \in V^*$.

If $x^* = 0$ and $x = 0$, then $x^*(y) = \langle y,x\rangle$, $y \in V$. Furthermore, $x = 0$ is the only such element of $V$ since $0 = x^*(x) = \langle x,x\rangle = \|x\|^2$.

Suppose then that $x^* \neq 0$ and let $W = N(x^*)$, the kernel of $x^*$. From Theorem 3.3.2 we see that $W$ is a proper closed linear subspace of $V$, and so, by the Orthogonal Decomposition Theorem, we have $V = W \oplus W^\perp$. Since $W^\perp \neq \{0\}$, there exists some $x_o \in W^\perp$, $x_o \neq 0$. Clearly $x^*(x_o) \neq 0$, as $x_o \notin N(x^*) = W$, and thus for each $y \in V$ we can form the element $y - [x^*(y)/x^*(x_o)]x_o \in V$.

Evidently $y - [x^*(y)/x^*(x_o)]x_o \in W$, and hence

$$\langle y - [\frac{x^*(y)}{x^*(x_o)}]x_o, x_o\rangle = 0,$$

that is,

$$\langle y, x_o\rangle = [\frac{x^*(y)}{x^*(x_o)}]\langle x_o, x_o\rangle.$$

Consequently we see that

$$x^*(y) = \langle y, \left[\frac{\overline{x^*(x_o)}}{\|x_o\|^2}\right]x_o\rangle \qquad (y \in V),$$

and we set $x = [\overline{x^*(x_o)}/\|x_o\|^2]x_o$.

The uniqueness of $x$ follows at once from the isometric nature of the mapping, and thus the first hypothesis of the theorem implies the second one. □

Thus, for example, suppose $V = L_2([-\pi,\pi],dt/2\pi)$. From Corollary 8.3.2 we know that for each $x^* \in L_2([-\pi,\pi],dt/2\pi)^*$ there corresponds a unique $g \in L_2([-\pi,\pi],dt/2\pi)$ such that

$$x^*(f) = \frac{1}{2\pi}\int_{-\pi}^{\pi} f(t)\overline{g(t)}\,dt \qquad (f \in L_2([-\pi,\pi],dt/2\pi)).$$

The mapping $g \to x^*$ is the one that appears in the Riesz Representation Theorem.

Combining Theorems 13.4.2 and 4.6.1, we easily obtain the next result. The details are left to the reader.

Corollary 13.4.1. Let $(V,\langle\cdot,\cdot\rangle)$ be a Hilbert space over $\Phi$. If $W \subset V$ is a closed linear subspace, then $W^{\perp\perp} = W$.

This corollary can, of course, also be proved by direct arguments without using Theorems 13.4.2 and 4.6.1.

13.5. Orthogonal Projections. In this section we wish to show that there exists a one-to-one correspondence between the closed linear subspaces of a Hilbert space and certain continuous linear transformations on the Hilbert space called orthogonal projections.

To begin we consider a Hilbert space $(V,\langle\cdot,\cdot\rangle)$ and a closed linear subspace $W$ of $V$. From the Orthogonal Decomposition Theorem (Theorem 13.4.1) we know that $V = W \oplus W^\perp$. Thus for each $x \in V$ there exist a unique $y \in W$ and a unique $z \in W^\perp$ such that $x = y + z$. We define $P : V \to V$ by $P(x) = y$. It is evident that $P$ is a linear transformation from $V$ to itself and that the range of $P$, denoted as usual by $R(P)$, is precisely $W$. Moreover, since $y \perp z$, we see from Proposition 13.4.1 that $\|x\|^2 = \|y + z\|^2 = \|y\|^2 + \|z\|^2$, and so

$$\|P(x)\|^2 = \|y\|^2 \le \|x\|^2,$$

whence $P \in L(V)$ and $\|P\| \le 1$. If $W \neq \{0\}$, then $\|P\| = 1$ since

$$\|P\| \ge \sup_{\substack{x\in W\\ \|x\|=1}} \|P(x)\|$$

$$= \sup_{\substack{x \in W \\ \|x\| = 1}} \|x\|$$

$$= 1.$$

The fact that $P(x) = x$, $x \in W$, utilized in the preceding inequalities, also shows at once that $P^2 = P$; that is, $P^2(x) = P(x)$, $x,y \in V$.

We claim furthermore that $\langle P(x),y\rangle = \langle x,P(y)\rangle$, $x,y \in V$, arguing as follows: Suppose $x,y \in V$ are given and write $x = u + v$ and $y = z + w$, where $u,z \in W$ and $v,w \in W^{\perp}$. Then, since $u \perp w$ and $z \perp v$, we see that, on the one hand,

$$\langle P(x),y\rangle = \langle P(u + v),z + w\rangle$$

$$= \langle u,z\rangle + \langle u,w\rangle$$

$$= \langle u,z\rangle,$$

and, on the other hand,

$$\langle x,P(y)\rangle = \langle u + v,P(z + w)\rangle$$

$$= \langle u,z\rangle + \langle v,z\rangle$$

$$= \langle u,z\rangle.$$

Hence $\langle P(x),y\rangle = \langle x,P(y)\rangle$, $x,y \in V$.

Finally, we claim that the transformation $P$ defined here is the only element of $L(V)$ such that $P^2 = P$, $R(P) = W$, and $\langle P(x),y\rangle = \langle x,P(y)\rangle$, $x,y \in V$. Indeed, suppose $Q \in L(V)$ also has these properties. Then we note first that

$$\{x \mid x \in V,\ P(x) = x\} = R(P) = W = R(Q) = \{x \mid x \in V,\ Q(x) = x\}.$$

This follows easily on noting, for example, that, if $x = P(y) \in R(P)$, then $P(x) = P^2(y) = P(y) = x$. In a similar vein we claim that

$$\{x \mid x \in V,\ P(x) = 0\} = R(P)^{\perp} = W^{\perp} = R(Q)^{\perp} = \{x \mid x \in V,\ Q(x) = 0\}.$$

The argument is as follows: If, for example, $P(x) = 0$, then $\langle P(y),x\rangle = \langle y,P(x)\rangle = 0$, $y \in V$, and hence $x \perp R(P)$; that is, $\{x \mid x \in V,\ P(x) = 0\} \subset W^{\perp}$. Conversely, if $x \in W^{\perp} = R(P)^{\perp}$, then $\langle y,P(x)\rangle = \langle P(y),x\rangle = 0$, $y \in V$, and so, in particular $\langle P(x),P(x)\rangle = \|P(x)\|^2 = 0$. Thus $W^{\perp} = \{x \mid x \in V,\ P(x) = 0\}$.

But then for any $x \in V$, if $x = y + z$, $y \in W$, $z \in W^{\perp}$, we have $P(x) = P(y + z) = y = Q(y + z) = Q(x)$, that is, $P = Q$.

We can summarize this discussion by saying that a closed linear subspace $W$ of a Hilbert space $V$ determines a unique $P \in L(V)$ such that $P^2 = P$, $R(P) = W$, and $\langle P(x),y\rangle = \langle x,P(y)\rangle$, $x,y \in V$. The converse of this observation is also valid, as will be demonstrated by Theorem 13.5.1. First, however, we introduce a name for such transformations.

<u>Definition 13.5.1</u>. Let $(V,\langle\cdot,\cdot\rangle)$ be an inner product space over $\Phi$. If $P \in L(V)$ is such that

(i) $P^2 = P$,

(ii) $\langle P(x),y\rangle = \langle x,P(y)\rangle$ $\qquad (x,y \in V)$,

then $P$ is said to be an <u>orthogonal projection</u>.

The source of the name will become apparent from Proposition 13.5.1.

<u>Theorem 13.5.1</u>. Let $(V,\langle\cdot,\cdot\rangle)$ be a Hilbert space over $\Phi$ and let $W \subset V$. Then the following are equivalent:

(i) $W$ is a closed linear subspace of $V$.

(ii) There exists a unique orthogonal projection $P \in L(V)$ such that $R(P) = W$.

Proof. The implication from part (i) to part (ii) is contained in the foregoing discussion. Suppose part (ii) holds. Clearly $W$ is then a linear subspace. We only need to show that $W$ is closed, and to this end, in view of Proposition 13.4.2(i), it suffices to prove that $W = X^{\perp}$ for some set $X \subset V$. If $I$ denotes the identity mapping on $V$, that is, $I(x) = x$, $x \in V$, then it is apparent that $I - P \in L(V)$. We set $X = R(I - P)$.

If $x \in W = R(P)$, then there exists some $u \in V$ such that $P(u) = x$, and if $y \in X = R(I - P)$, then there exists some $v \in V$ for which $(I - P)(v) = y$. Hence

$$\begin{aligned} \langle x,y\rangle &= \langle P(u),(I - P)(v)\rangle \\ &= \langle u,P(I - P)(v)\rangle \\ &= \langle u,(P - P^2)(v)\rangle \\ &= 0, \end{aligned}$$

as $P^2 = P$. Thus $W \subset X^{\perp}$.

Conversely, suppose $x \in X^{\perp}$. Then for any $y \in V$ we have $(I - P)(y) \in X$, and so

$$\begin{aligned} 0 &= \langle x,(I - P)(y)\rangle \\ &= \langle x,y - P(y)\rangle \\ &= \langle x,y\rangle - \langle x,P(y)\rangle \\ &= \langle x,y\rangle - \langle P(x),y\rangle \\ &= \langle (I - P)(x),y\rangle. \end{aligned}$$

In particular, if $y = (I - P)(x)$, then $0 = \|(I - P)(x)\|^2$; that is, $x = P(x) \in R(P) = W$.

Therefore $W = X^{\perp}$, and $W$ is closed. $\square$

It is easily verified that, if $V$ is a Hilbert space and $P \in L(V)$, then $P$ is an orthogonal projection if and only if $I - P$ is an orthogonal projection, and that if $P$ is an orthogonal projection, then $R(P) = R(I - P)^{\perp}$ and $R(I - P) = R(P)^{\perp}$. The details are left to the reader. These observations, combined with the Orthogonal Decomposition Theorem (Theorem 13.4.1), immediately yield the next proposition.

Proposition 13.5.1. Let $(V,\langle\cdot,\cdot\rangle)$ be a Hilbert space over $\Phi$. If $P \in L(V)$ is an orthogonal projection, then $V = R(P) \oplus R(I - P)$ and $R(P) \perp R(I - P)$.

This proposition should shed some light on the reason for the name "orthogonal projection".

There are several equivalent formulations of the notion of orthogonal projection. We collect some of these in the next proposition, whose proof is left to the reader.

Proposition 13.5.2. Let $(V,\langle\cdot,\cdot\rangle)$ be a Hilbert space over $\Phi$. Then the following are equivalent:

(i) $P \in L(V)$ is an orthogonal projection.

(ii) $P \in L(V)$ is such that $P^2 = P$ and

$$R(P)^{\perp} = \{x \mid x \in V,\ P(x) = 0\}.$$

(iii) $P \in L(V)$ is such that $P^2 = P$ and $R(P) \perp R(I - P)$.

13.6. Complete Orthonormal Sets. The Riesz Representation Theorem proved in Section 13.4 is obviously a generalization to Hilbert spaces of a standard result of linear algebra. In this section we wish to develop another generalization to Hilbert spaces of a fundamental result for finite-dimensional linear spaces: the notion of an orthonormal basis.

To be more concrete, suppose $V = \mathbb{R}^n$. Then it is well-known that there exists a basis for $V$ consisting of vectors $e_1, e_2, \ldots, e_n$ such that

$$\langle e_k, e_j \rangle = 1 \quad \text{for} \quad k = j,$$

$$\langle e_k, e_j \rangle = 0 \quad \text{for} \quad k \neq j.$$

Our goal is to obtain a counterpart in an arbitrary Hilbert space for such a basis in $\mathbb{R}^n$. The notion of basis that is most appropriate for our concerns is not the usual algebraic one, but rather a concept utilizing the fact that a Hilbert space comes equipped with a natural topology (the norm topology), which thus allows us to consider infinite linear combinations of elements of the space as well as finite linear combinations. The precise definition of basis we require will appear only after some preliminary development. We begin with a definition and a fundamental theorem.

Definition 13.6.1. Let $(V, \langle \cdot, \cdot \rangle)$ be an inner product space over $\Phi$. A set of nonzero vectors $\{x_\alpha \mid \alpha \in \Lambda\} \subset V$ is said to be an *orthogonal set* if $\langle x_\alpha, x_\beta \rangle = 0$, $\alpha, \beta \in \Lambda$, $\alpha \neq \beta$; it is said to be an *orthonormal set* if it is an orthogonal set such that $\langle x_\alpha, x_\alpha \rangle = 1$, $\alpha \in \Lambda$. An orthonormal set of vectors in $V$ that is properly contained in no other orthonormal set in $V$ is said to be a *complete orthonormal set*.

Theorem 13.6.1. Let $(V, \langle \cdot, \cdot \rangle)$ be an inner product space over $\Phi$ and let $V \neq \{0\}$. Then $V$ contains a complete orthonormal set.

Proof. Let $\mathcal{O}$ denote the collection of all orthonormal sets in $V$. Clearly $\mathcal{O} \neq \emptyset$ because, if $x \in V$, $\|x\| = 1$, then $\{x\} \in \mathcal{O}$. We define a partial ordering in $\mathcal{O}$ by setting $E_1 > E_2$ if $E_1 \supset E_2$, $E_1, E_2 \in \mathcal{O}$. If $\mathcal{U} = \{E_\alpha \mid \alpha \in \Lambda\}$ is a linearly ordered subset of $\mathcal{O}$, then it is easily verified that $E_0 = \bigcup_\alpha E_\alpha$ is an upper bound for $\mathcal{U}$. Consequently we may apply Zorn's Lemma [$DS_1$, p.6] to deduce the existence of a maximal element $E$ in $\mathcal{O}$ with respect to the partial ordering.

From the preceding definition we see at once that $E$ is a complete orthonormal set. □

Thus every inner product space contains complete orthonormal sets. Moreover, the proof of Theorem 13.6.1 can be modified in an obvious way to obtain the following more general result. The details are left to the reader.

Corollary 13.6.1. Let $(V,\langle\cdot,\cdot\rangle)$ be an inner product space over $\Phi$. If $F \subset V$ is an orthonormal set, then there exists a complete orthonormal set $E \subset V$ such that $E \supset F$.

An alternative description of complete orthonormal sets is provided by the next theorem.

Theorem 13.6.2. Let $(V,\langle\cdot,\cdot\rangle)$ be an inner product space over $\Phi$ and suppose $E \subset V$ is an orthonormal set. Then the following are equivalent:

(i) $E$ is a complete orthonormal set.
(ii) If $x \in V$ is such that $x \perp E$, then $x = 0$.

Proof. Suppose $E$ is a complete orthonormal set. If $x \in V$ is such that $x \perp E$ and $x \neq 0$, then evidently $E \cup \{x/\|x\|\}$ is an orthonormal set that properly contains $E$, thereby contradicting the fact that $E$ is a complete orthonormal set. Thus $x \perp E$ implies $x = 0$.

Conversely, suppose $x \perp E$ implies $x = 0$. If $E$ were not a complete orthonormal set, then there would exist some orthonormal set $F \subset V$ such that $F$ properly contained $E$. Hence, if $x \in F \sim E$, then $\|x\| = 1$ and $x \perp E$, contrary to the assumption that $x \perp E$ implies $x = 0$.

Therefore $E$ is a complete orthonormal set, and the proof is finished. □

We shall ultimately see that a complete orthonormal set in a Hilbert space is an orthonormal basis for the space. Before coming to grips with this, though, we need to review some facts concerning the notion of noncountable sums in a normed linear space.

Suppose $(V,\|\cdot\|)$ is a normed linear space over $\Phi$ and $\{x_\alpha \mid \alpha \in \Lambda\} \subset V$. We denote by $\Pi$ the collection of all finite subsets $\pi$ of the index set $\Lambda$, and partially order $\Pi$ by setting $\pi_1 > \pi_2$ if $\pi_1 \supset \pi_2$, $\pi_1,\pi_2 \in \Pi$. Clearly, with this partial ordering, $\Pi$ is a directed system, and $\{\Sigma_{\alpha\in\pi}x_\alpha\}_{\pi\in\Pi}$ is a net of elements in $V$. With these notational conventions agreed on, we make the following definition:

Definition 13.6.2. Let $(V,\|\cdot\|)$ be a normed linear space over $\Phi$ and suppose $\{x_\alpha \mid \alpha \in \Lambda\} \subset V$. Then $\Sigma_\Lambda x_\alpha$ is said to converge to $x \in V$ if the net $\{\Sigma_{\alpha\in\pi}x_\alpha\}_{\pi\in\Pi}$ converges to $x$; that is, given $\varepsilon > 0$, there exists some $\pi_0 \in \Pi$ such that, if $\pi > \pi_0$, then $\|x - \Sigma_{\alpha\in\pi}x_\alpha\| < \varepsilon$.

Standard arguments reveal that, if $\Sigma_\Lambda x_\alpha = x$ and $\Sigma_\Lambda y_\alpha = y$, then

(i) $\Sigma_\Lambda(x_\alpha + y_\alpha) = x + y$.
(ii) $\Sigma_\Lambda(ax_\alpha) = ax$ $\quad (a \in \Phi)$.

Although our main concern will be with sums, there is one general fact about nets in Banach spaces that we shall have need of: the sequential completeness of Banach spaces implies net completeness. More precisely, we have the following proposition:

Proposition 13.6.1. Let $(V,\|\cdot\|)$ be a Banach space over $\Phi$. If $\{x_\alpha\}_{\alpha\in\Lambda}$ is a Cauchy net, then there exists some $x \in V$ such that $\lim_\alpha x_\alpha = x$.

Proof. For each positive integer $n$ let $\alpha_n \in \Lambda$ be such that, if $\alpha,\beta > \alpha_n$, then $\|x_\alpha - x_\beta\| < 1/n$. This is possible because

$\{x_\alpha\}_{\alpha\in\Lambda}$ is a Cauchy net. Then set $\alpha^1 = \alpha_1$ and choose $\alpha^n \in \Lambda$ such that $\alpha^n > \alpha^1$, $\alpha^n > \alpha^2, \ldots, \alpha^n > \alpha^{n-1}$, $\alpha^n > \alpha_n$, $n = 2,3,4,\ldots$ . It is then apparent that $\{x_{\alpha^n}\}$ is a Cauchy sequence in $V$, and so there exists some $x \in V$ such that $\lim_n x_{\alpha^n} = x$. Routine arguments now show that $\lim_\alpha x_\alpha = x$. □

An important fact about the sum $\Sigma_\Lambda x_\alpha$ is the following proposition:

<u>Proposition 13.6.2</u>. Let $(V,\|\cdot\|)$ be a normed linear space over $\Phi$ and suppose $\{x_\alpha \mid \alpha \in \Lambda\} \subset V$. If there exists some $x \in V$ such that $\Sigma_\Lambda x_\alpha = x$, then $x_\alpha = 0$ for all but a countable number of $\alpha$.

<u>Proof</u>. For each positive integer $n$ let $\pi_n \in \Pi$ be such that, if $\pi > \pi_n$, then $\|x - \Sigma_{\alpha\in\pi} x_\alpha\| < 1/2n$. Clearly $\pi_\infty = \bigcup_{n=1}^\infty \pi_n$ is a countable subset of $\Lambda$. If $\beta \in \Lambda$ and $\beta \notin \pi_\infty$, then $\beta \notin \pi_n$, $n = 1,2,3,\ldots$, whence we see that for each $n$

$$\begin{aligned}\|x_\beta\| &= \Big\|\sum_{\alpha\in\pi_n\cup\{\beta\}} x_\alpha - \sum_{\alpha\in\pi_n} x_\alpha\Big\| \\ &\le \Big\|\sum_{\alpha\in\pi_n\cup\{\beta\}} x_\alpha - x\Big\| + \Big\|x - \sum_{\alpha\in\pi_n} x_\alpha\Big\| \\ &< \frac{1}{2n} + \frac{1}{2n} \\ &= \frac{1}{n},\end{aligned}$$

as $\pi_n \cup \{\beta\} > \pi_n$ and $\pi_n > \pi_n$.

Therefore $x_\beta = 0$ if $\beta \notin \pi_\infty$. □

If the normed linear space in question is just the scalar field $\Phi$, then we have the following corollary. The first portion of the result is proved by noting that a series of numbers converges absolutely if and only if every rearrangement of the series converges.

Corollary 13.6.2.

(i) Let $\{a_k\} \subset \Phi$ be a sequence. Then $\Sigma_{k=1}^{\infty} a_k$ converges absolutely to $a$ if and only if $\Sigma_{\mathbb{Z}^+} a_k$ converges to $a$, where $\mathbb{Z}^+ = \{1,2,3,\ldots\}$.

(ii) Let $\{a_\alpha \mid \alpha \in \Lambda\} \subset \Phi$ and suppose there exists some $a \in \Phi$ such that $\Sigma_\Lambda a_\alpha = a$. Then $a_\alpha = 0$ for all but a countable number of $\alpha \in \Lambda$, and if this countable number of indices is enumerated as the sequence $\{\alpha_k\}$, then $\Sigma_{k=1}^{\infty} a_{\alpha_k}$ converges absolutely to a.

In the case that $V = \Phi$ we shall often write $\Sigma_\Lambda a_\alpha < \infty$ when $\{a_\alpha \mid \alpha \in \Lambda\} \subset \Phi$, $a_\alpha \geq 0$, and $\Sigma_\Lambda a_\alpha$ converges to some $a \in \Phi$.

We can now make a meaningful definition of orthonormal basis.

Definition 13.6.3. Let $(V,\langle\cdot,\cdot\rangle)$ be an inner product space over $\Phi$. A set of vectors $\{e_\alpha \mid \alpha \in \Lambda\} \subset V$ is said to be an orthonormal basis for $V$ if

(i) $\{e_\alpha \mid \alpha \in \Lambda\}$ is an orthornormal set.

(ii) For each $x \in V$ there exist some $a_\alpha(x) \in \Phi$, $\alpha \in \Lambda$, such that $x = \Sigma_\Lambda a_\alpha(x) e_\alpha$.

We need make no explicit mention of the linear independence in the definition, since an orthornormal set of vectors is always linearly independent. Indeed, every orthogonal set of vectors is linearly independent, as shown by the following: Suppose $\{x_\alpha \mid \alpha \in \Lambda\}$ is an orthogonal set and, for example, that $\Sigma_{k=1}^{n} a_k x_{\alpha_k} = 0$, where $a_k \in \Phi$. Then we see that

$$0 = \langle \sum_{k=1}^{n} a_k x_{\alpha_k}, x_{\alpha_j} \rangle = a_j \langle x_{\alpha_j}, x_{\alpha_j} \rangle \qquad (j = 1,2,\ldots,n),$$

whence $a_j = 0$, $j = 1,2,\ldots,n$. Thus $\{x_\alpha \mid \alpha \in \Lambda\}$ is linearly independent.

Furthermore, if $\{e_\alpha \mid \alpha \in \Lambda\}$ is an orthonormal basis for $V$, then it is easy to compute the coefficients $\{a_\alpha(x)\}$ for each $x \in V$: if $x = \Sigma_\Lambda a_\alpha(x) e_\alpha$, then it follows at once from the orthonormality of $\{e_\alpha \mid \alpha \in \Lambda\}$ and the continuity of the inner product that

$$\begin{aligned}\langle x, e_\beta\rangle &= \langle \sum_\Lambda a_\alpha(x) e_\alpha, e_\beta\rangle \\ &= \sum_\Lambda a_\alpha(x)\langle e_\alpha, e_\beta\rangle \\ &= a_\beta(x) \qquad (\beta \in \Lambda).\end{aligned}$$

Hence, if $\{e_\alpha \mid \alpha \in \Lambda\}$ is an orthonormal basis for an inner product space $V$, then

$$x = \sum_\Lambda \langle x, e_\alpha\rangle e_\alpha \qquad (x \in V).$$

The next three theorems are concerned with the fundamental properties of orthonormal sets, and the last of the three contains detailed descriptions of orthonormal bases for Hilbert spaces.

Theorem 13.6.3. Let $(V, \langle\cdot,\cdot\rangle)$ be an inner product space over $\Phi$ and let $\{e_\alpha \mid \alpha \in \Lambda\}$ be an orthonormal set. Then

(i) Given $x \in V$, if $\pi \in \Pi$, then

$$\inf_{\substack{a_\alpha \in \Phi \\ \alpha \in \pi}} \|x - \sum_{\alpha \in \pi} a_\alpha e_\alpha\|$$

occurs when $a_\alpha = \langle x, e_\alpha\rangle$.

(ii) $\Sigma_\Lambda |\langle x, e_\alpha\rangle|^2 \le \|x\|^2 \qquad (x \in V)$.

(iii) If $x \in V$, then $\langle x, e_\alpha\rangle \neq 0$ for at most a countable number of $\alpha \in \Lambda$.

(iv) $\Sigma_\Lambda |\langle x, e_\alpha\rangle \overline{\langle y, e_\alpha\rangle}| \le \|x\|\|y\| \qquad (x, y \in V)$.

Proof. Let $x \in V$. Then for any $a_\alpha \in \Phi$ we have

$$\begin{aligned}\|x - \sum_{\alpha \in \pi} a_\alpha e_\alpha\|^2 &= \langle x - \sum_{\alpha \in \pi} a_\alpha e_\alpha, x - \sum_{\beta \in \pi} a_\beta e_\beta\rangle \\ &= \langle x, x\rangle - \sum_{\alpha \in \pi} a_\alpha \langle e_\alpha x\rangle - \sum_{\beta \in \pi} \bar{a}_\beta \langle x, e_\beta\rangle + \sum_{\alpha,\beta \in \pi} a_\alpha \bar{a}_\beta \langle e_\alpha, e_\beta\rangle\end{aligned}$$

$$= \|x\|^2 - \sum_{\alpha\in\pi} a_\alpha\overline{\langle x,e_\alpha\rangle} - \sum_{\alpha\in\pi} \bar{a}_\alpha\langle x,e_\alpha\rangle + \sum_{\alpha\in\pi} |a_\alpha|^2$$

$$= \|x\|^2 + \sum_{\alpha\in\pi} |a_\alpha - \langle x,e_\alpha\rangle|^2 - \sum_{\alpha\in\pi} |\langle x,e_\alpha\rangle|^2.$$

It is now apparent that

$$\inf_{\substack{a_\alpha\in\Phi\\ \alpha\in\pi}} \|x - \sum_{\alpha\in\pi} a_\alpha e_\alpha\|$$

occurs when $a_\alpha = \langle x,e_\alpha\rangle$, thereby proving part (i), and that

$$\sum_{\alpha\in\pi} |\langle x,e_\alpha\rangle|^2 \le \|x\|^2.$$

Since this last estimate remains valid for each $\pi \in \Pi$, an easy argument shows that $\Sigma_\Lambda|\langle x,e_\alpha\rangle|^2 < \infty$ and

$$\sum_\Lambda |\langle x,e_\alpha\rangle|^2 \le \|x\|^2.$$

Thus part (ii) holds, and part (iii) follows immediately from Proposition 13.6.2 or Corollary 13.6.2(ii).

Furthermore, if $x,y \in V$, then, by part (iii), $\langle x,e_\alpha\rangle = 0$ and $\langle y,e_\alpha\rangle = 0$ for all but a countable number of $\alpha \in \Lambda$, say for the sequence $\{\alpha_k\}$. Then from Corollary 13.6.2(ii), the Cauchy-Schwarz Inequality for series and the second part of the present theorem we see that

$$\sum_\Lambda |\langle x,e_\alpha\rangle\overline{\langle y,e_\alpha\rangle}| = \sum_{k=1}^\infty |\langle x,e_{\alpha_k}\rangle\overline{\langle y,e_{\alpha_k}\rangle}|$$

$$\le \left[\sum_{k=1}^\infty |\langle x,e_{\alpha_k}\rangle|^2\right]^{1/2}\left[\sum_{k=1}^\infty |\langle y,e_{\alpha_k}\rangle|^2\right]^{1/2}$$

$$= \left[\sum_\Lambda |\langle x,e_\alpha\rangle|^2\right]^{1/2}\left[\sum_\Lambda |\langle y,e_\alpha\rangle|^2\right]^{1/2}$$

$$\leq \|x\|\|y\|.$$

Therefore part (iv) holds, and the proof is complete. □

The numbers $\{\langle x,e_\alpha\rangle \mid \alpha \in \Lambda\}$ are often referred to as the Fourier coefficients of $x$ with respect to the orthonormal set $\{e_\alpha \mid \alpha \in \Lambda\}$. The reason for this terminology will become apparent when we discuss the Hilbert space $L_2([-\pi,\pi],dt/2\pi)$ in greater detail in the next section. Theorem 13.6.3(ii), generally called Bessel's Inequality, implies that the Fourier coefficients of $x$ are the components of an element of $L_2(\Lambda)$, the measure being counting measure on $\Lambda$.

The next theorem says that the components of every element of $L_2(\Lambda)$ are the Fourier coefficients of some $x \in V$, provided $V$ is a Hilbert space.

Theorem 13.6.4 (Riesz-Fischer Theorem). Let $(V,\langle\cdot,\cdot\rangle)$ be a Hilbert space over $\Phi$ and suppose $\{e_\alpha \mid \alpha \in \Lambda\}$ is an orthonormal set. If $\{a_\alpha \mid \alpha \in \Lambda\} \subset \Phi$, then the following are equivalent:

(i) $\Sigma_\Lambda |a_\alpha|^2 < \infty$.

(ii) There exists some $x \in V$ such that $x = \Sigma_\Lambda a_\alpha e_\alpha$.

Moreover, if $x = \Sigma_\Lambda a_\alpha e_\alpha$, then $a_\alpha = \langle x,e_\alpha\rangle$, $\alpha \in \Lambda$.

Proof. Suppose that $\Sigma_\Lambda |a_\alpha|^2 < \infty$. Then, since only countably many of the $a_\alpha$ are nonzero (Proposition 13.6.2), it is easily verified that given $\epsilon > 0$ there exists some $\pi_o \in \Pi$ such that, if $\pi > \pi_o$ and $\pi' > \pi_o$, then $\Sigma_{\alpha \in \pi \Delta \pi'} |a_\alpha|^2 < \epsilon$, where $\pi \Delta \pi' = (\pi \sim \pi') \cup (\pi' \sim \pi)$ Consequently, if $\pi > \pi_o$ and $\pi' > \pi_o$, we have

$$\left\| \sum_{\alpha \in \pi} a_\alpha e_\alpha - \sum_{\alpha \in \pi'} a_\alpha e_\alpha \right\|^2 = \left\| \sum_{\alpha \in \pi \Delta \pi'} a_\alpha e_\alpha \right\|^2$$

$$= \sum_{\alpha \in \pi \Delta \pi'} |a_\alpha|^2$$

$$< \epsilon,$$

since $\{e_\alpha \mid \alpha \in \Lambda\}$ is an orthonormal set. Thus $\{\Sigma_{\alpha \in \pi} a_\alpha e_\alpha\}_{\pi \in \Pi}$ is a Cauchy net in $V$, and so from Proposition 13.6.1 we see that there exists some $x \in V$ such that $x = \Sigma_\Lambda a_\alpha e_\alpha$. Hence part (i) implies part (ii).

Conversely, suppose $x = \Sigma_\Lambda a_\alpha e_\alpha$. Then the continuity of the inner product (Corollary 13.1.2) and Bessel's Inequality (Theorem 13.6.3(ii)) combine to show that $\Sigma_\Lambda |a_\alpha|^2 < \infty$.

Therefore part (ii) implies part (i), and the proof is complete. □

Note that Theorem 13.6.4 does not assert that every $x \in V$ can be expressed as $x = \Sigma_\Lambda \langle x, e_\alpha \rangle e_\alpha$ for a given orthonormal set $\{e_\alpha \mid \alpha \in \Lambda\}$; that is, it does not assert that $\{e_\alpha \mid \alpha \in \Lambda\}$ is an orthonormal basis for $V$. All we can deduce at this point is that, given $x \in V$, there exists some $y \in V$ for which $y = \Sigma_\Lambda \langle x, e_\alpha \rangle e_\alpha$. From the continuity of the inner product we know, however, that $\langle x - y, e_\alpha \rangle = 0$, $\alpha \in \Lambda$, and so $x = y$ provided $\{e_\alpha \mid \alpha \in \Lambda\}$ is a complete orthonormal set (Theorem 13.6.2). This observation proves the first implication of the next theorem.

<u>Theorem</u> <u>13.6.5</u>. Let $(V, \langle \cdot, \cdot \rangle)$ be a Hilbert space over $\Phi$ and suppose $\{e_\alpha \mid \alpha \in \Lambda\}$ is an orthonormal set. Then the following are equivalent:

(i) $\{e_\alpha \mid \alpha \in \Lambda\}$ is a complete orthonormal set.

(ii) $\{e_\alpha \mid \alpha \in \Lambda\}$ is an orthonormal basis for $V$.

(iii) The linear subspace of $V$ spanned by $\{e_\alpha \mid \alpha \in \Lambda\}$ is norm dense in $V$.

(iv) If $x \in V$, then $\Sigma_\Lambda |\langle x, e_\alpha \rangle|^2 = \|x\|^2$.

(v) If $x, y \in V$, then $\Sigma_\Lambda \langle x, e_\alpha \rangle \overline{\langle y, e_\alpha \rangle} = \langle x, y \rangle$.

Proof. The implication from part (i) to part (ii) was established in the paragraph preceding the theorem, and that from part (ii) to part (iii) is obvious. Suppose that part (iii) holds and that $\{e_\alpha \mid \alpha \in \Lambda\}$ is not a complete orthonormal set. Then there would exist, by Theorem 13.6.2, some $x_o \in V$, $\|x_o\| = 1$, such that $x_o \perp e_\alpha$, $\alpha \in \Lambda$. However, by part (iii) there exist $a_\alpha \in \Phi$ and $\pi \in \Pi$ for which $\|x_o - \Sigma_{\alpha \in \pi} a_\alpha e_\alpha\| < 1/2$. But Theorem 13.6.3(i) then shows that

$$\begin{aligned}
\tfrac{1}{2} &> \|x_o - \sum_{\alpha \in \pi} a_\alpha e_\alpha\| \\
&\geq \|x_o - \sum_{\alpha \in \pi} \langle x_o, e_\alpha \rangle e_\alpha\| \\
&= \|x_o\| \\
&= 1,
\end{aligned}$$

which is clearly absurd. Hence part (iii) implies part (i), and parts (i) through (iii) are equivalent.

It is evident that the implication from part (iii) to part (i) can also be established by an appeal to the Riesz Representation Theorem (Theorem 13.4.2) and Corollary 4.2.8 to the Hahn-Banach Theorem.

Now suppose that $\{e_\alpha \mid \alpha \in \Lambda\}$ is an orthonormal basis for $V$. Then we know from the comments following Definition 13.6.3 that $x = \Sigma_\Lambda \langle x, e_\alpha \rangle e_\alpha$ for each $x \in V$. Appealing to the continuity of the inner product (Corollary 13.1.2), we conclude that

$$\begin{aligned}
\|x\|^2 = \langle x, x \rangle &= \langle \sum_\Lambda \langle x, e_\alpha \rangle e_\alpha, \sum_\Lambda \langle x, e_\beta \rangle e_\beta \rangle \\
&= \sum_\Lambda \langle x, e_\alpha \rangle \overline{\langle x, e_\beta \rangle} \langle e_\alpha, e_\beta \rangle \\
&= \sum_\Lambda |\langle x, e_\alpha \rangle|^2,
\end{aligned}$$

and so part (ii) implies part (iv).

A similar argument shows that part (ii) implies part (v), and obviously part (v) implies part (iv) on setting $x = y$ in (v).

Finally, suppose part (iv) holds and let $x_o \in V$ be such that $x_o \perp e_\alpha$, $\alpha \in \Lambda$. Then

$$\|x_o\|^2 = \sum_{\Lambda} |\langle x_o, e_\alpha \rangle|^2 = 0,$$

and so $x_o = 0$. Thus part (iv) implies part (i).

Therefore all five parts of the theorem are equivalent. □

In particular, we see from the preceding theorem that every Hilbert space has an orthonormal basis and that any complete orthonormal set can serve as such a basis.

Parts (iv) and (v) of the theorem are generally referred to as Parseval's Identity.

Let us now look more closely at the remark preceding the Riesz-Fischer Theorem (Theorem 13.6.4). Suppose that $(V, \langle \cdot, \cdot \rangle)$ is a Hilbert space over $\Phi$ and that $\{e_\alpha \mid \alpha \in \Lambda\} \subset V$ is a complete orthonormal set. Consider the Hilbert space $L_2(\Lambda, \mathcal{S}, \mu)$ over $\Phi$, where $\mathcal{S}$ is all subsets of $\Lambda$ and $\mu$ is counting measure on $\Lambda$; that is, if $E \in \mathcal{S}$, then $\mu(E)$ is equal to the number of points in $E$ when $E$ is finite, and $\mu(E)$ is plus infinity when $E$ is infinite. Then define the mapping $\varphi : V \to L_2(\Lambda, \mathcal{S}, \mu)$ by setting $\varphi(x) = \{\langle x, e_\alpha \rangle\}$, $x \in V$, that is, $\varphi(x)$ is that element of $L_2(\Lambda, \mathcal{S}, \mu)$ whose value at $\alpha \in \Lambda$ is $\langle x, e_\alpha \rangle$. That $\varphi(x) \in L_2(\Lambda, \mathcal{S}, \mu)$, $x \in V$, is an immediate consequence of Bessel's Inequality.

Clearly $\varphi$ is a linear mapping, and the two forms of Parseval's Identity show that $\varphi$ is an isometry and that $\varphi$ preserves inner products; that is, $\langle x, y \rangle = \langle \varphi(x), \varphi(y) \rangle$, $x, y \in V$. The Riesz-Fischer Theorem shows that $\varphi$ is surjective. Hence we have the following theorem:

<u>Theorem 13.6.6</u>. Let $(V,\langle\cdot,\cdot\rangle)$ be a Hilbert space over $\Phi$ and suppose $\{e_\alpha \mid \alpha \in \Lambda\} \subset V$ is a complete orthonormal set. Then $(V,\langle\cdot,\cdot\rangle)$ is isometrically isomorphic to $L_2(\Lambda,\mathcal{S},\mu)$, where $\mathcal{S}$ is all subsets of $\Lambda$ and $\mu$ is counting measure on $\Lambda$.

Thus we see that Hilbert spaces are essentially just $L_2$-spaces.

A number of other rather easily derived results about Hilbert spaces are obtainable from the foregoing development. We wish, however, merely to mention one of these. The details are left to the reader.

First, in analogy with the case of finite-dimensional spaces, we make the following definition:

<u>Definition 13.6.4</u>. Let $(V,\langle\cdot,\cdot\rangle)$ be a Hilbert space over $\Phi$. If $\{e_\alpha \mid \alpha \in \Lambda\} \subset V$ is a complete orthonormal set, then the <u>dimension</u> of $V$, denoted by $\dim(V)$, is defined to be the cardinality of $\Lambda$.

On the surface this definition would seem to depend on the choice of the complete orthonormal set. However, this is not the case, because we can show that, if $\{e_\alpha \mid \alpha \in \Lambda\}$ and $\{f_\beta \mid \beta \in \Lambda'\}$ are complete orthonormal sets in $V$, then the cardinalities of $\Lambda$ and $\Lambda'$ are equal. Thus the notion of dimension in a Hilbert space is well-defined.

One useful result involving this concept is the following theorem:

<u>Theorem 13.6.7</u>. Let $(V,\langle\cdot,\cdot\rangle)$ be a Hilbert space over $\Phi$. Then the following are equivalent:

(i) $\dim(V) = \aleph_0$.
(ii) $(V,\|\cdot\|)$ is separable.

Consequently a Hilbert space is separable if and only if it has a countable orthonormal basis.

<u>13.7. Fourier Analysis in $L_2([-\pi,\pi],dt/2\pi)$</u>. We now wish to see what the development of the preceding section can tell us about the Hilbert space $L_2([-\pi,\pi],dt/2\pi)$. Recall that the inner product in this space is defined by

$$\langle f,g\rangle = \frac{1}{2\pi}\int_{-\pi}^{\pi} f(t)\overline{g(t)}\,dt \qquad (f,g \in L_2([-\pi,\pi],dt/2\pi).$$

The first thing we do is to examine a particular orthonormal basis for $L_2([-\pi,\pi],dt/2\pi)$.

<u>Theorem 13.7.1</u>. The family of functions $\{e^{ik\cdot} \mid k \in \mathbb{Z}\}$ contained in $L_2([-\pi,\pi],dt/2\pi)$ is a complete orthonormal set.

<u>Proof</u>. The fact that the complex exponentials form an orthonormal set in $L_2([-\pi,\pi],dt/2\pi)$ is a routine exercise in calculus. To prove the completeness of this orthonormal set is, however, less trivial. Recalling Theorem 13.6.2, we see that it suffices to show that, if $f \in L_2([-\pi,\pi],dt/2\pi)$ and $\langle f,e^{ik\cdot}\rangle = 0$, $k \in \mathbb{Z}$, then $f(t) = 0$ almost everywhere.

Suppose $f$ is such a function and define

$$F(t) = \frac{1}{2\pi}\int_{-\pi}^{t} f(s)\,ds \qquad (t \in [-\pi,\pi]).$$

Since $L_2([-\pi,\pi],dt/2\pi) \subset L_1([-\pi,\pi],dt/2\pi)$, it is evident that $F$ is a well-defined absolutely continuous function on $[-\pi,\pi]$ [Ry, p. 106] In particular, $F \in L_2([-\pi,\pi],dt/2\pi)$. Define $G$ by

$$G(t) = F(t) - \frac{1}{2\pi}\int_{-\pi}^{\pi} F(s)\,ds \qquad (t \in [-\pi,\pi]).$$

Then $G \in L_2([-\pi,\pi],dt/2\pi)$, $G$ is continuous, and the derivative of $G$ exists and equals $f$ almost everywhere, because $F$ is the indefinite integral of $f$ [Ry, pp. 105-107]. Consequently, utilizing integration by parts, we see that

$$0 = \langle f,e^{ik\cdot}\rangle$$

$$= \langle G',e^{ik\cdot}\rangle$$

$$= \frac{1}{2\pi} G(t)e^{-ikt}\Big|_{-\pi}^{\pi} + \frac{ik}{2\pi}\int_{-\pi}^{\pi} G(t)e^{-ikt}\,dt$$

$$= \frac{\cos(k\pi)}{2\pi}[G(\pi) - G(-\pi)] + ik\langle G, e^{ik\cdot}\rangle \qquad (k \in \mathbb{Z}).$$

On setting $k = 0$, we see that $G(\pi) - G(-\pi) = 0$, whence we conclude that $\langle G, e^{ik\cdot}\rangle = 0$, $k \in \mathbb{Z}$, $k \neq 0$. However, when $k = 0$, we have

$$\langle G, 1\rangle = \frac{1}{2\pi}\int_{-\pi}^{\pi} G(t)\,dt$$

$$= \frac{1}{2\pi}\int_{-\pi}^{\pi} F(t)\,dt - \frac{1}{2\pi}\int_{-\pi}^{\pi} F(t)\,dt$$

$$= 0.$$

Thus $\langle G, e^{ik\cdot}\rangle = 0$, $k \in \mathbb{Z}$.

Now, considering $\{e^{ik\cdot} \mid k \in \mathbb{Z}\}$ as a family of continuous functions defined on $\Gamma = \{z \mid z \in \mathbb{C}, |z| = 1\}$, we denote by $W$ the linear subspace of $C(\Gamma)$ spanned by $\{e^{ik\cdot} \mid k \in \mathbb{Z}\}$. It can be routinely verified that $W$ is a subalgebra of $C(\Gamma)$ that separates points, is closed under complex conjugation, and is such that $Z(W) = \{z \mid z \in \Gamma, h(z) = 0, h \in W\} = \emptyset$. Hence by the Stone-Weierstrass Theorem (Theorem 11.4.1) we conclude that $W$ is dense in $(C(\Gamma), \|\cdot\|_\infty)$. It is apparent that this is equivalent to asserting that the linear subspace of $C([-\pi,\pi])$ spanned by $\{e^{ik\cdot} \mid k \in \mathbb{Z}\}$ is norm dense in the space of continuous functions $h$ on $[-\pi,\pi]$ such that $h(\pi) = h(-\pi)$.

Suppose that $\|G\|_2 \neq 0$ and let $\epsilon > 0$. Since $G \in C([-\pi,\pi])$ and $G(\pi) = G(-\pi)$, the preceding paragraph shows that there exists some finite linear combination of the elements of $\{e^{ik\cdot} \mid k \in \mathbb{Z}\}$, call it $h$, such that $\|G - h\|_\infty < \epsilon$. Evidently

$$\frac{1}{2\pi}\int_{-\pi}^{\pi} \overline{G(t)h(t)}\,dt = \overline{\langle G, h\rangle} = 0,$$

as $\langle G,e^{ik\cdot}\rangle = 0$, $k \in \mathbb{Z}$, and $h$ is a finite linear combination of the $e^{ik\cdot}$. Hence

$$
\begin{aligned}
(\|G\|_2)^2 &= \frac{1}{2\pi}\int_{-\pi}^{\pi} G(t)\overline{G(t)}\,dt \\
&= \frac{1}{2\pi}\int_{-\pi}^{\pi} \overline{G(t)}[G(t) - h(t)]\,dt \\
&\le \frac{1}{2\pi}\int_{-\pi}^{\pi} |\overline{G(t)}||G(t) - h(t)|\,dt \\
&< \frac{\epsilon}{2\pi}\int_{-\pi}^{\pi} |G(t)|\,dt \\
&\le \epsilon\|G\|_2,
\end{aligned}
$$

the last step being valid because of the Cauchy-Schwarz Inequality (Theorem 13.1.1). Consequently $\|G\|_2 < \epsilon$, and since $\epsilon$ is arbitrary, we must conclude that $\|G\|_2 = 0$, contrary to assumption.

Thus $\|G\|_2 = 0$, and so

$$
\begin{aligned}
G(t) &= F(t) - \frac{1}{2\pi}\int_{-\pi}^{\pi} F(s)\,ds \\
&= 0 \qquad\qquad (t \in [-\pi,\pi]),
\end{aligned}
$$

as $G$ is continuous. But, since $F$ is an indefinite integral, we have $F'(t) = f(t) = 0$ for almost all $t$.

Therefore $\{e^{ik\cdot} \mid k \in \mathbb{Z}\}$ is a complete orthonormal set. □

Theorem 13.7.1 has as a corollary the fact that the Fourier transform on $L_1([-\pi,\pi],dt/2\pi)$ is injective, a result remarked on in Section 7.8.

<u>Corollary 13.7.1</u>. Let $f \in L_1([-\pi,\pi],dt/2\pi)$. If $\hat{f}(k) = 0$, $k \in \mathbb{Z}$, then $f = 0$.

<u>Proof</u>. It is evident from the proof of Theorem 13.7.1 that

$$G(t) = F(t) - \frac{1}{2\pi}\int_{-\pi}^{\pi} F(s)\, ds \qquad (t \in [-\pi,\pi])$$

defines an element $G \in C([-\pi,\pi]) \subset L_2([-\pi,\pi],dt/2\pi)$, where, as before,

$$F(t) = \int_{-\pi}^{t} f(s)\, ds \qquad (t \in [-\pi,\pi]).$$

Since $G' = F' = f$ almost everywhere and $\hat{f}(k) = 0$, $k \in \mathbb{Z}$, we deduce once again that $\langle G, e^{ik\cdot}\rangle = 0$, $k \in \mathbb{Z}$, and so, by Theorem 13.7.1, $G$ is identically zero. Hence $f = 0$. □

The proof of the next theorem is obtained by quoting the appropriate result of the preceding section. We establish the notational convention that

$$\hat{f}(k) = \langle f, e^{ik\cdot}\rangle \qquad (k \in \mathbb{Z};\ f \in L_2([-\pi,\pi],dt/2\pi)).$$

This is consistent with previous uses of $\hat{f}$ in, for instance, Sections 3.1, 6.6, 7.8, and 9.7. It also provides a rationale for the Fourier coefficient terminology mentioned in the preceding section.

<u>Theorem</u> <u>13.7.2</u>. Consider the Hilbert space $L_2([-\pi,\pi],dt/2\pi)$ over $\mathbb{C}$.

(i) If $f \in L_2([-\pi,\pi],dt/2\pi)$, then $f = \Sigma_{\mathbb{Z}}\hat{f}(k)e^{ik\cdot}$ and $\|f\|_2 = [\Sigma_{k=-\infty}^{\infty}|\hat{f}(k)|^2]^{1/2}$.

(ii) If $f,g \in L_2([-\pi,\pi],dt/2\pi)$, then

$$\frac{1}{2\pi}\int_{-\pi}^{\pi} f(t)\overline{g(t)}\, dt = \sum_{k=-\infty}^{\infty} \hat{f}(k)\overline{\hat{g}(k)}$$

and

$$\frac{1}{2\pi}\int_{-\pi}^{\pi} f(t)g(t)\, dt = \sum_{k=-\infty}^{\infty} \hat{f}(k)\hat{g}(-k).$$

(iii) If $\{a_k \mid k \in \mathbb{Z}\} \subset \mathbb{C}$ is such that $\Sigma_{k=-\infty}^{\infty}|a_k|^2 < \infty$, then there exists a unique $f \in L_2([-\pi,\pi],dt/2\pi)$ such that $f = \Sigma_{\mathbb{Z}} a_k e^{ik\cdot}$ and $a_k = \hat{f}(k)$, $k \in \mathbb{Z}$.

(iv) The mapping $\varphi : L_2([-\pi,\pi],dt/2\pi) \to L_2(\mathbb{Z})$ defined for each $f \in L_2([-\pi,\pi],dt/2\pi)$ by $\varphi(f)(k) = \hat{f}(k)$, $k \in \mathbb{Z}$, is a surjective isometric isomorphism.

The results of this theorem have valid analogs in the more general situation in which $[-\pi,\pi]$ is replaced by an arbitrary compact Abelian topological group $G$, and one considers the Hilbert space $L_2(G,\mu)$, where $\mu$ is Haar measure on $G$. In this case a complete orthonormal set in $L_2(G,\mu)$ is provided by the continuous characters of $G$, that is, by all continuous homomorphisms of $G$ into the circle group $\Gamma = \{z \mid z \in \mathbb{C}, |z| = 1\}$. The proof that the continuous characters on $G$ form a complete orthonormal set hinges on the fact that these functions separate the points of $G$. This is a highly nontrivial fact (see, for instance, [$HR_1$, pp. 335-355; R, pp. 23-26]). Furthermore, the second portion of implication (i), as well as implications (ii) and (iv) of Theorem 13.7.2 can be generalized to the context of locally compact Abelian topological groups. Part (iv) of Theorem 13.7.2 is generally referred to as <u>Plancherel's Theorem</u>. The interested reader may consult, for instance, [$HR_2$, pp. 225 and 226; HSt, pp. 411-413; Ka, pp. 139-142; Lo, pp. 141-146; R, pp. 128-130; $Ru_1$, pp. 26 and 27].

In particular, Theorem 13.7.2(i) says that the Fourier series $\Sigma_{\mathbb{Z}}\hat{f}(k)e^{ik\cdot}$ of $f \in L_2([-\pi,\pi],dt/2\pi)$ converges to $f$ in the norm of $L_2([-\pi,\pi],dt/2\pi)$. It was conjectured by Lusin [Lu] in 1915 that the Fourier series of such a function actually converges almost everywhere. This conjecture was given an affirmative answer by Carleson [Ca] in 1966. Subsequently the result was extended to $L_p([-\pi,\pi],dt/2\pi)$, $1 < p \leq \infty$, by Hunt [Ht]. A discussion of these results is considerably beyond the scope of this book, and the reader is referred either to the original papers or to [M] for an exposition. However, we would like to prove a result of Kolmogorov

[Ko], which asserts that certain subsequences of the symmetric partial sums of the Fourier series of an $L_2$-function converge almost everywhere.

We need one fact that we do not prove: if $f \in L_2([-\pi,\pi], dt/2\pi)$, then, as indicated in Section 9.1, the nth Cesàro mean of the Fourier series of $f$ can be written as

$$\sigma_n(f)(t) = \sum_{k=-n}^{n} \left(1 - \frac{|k|}{n+1}\right)\hat{f}(k)e^{ikt} \qquad (t \in [-\pi,\pi]).$$

It is the case that the sequence $\{\sigma_n(f)\}$ converges almost everywhere to $f$. This follows at once from Carleson's result, as convergence almost everywhere implies convergence of the Cesàro means. However, a simpler and more direct proof is also available [$E_1$, p. 96].

One definition is necessary before we can state Kolmogorov's theorem.

Definition 13.7.1. A sequence $\{n_k\}$ of positive integers is said to be a Hadamard sequence if $\inf_k [n_{k+1}/n_k] > 1$.

Theorem 13.7.3 (Kolmogorov's Theorem). Consider the Hilbert space $L_2([-\pi,\pi], dt/2\pi)$ over $\mathbb{C}$. If $\{n_k\}$ is a Hadamard sequence, then for each $f \in L_2([-\pi,\pi], dt/2\pi)$ the sequence $\{s_{n_k}(f)\}$ converges almost everywhere to $f$ where

$$s_{n_k}(f)(t) = \sum_{m=-n_k}^{n_k} \hat{f}(m)e^{imt} \qquad (t \in [-\pi,\pi]).$$

Proof. We first observe the following immediate consequence of the Monotone Convergence Theorem [Ry, pp. 84 and 227]: Suppose $\{g_k\}$ is a sequence of nonnegative integrable functions on $[-\pi,\pi]$ such that

$$\sum_{k=1}^{\infty} \int_{-\pi}^{\pi} g_k(t)\, dt < \infty.$$

Then, since $\{\Sigma_{k=1}^{n} g_k\}$ is an increasing sequence of functions that converges to $\Sigma_{k=1}^{\infty} g_k$, we conclude via the Monotone Convergence Theorem that

$$\int_{-\pi}^{\pi} \sum_{k=1}^{\infty} g_k(t)\, dt = \sum_{k=1}^{\infty} \int_{-\pi}^{\pi} g_k(t)\, dt,$$

and hence $\Sigma_{k=1}^{\infty} g_k$ is integrable. In particular, $\Sigma_{k=1}^{\infty} g_k$ must converge almost everywhere, and so $\lim_k g_k(t) = 0$ for almost all $t \in [-\pi,\pi]$.

Now suppose $f \in L_2([-\pi,\pi],dt/2\pi)$. Clearly $\{|\sigma_{n_k}(f) - s_{n_k}(f)|^2\}$ is a sequence of nonnegative integrable functions on $[-\pi,\pi]$. Suppose for the moment that

$$\sum_{k=1}^{\infty} (\|\sigma_{n_k}(f) - s_{n_k}(f)\|_2)^2 = \sum_{k=1}^{\infty} \frac{1}{2\pi} \int_{-\pi}^{\pi} |\sigma_{n_k}(f)(t) - s_{n_k}(f)(t)|^2\, dt < \infty .$$

From our previous observations we know that

$$\lim_k |\sigma_{n_k}(f)(t) - s_{n_k}(f)(t)| = 0$$

for almost all $t \in [-\pi,\pi]$. Since $\{\sigma_{n_k}(f)\}$ converges almost everywhere to $f$, it follows at once that $\{s_{n_k}(f)\}$ converges almost everywhere to $f$, which is what we wish to prove. Thus it suffices to show that

$$\sum_{k=1}^{\infty} (\|\sigma_{n_k}(f) - s_{n_k}(f)\|_2)^2 < \infty.$$

Now we see that

$$(\|\sigma_{n_k}(f) - s_{n_k}(f)\|_2)^2 = \frac{1}{2\pi} \int_{-\pi}^{\pi} \left| \sum_{m=-n_k}^{n_k} \left[\left(1 - \frac{|m|}{n_k+1}\right) - 1\right] \hat{f}(m) e^{imt} \right|^2 dt$$

$$= \frac{1}{(n_k+1)^2} \sum_{m=-n_k}^{n_k} m^2 |\hat{f}(m)|^2,$$

since $\{e^{im\cdot} \mid m \in \mathbb{Z}\}$ forms an orthonormal set in $L_2([-\pi,\pi],dt/2\pi)$. Hence, setting $n_0 = 0$ and interchanging the order of summation, we obtain

$$\sum_{k=1}^{N} (\|\sigma_{n_k}(f) - s_{n_k}(f)\|_2)^2 = \sum_{k=1}^{N} \left[\frac{1}{(n_k+1)^2} \sum_{m=-n_k}^{n_k} m^2|\hat{f}(m)|^2\right]$$

$$\leq \sum_{k=1}^{N} \left[\frac{1}{n_k^2} \sum_{j=1}^{k} \Big(\sum_{n_{j-1}<|m|\leq n_j} m^2|\hat{f}(m)|^2\Big)\right]$$

$$= \sum_{j=1}^{N} \left[\Big(\sum_{n_{j-1}<|m|\leq n_j} m^2|\hat{f}(m)|^2\Big)\Big(\sum_{k=j}^{N} \frac{1}{n_k^2}\Big)\right].$$

However, $\{n_k\}$ is a Hadamard sequence, and so $\inf_k(n_{k+1}/n_k) = c > 1$. It is easily seen that $n_{j+r} \geq c^r n_j$, $j,r = 1,2,3,\ldots$ . Consequently

$$\sum_{k=j}^{N} \frac{1}{n_k^2} = \sum_{r=0}^{N-j} \frac{1}{n_{j+r}^2}$$

$$\leq \sum_{r=0}^{N-j} \frac{1}{c^{2r} n_j^2}$$

$$= \frac{1}{n_j^2} \sum_{r=0}^{N-j} \frac{1}{c^{2r}}$$

$$\leq \frac{1}{n_j^2} \sum_{r=0}^{\infty} \frac{1}{c^{2r}}$$

$$= \frac{C}{n_j^2} \qquad (N = 1,2,3,\ldots),$$

where $C = \Sigma_{r=0}^{\infty}(1/c^2)^r$ converges because $1/c^2 < 1$.

Hence we have

$$\sum_{k=1}^{N} (\|\sigma_{n_k}(f) - s_{n_k}(f)\|_2)^2 \le \sum_{j=1}^{N} \Big[ (n_j^2 \sum_{n_{j-1} < |m| \le n_j} |\hat{f}(m)|^2)(\frac{C}{n_j^2}) \Big]$$

$$= C\Big[ \sum_{m=-n_N}^{n_N} |\hat{f}(m)|^2 \Big]$$

$$\le C\Big[ \sum_{m=-\infty}^{\infty} |\hat{f}(m)|^2 \Big]$$

$$= C(\|f\|_2)^2 \qquad (N = 1,2,3,\ldots).$$

The penultimate step is valid by Theorem 13.7.2(i) or Parseval's Identity (Theorem 13.6.5(iv)).

It is then apparent that

$$\sum_{k=1}^{\infty} (\|\sigma_{n_k}(f) - s_{n_k}(f)\|_2)^2 < \infty,$$

which completes the proof. □

<u>13.8. Rademacher Functions.</u> As should be evident from the preceding development, a Hilbert space contains many different orthonormal sets. The choice of a particular orthonormal set in a given Hilbert space is often extremely helpful in investigating certain problems. In Section 13.7 we saw how the complete orthonormal set $\{e^{ik\cdot} \mid k \in \mathbb{Z}\}$ can be used to study Fourier analysis in the space $L_2([-\pi,\pi],dt/2\pi)$. In this section we examine an orthonormal set that plays an important role in probability theory, namely, the Rademacher functions.

<u>Definition 13.8.1.</u> For each $n = 1,2,3,\ldots$ we define the nth <u>Rademacher function</u> $r_n$ on $[0,1]$ by

$$r_n(t) = 1 \quad \text{for } \frac{k-1}{2^n} \le t < \frac{k}{2^n}, \ k \text{ odd}, \ 1 \le k < 2^n;$$

$$r_n(t) = -1 \quad \text{for } \frac{k-1}{2^n} \le t < \frac{k}{2^n}, \ k \text{ even}, \ 1 < k \le 2^n;$$

$$r_n(t) = -1 \quad \text{for } t = 1.$$

Thus $r_n$ assumes the values +1 and -1 alternately on the $2^n$ subintervals of $[0,1]$:

$$[0,\frac{1}{2^n}),\ [\frac{1}{2^n}, \frac{2}{2^n}),\ldots,\ [\frac{2^n-2}{2^n}, \frac{2^n-1}{2^n}),\ [\frac{2^n-1}{2^n}, 1].$$

One can also describe the values of $r_n$ in terms of the nth digit of the binary expansion of any $t$, $0 \le t \le 1$. Namely, if $t = \Sigma_{n=1}^{\infty} b_n(t)/2^n$, $0 \le t \le 1$, where $b_n(t) = 0$ or 1, $n = 1,2,3,\ldots$, then it is easily seen that

$$r_n(t) = 1 - 2b_n(t) \qquad (n = 1,2,3,\ldots;\ t \in [0,1]),$$

provided that in the case of two valid binary expansions for $t$ we agree always to choose the one for which all but a finite number of the $b_n(t)$ are zero. Thus, for example, we write $1/2 = \Sigma_{n=1}^{\infty} b_n(t)/2^n$ where $b_1(t) = 1$ and $b_n(t) = 0$, $n > 1$, instead of where $b_1(t) = 0$ and $b_n(t) = 1$, $n > 1$. We shall return shortly to this connection between Rademacher functions and binary digits.

First, however, we have the following fundamental proposition, whose proof is left to the reader:

Proposition 13.8.1. The Rademacher functions $\{r_n\}$ form an orthonormal set in $L_2([0,1],dt)$, where $dt$ denotes Lebesgue measure on $[0,1]$.

The Rademacher functions do not form a complete orthonormal set because, as is easily verified, the function $r_0(t) = 1$, $t \in [0,1]$, is orthogonal to each $r_n$, $n = 1,2,3,\ldots$ . However, one can use the Rademacher functions to construct a complete orthonormal set in $L_2([0,1],dt)$. Indeed, let $w_0(t) = 1$, $t \in [0,1]$. If $n \ge 1$ and

$n = 2^{k_1} + 2^{k_2} + \cdots + 2^{k_{m(n)}}$, where $k_1 > k_2 > \cdots > k_{m(n)} \geq 0$,

$$w_n = r_{k_1+1} r_{k_2+1} \cdots r_{k_{m(n)}+1}.$$

It can then be shown that the set of functions $\{w_n \mid n = 0,1,2,\ldots\}$ is a complete orthonormal set in $L_2([0,1],dt)$. This set of functions is called the Walsh functions. It contains the Rademacher functions as a subset since $w_{2^n} = r_{n+1}$, $n = 0,1,2,\ldots$ . We do not prove these assertions, as we shall make no use of them, but instead refer the interested reader to [A1, pp. 59-62].

Our first use of the orthonormality of the Rademacher functions is to prove a special instance of the strong law of large numbers. Such laws assert that the arithmetic averages of certain functions converge almost everywhere. They play a central role in probability theory and ergodic theory (see, for example [Ei, pp. 439-443; F, pp. 1-29; $\text{Ha}_1$, pp. 13-24; Hs, pp. 221-242; HSt, pp. 447-449; Pa, pp. 417-422]).

Theorem 13.8.1. Let $\{r_n\}$ be the Rademacher functions. Then the sequence of arithmetice averages $\{(\Sigma_{k=1}^n r_k)/n\}$ converges to zero almost everywhere with respect to Lebesgue measure on $[0,1]$.

Proof. For each $n = 1,2,3,\ldots$ consider $f_n = [(\Sigma_{k=1}^n r_k)/n]^4$, which belongs to $L_1([0,1],dt)$. Some elementary, but tedious, computation reveals that

$$n^4 f_n = [(\sum_{k=1}^n r_k)^2]^2$$

$$= (\sum_{k=1}^n r_k^2 + 2 \sum_{\substack{k,m=1 \\ k<m}}^n r_k r_m)^2$$

$$= \Big(n + 2 \sum_{\substack{k,m=1 \\ k<m}}^{n} r_k r_m\Big)^2$$

$$= n^2 + 4n \sum_{\substack{k,m=1 \\ k<m}}^{n} r_k r_m + 4\Big(\sum_{\substack{k,m=1 \\ k<m}}^{n} r_k r_m\Big)\Big(\sum_{\substack{i,j=1 \\ i<j}}^{n} r_i r_j\Big)$$

$$= n^2 + 4n \sum_{\substack{k,m=1 \\ k<m}}^{n} r_k r_m + 4 \sum_{\substack{k,m=1 \\ k<m}}^{n} r_k^2 r_m^2 + 4\Big[\sum_{j=1}^{n} 2r_j^2\Big(\sum_{\substack{k,m=1 \\ k<m \\ k,m \neq j}}^{n} r_k r_m\Big)\Big]$$

$$= n^2 + 4n \sum_{\substack{k,m=1 \\ k<m}}^{n} r_k r_m + 4[(n-1) + (n-2) + \cdots + 2 + 1]$$

$$+ 8 \sum_{j=1}^{n} \Big(\sum_{\substack{k,m=1 \\ k<m \\ k,m \neq j}}^{n} r_k r_m\Big)$$

$$= n^2 + 2n(n-1) + 4n \sum_{\substack{k,m=1 \\ k<m}}^{n} r_k r_m + 8 \sum_{j=1}^{n} \Big(\sum_{\substack{k,m=1 \\ k<m \\ k,m \neq j}}^{n} r_k r_m\Big),$$

where we have used several times that $r_k^2 = 1$, $k = 1,2,3,\ldots$ . Dividing both sides of the preceding equation by $n^4$, integrating over $[0,1]$, and using the orthogonality of the Rademacher functions, we deduce that

$$\int_0^1 f_n(t)\,dt = \frac{1}{n^2} + \frac{2n(n-1)}{n^4}$$

$$\leq \frac{3}{n^2} \qquad (n = 1,2,3,\ldots).$$

Consequently we see that

$$\sum_{n=1}^{\infty} \int_0^1 f_n(t)\, dt < \infty.$$

Thus, by the same argument, _mutatis mutandis_, involving the Monotone Convergence Theorem as used in the proof of Kolmogorov's Theorem (Theorem 13.7.3), we conclude that the sequence $\{f_n\}$ converges to zero almost everywhere with respect to Lebesgue measure on $[0,1]$. It follows immediately that $\{(\Sigma_{k=1}^n r_k)/n\}$ also converges almost everywhere to zero, and the theorem is proved. □

As an immediate corollary we have the next result.

_Corollary 13.8.1._ For each $t \in [0,1]$ let $b_n(t)$ denote the nth digit in the binary expansion of $t$, $n = 1,2,3,\ldots$ . Then the sequence $\{(\Sigma_{k=1}^n b_k(t))/n\}$ converges to $1/2$ almost everywhere with respect to Lebesgue measure on $[0,1]$.

_Proof._ $r_k = 1 - 2b_k$, $k = 1,2,3,\ldots,$ and Theorem 13.8.1. □

In particular, the corollary asserts that the binary expansion of almost every $t \in [0,1]$ contains the same number of zeros and ones, as $\Sigma_{k=1}^n b_k(t)$ is clearly equal to the number of ones that appear in the first $n$ places of the binary expansion of $t$.

Next we wish to employ the orthonormality of the Rademacher functions to obtain a result about the convergence of random series. To be precise, suppose $\{a_k\}$ is a sequence of complex numbers and consider the formal infinite series $\Sigma_{k=1}^\infty \pm a_k$, where the plus and minus signs are chosen at random; that is, for a given $k$ the probabilities that $+a_k$ or $-a_k$ will appear in the series are each 1/2. Under what conditions on the sequence $\{a_k\}$ and for which choices of the plus and minus signs will $\Sigma_{k=1}^\infty \pm a_k$ converge? We shall see that convergence will result for almost all choices of the signs provided that $\Sigma_{k=1}^\infty |a_k|^2 < \infty$. It is evident for each

$n = 1,2,3,\ldots$ that the Lebesgue measure of both the set $\{t \mid r_n(t) = 1\}$ and the set $\{t \mid r_n(t) = -1\}$ is equal to $1/2$. Thus we see that our question is equivalent to examining the convergence of the series $\Sigma_{k=1}^{\infty} a_k r_k$ for a given sequence $\{a_k\} \subset \mathbb{C}$. We then have the following theorem due to Rademacher:

Theorem 13.8.2. Let $\{a_k\} \subset \mathbb{C}$. If $\Sigma_{k=1}^{\infty} |a_k|^2 < \infty$, then $\Sigma_{k=1}^{\infty} a_k r_k$ converges almost everywhere with respect to Lebesgue measure on $[0,1]$.

Proof. Since the Rademacher functions form an orthonormal set in $L_2([0,1],dt)$ we see from the Riesz-Fischer Theorem (Theorem 13.6.4) that there exists some function $f \in L_2([0,1],dt) \subset L_1([0,1],dt)$ such that

$$\lim_n \|f - \sum_{k=1}^{n} a_k r_k\|_2 = 0.$$

Moreover, for any $0 \leq s < t \leq 1$ we have from the Cauchy-Schwarz Inequality (Theorem 13.1.1)

$$\left|\int_s^t [f(u) - \sum_{k=1}^{n} a_k r_k(u)]\, du\right| \leq \int_s^t \left|f(u) - \sum_{k=1}^{n} a_k r_k(u)\right| du$$

$$\leq (t - s)^{1/2} \|f - \sum_{k=1}^{n} a_k r_k\|_2 \quad (n = 1,2,3,\ldots),$$

from which it follows at once that

$$\int_s^t f(u)\, du = \sum_{k=1}^{\infty} a_k \int_s^t r_k(u)\, du.$$

Now let $t_o \in [0,1]$ be any point such that $f(t_o)$ is defined, $t_o$ is not a dyadic rational ( that is, $t_o$ is not of the form $j/2^r$ for any nonnegative integers $j$ and $r$ for which $0 \leq j \leq 2^r$, $r \geq 0$), and such that $t_o$ is a Lebesgue point of $f$. The last assumption says, by definition, that

$$\lim_{\substack{(t-s)\to 0 \\ s<t_o<t}} \frac{1}{t-s}\int_s^t |f(u) - f(t_o)|\, du = 0.$$

These assumptions on $t_o$ are compatible because the set of dyadic rationals and the complement of the set of Lebesgue points for $f$ are both sets of Lebesgue measure zero. A proof of the latter assertion can be found, for instance, in [HSt, pp. 476 and 477]. Note that we are dealing here with a function in $L_2([0,1],dt)$, and not an equivalence class.

Since $t_o$ is not a dyadic rational, we see that there exist points $s_m = j_m/2^m$ and $t_m = (j_m + 1)/2^m$ such that

$$0 < s_m < t_o < t_m < 1.$$

It is then easily seen from the definition of the Rademacher functions that

$$\int_{s_m}^{t_m} r_k(u)\, du = 0 \qquad (k > m),$$

and

$$\int_{s_m}^{t_m} r_k(u)\, du = (t_m - s_m) r_k(t_o) \qquad (k \le m).$$

Consequently

$$\begin{aligned}\int_{s_m}^{t_m} f(u)\, du &= \sum_{k=1}^{\infty} a_k \int_{s_m}^{t_m} r_k(u)\, du \\ &= \sum_{k=1}^{m} a_k (t_m - s_m) r_k(t_o),\end{aligned}$$

from which it follows that

$$\frac{1}{t_m - s_m}\int_{s_m}^{t_m} f(u)\, du = \sum_{k=1}^{m} a_k r_k(t_o).$$

But then, if we similarly choose sequences $\{s_m\}$ and $\{t_m\}$ such that $\lim_m s_m = \lim_m t_m = t_o$, we see at once that

$$f(t_o) = \lim_m \frac{1}{t_m - s_m} \int_{s_m}^{t_m} f(u)\,du$$

$$= \lim_m \sum_{k=1}^{m} a_k r_k(t_o)$$

$$= \sum_{k=1}^{\infty} a_k r_k(t_o).$$

Thus for each point $t \in [0,1]$ that is a Lebesgue point for $f$ and is not a dyadic rational the series $\Sigma_{k=1}^{\infty} a_k r_k(t)$ converges. In particular, $\Sigma_{k=1}^{\infty} a_k r_k$ converges for almost all $t \in [0,1]$. □

Hence we see that $\Sigma_{k=1}^{\infty} \pm a_k$ converges for almost all choices of the plus and minus signs provided that $\Sigma_{k=1}^{\infty} |a_k|^2 < \infty$. The converse of this result is also valid: if $\Sigma_{k=1}^{\infty} |a_k|^2 = \infty$, then $\Sigma_{k=1}^{\infty} a_k r_k$ diverges almost everywhere. We do not give the proof of this result. For this and other results about random series the reader is referred to [A1, pp. 51-62; Dn, pp. 221-229; E$_2$, pp. 195-210; Ei, pp. 403 and 404; Kh; Z, pp. 212-222].

We hope that the discussion in this and the preceding section has convinced the reader that an appropriate choice of an orthonormal set can be extremely helpful in studying certain problems. There are many other more or less standard orthonormal sets that appear in various areas of mathematical investigation, such as the Laguerre, Legendre, and Jacobi polynomials and the Hermite functions, to mention a few. We do not pursue the subject further, but instead refer the reader to [A1, C, Sa, Sn, Sz] for additional material and applications.

<u>13.9</u>. <u>The Hilbert Space Adjoint</u>. In this section we wish to introduce the notion of the Hilbert space adjoint of a continuous linear transformation on a Hilbert space. This could be accomplished by combining the discussion of adjoints in Section 4.4 with the

Riesz Representation Theorem (Theorem 13.4.2), but instead we shall give an independent development.

Suppose $(V,\langle\cdot,\cdot\rangle)$ is a Hilbert space over $\Phi$ and let $T \in L(V)$. For a given fixed $x \in V$ we define $x^*$ by $x^*(y) = \langle T(y),x\rangle$, $y \in V$. Clearly $x^* \in V'$, and

$$\begin{aligned} |x^*(y)| &= |\langle T(y),x\rangle| \\ &\le \|T(y)\|\|x\| \\ &\le \|T\|\|x\|\|y\| \qquad (y \in V) \end{aligned}$$

shows that $x^* \in V^*$ and $\|x^*\| \le \|T\|\|x\|$. Thus by the Riesz Representation Theorem ( Theorem 13.4.2), there exists a unique $z_x \in V$ such that

$$x^*(y) = \langle T(y),x\rangle = \langle y,z_x\rangle \qquad (y \in V).$$

This leads to the following definition:

<u>Definition 13.9.1</u>. Let $(V,\langle\cdot,\cdot\rangle)$ be a Hilbert space over $\Phi$ and let $T \in L(V)$. Then the <u>Hilbert space adjoint</u> $T^*$ of $T$ is the mapping $T^* : V \to V$, defined by $T^*(x) = z_x$, $x \in V$, where $z_x$ is the unique vector in $V$ such that

$$\langle T(y),x\rangle = \langle y,z_x\rangle \qquad (y \in V).$$

It is easily seen that this definition of adjoint transformation is equivalent to that given in Section 4.4 in the context of Hilbert spaces. We shall generally speak of $T^*$ as the <u>adjoint</u> of $T$.

<u>Proposition 13.9.1</u>. Let $(V,\langle\cdot,\cdot\rangle)$ be a Hilbert space over $\Phi$ and let $T \in L(V)$. Then $T^* \in L(V)$, $\|T^*\| \le \|T\|$, and

$$\langle T^*(x),y\rangle = \langle x,T(y)\rangle \qquad (x,y \in V).$$

Proof. For any $x,y,z \in V$ and $a,b \in \Phi$ we have

$$\begin{aligned}\langle y,T^*(ax + bz)\rangle &= \langle T(y),ax + bz\rangle \\ &= \bar{a}\langle T(y),x\rangle + \bar{b}\langle T(y),z\rangle \\ &= \bar{a}\langle y,T^*(x)\rangle + \bar{b}\langle y,T^*(z)\rangle \\ &= \langle y,aT^*(x) + bT^*(z)\rangle.\end{aligned}$$

Hence $T^*$ is linear.

Second, recalling the definition of $T^*$ and the Riesz Representation Theorem (Theorem 13.4.2), we see that

$$\begin{aligned}\|T^*(x)\| &= \|z_x\| \\ &= \|x^*\| \\ &\leq \|T\|\|x\| \qquad (x \in V),\end{aligned}$$

whence $T^* \in L(V)$ and $\|T^*\| \leq \|T\|$.

The proof of the last assertion is left to the reader. □

Actually $\|T^*\| = \|T\|$, as is seen in the next result.

Proposition 13.9.2. Let $(V,\langle\cdot,\cdot\rangle)$ be a Hilbert space over $\Phi$.

(i) If $T,S \in L(V)$ and $a \in \Phi$, then $(T + S)^* = T^* + S^*$, $(aT)^* = \bar{a}T^*$, and $T^{**} = (T^*)^* = T$.

(ii) If $T \in L(V)$, then $\|T\| = \|T^*\|$.

Proof. The first two parts of part (i) are routine, and are left to the reader, while by Proposition 13.9.1

$$\langle y,T^{**}(x)\rangle = \langle T^*(y),x\rangle = \langle y,T(x)\rangle \qquad (x,y \in V),$$

whence $\|T^{**}(x) - T(x)\| = 0$, $x \in V$. Thus $T = T^{**}$.

The proof of implication (ii) is contained in the following inequalities:

$$\|T^*\| \leq \|T\| = \|T^{**}\| \leq \|T^*\|. \qquad \square$$

Evidently the preceding two propositions can be summarized in the following theorem:

Theorem 13.9.1. Let $(V,\langle\cdot,\cdot\rangle)$ be a Hilbert space over $\Phi$. Then the mapping $* : L(V) \to L(V)$, defined by $* : T \to T^*$, $T \in L(V)$, is a bijective antilinear isometric mapping from $L(V)$ to itself.

Let us look at a concrete example of a transformation $T$ and its adjoint $T^*$. Suppose $V = L_2([0,1],dt)$ with the usual inner product

$$\langle f,g\rangle = \int_0^1 f(t)\overline{g(t)}\,dt \qquad (f,g \in L_2([0,1],dt)$$

and let $K \in L_2([0,1] \times [0,1], ds\,dt)$, where $ds\,dt$ denotes Lebesgue measure on $[0,1] \times [0,1]$. Then we define $T$ by

$$T(f)(s) = \int_0^1 K(s,t)f(t)\,dt \qquad (f \in L_2([0,1],dt).$$

The formula, of course, is required to hold only almost everywhere. Clearly $T$ is linear, and the estimates

$$\begin{aligned}\|T(f)\|_2 &= [\int_0^1|\int_0^1 K(s,t)f(t)\,dt|^2\,ds]^{1/2}\\ &\leq [\int_0^1(\int_0^1 |K(s,t)|^2\,dt)(\int_0^1 |f(t)|^2\,dt)\,ds]^{1/2}\\ &= [\int_0^1(\int_0^1 |K(s,t)|^2\,dt)\,ds]^{1/2}[\int_0^1 |f(t)|^2\,dt]^{1/2}\\ &= \|K\|_2\|f\|_2 \qquad (f \in L_2([0,1],dt)\end{aligned}$$

show that $T \in L(L_2([0,1],dt))$.

Now $T^*$ is defined by the relation $\langle T(f),g\rangle = \langle f,T^*(g)\rangle$, $f,g \in L_2([0,1],dt)$, and so we have

$$\int_0^1 [\int_0^1 K(s,t)f(t)\,dt]\overline{g(s)}\,ds = \int_0^1 f(s)\overline{T^*(g)(s)}\,ds \quad (f,g \in L_2([0,1],dt).$$

Applying Fubini's Theorem [Ry, p. 269] to the first integral, we deduce that

$$\int_0^1 f(s)\overline{T^*(g)(s)}\,ds = \int_0^1 f(s)[\int_0^1 K(t,s)\overline{g(t)}\,dt]\,ds$$

$$= \int_0^1 f(s)\overline{[\int_0^1 \overline{K(t,s)}g(t)\,dt]}\,ds \quad (f,g \in L_2([0,1],dt),$$

from which we conclude that

$$T^*(g)(s) = \int_0^1 \overline{K(t,s)}g(t)\,dt \qquad (g \in L_2([0,1],dt).$$

As another example consider $V = \ell_2$ with the usual inner product

$$\langle\{a_k\},\{b_k\}\rangle = \sum_{k=1}^{\infty} a_k\overline{b}_k \qquad (\{a_k\},\{b_k\} \in \ell_2)$$

and define $T \in L(\ell_2)$ by $T(\{a_k\}) = \{b_k\}$, where $b_1 = 0$ and $b_k = a_{k-1}$, $k = 2,3,\ldots$ . Then to describe $T^*$ we observe that

$$\langle\{a_k\},T^*(\{b_k\})\rangle = \langle T(\{a_k\}),\{b_k\}\rangle$$

$$= \sum_{k=2}^{\infty} a_{k-1}\overline{b}_k$$

$$= \sum_{k=1}^{\infty} a_k\overline{b}_{k+1} \quad (\{a_k\},\{b_k\} \in \ell_2).$$

Consequently we see that $T^*(\{b_k\}) = \{c_k\}$, where $c_k = b_{k+1}$, $k = 1,2,3,\ldots$ .

As is well known from linear algebra, linear transformations between finite-dimensional linear spaces can be described in terms of matrices. An analog of this fact for continuous linear transformations $T$ on a Hilbert space is often useful in studying $T$ and $T^*$.

For example, suppose $(V,\langle\cdot,\cdot\rangle)$ is a Hilbert space over $\Phi$ and $\{e_\alpha \mid \alpha \in \Lambda\}$ is a complete orthonormal set in V. Then we see from the comments following Definition 13.6.3 that for any $T \in L(V)$

$$T(e_\alpha) = \sum_\Lambda \langle T(e_\alpha),e_\beta\rangle e_\beta \qquad (\alpha \in \Lambda).$$

It is easily seen that the <u>matrix</u>

$$(c_{\alpha\beta})_{\alpha,\beta\in\Lambda} = (\langle T(e_\alpha),e_\beta\rangle)_{\alpha,\beta\in\Lambda}$$

completely determines T. However, the matrix $(c_{\alpha\beta})_{\alpha,\beta\in\Lambda}$ clearly depends on the choice of the orthonormal basis $\{e_\alpha \mid \alpha \in \Lambda\}$.

Moreover, if $(c_{\alpha\beta})_{\alpha,\beta\in\Lambda}$ is a matrix corresponding to T and $(d_{\alpha\beta})_{\alpha,\beta\in\Lambda}$ is the matrix corresponding to $T^*$, then

$$\begin{aligned} d_{\alpha\beta} &= \langle T^*(e_\alpha),e_\beta\rangle \\ &= \langle e_\alpha,T(e_\beta)\rangle \\ &= \overline{\langle T(e_\beta),e_\alpha\rangle} \\ &= \overline{c}_{\beta\alpha} \qquad (\alpha,\beta \in \Lambda). \end{aligned}$$

Hence the matrix corresponding to $T^*$ is the conjugate transpose of the matrix corresponding to T, just as in the finite-dimensional case.

<u>13.10. Self-adjoint and Unitary Transformations.</u> Subsequently we shall restrict our attention, for the most part, to two special classes of Hilbert space transformations: the self-adjoint and unitary transformations. In this section we define these types of transformations and examine some of their fundamental properties.

<u>Definition 13.10.1.</u> Let $(V,\langle\cdot,\cdot\rangle)$ be a Hilbert space over $\Phi$ and let $T \in L(V)$. The transformation T is said to be <u>self-adjoint</u> if $T = T^*$, that is, $\langle T(x),y\rangle = \langle x,T(y)\rangle$, $x,y \in V$; and

T is said to be <u>unitary</u> if $TT^* = T^*T = I$, where $I$ denotes the identity mapping $I(x) = x$, $x \in V$.

In the case that $\Phi = \mathbb{R}$ unitary transformations are often called <u>orthogonal</u>. Let us first look at some examples and properties of self-adjoint transformations.

From Definition 13.5.1 we see that every orthogonal projection is a self-adjoint transformation. In general, if $T \in L(V)$ and the matrix corresponding to $T$ with respect to some orthonormal basis of $V$ is $(c_{\alpha\beta})_{\alpha,\beta \in \Lambda}$, then, since the matrix corresponding to $T^*$ is $(\overline{c}_{\beta\alpha})_{\alpha,\beta \in \Lambda}$, we see at once that $T$ is self-adjoint if and only if $c_{\alpha\beta} = \overline{c}_{\beta\alpha}$, $\alpha,\beta \in \Lambda$. This again generalizes a well known fact from linear algebra.

Finally, if we consider again the transformation $T$ on $L_2([0,1],dt)$, defined by

$$T(f)(s) = \int_0^1 K(s,t)f(t)\,dt \qquad (f \in L_2([0,1],dt),$$

where $K \in L_2([0,1] \times [0,1], ds\,dt)$, then we see that $T$ is self-adjoint if and only if

$$\int_0^1 K(s,t)f(t)\,dt = \int_0^1 \overline{K(t,s)}f(t)\,dt \qquad (f \in L_2([0,1],dt),$$

that is, if and only if $K(s,t) = \overline{K(t,s)}$ almost everywhere with respect to Lebesgue measure on the square $[0,1] \times [0,1]$.

Our first general result about self-adjoint transformations is the following proposition:

<u>Proposition 13.10.1</u>. Let $(V,\langle\cdot,\cdot\rangle)$ be a Hilbert space over $\Phi$. If $T \in L(V)$ is self-adjoint, then $\langle T(x),x\rangle \in \mathbb{R}$, $x \in V$.

<u>Proof</u>. $\langle T(x),x\rangle = \langle x,T(x)\rangle = \overline{\langle T(x),x\rangle}$, $x \in V$. □

Evidently, if $V$ is a Hilbert space over $\mathbb{R}$, then $\langle T(x),x\rangle$ is real for each $x \in V$ and any $T \in L(V)$. But since not all real matrices are symmetric about the diagonal, it is apparent that there exist $T \in L(V)$ that are not self-adjoint. However, if $V$ is a Hilbert space over $\mathbb{C}$, then the reality of $\langle T(x),x\rangle$, $x \in V$, does imply the self-adjointness of $T$, as shown in the next theorem. This difference between Hilbert spaces over $\mathbb{R}$ and over $\mathbb{C}$ will recur at several points in the next sections.

Theorem 13.10.1. Let $(V,\langle\cdot,\cdot\rangle)$ be a Hilbert space over $\mathbb{C}$ and let $T \in L(V)$. Then the following are equivalent:

(i) $T$ is self-adjoint.

(ii) $\langle T(x),x\rangle \in \mathbb{R}$, $x \in V$.

Proof. The implication from part (i) to part (ii) is contained in Proposition 13.10.1. Suppose that $\langle T(x),x\rangle \in \mathbb{R}$, $x \in V$. Then $\langle T^*(x),x\rangle \in \mathbb{R}$, $x \in V$, since $\langle T^*(x),x\rangle = \langle x,T(x)\rangle = \overline{\langle T(x),x\rangle}$. Consequently

$$\begin{aligned}\langle (T - T^*)(x),x\rangle &= \langle x,(T^* - T)(x)\rangle \\ &= \overline{\langle (T^* - T)(x),x\rangle} \\ &= \overline{\langle T^*(x),x\rangle} - \overline{\langle T(x),x\rangle} \\ &= \langle T^*(x),x\rangle - \langle T(x),x\rangle \\ &= -\langle (T - T^*)(x),x\rangle \qquad (x \in V).\end{aligned}$$

Hence $\langle (T - T^*)(x),x\rangle = 0$, $x \in V$.

However, a straightforward computation reveals that

$$\begin{aligned}4\langle (T - T^*)(x),y\rangle &= \langle (T - T^*)(x + y),x + y\rangle \\ &\quad -\langle (T - T^*)(x - y),x - y\rangle \\ &\quad +i\langle (T - T^*)(x + iy),x + iy\rangle \\ &\quad -i\langle (T - T^*)(x - iy),x - iy\rangle \qquad (x,y \in V),\end{aligned}$$

whence we conclude that $\langle (T - T^*)(x), y \rangle = 0$, $x, y \in V$.

In particular, $\|(T - T^*)(x)\| = 0$, $x \in V$, and so $T = T^*$. □

The computation utilized at the end of the proof of the theorem is a special case of a useful general result that we isolate in the next proposition. The proof is left to the reader.

Proposition 13.10.2. Let $(V, \langle \cdot, \cdot \rangle)$ be an inner product space over $\mathbb{C}$. If $T \in L(V)$ is such that $\langle T(x), x \rangle = 0$, $x \in V$, then $T = 0$.

The proposition fails if the scalar field is $\mathbb{R}$.

A final simple, but useful, proposition about self-adjoint transformations is the next result.

Proposition 13.10.3. Let $(V, \langle \cdot, \cdot \rangle)$ be a Hilbert space over $\mathbb{C}$. If $T \in L(V)$, then there exist unique self-adjoint transformations $T_1, T_2 \in L(V)$ such that $T = T_1 + iT_2$.

Proof. Set $T_1 = (T + T^*)/2$ and $T_2 = (T - T^*)/2i$. □

Now let us turn our attention to unitary transformations. The next proposition provides some equivalent descriptions of such transformations.

Proposition 13.10.4. Let $(V, \langle \cdot, \cdot \rangle)$ be a Hilbert space over $\mathbb{C}$ and let $T \in L(V)$. Then the following are equivalent:

(i) $T$ is unitary.
(ii) $T$ is surjective and $\|T(x)\| = \|x\|$, $x \in V$.
(iii) $T$ is bijective and $T^{-1} = T^*$.

Proof. If $T$ is unitary, then

$$\begin{aligned} \|T(x)\|^2 &= \langle T(x), T(x) \rangle \\ &= \langle x, T^*T(x) \rangle \end{aligned}$$

$$= \langle x,x \rangle$$

$$= \|x\|^2 \qquad (x \in V),$$

and if $x \in V$, then $x = I(x) = TT^*(x)$. Hence part (i) implies part (ii).

Moreover, if part (ii) holds, then from the Polarization Identity (Theorem 13.2.2) and the fact that $\|T(x)\| = \|x\|$, $x \in V$, we deduce at once that $\langle T(x),T(y) \rangle = \langle x,y \rangle$, $x,y \in V$. Thus

$$\langle T^*T(x),y \rangle = \langle T(x),T(y) \rangle$$

$$= \langle x,y \rangle \qquad (x,y \in V),$$

whence $T^*T = I$. However, $T$ is bijective, because $\|T(x)\| = \|x\|$, $x \in V$, implies that $T$ is injective, and so $T^{-1}$ exists. Clearly then $T^{-1} = T^*$, and part (ii) of the proposition implies part (iii).

Finally, suppose $T$ is bijective and $T^{-1} = T^*$. Then $TT^* = TT^{-1} = I = T^{-1}T = T^*T$, and so $T$ is unitary.

Therefore part (iii) implies part (i), and all three parts of the proposition are equivalent. □

It is perhaps worth noting explicitly that a unitary transformation preserves inner products; that is, $\langle T(x),T(y) \rangle = \langle x,y \rangle$, $x,y \in V$. In particular, every unitary transformation is an isometry.

Note also that in defining a unitary transformation it is not sufficient to require only that either $TT^* = I$ or $T^*T = I$. For instance, the shift transformation on $\ell_2$ discussed after Theorem 13.9.1 is a transformation $T \in L(\ell_2)$ such that $T^*T = I$. However, it is not unitary since it is not surjective. The general reason for each of the equations $TT^* = I$ or $TT^* = I$ alone not being enough to imply that $T$ is unitary is as follows: for infinite-dimensional spaces $T \in L(V)$ must be bijective for $T^{-1}$ to exist and belong to $L(V)$. This is in contrast to the situation when $V$ is a finite-dimensional space, as then injectivity and surjectivity of linear transformations are equivalent.

Proposition 13.10.4 allows us immediately to characterize unitary $T \in L(V)$ in terms of their matrix representations. We leave it to the reader to show that, if $T \in L(V)$ and $(c_{\alpha\beta})_{\alpha,\beta \in \Lambda}$ is the matrix corresponding to $T$, then $T$ is unitary if and only if $(d_{\alpha\beta})_{\alpha,\beta \in \Lambda} = (\overline{c}_{\beta\alpha})_{\alpha,\beta \in \Lambda}$, where $(d_{\alpha\beta})_{\alpha,\beta \in \Lambda}$ corresponds to $T^{-1}$. Once more this generalizes a standard result from linear algebra.

13.11. The Mean Ergodic Theorem. As mentioned in Section 13.8 one aspect of ergodic theory is the study of the convergence of certain arithmetic averages. We now wish to examine such questions in somewhat greater detail and see how unitary transformations enter into the subject in a natural manner. Some preliminaries are in order.

Definition 13.11.1. Let $(X,S,\mu)$ be a positive measure space. A mapping $\varphi : X \to X$ is said to be measure preserving if

(i) $\varphi$ is bijective.
(ii) $E \in S$ if and only if $\varphi(E) \in S$.
(iii) $\mu[\varphi(E)] = \mu(E)$, $E \in S$.

It is not difficult to verify that, if $\varphi$ is a measure preserving mapping on $X$, then $\mu[\varphi^{-1}(E)] = \mu[\varphi(E)] = \mu(E)$, $E \in S$.

One of the simplest examples of a measure preserving mapping is to take $X = \mathbb{R}$, $\mu$ equal to Lebesgue measure on $\mathbb{R}$ and $\varphi(t) = t + s$, $t \in \mathbb{R}$, where $s \in \mathbb{R}$ is fixed. There are, of course, other considerably more complicated measure preserving mappings (see, for example, [F, pp. 3 and 4, 75-99; $Ha_1$, pp. 5-9]).

Suppose now that $(X,S,\mu)$ is a positive measure space and $\varphi$ is a measure preserving mapping on $X$. If $f$ is any $\mu$-measurable function on $X$, we define $T(f)$ by the formula $T(f)(t) = f[\varphi(t)]$, which is to be valid for $\mu$-almost all $t \in X$. In other words, $T(f) = f \circ \varphi$. Clearly $T(f)$ is again a $\mu$-measurable function on $X$, and $T$ is a linear transformation on the linear space of $\mu$-measurable functions on $X$ to itself. If $f$ and $g$ are $\mu$-measurable and $T(f) = T(g)$ almost everywhere with respect to $\mu$, then $f = g$ almost everywhere. Indeed, $f(s) - g(s) \neq 0$ if and only if $s$ belongs to the set

$$\varphi(E) = \varphi(\{t \mid T(f)(t) - T(g)(t) \neq 0\}).$$

But since $\mu(E) = 0$ and $\varphi$ is measure preserving, we see that $\mu[\varphi(E)] = 0$. Thus $f = g$ almost everywhere. Consequently $T$ is injective. A similar argument shows that $T$ is surjective.

Given a positive measure space $(X,\mathcal{S},\mu)$ and a measure preserving mapping $\varphi$ on $X$, one of the fundamental concerns of ergodic theory is to investigate the convergence of the arithmetic averages

$$\frac{1}{n+1} \sum_{k=0}^{n} T^k(f)$$

for various classes of $\mu$-measurable functions $f$. The term $T^0$ is, of course, the identity transformation $I$.

If we restrict our attention to $f \in L_p(X,\mathcal{S},\mu)$, $1 \leq p \leq \infty$, then the transformation $T$ possesses some additional pleasant properties. The proof of the next proposition is left to the reader.

<u>Proposition 13.11.1</u>. Let $(X,\mathcal{S},\mu)$ be a positive measure space and let $\varphi$ be a measure preserving mapping on $X$. If $1 \leq p \leq \infty$ and $T(f) = f \circ \varphi$, $f \in L_p(X,\mathcal{S},\mu)$, then $T$ is a bijective linear isometry from $L_p(X,\mathcal{S},\mu)$ to $L_p(X,\mathcal{S},\mu)$.

In the particular case that $p = 2$, we see from Proposition 13.10.4 that $T$ is a unitary transformation on $L_2(X,\mathcal{S},\mu)$. This observation leads us to consider the convergence behavior of the arithmetic averages of certain unitary transformations on Hilbert space -- that is, the Mean Ergodic Theorem.

At this point the reader should, perhaps, recall the definition and basic properties of the strong operator topology $so(\mathcal{T})$ on $L(V)$, as discussed in Section 6.3.

Theorem 13.11.1 (Mean Ergodic Theorem). Let $(V,\langle\cdot,\cdot\rangle)$ be a Hilbert space over $\Phi$. If $T \in L(V)$ is unitary, then the sequence $\{(\Sigma_{k=0}^{n} T^k)/n+1\}$ converges in $(L(V),so(\mathcal{T}))$ to P, the orthogonal projection of V onto the closed linear subspace

$$W = \{x \mid x \in V,\ T(x) = x\}.$$

Proof. Since $T \in L(V)$, it is evident that W is indeed a closed linear subspace of V. Moreover, since T is unitary and hence an isometry, we see that $\|T^k(x)\| = \|x\|$, $x \in V$, $k = 0,1,2,\ldots,$ and so

$$\begin{aligned}\left\| \frac{1}{n+1} \sum_{k=0}^{n} T^k(x)\right\| &\le \frac{1}{n+1} \sum_{k=0}^{n} \|T^k(x)\| \\ &= \|x\| \qquad\qquad (x \in V).\end{aligned}$$

Now set $W_1 = \{T(x) - x \mid x \in V\}$. Clearly $W_1$ is a linear subspace of V, but it is not necessarily closed. However, we claim that $W = W_1^{\perp}$, the argument being as follows: If $x \in V$ and $y \in W$, then, since T is unitary, we have

$$\begin{aligned}\langle T(x) - x,y\rangle &= \langle T(x),y\rangle - \langle x,y\rangle \\ &= \langle x,T^*(y)\rangle - \langle x,y\rangle \\ &= \langle x,T^{-1}(y) - y\rangle \\ &= \langle x,T^{-1}[T(y)] - y\rangle \\ &= \langle x,y - y\rangle \\ &= 0.\end{aligned}$$

Thus $W \subset W_1^{\perp}$. Conversely, if $y \in W_1^{\perp}$, then, since T is unitary, we have for each $x \in V$

$$\begin{aligned} 0 &= \langle T(x) - x, y \rangle \\ &= \langle x, T^*(y) \rangle - \langle x, y \rangle \\ &= \langle x, T^{-1}(y) - y \rangle, \end{aligned}$$

whence $T^{-1}(y) = y$, $y \in W_1^\perp$. Consequently, $T(y) = T[T^{-1}(y)] = y$, $y \in W_1^\perp$, and $W_1^\perp \subset W$. Hence $W = W_1^\perp$.

Furthermore, from the continuity of the inner product (Corollary 13.1.2) it is readily deduced that $W_1^\perp = [cl(W_1)]^\perp$, where $cl(W_1)$, as usual, denotes the closure of $W_1$ in $V$. From Corollary 13.4.1 we then see that $cl(W_1) = [cl(W_1)]^{\perp\perp} = W^\perp$, since $cl(W_1)$ is a closed linear subspace of $V$. Thus from the Orthogonal Decomposition Theorem (Theorem 13.4.1) we conclude that

$$V = W \oplus W^\perp = W \oplus cl(W_1).$$

Recalling the definitions of the strong operator topology and of the orthogonal projection of $V$ onto $W$, we see that the proof will be complete if we can show that for each $x = x_1 + x_2 \in V$, where $x_1 \in W$ and $x_2 \in cl(W_1)$,

$$\lim_n \left\| \frac{1}{n+1} \sum_{k=0}^{n} T^k(x_1) - x_1 \right\| = 0$$

and

$$\lim_n \left\| \frac{1}{n+1} \sum_{k=0}^{n} T^k(x_2) \right\| = 0.$$

The first assertion is evident, because if $x_1 \in W$, then $T^k(x_1) = x_1$, $k = 0,1,2,\ldots$, and so $[\Sigma_{k=0}^n T^k(x_1)]/n + 1 = x_1$, $n = 0,1,2,\ldots$ .

On the other hand, suppose $x_2 \in W_1$, say $x_2 = T(y) - y$. Then

$$\left\| \frac{1}{n+1} \sum_{k=0}^{n} T^k(x_2) \right\| = \left\| \frac{1}{n+1} \left[ \sum_{k=0}^{n} T^{k+1}(y) - \sum_{k=0}^{n} T^k(y) \right] \right\|$$

$$= \frac{1}{n+1} \|T^{n+1}(y) - y\|$$

$$\leq \frac{2\|y\|}{n+1} \qquad (n = 0,1,2,\ldots),$$

$T$ being unitary and hence an isometry. It follows at once that

$$\lim_n \left\|\frac{1}{n+1} \sum_{k=0}^{n} T^k(x_2)\right\| = 0 \qquad (x_2 \in W_1).$$

However, if $x_2 \in \text{cl}(W_1)$ and $\epsilon > 0$ is given, then there exists some $y_2 \in W_1$ such that $\|x_2 - y_2\| < \epsilon$. Hence

$$\left\|\frac{1}{n+1} \sum_{k=0}^{n} T^k(x_2)\right\| \leq \left\|\frac{1}{n+1} \sum_{k=0}^{n} T^k(x_2 - y_2)\right\| + \left\|\frac{1}{n+1} \sum_{k=0}^{n} T^k(y_2)\right\|$$

$$\leq \frac{1}{n+1} \sum_{k=0}^{n} \|T^k(x_2 - y_2)\| + \left\|\frac{1}{n+1} \sum_{k=0}^{n} T^k(y_2)\right\|$$

$$= \|x_2 - y_2\| + \left\|\frac{1}{n+1} \sum_{k=0}^{n} T^k(y_2)\right\|$$

$$< \epsilon + \left\|\frac{1}{n+1} \sum_{k=0}^{n} T^k(y_2)\right\| \qquad (n = 0,1,2,\ldots),$$

$T$ being an isometry. An immediate consequence of these estimates is that

$$\limsup_n \left\|\frac{1}{n+1} \sum_{k=0}^{n} T^k(x_2)\right\| \leq \epsilon \qquad (x_2 \in \text{cl}(W_1)),$$

which implies, $\epsilon > 0$ being arbitrary, that

$$\lim_n \left\|\frac{1}{n+1} \sum_{k=0}^{n} T^k(x_2)\right\| = 0 \qquad (x \in \text{cl}(W_1)).$$

Therefore $\{[\Sigma_{k=0}^{n} T^k(x)]/n + 1\}$ converges to $P(x)$, $x \in V$, and the proof is complete. □

It should be noted that the sequence $\{(\Sigma_{k=0}^{n} T^k)/n + 1\}$ does not, in general, converge in the norm topology to $P$.

If we combine the Mean Ergodic Theorem with our discussion of the unitary transformation $T$ on $L_2(X,\mathcal{S},\mu)$ induced by a measure preserving mapping $\varphi$ on $X$, we easily deduce the validity of the next theorem.

<u>Theorem</u> <u>13.11.2</u>. Let $(X,\mathcal{S},\mu)$ be a positive measure space and let $\varphi$ be a measure preserving mapping on $X$. If $T(f) = f \circ \varphi$, $f \in L_2(X,\mathcal{S},\mu)$, then for each $f \in L_2(X,\mathcal{S},\mu)$ there exists some $f_o \in L_2(X,\mathcal{S},\mu)$ such that

(i) $\lim_n \left\|\frac{1}{n+1} \sum_{k=0}^{n} T^k(f) - f_o\right\|_2 = 0.$

(ii) $T(f_o) = f_o$.

Clearly $f_o$ is the image of $f$ under the orthogonal projection of $L_2(X,\mathcal{S},\mu)$ onto $W = \{g \mid g \in L_2(X,\mathcal{S},\mu),\ T(g) = g \circ \varphi = g\}$.

It is not difficult to extend this result to $L_1(X,\mathcal{S},\mu)$ provided we assume that $\mu$ has finite total mass.

<u>Theorem</u> <u>13.11.3</u>. Let $(X,\mathcal{S},\mu)$ be a positive measure space such that $\mu(X) = 1$ and let $\varphi$ be a measure preserving mapping on $X$. If $T(f) = f \circ \varphi$, $f \in L_1(X,\mathcal{S},\mu)$, then for each $f \in L_1(X,\mathcal{S},\mu)$ there exists some $f_o \in L_1(X,\mathcal{S},\mu)$ such that

(i) $\lim_n \left\|\frac{1}{n+1} \sum_{k=0}^{n} T^k(f) - f_o\right\|_1 = 0.$

(ii) $T(f_o) = f_o$.

<u>Proof</u>. Let $f \in L_1(X,\mathcal{S},\mu)$ and $\varepsilon > 0$ be given. Now, since $\mu(X) = 1$, we see that $L_2(X,\mathcal{S},\mu) \subset L_1(X,\mathcal{S},\mu)$, and moreover $L_2(X,\mathcal{S},\mu)$ is norm dense in $L_1(X,\mathcal{S},\mu)$ [Ry, p.244]. Thus there exists some $g \in L_2(X,\mathcal{S},\mu)$ such that $\|f - g\|_1 < \varepsilon/3$. Since $\mu(X) = 1$, we see from the Cauchy-Schwarz Inequality (Theorem 13.1.1) that $\|h\|_1 \le \|h\|_2$, $h \in L_2(X,\mathcal{S},\mu)$, and hence for each $n,m = 0,1,2,\ldots$ we have

$$\left\|\frac{1}{n+1}\sum_{k=0}^{n} T^k(f) - \frac{1}{m+1}\sum_{k=0}^{m} T^k(f)\right\|_1$$

$$\leq \left\|\frac{1}{n+1}\left[\sum_{k=0}^{n} T^k(f) - \sum_{k=0}^{n} T^k(g)\right]\right\|_1$$

$$+ \left\|\frac{1}{n+1}\sum_{k=0}^{n} T^k(g) - \frac{1}{m+1}\sum_{k=0}^{m} T^k(g)\right\|_1$$

$$+ \left\|\frac{1}{m+1}\left[\sum_{k=0}^{m} T^k(g) - \sum_{k=0}^{m} T^k(f)\right]\right\|_1$$

$$\leq \frac{1}{n+1}\sum_{k=0}^{n} \|T^k(f-g)\|_1$$

$$+ \left\|\frac{1}{n+1}\sum_{k=0}^{n} T^k(g) - \frac{1}{m+1}\sum_{k=0}^{m} T^k(g)\right\|_2$$

$$+ \frac{1}{m+1}\sum_{k=0}^{m} \|T^k(g-f)\|_1$$

$$< \frac{2\epsilon}{3} + \left\|\frac{1}{n+1}\sum_{k=0}^{n} T^k(g) - \frac{1}{m+1}\sum_{k=0}^{m} T^k(g)\right\|_2,$$

since $T$ is an isometry on $L_1(X,\mathcal{S},\mu)$ by Proposition 13.11.1. From the Mean Ergodic Theorem we see that $\{[\Sigma_{k=0}^{n} T^k(g)]/n+1\}$ is a Cauchy sequence in $L_2(X,\mathcal{S},\mu)$, from which it follows at once that $\{[\Sigma_{k=0}^{n} T^k(f)]/n+1\}$ is a Cauchy sequence in $L_1(X,\mathcal{S},\mu)$

Consequently, since $L_1(X,\mathcal{S},\mu)$ is a Banach space, there exists some $f_o \in L_1(X,\mathcal{S},\mu)$ such that

$$\lim_n \left\|\frac{1}{n+1}\sum_{k=0}^{n} T^k(f) - f_o\right\|_1 = 0.$$

Moreover,

$$\|T(f_o) - f_o\|_1 = \lim_n \frac{1}{n+1}\left\|T\left[\sum_{k=0}^{n} T^k(f)\right] - \sum_{k=0}^{n} T^k(f)\right\|_1$$

$$= \lim_n \frac{1}{n+1}\|T^{n+1}(f) - f\|_1$$

$$= 0,$$

as $T$ is an isometry.

Therefore $T(f_o) = f_o$, and the proof is complete. □

Obviously, if $f \in L_1(X,S,\mu)$, then from Theorem 13.11.3 we can deduce the existence of a subsequence of $\{[\Sigma_{k=0}^{n} T^k(f)]/n+1\}$ that converges almost everywhere to $f_o$. Actually it can be shown that the sequence itself converges almost everywhere to $f_o$. This is an immediate consequence of the following general theorem due to G.D.Birkhoff:

Theorem 13.11.4 (Individual Ergodic Theorem). Let $(X,S,\mu)$ be a $\sigma$-finite positive measure space and let $\varphi$ be a measure preserving mapping on $X$. If $T(f) = f \circ \varphi$, $f \in L_1(X,S,\mu)$, then for each $f \in L_1(X,S,\mu)$ there exists some $f_o \in L_1(X,S,\mu)$ such that

(i) $\lim_n \frac{1}{n+1} \sum_{k=0}^{n} T^k(f)(t) = f_o(t)$ for $\mu$-almost all $t \in X$.

(ii) $T(f_o) = f_o$.

We do not give the proof of this theorem, as it would involve too long a digression from our main interests. For proofs of the theorem and other material on ergodic theory we refer the reader to [F, Gr, $Ha_1$]. It is also of interest to compare the development of this section with the discussion of fixed point theorems in Chapter 12.

## 13.12. A Theorem About $H_2$.

In this section we wish to establish an analog of a classical theorem about analytic functions for the elements of a certain subspace of the Hilbert space $L_2([-\pi,\pi],dt/2\pi)$. The main Hilbert space tool we shall employ will be the Riesz Representation Theorem (Theorem 13.4.2). Before stating the theorem we discuss some preliminary material that we hope will illustrate the connection between a certain subspace $H_2$ of $L_2([-\pi,\pi],dt/2\pi)$ and the theory of analytic functions. We do not prove any of the results mentioned in this discussion.

<u>Definition 13.12.1</u>. The Hardy space $H_2$ is defined by

$$H_2 = \{f \mid f \in L_2([-\pi,\pi],dt/2\pi),\ \hat{f}(k) = \langle f,e^{ik\cdot}\rangle = 0,\ k \in \mathbb{Z},\ k < 0\}.$$

It is easily verified that $H_2$ is a closed linear subspace of the Hilbert space $L_2([-\pi,\pi],dt/2\pi)$ and so is itself a Hilbert space. There are other descriptions of $H_2$ that exhibit the relationship between $H_2$ and certain analytic functions on the open unit disk in the complex plane. We outline these here in order to further motivate the theorem to be established later in the section. No proofs are given. The interested reader is referred instead to [Dn, pp. 1-38; Ho, pp. 27-39; R, pp. 84-89; Z, pp. 271-285].

Consider the linear space $H_2'$ of functions $g$ that are analytic in the open unit disk $\{re^{it} \mid t \in [-\pi,\pi],\ 0 \leq r < 1\}$ and such that there exists a constant $M_g$, depending on $g$, for which

$$\frac{1}{2\pi}\int_{-\pi}^{\pi} |g(re^{it})|^2\, dt \leq M_g \qquad (0 \leq r < 1).$$

It can be shown that

$$\lim_{\substack{r\to 1\\ r<1}} \frac{1}{2\pi}\int_{-\pi}^{\pi} |g(re^{it})|^2\, dt$$

exists for each $g \in H_2'$ and that $H_2'$ is a Banach space with the norm

$$\|g\| = \lim_{\substack{r\to 1\\ r<1}} \left[\frac{1}{2\pi}\int_{-\pi}^{\pi} |g(re^{it})|^2\, dt\right]^{1/2} \qquad (g \in H_2').$$

Furthermore, if $g \in H_2'$, then one can also prove that

$$\lim_{\substack{r\to 1\\ r<1}} g(re^{it}) = \tilde{g}(e^{it})$$

exists for almost all $e^{it}$ with respect to Lebesgue measure on the unit circle (equivalently, for almost all $t \in [-\pi,\pi]$ with respect to Lebesgue measure on $[-\pi,\pi]$) and that the formula $f(t) = \tilde{g}(e^{it})$ defines an element $f \in H_2$. The correspondence so defined from $H_2'$ to $H_2$ can be shown to be a surjective isometric isomorphism of the Banach space $H_2'$ to the Banach space $H_2$. Thus the elements of $H_2$ can be thought of as the boundary values of certain analytic functions defined on the open unit disk.

It can also be shown that, if $g$ is analytic on the open unit disk and has the power series expansion

$$g(re^{it}) = \sum_{k=0}^{\infty} a_k r^k e^{ikt} \qquad (t \in [-\pi,\pi];\ 0 \le r < 1),$$

then $g \in H_2'$ if and only if $\Sigma_{k=0}^{\infty}|a_k|^2 < \infty$. Moreover, if $f \in H_2$ corresponds to $g \in H_2'$ under the isomorphism described in the preceding paragraph, then $\hat{f}(k) = a_k$, $k \in \mathbb{Z}$, $k \ge 0$, and

$$\|f\|_2 = [\sum_{k=0}^{\infty} |a_k|^2]^{1/2}.$$

These observations should convince the reader that there is an intimate connection between the space $H_2$ and certain analytic functions on the open unit disk. Now it is a standard result of analytic function theory that, if $g$ is an analytic function defined on some connected open set $\Omega$ and $g(z) = 0$ for all $z$ in some infinite subset $E \subset \Omega$ that has a cluster point in $\Omega$, then $g(z) = 0$, $z \in \Omega$. We wish to prove an analog of this result in $H_2$ by showing that, if $f \in H_2$ and $f$ vanishes almost everywhere on some subset $E \subset [-\pi,\pi]$ of positive Lebesgue measure, then $f$ vanishes almost everywhere. Before we can state the theorem we need two lemmas that will be used in the proof. The proof of the first lemma is left to the reader.

<u>Lemma 13.12.1</u>. If $f \in L_1([-\pi,\pi],dt/2\pi)$, then for each $n \in \mathbb{Z}$

$$(e^{in\cdot}f)^{\wedge}(k) = \hat{f}(k-n) \qquad (k \in \mathbb{Z}).$$

<u>Lemma 13.12.2</u>. If $f \in L_1([-\pi,\pi],dt/2\pi)$ and $\hat{f}(k) = 0$, $k \in \mathbb{Z}$, $k \neq 0$, then $f$ is a constant.

Proof. Elementary computations show that $[f - \hat{f}(0)]^\wedge(k) = 0$, $k \in \mathbb{Z}$, whence by Corollary 13.7.1 we conclude that $f = \hat{f}(0)$. □

For $f \in L_2([-\pi,\pi],dt/2\pi)$ the last lemma follows at once from the fact that $\{e^{ik\cdot} \mid k \in \mathbb{Z}\}$ is a complete orthonormal set in $L_2([-\pi,\pi],dt/2\pi)$. But we shall have need of the more general result.

Theorem 13.12.1. Let $E \subset [-\pi,\pi]$ be a set of positive Lebesgue measure. If $f \in H_2$ is such that $f$ vanishes almost everywhere on $E$, then $f$ vanishes almost everywhere on $[-\pi,\pi]$.

Proof. Let $W$ be the linear subspace of $H_2$ defined as

$$\begin{aligned} W &= \{g \mid g \in H_2,\ g \text{ vanishes almost everywhere on } E\} \\ &= \{g \mid g \in H_2,\ \chi_E g \text{ vanishes almost everywhere}\}, \end{aligned}$$

where $\chi_E$ is, as usual, the characteristec function of the set $E$. It is easily seen that $W$ is a closed linear subspace of $H_2$ and hence a Hilbert space. The theorem will be established if we can show that $W = \{0\}$.

To prove this it clearly suffices, either by Corollary 13.7.1 or the complete orthonormality of $\{e^{ik\cdot} \mid k \in \mathbb{Z}\}$, to show that $\hat{g}(k) = 0$, $k \in \mathbb{Z}$, for each $g \in W$. However, to establish this we claim it is sufficient to show that $\hat{g}(0) = 0$, $g \in W$.

Indeed, suppose that $\hat{g}(0) = 0$, $g \in W$. Then for each $g \in W$ we have $\hat{g}(k) = 0$, $k \in \mathbb{Z}$, $k < 0$, as $g \in H_2$. On the other hand, if $g \in W$, then from Lemma 13.12.1 we see that

$$(e^{-i\cdot}g)^\wedge(k) = \hat{g}(k + 1) = 0 \qquad (k \in \mathbb{Z},\ k < 0),$$

since $g \in H_2$ and $\hat{g}(0) = 0$. Thus $e^{-i\cdot}g \in H_2$. Moreover, since $\chi_E g$ vanishes almost everywhere, it follows at once that $e^{-i\cdot}g \in W$. But then

$$(e^{-i2\cdot}g)^\wedge(k) = (e^{-i\cdot}g)^\wedge(k + 1) = 0 \qquad (k \in \mathbb{Z},\ k < 0),$$

as $e^{-i\cdot}g \in H_2$ and $(e^{-i\cdot}g)^{\wedge}(0) = 0$. Hence, as before, we conclude that $e^{-i2\cdot}g \in W$. Continuing in this fashion, we deduce that $e^{-ik\cdot}g \in W$ for each $k \in \mathbb{Z}$, $k > 0$, and for each $g \in W$. However, we then have

$$0 = (e^{-ik\cdot}g)^{\wedge}(0) = \hat{g}(k) \qquad (k \in \mathbb{Z},\ k > 0),$$

which, coupled with our previous observation, shows that, if $\hat{g}(0) = 0$, $g \in W$, then $\hat{g}(k) = 0$, $k \in \mathbb{Z}$, $g \in W$.

Consequently, we need only show that $\hat{g}(0) = 0$, $g \in W$, in order to prove the theorem.

Define the linear functional $x^*$ on $W$ by

$$x^*(g) = \hat{g}(0) = \frac{1}{2\pi}\int_{-\pi}^{\pi} g(t)\, dt \qquad (g \in W).$$

It follows at once from the Cauchy-Schwarz Inequality (Theorem 13.1.1) that $x^* \in W^*$. Thus, since $W$ is a Hilbert space, from the Riesz Representation Theorem (Theorem 13.4.2) we see that there exists some $h \in W$ such that $x^*(g) = \langle g,h\rangle$, $g \in W$. We claim that $h = 0$, that is, $h$ vanishes almost everywhere.

Since $h \in W$, that is, $\chi_E h$ vanishes almost everywhere, it follows at once that $e^{ik\cdot}h \in W$ for each $k \in \mathbb{Z}$, $k > 0$, and so, by Lemma 13.12.1,

$$\begin{aligned}\langle e^{ik\cdot}h,h\rangle &= x^*(e^{ik\cdot}h)\\ &= (e^{ik\cdot}h)^{\wedge}(0)\\ &= \hat{h}(-k)\\ &= 0 \qquad (k \in \mathbb{Z},\ k > 0),\end{aligned}$$

as $h \in H_2$. Consequently

$$(|h|^2)^{\wedge}(-k) = \frac{1}{2\pi}\int_{-\pi}^{\pi} |h(t)|^2 e^{ikt}\, dt$$

$$= \frac{1}{2\pi}\int_{-\pi}^{\pi} e^{ikt} h(t)\overline{h(t)}\,dt$$

$$= \langle e^{ik\cdot}h,h\rangle$$

$$= 0 \qquad (k \in \mathbb{Z},\ k > 0).$$

But we also know that

$$(|h|^2)\hat{}(k) = \frac{1}{2\pi}\int_{-\pi}^{\pi} |h(t)|^2 e^{-ikt}\,dt$$

$$= \overline{\frac{1}{2\pi}\int_{-\pi}^{\pi} e^{ikt} h(t)\overline{h(t)}\,dt}$$

$$= \overline{\langle e^{ik\cdot}h,h\rangle}$$

$$= \overline{\hat{h}(-k)}$$

$$= 0 \qquad (k \in \mathbb{Z},\ k > 0).$$

Thus $(|h|^2)\hat{}(k) = 0$, $k \in \mathbb{Z}$, $k \neq 0$, whence we conclude from Lemma 13.12.2 that $|h|^2$ is a constant. The lemma is applicable because $|h| \in L_2([-\pi,\pi],dt/2\pi)$ implies $|h|^2 \in L_1([-\pi,\pi],dt/2\pi)$ by the Cauchy-Schwarz Inequality (Theorem 13.1.1).

However, $h \in W$ implies that $h$, and hence also $|h|^2$, vanishes almost everywhere on $E$, a set of positive measure. Hence $h$ must vanish almost everywhere, and so therefore $\hat{g}(0) = \langle g,h\rangle = 0$, $g \in W$.

Consequently $W = \{0\}$, and the proof is complete. □

<u>13.13. Some Basic Results in Spectral Theory</u>. As already indicated (for example, in Section 3.1), the action of a linear transformation $T$ from a finite-dimensional linear space $V$ to itself can always be realized as matrix multiplication of some matrix $A = (a_{jk})$ and the coefficient vectors of the elements of $V$. The matrix $A$, of course, depends on the choice of a basis for $V$. For certain

types of such transformations one can choose the basis of $V$ in such a way that the matrix $A$ is a diagonal matrix -- that is, $A$ only has nonzero entries on the diagonal. This means that, for example, in the case where $V$ is either $\mathbb{R}^n$ or $\mathbb{C}^n$, one can write

$$T(x) = \sum_{k=1}^{m} \langle x, e_k \rangle \lambda_k e_k \qquad (x \in V),$$

where $e_1, e_2, \ldots, e_m$ are certain linearly independent orthonormal vectors in $V$; $\lambda_1, \lambda_2, \ldots, \lambda_m$ are the not necessarily distinct nonzero diagonal entries of the matrix $A$; $\langle \cdot, \cdot \rangle$ denotes the usual inner product; and $m \leq n$. This is the case, for instance, whenever $T$ is self-adjoint -- that is, whenever the matrix corresponding to $T$ is equal to its conjugate transpose.

Our ultimate goal in the remaining sections is to establish a counterpart of this result for certain self-adjoint linear transformations on an arbitrary Hilbert space over $\mathbb{C}$. This result will appear in Section 13.15 when we prove a spectral decomposition theorem for compact self-adjoint linear transformations. The result for finite-dimensional linear spaces discussed at the beginning of this section will be a special case of this theorem. However, before we can establish the indicated result, we must develop some aspects of the general spectral theory of linear transformations on Hilbert spaces. For reasons that will subsequently become apparent we restrict our attention to Hilbert spaces over $\mathbb{C}$.

_Definition 13.13.1._ Let $(V, \langle \cdot, \cdot \rangle)$ be a Hilbert space over $\mathbb{C}$ and let $T \in \mathcal{L}(V)$. Then $\lambda \in \mathbb{C}$ is said to be an _eigenvalue_ for $T$ if there exists some $x \in V$, $x \neq 0$, such that $T(x) = \lambda x$, that is, $(\lambda I - T)(x) = 0$; and $x \in V$, $x \neq 0$, is said to be an _eigenvector_ for $T$ if there exists some $\lambda \in \mathbb{C}$ such that $T(x) = \lambda x$.

Thus we see that $\lambda \in \mathbb{C}$ is an eigenvalue for $T \in \mathcal{L}(V)$ if and only if $\lambda I - T$ is not injective. Evidently, if $\lambda$ is an eigenvalue for $T$, then one can always find a vector $x \in V$, $\|x\| = 1$, such that $T(x) = \lambda x$.

Definition 13.13.2. Let $(V,\langle\cdot,\cdot\rangle)$ be a Hilbert space over $\mathbb{C}$ and let $T \in L(V)$. Then

(i) $\sigma(T)$, the spectrum of $T$, is the set of $\lambda \in \mathbb{C}$ such that $\lambda I - T$ is either not injective or not surjective.

(ii) $P_\sigma(T)$, the point spectrum of $T$, is the set of $\lambda \in \mathbb{C}$ such that $\lambda I - T$ is not injective.

(iii) $C_\sigma(T)$, the continuous spectrum of $T$, is the set of $\lambda \in \mathbb{C}$ such that $\lambda I - T$ is injective, but not surjective, and such that $R(\lambda I - T)$ is dense in $V$.

(iv) $R_\sigma(T)$, the residual spectrum of $T$, is the set of $\lambda \in \mathbb{C}$ such that $\lambda I - T$ is injective, but not surjective, and such that $R(\lambda I - T)$ is not dense in $V$.

(v) $\rho(T)$, the resolvent set of $T$, is the set of $\lambda \in \mathbb{C}$ such that $\lambda I - T$ is bijective.

It is evident that $\mathbb{C} = \sigma(T) \cup \rho(T)$ and that

$$\sigma(T) = P_\sigma(T) \cup C_\sigma(T) \cup R_\sigma(T)$$

for each $T \in L(V)$. Moreover, the various sets defined here are pairwise disjoint. Clearly $P_\sigma(T)$ consists precisely of the eigenvalues of $T$. If $\lambda \in \rho(T)$, then $\lambda I - T$ is a bijective continuous linear mapping from $V$ to itself, and so by Corollary 7.2.2 to the Open Mapping Theorem we see that $(\lambda I - T)^{-1}$ exists and belongs to $L(V)$. Thus $\rho(T)$ is precisely the set of points $\lambda \in \mathbb{C}$ such that $\lambda I - T$ has a bounded inverse defined on all of $V$. When $\lambda \in \rho(T)$, we shall often write $(\lambda I - T)^{-1} = R(\lambda,T)$ and call $R(\lambda,T)$ the resolvent of $T$. Note that the notation $R(\lambda,T)$ for $(\lambda I - T)^{-1}$ is used only when $(\lambda I - T)^{-1} \in L(V)$.

It should be pointed out that the equality $P_\sigma(T) = \sigma(T)$ does not generally hold. For example, the shift transformation $T \in L(\ell_2)$, defined by $T(\{a_k\}) = \{b_k\}$ where $b_1 = 0$ and $b_k = a_{k-1}$, $k = 2,3,\ldots$, is clearly injective, but not surjective. Hence $0 \in \sigma(T) \sim P_\sigma(T)$. Actually it is possible to construct examples of continuous linear transformations $T$ such that each or any pair of the sets $P_\sigma(T)$, $R_\sigma(T)$, and $C_\sigma(T)$ is empty, with the one exception that not all can be simultaneously empty -- that is, it cannot be the case that $\sigma(T) = \emptyset$. We shall prove this last assertion shortly. Since ultimately we shall be mainly concerned with compact self-adjoint linear transformations, where $\sigma(T) \sim P_\sigma(T)$ is either $\emptyset$ or $\{0\}$, we do not pursue further an investigation of the sets $C_\sigma(T)$ and $R_\sigma(T)$.

Our first concern is to examine $\rho(T)$ and $R(\lambda,T)$ a bit more thoroughly.

Lemma 13.13.1. Let $(V,\langle\cdot,\cdot\rangle)$ be a Hilbert space over $\mathbb{C}$, let $T \in L(V)$, and suppose $\lambda_o \in \rho(T)$. If $\lambda \in \mathbb{C}$ is such that $|\lambda - \lambda_o| < 1/\|R(\lambda_o,T)\|$, then

(i) $I - (\lambda_o - \lambda)R(\lambda_o,T)$ is bijective.
(ii) $[I - (\lambda_o - \lambda)R(\lambda_o,T)]^{-1} \in L(V)$.
(iii) $[I - (\lambda_o - \lambda)R(\lambda_o,T)]^{-1} = \Sigma_{k=0}^{\infty}[(\lambda_o - \lambda)R(\lambda_o,T)]^k$.

Proof. The proof is a straightforward application of the remarks following Theorem 6.3.3, combined with the observation that the series $\Sigma_{k=0}^{\infty}[(\lambda_o - \lambda)R(\lambda_o,T)]^k$ converges in $L(V)$ since $|\lambda_o - \lambda| < 1/\|R(\lambda_o,T)\|$. The details are left to the reader. □

This lemma allows us to establish the following important result concerning the existence of $R(\lambda,T)$:

Theorem 13.13.1. Let $(V,\langle\cdot,\cdot\rangle)$ be a Hilbert space over $\mathbb{C}$, let $T \in L(V)$, and suppose $\lambda_o \in \rho(T)$. If $\lambda \in \mathbb{C}$ is such that $|\lambda - \lambda_o| < 1/\|R(\lambda_0,T)\|$, then

(i) $\lambda \in \rho(T)$.
(ii) $R(\lambda,T) = R(\lambda_o,T)\Sigma_{k=0}^{\infty}[(\lambda_o - \lambda)R(\lambda_o,T)]^k$.

Proof. By Lemma 13.13.1 we see that

$$R(\lambda_o,T)\sum_{k=0}^{\infty}[(\lambda_o - \lambda)R(\lambda_o,T)]^k = R(\lambda_o,T)[I - (\lambda_o - \lambda)R(\lambda_o,T)]^{-1}.$$

Moreover,

$$(\lambda I - T)R(\lambda_0,T)[I - (\lambda_0 - \lambda)R(\lambda_0,T)]^{-1}$$
$$= [(\lambda - \lambda_0)I + \lambda_0 I - T]R(\lambda_0,T)[I - (\lambda_0 - \lambda)R(\lambda_0,T)]^{-1}$$
$$= [(\lambda - \lambda_0)R(\lambda_0,T) + I][I - (\lambda_0 - \lambda)R(\lambda_0,T)]^{-1}$$
$$= I.$$

Similarly, since

$$R(\lambda_0,T)[I - (\lambda_0 - \lambda)R(\lambda_0,T)]^{-1} = \sum_{k=0}^{\infty} (\lambda_0 - \lambda)^k [R(\lambda_0,T)]^{k+1}$$
$$= [I - (\lambda_0 - \lambda)R(\lambda_0,T)]^{-1}R(\lambda_0,T),$$

we see that

$$R(\lambda_0,T)[I - (\lambda_0 - \lambda)R(\lambda_0,T)]^{-1}(\lambda I - T) = I.$$

It then follows immediately that

$$R(\lambda,T) = (\lambda I - T)^{-1} = R(\lambda_0,T)\sum_{k=0}^{\infty} [(\lambda_0 - \lambda)R(\lambda_0,T)]^k,$$

and $\lambda \in \rho(T)$. □

Corollary 13.13.1. Let $(V,\langle\cdot,\cdot\rangle)$ be a Hilbert space over $\mathbb{C}$. If $T \in L(V)$, then $\rho(T)$ is an open subset of $\mathbb{C}$, and $\sigma(T)$ is a closed subset of $\mathbb{C}$.

We are now in a position to prove that $\sigma(T)$ is always nonempty.

Theorem 13.13.2. Let $(V,\langle\cdot,\cdot\rangle)$ be a Hilbert space over $\mathbb{C}$. If $T \in L(V)$, then $\sigma(T) \neq \emptyset$.

Proof. We first make some preliminary observations based on Theorem 13.13.1. For each $x \in V$ and $x^* \in V^*$ we define the complex-valued function $f_{x,x^*}$ on $\rho(T)$ by

$$f_{x,x^*}(\lambda) = x^*[R(\lambda,T)(x)] \qquad (\lambda \in \rho(T)).$$

We claim that each such function $f_{x,x^*}$ is analytic on $\rho(T)$. Since $\rho(T)$ is open, to prove this assertion it suffices to show that $f_{x,x^*}$ has a power series expansion about each point in $\rho(T)$. But, if $\lambda_o \in \rho(T)$, then, from Theorem 13.13.1 we see that

$$f_{x,x^*}(\lambda) = \sum_{k=0}^{\infty} x^*([R(\lambda_o,T)]^{k+1}(x))(\lambda_o - \lambda)^k$$

for all $\lambda \in \mathbb{C}$ such that $|\lambda - \lambda_o| < 1/\|R(\lambda_o,T)\|$. Thus each $f_{x,x^*}$ is analytic on $\rho(T)$.

Furthermore, we note that, if $\lambda \in \mathbb{C}$ is such that $|\lambda| > \|T\|$, then $\lambda \in \rho(T)$, as, in this case, we see at once that the series $\Sigma_{k=0}^{\infty} T^k/\lambda^{k+1}$ converges in $L(V)$, as $\|T\|/|\lambda| < 1$. Hence

$$\begin{aligned}(\lambda I - T) \sum_{k=0}^{\infty} \frac{T^k}{\lambda^{k+1}} &= \lim_n (\lambda I - T) \sum_{k=0}^{n} \frac{T^k}{\lambda^{k+1}} \\ &= \lim_n \left[\sum_{k=0}^{n} \frac{T^k}{\lambda^k} - \sum_{k=0}^{n} \frac{T^{k+1}}{\lambda^{k+1}}\right] \\ &= \lim_n \left[I - \frac{T^{n+1}}{\lambda^{n+1}}\right] \\ &= I,\end{aligned}$$

since $\lim_n (\|T\|/|\lambda|)^{n+1} = 0$. Similarly

$$\sum_{k=0}^{\infty} \frac{T^k}{\lambda^{k+1}} (\lambda I - T) = I,$$

and so $R(\lambda,T) = \Sigma_{k=0}^{\infty} T^k/\lambda^{k+1}$. Consequently

$$\begin{aligned}\|R(\lambda,T)\| &\le \sum_{k=0}^{\infty} \frac{\|T\|^k}{|\lambda|^{k+1}} \\ &= \frac{1}{|\lambda|(1 - \|T\|/|\lambda|)},\end{aligned}$$

from which it follows easily that $\lim_{\lambda\to\infty} |f_{x,x^*}(\lambda)| = 0$ for each $x \in V$ and each $x^* \in V^*$.

Now suppose that $\sigma(T) = \emptyset$. We see that for each $x \in V$ and $x^* \in V^*$, the function $f_{x,x^*}$ is analytic on $\rho(T) = \mathbb{C}$, that is, $f_{x,x^*}$ is an entire function. Moreover, the conclusion of the preceding paragraph shows that each $f_{x,x^*}$ is bounded. Hence by Liouville's Theorem [$Ru_2$, p. 213] we conclude that each $f_{x,x^*}$ is identically zero on $\mathbb{C}$. Thus by a consequence of the Hahn-Banach Theorem (Corollary 4.2.6) we have $R(\lambda,T)(x) = 0$, $\lambda \in \mathbb{C}$ and $x \in V$, and hence $R(\lambda,T) = 0$, $\lambda \in \mathbb{C}$, which is clearly absurd.

Therefore $\sigma(T) \neq \emptyset$. □

We have restricted our attention in the present discussion to Hilbert spaces over the complex numbers $\mathbb{C}$ in order to ensure the validity of the preceding result. If one considers Hilbert spaces over $\mathbb{R}$, then the theorem may very well be false -- provided, of course, one recognizes that in any case $\sigma(T)$ must be a subset of the scalar field of the Hilbert space in question. Otherwise, one could not meaningfully discuss the transformation $\lambda I - T$. Thus, for instance, if one considers the linear transformation $T$ on $\mathbb{R}^2$ determined by the matrix

$$A = \begin{bmatrix} 0 & -1 \\ 1 & 0 \end{bmatrix}$$

then it is readily verified that $\lambda I - T$ is not injective if and only if the determinant of the matrix $\lambda I - A$ vanishes at $\lambda$ (that is, $\lambda$ is a root of $\lambda^2 + 1 = 0$). Since $\mathbb{R}^2$ is a finite-dimensional Hilbert space over $\mathbb{R}$, it is apparent that $\sigma(T) = \emptyset$.

<u>13.14.</u> <u>Some Spectral Theory Results for Self-adjoint Transformations.</u> In this section we wish to discuss some special results in the spectral theory of self-adjoint continuous linear transformations. The utility of these results will become apparent in the next section. We begin by defining the notion of the numerical range of an element of $L(V)$.

Definition 13.14.1. Let $(V,\langle\cdot,\cdot\rangle)$ be a Hilbert space over $\mathbb{C}$ and let $T \in L(V)$. Then the numerical range $\omega(T)$ of $T$ is defined by

$$\omega(T) = \{\langle T(x),x\rangle \mid x \in V, \|x\| = 1\}.$$

From Proposition 13.10.1 we see that, if $T \in L(V)$ is self-adjoint, then $\omega(T) \subset \mathbb{R}$, since $\langle T(x),x\rangle \in \mathbb{R}$, $x \in V$. An immediate consequence of this observation and the next theorem is that $\sigma(T) \subset \mathbb{R}$ whenever $T$ is self-adjoint.

Theorem 13.14.1. Let $(V,\langle\cdot,\cdot\rangle)$ be a Hilbert space over $\mathbb{C}$. If $T \in L(V)$ is self-adjoint, then $\sigma(T) \subset \mathrm{cl}[\omega(T)]$.

Proof. Suppose $\lambda \notin \mathrm{cl}[\omega(T)]$ and let $d = \inf_{\xi \in \mathrm{cl}[\omega(T)]}|\lambda - \xi|$. Clearly $d > 0$. Moreover, if $x \in V$ and $\|x\| = 1$, then by the Cauchy-Schwarz Inequality (Theorem 13.1.1) and the definition of $\omega(T)$ we see that

$$\begin{aligned} d &\le |\lambda - \langle T(x),x\rangle| \\ &= |\langle(\lambda I - T)(x),x\rangle| \\ &\le \|(\lambda I - T)(x)\|\|x\| \\ &= \|(\lambda I - T)(x)\|. \end{aligned}$$

Thus it follows easily that for any $x \in V$

$$d\|x\| \le \|(\lambda I - T)(x)\|.$$

Consequently by Proposition 3.2.3 we have $(\lambda I - T)^{-1} \in L(R(\lambda I - T),V)$. An elementary argument, whose details we leave to the reader, shows that $R(\lambda I - T)$ is a closed linear subspace of $V$. We claim that $R(\lambda I - T) = V$ by the following argument: If $(\lambda I - T) \neq V$, then by a consequence of the Hahn-Banach Theorem (Theorem 4.2.3) and the Riesz Representation Theorem (Theorem 13.4.2) there exists some $y \in V$, $\|y\| = 1$, such that $\langle(\lambda I - T)(x),y\rangle = 0$, $x \in V$. Then, in particular, we would have $\langle(\lambda I - T)(y),y\rangle = 0$, from which we deduce at once that $\lambda = \langle T(y),y\rangle \in \omega(T)$, contrary to the assumption that $\lambda \notin \mathrm{cl}[\omega(T)]$. Hence $(\lambda I - T) = V$, and $(\lambda I - T)^{-1} = R(\lambda,T) \in L(V)$.

Therefore $\lambda \in \rho(T)$, whence $\sigma(T) \subset \mathrm{cl}[\omega(T)]$. □

Corollary 13.14.1. Let $(V,\langle\cdot,\cdot\rangle)$ be a Hilbert space over $\mathbb{C}$. If $T \in L(V)$ is self-adjoint, then $\sigma(T) \subset \mathbb{R}$.

An easy consequence of Corollary 13.14.1 is that the eigenvectors corresponding to distinct eigenvalues of a self-adjoint continuous linear transformation are orthogonal.

Corollary 13.14.2. Let $(V,\langle\cdot,\cdot\rangle)$ be a Hilbert space over $\mathbb{C}$ and let $T \in L(V)$ be self-adjoint. If $\lambda_1,\lambda_2 \in P_\sigma(T)$ are such that $\lambda_1 \neq \lambda_2$ and if $x_1,x_2 \in V$ are nonzero vectors such that $T(x_k) = \lambda_k x_k$, $k = 1,2$, then $x_1 \perp x_2$.

Proof. Clearly we have

$$\begin{aligned}\lambda_1\langle x_1,x_2\rangle &= \langle\lambda_1 x_1,x_2\rangle \\ &= \langle T(x_1),x_2\rangle \\ &= \langle x_1,T(x_2)\rangle \\ &= \langle x_1,\lambda_2 x_2\rangle \\ &= \lambda_2\langle x_1,x_2\rangle,\end{aligned}$$

since $\sigma(T) \subset \mathbb{R}$ by Corollary 13.14.1. Thus $(\lambda_1 - \lambda_2)\langle x_1,x_2\rangle = 0$, and so $x_1 \perp x_2$. □

If $T \in L(V)$, then it is apparent that for each $x \in V$, $\|x\| = 1$, we have $|\langle T(x),x\rangle| \leq \|T\|$, and hence $\sup_{\xi \in \mathrm{cl}[\omega(T)]}|\xi| \leq \|T\|$. Thus if $T \in L(V)$ is self-adjoint, Theorem 13.14.1 shows that

$$\sup_{\lambda \in \sigma(T)} |\lambda| \leq \|T\|.$$

Ultimately we shall prove in this case that

$$\sup_{\lambda \in \sigma(T)} |\lambda| = \|T\|.$$

As a first step in this direction we prove the following theorem:

Theorem 13.14.2. Let $(V,\langle\cdot,\cdot\rangle)$ be a Hilbert space over $\mathbb{C}$. If $T \in L(V)$ is self-adjoint, then

$$\|T\| = \sup\{|\langle T(x),x\rangle| \mid x \in V, \|x\| = 1\}$$

$$= \sup_{\xi \in \omega(T)} |\xi|.$$

Proof. By the foregoing remarks we know that $|\xi| \leq \|T\|$, $\xi \in \omega(T)$. Let

$$a = \sup_{\xi \in \omega(T)} |\xi| = \sup\{|\langle T(x),x\rangle| \mid x \in V, \|x\| = 1\}.$$

Clearly $a \leq \|T\|$. Some elementary computations and the self-adjointness of $T$ reveal that

$$\langle T(x + y),x + y\rangle - \langle T(x - y),x - y\rangle = 4\mathrm{Re}\langle T(y),x\rangle \qquad (x,y \in V),$$

whence, if $x \neq y$ and $x \neq -y$, we have

$$\begin{aligned}
4\mathrm{Re}\langle T(y),x\rangle &\leq |\langle T(x + y),x + y\rangle| + |\langle T(x - y),x - y\rangle| \\
&= \left|\left\langle \frac{T(x + y)}{\|x + y\|},\frac{x + y}{\|x + y\|}\right\rangle\right| \|x + y\|^2 \\
&\quad + \left|\left\langle \frac{T(x - y)}{\|x - y\|},\frac{x - y}{\|x - y\|}\right\rangle\right| \|x - y\|^2 \\
&\leq a(\|x + y\|^2 + \|x - y\|^2) \\
&= 2a(\|x\|^2 + \|y\|^2) \qquad (x,y \in V).
\end{aligned}$$

The last step is, of course, just the Parallelogram Identity (Theorem 13.2.1). A similar argument reveals that the estimate is also valid in the case that $x = y$ or $x = -y$.

If $y \in V$ is such that $\|y\| = 1$ and $\|T(y)\| \neq 0$, then, on setting $x = T(y)/\|T(y)\|$ in the preceding estimate, we obtain

$$\begin{aligned}
\|T(y)\| &= \langle T(y), \frac{T(y)}{\|T(y)\|}\rangle \\
&= \operatorname{Re}\langle T(y), \frac{T(y)}{\|T(y)\|}\rangle \\
&\leq \frac{a}{2}(\|y\|^2 + \frac{\|T(y)\|^2}{\|T(y)\|^2}) \\
&= a.
\end{aligned}$$

Since this inequality obviously remains valid when $\|T(y)\| = 0$, $\|y\| = 1$, we conclude that

$$\|T\| = \sup_{\|y\| = 1} \|T(y)\| \leq a,$$

which completes the proof. □

To show that $\|T\| = \sup_{\lambda \in \sigma(T)} |\lambda|$ it clearly suffices, in view of the preceding result, to prove that the supremum and infimum of $\omega(T)$ both lie in $\sigma(T)$. With this in mind we make the following definition:

<u>Definition 13.14.2</u>. Let $(V, \langle\cdot,\cdot\rangle)$ be a Hilbert space over $\mathbb{C}$. If $T \in L(V)$ is self-adjoint, then we set $M_T = \sup \omega(T)$ and $m_T = \inf \omega(T)$.

<u>Theorem 13.14.3</u>. Let $(V, \langle\cdot,\cdot\rangle)$ be a Hilbert space over $\mathbb{C}$. If $T \in L(V)$ is self-adjoint, then $M_T \in \sigma(T)$ and $m_T \in \sigma(T)$.

<u>Proof</u>. We shall show only that $m_T \in \sigma(T)$, leaving the analogous proof that $M_T \in \sigma(T)$ to the reader. Let $S = T - m_T I$. It follows immediately from the definition of $m_T$ and the equation

$\langle S(x),x\rangle = \langle T(x),x\rangle - m_T\langle x,x\rangle$, $x \in V$, that $\langle S(x),x\rangle \geq 0$, $x \in V$, and that $\inf_{\|x\|=1}\langle S(x),x\rangle = 0$. A routine argument then shows that the formula

$$\psi(x,y) = \langle S(x),y\rangle \qquad (x,y \in V)$$

defines a nonnegative symmetric bilinear form $\psi$ on V. From the comment following Theorem 13.1.1 we see that the Cauchy-Schwarz Inequality is valid for $\psi$, and so, in particular, we have

$$\begin{aligned}|\psi[S(x),x]|^2 &\leq \psi[S(x),S(x)]\psi(x,x)\\ &= \langle S^2(x),S(x)\rangle\langle S(x),x\rangle \qquad (x \in V).\end{aligned}$$

However, since T is self-adjoint and $m_T \in \mathbb{R}$, we see at once that S is self-adjoint, and so

$$\begin{aligned}\psi[S(x),x] &= \langle S^2(x),x\rangle\\ &= \langle S(x),S(x)\rangle\\ &\geq 0 \qquad (x \in V).\end{aligned}$$

Hence

$$\begin{aligned}\|S(x)\|^2 &\leq \langle S^2(x),S(x)\rangle\langle S(x),x\rangle\\ &= \langle S^3(x),x\rangle\langle S(x),x\rangle\\ &\leq \|S^3\|\|x\|^2\langle S(x),x\rangle \qquad (x \in V).\end{aligned}$$

Now, if S is not injective, then obviously $m_T \in P_\sigma(T) \subset \sigma(T)$. So suppose that S is injective. If S were also surjective, then by Corollary 7.2.2 to the Open Mapping Theorem we should have $S^{-1} \in L(V)$. Consequently by Proposition 3.2.3 there exists some $m > 0$ such that $m \leq \|S(x)\|$, $x \in V$, $\|x\| = 1$. However, $\|S\| \neq 0$ because S is injective, and so from the previously established inequality and the fact that $\inf_{\|x\|=1}\langle S(x),x\rangle = 0$, we deduce that $\inf_{\|x\|=1}\|S(x)\| = 0$, which contradicts the fact that $m > 0$.

Thus, if $S$ is injective, it cannot be surjective, and so $m_T \in \sigma(T)$. □

Corollary 13.14.3. Let $(V, \langle \cdot , \cdot \rangle)$ be a Hilbert space over $\mathbb{C}$. If $T \in L(V)$ is self-adjoint, then $\|T\| = \sup_{\lambda \in \sigma(T)} |\lambda|$.

13.15. A Spectral Decomposition Theorem for Compact Self-adjoint Transformations. We are now almost ready to extend the result of linear algebra indicated in the introductory remarks to Section 13.13 to a certain class of linear transformations on Hilbert spaces: the compact self-adjoint continuous linear transformations.

Definition 13.15.1. Let $(V, \langle \cdot , \cdot \rangle)$ be a Hilbert space over $\mathbb{C}$. The transformation $T \in L(V)$ is said to be compact if, whenever $E \subset V$ is norm bounded, $\mathrm{cl}[T(E)]$ is compact in the norm topology on $V$.

Thus $T \in L(V)$ is compact if and only if the image of a norm bounded set under $T$ has compact closure. It is easily verified that $T \in L(V)$ is compact if and only if $\mathrm{cl}[T(B_1)]$ is compact, where, as usual, $B_1 = \{x \mid x \in V, \|x\| \leq 1\}$.

Some examples of compact transformations are obviously desirable. First, suppose $(V, \langle \cdot , \cdot \rangle)$ is any Hilbert space over $\mathbb{C}$ and let $T \in L(V)$ be such that the range of $T$ is a finite-dimensional linear subspace of $V$. Since by Theorem 1.3.3 every closed bounded set in the range of $T$ is compact, it follows at once that $T$ is compact. In particular, every linear transformation from $\mathbb{C}^n$ to itself is compact. Thus the familiar linear transformations of linear algebra are all compact continuous linear transformations.

Not all compact transformations, however, have a finite-dimensional range. Another important class of compact transformations forms a subset of the so-called integral transformations or integral operators. For example, let $V = L_2([0,1],dt)$ and suppose $K \in C([0,1] \times [0,1])$. We define $T \in L(L_2([0,1],dt))$ by

$$T(f)(s) = \int_0^1 K(s,t)f(t)\,dt \qquad (f \in L_2([0,1],dt)).$$

That $T$ indeed belongs to $\mathcal{L}(L_2([0,1],dt))$ was indicated in the remarks following Theorem 13.9.1. We claim, however, that $T$ is also compact. To this end it suffices to show that $\mathrm{cl}[T(B_1)]$ is compact in $L_2([0,1],dt)$. With this in mind, suppose $f \in L_2([0,1],dt)$ and $\|f\|_2 \leq 1$. It is apparent, $K$ being continuous, that $T(f) \in C([0,1])$ and that

$$\begin{aligned}\|T(f)\|_\infty &\leq \|K\|_\infty \int_0^1 |f(t)|\,dt \\ &\leq \|K\|_\infty \|f\|_2 \\ &\leq \|K\|_\infty.\end{aligned}$$

Thus we see that $T(B_1)$ is a bounded subset of $(C([0,1]),\|\cdot\|_\infty)$. Furthermore, we claim that $T(B_1)$ is an equicontinuous set. Indeed, suppose $\epsilon > 0$ is given and let $\delta > 0$ be such that, if $s_1, s_2 \in [0,1]$ and $|s_1 - s_2| < \delta$, then $|K(s_1,t) - K(s_2,t)| < \epsilon$, $t \in [0,1]$. This is possible, of course, because $K$ is continuous, and hence uniformly continuous, on $[0,1] \times [0,1]$. Therefore, if $s_1, s_2 \in [0,1]$ and $|s_1 - s_2| < \delta$, then

$$\begin{aligned}|T(f)(s_1) - T(f)(s_2)| &\leq \int_0^1 |K(s_1,t) - K(s_2,t)||f(t)|\,dt \\ &< \epsilon \int_0^1 |f(t)|\,dt \\ &\leq \epsilon \|f\|_2 \\ &\leq \epsilon \qquad (f \in B_1),\end{aligned}$$

that is, $T(B_1)$ is equicontinuous. Consequently from the Ascoli-Arzelà Theorem [Y, pp. 85 and 86] it follows that $T(B_1)$ has compact closure in $(C([0,1]),\|\cdot\|_\infty)$.

Hence, if $\{f_n\} \subset B_1$, then there exist a subsequence $\{T(f_{n_k})\}$ of $\{T(f_n)\}$ and some $f \in C([0,1])$ such that $\lim_k \|T(f_{n_k}) - f\|_\infty = 0$, which leads at once to $\lim_k \|T(f_{n_k}) - f\|_2 = 0$. Therefore $T(B_1)$ has compact closure in $L_2([0,1],dt)$, and $T$ is a compact continuous linear transformation on $L_2([0,1],dt)$.

Actually one can even show that the formula

$$T(f)(s) = \int_0^1 K(s,t)f(t)\,dt \qquad (f \in L_2([0,1],dt))$$

defines a compact $T \in L(L_2([0,1],dt))$ whenever $K$ belongs to $L_2([0,1] \times [0,1], ds\,dt)$ (see, for instance [Y, pp. 277 and 278]). Such transformations are often called <u>Hilbert-Schmidt transformations</u>, and they play an important role in the study of differential and integral equations. The interested reader might consult, for instance, [CoHi, pp. 112-162; RzNg, pp. 143-194; Tr].

Next let us establish some preliminary results about the spectral theory of compact transformations.

<u>Proposition 13.15.1</u>. Let $(V,\langle\cdot,\cdot\rangle)$ be a Hilbert space over $\mathbb{C}$ and let $T \in L(V)$ be compact. If $V$ is not finite-dimensional, then $0 \in \sigma(T)$.

<u>Proof</u>. If $0 \in \rho(T)$, then $T^{-1} = R(0,T) \in L(V)$. Consequently $T^{-1}(B_1)$ is a norm bounded subset of $V$, and hence $T[T^{-1}(B_1)] = B_1$ is compact. Thus by the comments following Theorem 1.3.3 we conclude that $V$ is finite-dimensional, contrary to the hypothesis of the proposition.

Therefore $0 \in \sigma(T)$. □

Of course, if $V$ is finite-dimensional, then zero certainly may belong to the resolvent set. This will be the case if and only if the transformation is injective.

<u>Proposition 13.15.2</u>. Let $(V,\langle\cdot,\cdot\rangle)$ be a Hilbert space over $\mathbb{C}$ and let $T \in L(V)$ be compact and self-adjoint. Then $m_T$ and $M_T$ belong to $P_\sigma(T)$ whenever they are not zero.

Proof. Suppose, for instance, that $m_T \neq 0$. Recalling the definition of $m_T$ as

$$m_T = \inf\{\langle T(x),x\rangle \mid x \in V, \|x\| = 1\}$$

and the proof of Theorem 13.14.3 where we saw that

$$\|T(x) - m_T x\|^2 \leq \|(T - m_T I)^3\|\|x\|^2\langle T(x) - m_T x, x\rangle \qquad (x \in V),$$

we deduce at once the existence of some sequence $\{x_n\} \subset V$ such that $\|x_n\| = 1$, $n = 1,2,3,\ldots,$ and for which $\lim_n\|T(x_n) - m_T x_n\| = 0$. However, since $T$ is compact, there exists a subsequence $\{x_{n_k}\}$ of $\{x_n\}$ and some $y_o \in V$ such that $\lim_k\|T(x_{n_k}) - y_o\| = 0$. Thus from the estimates

$$\left\|\frac{y_o}{m_T} - x_{n_k}\right\| \leq \left\|\frac{y_o}{m_T} - \frac{T(x_{n_k})}{m_T}\right\| + \left\|\frac{T(x_{n_k})}{m_T} - x_{n_k}\right\|$$

$$= \frac{1}{|m_T|}\|y_o - T(x_{n_k})\| + \frac{1}{|m_T|}\|T(x_{n_k}) - m_T x_{n_k}\|$$

we deduce that $\lim_k\|x_{n_k} - y_o/m_T\| = 0$. Hence, on setting $x_o = y_o/m_T$ and using the continuity of $T$, we conclude that

$$\begin{aligned} T(x_o) &= \lim_k T(x_{n_k}) \\ &= \lim_k m_T x_{n_k} \\ &= m_T x_o. \end{aligned}$$

Therefore $m_T \in P_\sigma(T)$, as $\|x_o\| = \lim_k\|x_{n_k}\| = 1$. A similar argument shows that $M_T \in P_\sigma(T)$ whenever $M_T \neq 0$. □

Finally we can give the spectral decomposition theorem promised in Section 13.13. The statement and proof of the theorem are both rather long.

Theorem 13.15.1. Let $(V,\langle\cdot,\cdot\rangle)$ be a Hilbert space over $\mathbb{C}$ and let $T \in L(V)$ be compact and self-adjoint. Then

(i) There exists a finite or infinite sequence $\{\lambda_k\} \subset \mathbb{R}$ and a corresponding finite or infinite sequence $\{e_k\} \subset V$ such that

(a) $|\lambda_{k+1}| \leq |\lambda_k|$, $k = 1,2,3,\ldots$ .
(b) The sequence $\{e_k\}$ is an orthonormal set in $V$.
(c) $T(e_k) = \lambda_k e_k$, $k = 1,2,3,\ldots$ .

(ii) $T(x) = \Sigma_{k=1}^{\infty}\langle x,e_k\rangle\lambda_k e_k \qquad (x \in V)$.

(iii) If $V$ is finite-dimensional, then $\sigma(T) = \{\lambda_k\}$; if $V$ is not finite-dimensional, then $\sigma(T) = \{\lambda_k\} \cup \{0\}$.

(iv) If $\{\lambda_k\}$ is infinite, then $\lim_k \lambda_k = 0$.

(v) If $0 \notin P_\sigma(T)$, then $\{e_k\}$ is a complete orthonormal set in $V$.

Proof. Without loss of generality we may assume that $T \neq 0$, as otherwise the theorem is trivial. From Corollaries 13.13.1, 13.14.1, and 13.14.3, as well as Theorems 13.13.2 and 13.14.2, we know that $\sigma(T)$ is a nonempty closed bounded subset of $\mathbb{R}$ such that

$$\|T\| = \sup_{\lambda \in \sigma(T)} |\lambda| = \sup_{\xi \in \omega(T)} |\xi|.$$

Let $\lambda_1$ be that one of $m_T$ and $M_T$ which has the greatest absolute value. If $|m_T| = |M_T|$, then $\lambda_1$ can be either one. Since $T \neq 0$, it follows that $\lambda_1 \neq 0$, and so by Proposition 13.15.2 we deduce that $\lambda_1 \in P_\sigma(T)$. Let $e_1 \in V$ be so chosen that $\|e_1\| = 1$ and $T(e_1) = \lambda_1 e_1$. Clearly the choice of $\lambda_1$ also entails that $|\lambda_1| = \|T\|$.

Next let $V_1 = \{x \mid x \in V, x \perp e_1\}$. Then $V_1$ is a closed linear subspace of $V$. Moreover, $T(V_1) \subset V_1$ because, if $x \in V_1$, then

$$\begin{aligned}\langle T(x),e_1\rangle &= \langle x,T(e_1)\rangle \\ &= \lambda_1\langle x,e_1\rangle = 0,\end{aligned}$$

as $T$ is self-adjoint and $\lambda_1 \in \mathbb{R}$. Consequently it is evident that the restriction of $T$ to $V_1$, call it $T_1$, is a compact self-adjoint continuous linear transformation on the Hilbert space $V_1$. Thus we may repeat the foregoing argument with $T_1$ and $V_1$. More specifically, let $\lambda_2$ be that one of $m_{T_1}$ and $M_{T_1}$ which has the greatest absolute value. If $\lambda_2 = 0$, then $\|T_1\| = |\lambda_2| = 0$, and if $\lambda_2 \neq 0$, then, as before, $\lambda_2 \in P_\sigma(T_1)$, and there exists some $e_2 \in V_1$, $\|e_2\| = 1$, such that $T(e_2) = T_1(e_2) = \lambda_2 e_2$. Clearly $e_1 \perp e_2$, and

$$\begin{aligned} |\lambda_2| &= \|T_1\| \\ &= \sup_{\substack{x \in V_1 \\ \|x\| = 1}} \|T(x)\| \\ &\leq \sup_{\substack{x \in V \\ \|x\| = 1}} \|T(x)\| \\ &= \|T\| \\ &= |\lambda_1|. \end{aligned}$$

For the next step, in the case $\lambda_2 \neq 0$, we set $V_2 = \{x \mid x \in V_1, x \perp e_2\}$ and repeat the preceding arguments.

In this way we obtain a finite or infinite sequence $\{\lambda_k\} \subset P_\sigma(T) \subset \mathbb{R}$ and corresponding vectors $\{e_k\} \subset V$ such that

(a) $|\lambda_{k+1}| \leq |\lambda_k|$, $k = 1,2,3,\ldots$ .
(b) $\{e_k\}$ is an orthonormal set.
(c) $T(e_k) = \lambda_k e_k$, $k = 1,2,3,\ldots$,

and thereby have proved part (i) of the theorem. Furthermore, the preceding construction shows that for each $n$ we have

$$V = [e_1] \oplus [e_2] \oplus \cdots \oplus [e_n] \oplus V_n,$$

where $[e_k]$ denotes the closed one-dimensional linear subspace of $V$ spanned by $e_k$, $k = 1,2,3,\ldots$, and $V_n = \{x \mid x \in V_{n-1}, x \perp e_n\}$, $n = 1,2,3,\ldots$, where $V_0 = V$. Of course, if $\{\lambda_k\}$ is only a finite sequence, then there is only a finite number of such orthogonal decompositions of $V$.

Let us look first at the case where $V$ is finite-dimensional. The sequence $\{\lambda_k\}$ being clearly finite, a little reflection shows that there are two possibilities. First, there exists some $n$ such that $V_n \neq \{0\}$, $\lambda_{n+1} = 0$, and $\lambda_n \neq 0$. In this case it is evident that $0 \in P_\sigma(T)$, $T$ restricted to $V_n$ is the zero transformation, and

$$V = [e_1] \oplus [e_2] \oplus \cdots \oplus [e_n] \oplus V_n.$$

Thus if we set $x_\infty = x - \Sigma_{k=1}^n \langle x,e_k\rangle e_k$, $x \in V$, we see at once that $x_\infty \perp e_k$, $k = 1,2,\ldots,n$, and so $x_\infty \in V_n$. Consequently

$$\begin{aligned} T(x) &= \sum_{k=1}^{n} \langle x,e_k\rangle T(e_k) \\ &= \sum_{k=1}^{n} \langle x,e_k\rangle \lambda_k e_k \qquad (x \in V). \end{aligned}$$

In the second case there exists some $n$ such that $V_n = \{0\}$ and $\lambda_n \neq 0$. As a result we have $0 \notin P_\sigma(T)$ and

$$V = [e_1] \oplus [e_2] \oplus \cdots \oplus [e_n],$$

from which it follows at once that $e_1, e_2, \ldots, e_n$ form a complete orthonormal set in $V$, that is, a basis for the finite-dimensional space $V$, and

$$T(x) = \sum_{k=1}^{n} \langle x,e_k\rangle \lambda_k e_k \qquad (x \in V).$$

Moreover, since for finite-dimensional spaces we have $P_\sigma(T) = \sigma(T)$, it follows easily that in the first case, $\sigma(T) = \{\lambda_1, \lambda_2, \ldots, \lambda_n, 0\}$, whereas in the second case $\sigma(T) = \{\lambda_1, \lambda_2, \ldots, \lambda_n\}$.

Thus the theorem is completely proved for finite-dimensional $V$. Consequently for the remainder of the proof we assume that $V$ is not finite-dimensional. We have again two cases to consider. First, let us assume that $\{\lambda_k\}$ is finite. Since $V$ is not finite-dimensional, this entails that there exists some $n$ such that $V_n \neq \{0\}$, $\lambda_{n+1} = 0$, and $\lambda_n \neq 0$. By essentially the same argument as in the finite-dimensional case we see that $0 \in P_\sigma(T)$,

$$V = [e_1] \oplus [e_2] \oplus \cdots \oplus [e_n] \oplus V_n,$$

and

$$T(x) = \sum_{k=1}^{n} \langle x,e_k\rangle\lambda_k e_k \qquad (x \in V).$$

The theorem is thus proved for the first case, although we have not yet shown that $\sigma(T) = \{\lambda_1,\lambda_2,\ldots,\lambda_n,0\}$. We shall return to this point in a moment.

Now let us examine the case where $\{\lambda_k\}$ is infinite. In this case we claim that $\lim_k\lambda_k = 0$. If this were not true, then there would exist some $\epsilon > 0$ such that $|\lambda_k| \geq \epsilon$, $k = 1,2,3,\ldots,$ and for all $n,m = 1,2,3,\ldots,$ $n \neq m$, we would have

$$\begin{aligned} \|T(e_n) - T(e_m)\|^2 &= \|\lambda_n e_n - \lambda_m e_m\|^2 \\ &= \lambda_n^2 + \lambda_m^2 \\ &\geq 2\epsilon^2 \\ &> 0, \end{aligned}$$

as $\{e_k\}$ is an orthonormal set. This, however, clearly contradicts the compactness of $T$, since the sequence $\{T(e_k)\}$ could have no convergent subsequence. Hence $\lim_k\lambda_k = 0$, which proves part (iv) of the theorem.

If $x \in V$, then by Bessel's Inequality (Theorem 13.6.3(ii)) and the Riesz-Fischer Theorem (Theorem 13.6.4) it is apparent that the infinite sum $\Sigma_{k=1}^{\infty}\langle x,e_k\rangle e_k$ converges to some element of V. Consequently for each $x \in V$ we can define $x_\infty = x - \Sigma_{k=1}^{\infty}\langle x,e_k\rangle e_k$. It is obvious that $x_\infty \perp e_k$, $k = 1,2,3,\ldots$, and so $x_\infty \in V_k$, $k = 1,2,3,\ldots$ . Thus $T(x_\infty) = T_k(x_\infty)$, where $T_k$ is the restriction to $V_k$ of T. But then, since $\|T_k\| = |\lambda_{k+1}|$, $k = 1,2,3,\ldots$, by the construction of the sequence $\{\lambda_k\}$, we see that

$$\begin{aligned}\|T(x_\infty)\| &= \|T_k(x_\infty)\| \\ &\le \|T_k\|\|x_\infty\| \\ &= |\lambda_{k+1}|\|x_\infty\| \qquad (k = 1,2,3,\ldots).\end{aligned}$$

Hence $T(x_\infty) = 0$, as $\lim_k \lambda_k = 0$. Thus we conclude that

$$T(x) = \sum_{k=1}^{\infty} \langle x,e_k\rangle \lambda_k e_k \qquad (x \in V),$$

as $T \in L(V)$.

Moreover, since V is not finite-dimensional, we know from Proposition 13.15.1 that $0 \in \sigma(T)$. If $0 \notin P_\sigma(T)$, then T is injective, whence we deduce from $T(x_\infty) = 0$, $x \in V$, that $x = \Sigma_{k=1}^{\infty}\langle x,e_k\rangle e_k$, $x \in V$. Hence if follows from Theorem 13.6.5 that $\{e_k\}$ is a complete orthonormal set in V when $0 \notin P_\sigma(T)$.

It remains only to show that, if V is not finite dimensional, then $\sigma(T) = \{\lambda_k\} \cup \{0\}$. Clearly $\{\lambda_k\} \cup \{0\} \subset \sigma(T)$. So suppose $\lambda_o \in \mathbb{C} \sim (\{\lambda_k\} \cup \{0\})$. Since $\lim_k \lambda_k = 0$ when $\{\lambda_k\}$ is infinite, it is apparent that $d = \inf\{|\lambda_o - \lambda_k|, |\lambda_o| \mid k = 1,2,3,\ldots\} > 0$. We claim that $\lambda_o \notin P_\sigma(T)$. Were $\lambda_o \in P_\sigma(T)$, then there would exist some $x_o \in V$, $x_o \neq 0$, such that $T(x_o) = \lambda_o x_o$. However, since $\lambda_o \neq \lambda_k$, $k = 1,2,3,\ldots$, we know from Corollary 13.14.2 that $x_o \perp e_k$, $k = 1,2,3,\ldots$ . Consequently

$$\lambda_o x_o = T(x_o) = \sum_{k=1}^{\infty} \langle x_o,e_k\rangle \lambda_k e_k = 0,$$

which is a contradiction because $\lambda_o \neq 0$. Thus $\lambda_o \notin P_\sigma(T)$, that is, $\lambda_o I - T$ is injective.

We claim, moreover, that $\lambda_o I - T$ is even surjective. To see this we shall construct, given $y \in V$, some $x \in V$ such that $(\lambda_o I - T)(x) = y$. We note that, if such an equation were valid, then we would have

$$\begin{aligned}\langle y,e_k\rangle &= \langle(\lambda_o I - T)(x),e_k\rangle \\ &= \lambda_o\langle x,e_k\rangle - \langle T(x),e_k\rangle \\ &= \lambda_o\langle x,e_k\rangle - \langle x,T(e_k)\rangle \\ &= (\lambda_o - \lambda_k)\langle x,e_k\rangle \qquad (k = 1,2,3,\ldots),\end{aligned}$$

and so $\langle x,e_k\rangle = \langle y,e_k\rangle/(\lambda_o - \lambda_k)$, $k = 1,2,3,\ldots$, which is valid because $T$ is self-adjoint and $\lambda_o \neq \lambda_k$, $k = 1,2,3,\ldots$ . With this observation in mind, let $y \in V$ and consider the sum

$$\sum_{k=1}^{\infty} \left|\frac{\langle y,e_k\rangle}{\lambda_o - \lambda_k}\right|^2.$$

Since $|\lambda_o - \lambda_k| \geq d > 0$, it follows at once from the estimate

$$\sum_{k=1}^{\infty} \left|\frac{\langle y,e_k\rangle}{\lambda_o - \lambda_k}\right|^2 \leq \sum_{k=1}^{\infty} \left|\frac{\langle y,e_k\rangle}{d}\right|^2,$$

Bessel's Inequality (Theorem 13.6.3(ii)), and the Riesz-Fischer Theorem (Theorem 13.6.4) that $\Sigma_{k=1}^{\infty}[\langle y,e_k\rangle/(\lambda_o - \lambda_k)]e_k$ converges to some element in $V$. Consequently, since $\lambda_o \neq 0$,

$$x = \frac{y_\infty}{\lambda_o} + \sum_{k=1}^{\infty}\left[\frac{\langle y,e_k\rangle}{\lambda_o - \lambda_k}\right] e_k$$

is a well-defined element of $V$, where in accordance with our previous notation $y_\infty = y - \Sigma_{k=1}^{\infty}\langle y,e_k\rangle e_k$.

Since $T(y_\infty) = 0$, we see that

$$(\lambda_o I - T)(x) = y_\infty + \sum_{k=1}^{\infty} \frac{\langle y,e_k\rangle}{\lambda_o - \lambda_k} (\lambda_o I - T)(e_k)$$

$$= y_\infty + \sum_{k=1}^{\infty} \langle y, e_k \rangle e_k$$

$$= y,$$

whence we conclude that $\lambda_o I - T$ is surjective. Of course, if $\{\lambda_k\}$ is finite, then the infinite sums in the preceding argument are really only finite sums, and the discussion is even simpler.

Hence, if $\lambda_o \neq \lambda_k$, $k = 1,2,3,\ldots$, $\lambda_o \neq 0$, then $\lambda_o I - T$ is bijective -- that is, $\lambda_o \in \rho(T)$. Therefore $\sigma(T) = \{\lambda_k\} \cup \{0\}$.

This now completes the proof of the theorem. □

The theorem clearly contains as a special case the result on the diagonalization of self-adjoint linear transformations on finite-dimensional linear spaces mentioned at the beginning of Section 13.13.

The representation of $T$ described in part (ii) of the theorem can also be strengthened. To see this, suppose $T \in L(V)$ is compact and self-adjoint, and let $\{\lambda_k\} \subset \mathbb{R}$ and $\{e_k\} \subset V$ be as in part (i) of Theorem 13.15.1. Let $P_k$ denote the orthogonal projection of $V$ onto $[e_k]$, the closed linear subspace of $V$ spanned by $e_k$, $k = 1,2,3,\ldots$ . It is evident that $P_k(x) = \langle x, e_k \rangle e_k$, $k = 1,2,3,\ldots$, $x \in V$. Now, if $\{\lambda_k\} = \{\lambda_1, \lambda_2, \ldots, \lambda_n\}$, then it is apparent from Theorem 13.15.1(ii) that

$$T = \sum_{k=1}^{n} \lambda_k P_k,$$

whereas if $\{\lambda_k\}$ is infinite, then for each $x \in V$, $\|x\| = 1$, we have

$$\|(T - \sum_{k=1}^{n} \lambda_k P_k)(x)\|^2 = \|T(x) - \sum_{k=1}^{n} \lambda_k \langle x, e_k \rangle e_k\|^2$$

$$= \|\sum_{k=n+1}^{\infty} \lambda_k \langle x, e_k \rangle e_k\|^2$$

$$= \sum_{k=n+1}^{\infty} |\lambda_k|^2 |\langle x, e_k \rangle|^2$$

$$\leq |\lambda_{n+1}|^2 \sum_{k=n+1}^{\infty} |\langle x, e_k \rangle|^2$$

$$\leq |\lambda_{n+1}|^2 \|x\|^2$$

$$= |\lambda_{n+1}|^2 \qquad (n = 1,2,3,\ldots)$$

because the vectors $\{e_k\}$ form an orthonormal set. Bessel's Inequality (Theorem 13.6.3(ii)) ensures the validity of the penultimate estimate. Consequently we see that

$$\|T - \sum_{k=1}^{n} \lambda_k P_k\| = \sup_{\|x\|=1} \|(T - \sum_{k=1}^{n} \lambda_k P_k)(x)\|$$

$$\leq |\lambda_{n+1}| \qquad (n = 1,2,3,\ldots),$$

whence it follows at once that

$$T = \sum_{k=1}^{\infty} \lambda_k P_k,$$

as $\lim_n |\lambda_{n+1}| = 0$ by Theorem 13.15.1(iv).

We summarize these observations in the following corollary:

Corollary 13.15.1. Let $(V, \langle\cdot,\cdot\rangle)$ be a Hilbert space over $\mathbb{C}$ and let $T \in L(V)$ be compact and self-adjoint. Then

(i) There exists a finite or infinite sequence $\{\lambda_k\} \subset \mathbb{R}$ and a corresponding finite or infinite sequence $\{e_k\} \subset V$ such that

(a) $T(e_k) = \lambda_k e_k$, $k = 1,2,3,\ldots$ .
(b) The sequence $\{e_k\}$ is an orthonormal set in $V$.

(ii) If $P_k$ denotes the orthogonal projection of $V$ onto $[e_k]$, the closed linear subspace of $V$ spanned by $e_k$, $k = 1,2,3,\ldots$, then $T = \sum_{k=1}^{\infty} \lambda_k P_k$ where convergence is in the norm topology on $L(V)$.

The material of the last three sections is only a very small portion of the general study of spectral theory and spectral decomposition theorems for linear transformations on Hilbert and Banach spaces. The subject is vast, and for a further sampling of results the reader is referred to [AkG; BaNr; $DS_1$, pp. 555-730; $DS_2$; $E_1$, pp. 676-706; $Ha_2$; He; Lo; N, pp. 84-114, 441-511; RzNg, pp. 195-445; So; T, pp. 253-364; V; Y, pp. 193-362].

## 13.16. Problems.

1. (Theorem 13.2.2) Let $(V,\langle\cdot,\cdot\rangle)$ be an inner product space over $\Phi$. Prove each of the following:

(a) If $\Phi = \mathbb{R}$, then

$$\langle x,y\rangle = \left\|\frac{x+y}{2}\right\|^2 - \left\|\frac{x-y}{2}\right\|^2 \qquad (x,y \in V).$$

(b) If $\Phi = \mathbb{C}$, then

$$\langle x,y\rangle = \left\|\frac{x+y}{2}\right\|^2 - \left\|\frac{x-y}{2}\right\|^2 + i\left\|\frac{x+iy}{2}\right\|^2 - i\left\|\frac{x-iy}{2}\right\|^2 \qquad (x,y \in V).$$

2. Let $(V,\|\cdot\|) = (L_1([0,1],dt),\|\cdot\|_1)$ where

$$\|f\|_1 = \int_0^1 |f(t)|\,dt \qquad (f \in L_1([0,1],dt)).$$

Prove that $\|\cdot\|_1$ does not satisfy the Parallelogram Identity and conclude that $(L_1([0,1],dt),\|\cdot\|_1)$ is not a Hilbert space.

3. (Proposition 13.4.2) Let $(V,\langle\cdot,\cdot\rangle)$ be an inner product space over $\Phi$ and let $E \subset V$. Prove each of the following:

(a) $E^\perp$ is a closed linear subspace of $V$.

(b) $E \subset E^{\perp\perp}$.

(c) If $E$ is a linear subspace, then $E \cap E^\perp = \{0\}$.

4. (Corollary 13.4.1) Let $(V,\langle\cdot,\cdot\rangle)$ be a Hilbert space over $\Phi$. Prove that, if $W \subset V$ is a closed linear subspace, then $W = W^{\perp\perp}$.

5. Let $(V,\langle\cdot,\cdot\rangle)$ be a Hilbert space over $\Phi$ and let $W \subset V$ be a linear subspace. If $\mathrm{cl}(W) \neq V$, prove that there exists some $y \in V$, $y \neq 0$, such that $y \perp \mathrm{cl}(W)$. Conclude that $W$ is dense if and only if $y \perp W$ implies that $y = 0$.

6. Let $(V,\langle\cdot,\cdot\rangle)$ be a Hilbert space over $\Phi$ and let $W \subset V$ be a closed linear subspace. If $x_o \in V$, prove that

$$\min\{\|x - x_o\| \mid x \in W\} = \max\{|\langle x_o,y\rangle| \mid y \in W^{\perp}, \|y\| = 1\}.$$

7. (Proposition 13.5.1) Let $(V,\langle\cdot,\cdot\rangle)$ be a Hilbert space over $\Phi$. Prove that, if $P \in L(V)$ is an orthogonal projection, then $V = R(P) \oplus R(I - P)$ and $R(P) \perp R(I - P)$.

8. (Proposition 13.5.2) Let $(V,\langle\cdot,\cdot\rangle)$ be a Hilbert space over $\Phi$. Prove that the following are equivalent:

(i) $P \in L(V)$ is an orthogonal projection.

(ii) $P \in L(V)$ is such that $P^2 = P$ and

$$R(P)^{\perp} = \{x \mid x \in V, P(x) = 0\}.$$

(iii) $P \in L(V)$ is such that $P^2 = P$ and $R(P) \perp R(I - P)$.

9. Let $V = \ell_2$ and define $e_n \in \ell_2$ for $n = 1,2,3,\ldots$ by $e_n = \{\delta_{nk}\}_{k=1}^{\infty}$; that is, the nth coordiante of $e_n$ is $1$ and all other coordinates are zero. Let $M = \{e_{2n-1} + e_{2n} \mid n = 1,2,3,\ldots\}$.

(a) Identify the closed linear subspaces $M^{\perp}$ and $M^{\perp\perp}$ in $\ell_2$.

(b) Let $P$ and $Q$ in $L(\ell_2)$ be the orthogonal projections such that $R(P) = M^{\perp}$ and $R(Q) = M^{\perp\perp}$. If $a = \{a_k\} \in \ell_2$, prove that $P(a) = \{b_k\}$, where $b_{2k-1} = -b_{2k} = (a_{2k-1} - a_{2k})/2$, $k = 1,2,3,\ldots$; and that $Q(a) = \{c_k\}$, where $c_{2k-1} = c_{2k} = (a_{2k-1} + a_{2k})/2$, $k = 1,2,3,\ldots$ .

10. Let $(V,\langle\cdot,\cdot\rangle)$ be a Hilbert space over $\Phi$ and let $P$ and $Q$ in $L(V)$ be orthogonal projections.

(a) Prove that the following are equivalent:

(i) $PQ = QP$.

(ii) $PQ$ is an orthogonal projection.

(iii) $QP$ is an orthogonal projection.

(b) If $PQ$ is an orthogonal projection, prove that $R(PQ) = R(P) \cap R(Q)$.

11. Let $(V, \langle \cdot, \cdot \rangle)$ be a Hilbert space over $\Phi$ and let $P$ and $Q$ in $L(V)$ be orthogonal projections.

(a) Prove that the following are equivalent:

(i) $R(P) \perp R(Q)$.

(ii) $P[R(Q)] = 0$.

(iii) $Q[R(P)] = 0$.

(iv) $PQ = 0$.

(v) $QP = 0$.

(vi) $P + Q$ is an orthogonal projection.

(b) If $P + Q$ is an orthogonal projection, prove that $R(P + Q) = R(P) + R(Q)$.

12. Let $(V, \langle \cdot, \cdot \rangle)$ be a Hilbert space over $\Phi$ and let $P$ and $Q$ in $L(V)$ be orthogonal projections. If $\|P - Q\| < 1$, prove that $\dim[R(P)] = \dim[R(Q)]$, that is, $R(P)$ is isometrically linearly isomorphic to $R(Q)$.

13. Let $(V, \langle \cdot, \cdot \rangle)$ be a Hilbert space over $\Phi$, let $W \subset V$ be a closed linear subspace, and let $T \in L(V)$. The subspace $W$ is said to be <u>invariant</u> under $T$ if $T(x) \in W$ for all $x \in W$, and $W$ is said to <u>reduce</u> $T$ if both $W$ and $W^{\perp}$ are invariant under $T$. Let $P \in L(V)$ be the orthogonal projection such that $R(P) = W$. Prove that the following are equivalent:

(i) $W$ reduces $T$.

(ii) $PT = TP$.

(iii) $W$ is invariant under both $T$ and $T^*$.

14. (Corollary 13.6.1) Let $(V,\langle\cdot,\cdot\rangle)$ be an inner product space over $\Phi$. Prove that, if $F \subset V$ is an orthonormal set, then there exists a complete orthonormal set $E \subset V$ such that $E \supset F$.

15. (Corollary 13.6.2) Prove each of the following:

(a) Let $\{a_k\} \subset \Phi$ be a sequence. $\Sigma_{k=1}^{\infty} a_k$ converges absolutely to $a$ if and only if $\Sigma_{\mathbb{Z}^+} a_k$ converges to $a$ where $\mathbb{Z}^+$ denotes the set of positive integers.

(b) Let $\{a_\alpha \mid \alpha \in \Lambda\} \subset \Phi$ and suppose there exists some $a \in \Phi$ such that $\Sigma_\Lambda a_\alpha = a$. Then $a_\alpha = 0$ for all but a countable number of $\alpha \in \Lambda$, and if this countable number of indices is enumerated as the sequence $\{\alpha_k\}$, then $\Sigma_{k=1}^{\infty} a_{\alpha_k}$ converges absolutely to $a$.

16. Let $(V,\langle\cdot,\cdot\rangle)$ be a Hilbert space over $\Phi$ and let $\{e_\alpha \mid \alpha \in \Lambda\}$ and $\{f_\beta \mid \beta \in \Lambda'\}$ be complete orthonormal sets in $V$. Prove that the cardinalities of $\Lambda$ and $\Lambda'$ are equal.

17. (Theorem 13.6.7) Let $(V,\langle\cdot,\cdot\rangle)$ be a Hilbert space over $\Phi$. Prove that the following are equivalent:

(i) $\dim(V) = \aleph_0$.

(ii) $(V,\|\cdot\|)$ is separable.

18. Let $V_1$ and $V_2$ be two Hilbert spaces over $\Phi$. Prove that one of them is isometrically linearly isomorphic to a closed linear subspace of the other.

19. Let $(V,\langle\cdot,\cdot\rangle)$ be a Hilbert space over $\Phi$ and let $\{e_\alpha \mid \alpha \in \Lambda\}$ be a complete orthonormal set in $V$. If $\{x_n\}$ is a sequence in $V$, prove that $\{x_n\}$ converges weakly to $x \in V$ if and only if $\{\|x_n\|\}$ is a bounded sequence and $\lim_n \langle x_n, e_\alpha\rangle = \langle x, e_\alpha\rangle$ for each $\alpha \in \Lambda$.

*20. Let $(V, \langle\cdot,\cdot\rangle)$ be a Hilbert space over $\Phi$ and let $E = \{x_1, x_2, x_3, \ldots\}$ be a countable set of linearly independent vectors in $V$. Prove that there exists a countable orthonormal set $F = \{y_1, y_2, y_3, \ldots\}$ such that the smallest linear subspace containing $E$ is the same as the smallest linear subspace containing $F$. The process of constructing $F$ from $E$ is called Gram-Schmidt orthonormalization.

21. (a) Prove that $E = \{t^n \mid n = 0,1,2,\ldots\}$ is a linearly independent set in $L_2([a,b],dt)$, where $a$ and $b$ are finite real numbers and $dt$ is Lebesgue measure. Apply the Gram-Schmidt orthonormalization (Problem 20) to $E$ and obtain the polynomials

$$p_n = \frac{1}{c_n}\frac{d^n([(t-a)(t-b)]^n)}{dt^n} \qquad (n = 1,2,3,\ldots),$$

where $c_n \in \mathbb{R}$ is so chosen that $\|p_n\|_2 = 1$. When $a = -1$ and $b = 1$, these are called the Legendre polynomials.

(b) Prove the the Legendre polynomials in $L_2([-1,1],dt)$ are a complete orthonormal set.

*22. Let $W$ be a linear subspace of $C([0,1])$ that is closed as a subspace of $L_2([0,1],dt)$ and suppose there exists a constant $M$ such that $|f(t)| \leq M\|f\|_2$ for all $t \in [0,1]$, $f \in W$. Prove that $\dim(W) \leq M^2$.

23. Let $(V, \langle\cdot,\cdot\rangle)$ be a separable Hilbert space over $\Phi$, let $\{e_1, e_2, \ldots\}$ be an orthonormal basis for $V$, and let $T \in L(V)$. If $\|T\| \leq 1$, then certainly $\{\|T(e_n)\|\}$ is bounded by one. Prove that the converse of this is false by constructing linear transformations of arbitrarily large norms that are bounded by one on a basis.

24. (Theorem 13.7.2) Consider the Hilbert space $L_2([-\pi,\pi],dt/2\pi)$ over $\mathbb{C}$. Prove each of the following:

(a) If $f \in L_2([-\pi,\pi],dt/2\pi)$, then $f = \Sigma_{\mathbb{Z}}\hat{f}(k)e^{ik\cdot}$ and

$$\|f\|_2 = \left[\sum_{k=-\infty}^{\infty} |\hat{f}(k)|^2\right]^{1/2}.$$

(b) If $f,g \in L_2([-\pi,\pi],dt/2\pi)$, then

$$\frac{1}{2\pi}\int_{-\pi}^{\pi} f(t)\overline{g(t)}\, dt = \sum_{k=-\infty}^{\infty} \hat{f}(k)\overline{\hat{g}(k)}$$

and

$$\frac{1}{2\pi}\int_{-\pi}^{\pi} f(t)g(t)\, dt = \sum_{k=-\infty}^{\infty} \hat{f}(k)\hat{g}(-k).$$

(c) If $\{a_k \mid k \in \mathbb{Z}\} \subset \mathbb{C}$ is such that $\Sigma_{k=-\infty}^{\infty} |a_k|^2 < \infty$, then there exists a unique $f \in L_2([-\pi,\pi],dt/2\pi)$ such that $f = \Sigma_{\mathbb{Z}} a_k e^{ik\cdot}$ and $a_k = \hat{f}(k)$, $k \in \mathbb{Z}$.

(d) If the mapping $\varphi : L_2([-\pi,\pi],dt/2\pi) \to L_2(\mathbb{Z})$ is defined for each $f \in L_2([-\pi,\pi],dt/2\pi)$ by $\varphi(f)(k) = \hat{f}(k)$, $k \in \mathbb{Z}$, then $\varphi$ is a surjective isometric isomorphism.

25. (Proposition 13.8.1) Prove that the Rademacher functions $\{r_n\}$ form an orthonormal set in $L_2([0,1],dt)$, where dt denotes Lebesgue measure on $[0,1]$.

*26. Prove that, if $\{a_k\} \subset \mathbb{C}$ is such that $\Sigma_{k=1}^{\infty} |a_k|^2 = \infty$, then $\Sigma_{k=1}^{\infty} a_k r_k(t)$ diverges almost everywhere on $[0,1]$, where $\{r_k\}$ are the Rademacher functions in $L_2([0,1],dt)$.

27. Prove that the Walsh functions $\{w_n\}$ form a complete orthonormal set in $L_2([0,1],dt)$, where dt denotes Lebesgue measure on $[0,1]$.

28. (Proposition 13.9.2) Let $(V,\langle\cdot,\cdot\rangle)$ be a Hilbert space over $\Phi$ and let $T,S \in L(V)$ and $a \in \Phi$. Prove that $(T + S)^* = T^* + S^*$ and $(aT)^* = \bar{a}T^*$.

29. Let $(V,\langle\cdot,\cdot\rangle)$ be a Hilbert space over $\Phi$ and let $T \in L(V)$. Prove that, if $T$ is invertible, that is, $T^{-1} \in L(V)$, then so is $T^*$ and $(T^*)^{-1} = (T^{-1})^*$.

30. Let $(V,\langle\cdot,\cdot\rangle)$ be a Hilbert space over $\Phi$ and let $T \in L(V)$. Prove that $\|T^*T\| = \|T\|^2$.

31. Let $(V,\langle\cdot,\cdot\rangle)$ be a Hilbert space over $\Phi$ and let $T \in L(V)$. Prove that $N(T^*T) = N(T)$, where $N(T) = \{x \mid x \in V, T(x) = 0\}$.

32. Let $(V,\langle\cdot,\cdot\rangle)$ be a Hilbert space over $\Phi$ and let $T \in L(V)$. Prove that $[R(T)]^{\perp} = N(T^*)$.

33. Let $(V,\langle\cdot,\cdot\rangle)$ be a Hilbert space over $\Phi$ and let $T \in L(V)$. Then $T$ is said to be an <u>isometry</u> if $\|T(x)\| = \|x\|$, $x \in V$. Prove that $T$ is an isometry if and only if $T^*T = I$, where $I(x) = x$, $x \in V$, is the identity mapping.

34. Let $(V,\langle\cdot,\cdot\rangle)$ be a Hilbert space over $\Phi$ and let $T \in L(V)$. Prove that $I + T^*T$ has an inverse in $L(V)$ and that $\|(I + T^*T)^{-1}\| \leq 1$.

*35. Let $(V,\langle\cdot,\cdot\rangle)$ be a Hilbert space over $\Phi$. We have seen two different topologies that can be put on $L(V)$: the uniform, or norm topology $un(\mathcal{T})$, in which $\{T_n\}$ converges to $T$ if $\lim_n\|T_n - T\| = 0$; and the strong operator topology $so(\mathcal{T})$, in which $\{T_n\}$ converges to $T$ if $\lim_n\|T_n(x) - T(x)\| = 0$, $x \in V$. A third topology that is sometimes useful is the <u>weak operator topology</u> $wo(\mathcal{T})$, in which $\{T_n\}$ converges to $T$ if $\lim_n\langle T_n(x),y\rangle = \langle T(x),y\rangle$, $x,y \in V$.

(a) Consider the mapping $\varphi : (L(V),\mathcal{T}) \to \mathbb{R}$, defined by $\varphi(T) = \|T\|$. Prove that $\varphi$ is continuous if $\mathcal{T} = un(\mathcal{T})$, but discontinuous if $\mathcal{T} = so(\mathcal{T})$ or $\mathcal{T} = wo(\mathcal{T})$.

(b) Consider the mapping $\psi : (L(V),\mathcal{T}) \to (L(V),\mathcal{T})$, defined by $\psi(T) = T^*$. Prove that $\psi$ is continuous if $\mathcal{T} = un(\mathcal{T})$ or $\mathcal{T} = wo(\mathcal{T})$, but discontinuous if $\mathcal{T} = so(\mathcal{T})$.

36. (Proposition 13.10.2) Let $(V,\langle\cdot,\cdot\rangle)$ be an inner product space over $\mathbb{C}$. Prove that, if $T \in L(V)$ is such that $\langle T(x),x\rangle = 0$, $x \in V$, then $T = 0$.

37. Let $(V,\langle\cdot,\cdot\rangle)$ be a Hilbert space over $\Phi$. The transformation $T \in L(V)$ is said to be normal if $TT^* = T^*T$. Prove that $T$ is normal if and only if $\|T(x)\| = \|T^*(x)\|$, $x \in V$.

38. Let $(V,\langle\cdot,\cdot\rangle)$ be a Hilbert space over $\Phi$. If $T \in L'(V)$ is self-adjoint, prove that $T \in L(V)$. (Note: Self-adjointness for $T \in L'(V)$ is defined by $\langle T(x),y\rangle = \langle x,T(y)\rangle$, $x,y \in V$. It is necessary that $T$ be defined on all of $V$.)

39. Let $(V,\langle\cdot,\cdot\rangle)$ be a Hilbert space over $\Phi$.

(a) If $T \in L'(V)$ is self-adjoint, define $U = (T - iI)(T + iI)^{-1}$. Prove that $U \in L(V)$ and that $U$ is unitary ($U$ is called the Cayley transform of $T$).

(b) Let $U \in L(V)$ be unitary and suppose $U - I$ has an inverse in $L(V)$. If $T$ is defined by $T = i(I + U)(I - U)^{-1}$, prove that $T$ is self-adjoint and that $U$ is the Cayley transform of $T$.

40. Let $V = L_2([a,b],dt)$, where $-\infty \leq a < b \leq \infty$, and define $D = \{f(t) \mid f(t) \in L_2([a,b],dt) \text{ and } tf(t) \in L_2([a,b],dt)\}$. Let $T : D \to V$ be the multiplication operator $T(f)(t) = tf(t)$, $t \in [a,b]$.

(a) Prove that $D$ is a linear subspace of $L_2([a,b],dt)$ that is norm dense in $L_2([a,b],dt)$.

(b) Prove that $T \in L'(D,V)$ is self-adjoint.

(c) Prove that $T$ is bounded and $D = V$ if and only if the interval $[a,b]$ is finite.

(Note: The definition of the adjoint transformation and self-adjoint transformations are formally the same as when $D$ is a Hilbert space.)

41. Let $(V,\langle\cdot,\cdot\rangle)$ be a Hilbert space over $\Phi$. If $T \in L(V)$ is self-adjoint, then we know from Proposition 13.10.1 that $\langle T(x),x\rangle \in \mathbb{R}$, $x \in V$. So we may define an order relation for self-adjoint $S,T \in L(V)$ by setting $S \leq T$ if and only if

$$\langle S(x),x\rangle \leq \langle T(x),x\rangle \qquad (x \in V).$$

$T$ is said to be <u>positive</u> if $T \geq 0$, that is, $\langle T(x),x\rangle \geq 0$, $x \in V$. Prove that for any $T \in L(V)$ the transformations $T^*T$ and $TT^*$ are always positive.

42. Let $(V,\langle\cdot,\cdot\rangle)$ be a Hilbert space over $\Phi$ and let $P$ and $Q$ in $L(V)$ be orthogonal projections.

(a) Prove that the following are equivalent:

(i) $Q - P \geq 0$ ($\geq$ as defined in Problem 41).
(ii) $\|P(x)\| \leq \|Q(x)\|$, $x \in V$.
(iii) $R(P) \subset R(Q)$.
(iv) $QP = P$.
(v) $PQ = P$.
(vi) $Q - P$ is an orthogonal projection.

(b) If $Q - P$ is an orthogonal projection, prove that $R(Q - P) = R(Q) \cap [R(P)]^{\perp}$.

*43. Let $(V,\langle\cdot,\cdot\rangle)$ be a Hilbert space over $\Phi$ and suppose $\{T_n\} \subset L(V)$ and $S \in L(V)$ are self-adjoint transformations that satisfy $T_1 \leq T_2 \leq \cdots \leq T_n \leq \cdots \leq S$. Prove that there exists a self-adjoint $T \in L(V)$ such that $T_n \leq T \leq S$, $n = 1,2,3,\ldots$, and such that $\{T_n\}$ converges to $T$ in $(L(V),so(T))$. In the particular case where the $T_n$ are orthogonal projections and $S = I$, prove that $T$ is an orthogonal projection and that $R(T)$ is the smallest closed linear subspace containing $R(T_n)$, $n = 1,2,3,\ldots$ .

*44. Let $(V,\langle\cdot,\cdot\rangle)$ be a Hilbert space over $\Phi$ and let $T \in L(V)$ be a self-adjoint transformation such that $0 \leq T \leq I$ (see Problem 41 for a definition of $\leq$). Prove that there exists a unique positive self-adjoint $S \in L(V)$ such that $S^2 = T$.

45. (Proposition 13.11.1) Let $(X,S,\mu)$ be a positive measure space and let $\varphi$ be a measure preserving mapping on $X$. Prove that, if $1 \leq p \leq \infty$ and $T(f) = f \circ \varphi$, $f \in L_p(X,S,\mu)$, then $T$ is a bijective linear isometry from $L_p(X,S,\mu)$ to $L_p(X,S,\mu)$.

46. (Theorem 13.11.2) Let $(X,S,\mu)$ be a positive measure space and let $\varphi$ be a measure preserving mapping on $X$. Prove that, if $T(f) = f \circ \varphi$, $f \in L_2(X,S,\mu)$, then for each $f \in L_2(X,S,\mu)$ there exists some $f_o \in L_2(X,S,\mu)$ such that

(i) $\lim_n \left\| \frac{1}{n+1} \sum_{k=0}^{n} T^k(f) - f_o \right\|_2 = 0.$

(ii) $T(f_o) = f_o.$

*47. (Theorem 13.11.4) Let $(X,S,\mu)$ be a $\sigma$-finite positive measure space and let $\varphi$ be a measure preserving mapping on $X$. Prove that, if $T(f) = f \circ \varphi$, $f \in L_1(X,S,\mu)$, then for each $f \in L_1(X,S,\mu)$ there exists some $f_o \in L_1(X,S,\mu)$ such that

(i) $\lim_n \frac{1}{n+1} \sum_{k=0}^{n} T^k(f)(t) = f_o(t)$ for $\mu$-almost all $t \in X$.

(ii) $T(f_o) = f_o.$

48. (Lemma 13.12.1) If $f \in L_1([-\pi,\pi],dt/2\pi)$, prove that for each $n \in \mathbb{Z}$

$$(e^{in\cdot}f)^\wedge(k) = \hat{f}(k - n) \qquad (k \in \mathbb{Z}).$$

49. If $f \in H_2$ is real-valued, prove that $f$ is a constant.

50. (Lemma 13.13.1) Let $(V,\langle\cdot,\cdot\rangle)$ be a Hilbert space over $\mathbb{C}$, let $T \in L(V)$, and suppose $\lambda_o \in \rho(T)$. Let $\lambda \in \mathbb{C}$ be such that $|\lambda - \lambda_o| < 1/\|R(\lambda_o,T)\|$. Prove each of the following:

(a) $I - (\lambda_o - \lambda)R(\lambda_o,T)$ is bijective.

(b) $[I - (\lambda_o - \lambda)R(\lambda_o,T)]^{-1} \in L(V)$.

(c) $[I - (\lambda_o - \lambda)R(\lambda_o,T)]^{-1} = \Sigma_{k=0}^{\infty}[(\lambda_o - \lambda)R(\lambda_o,T)]^k$.

51. (Corollary 13.13.1) Let $(V,\langle\cdot,\cdot\rangle)$ be a Hilbert space over $\mathbb{C}$ and let $T \in L(V)$. Prove that $\rho(T)$ is an open subset of $\mathbb{C}$ and that $\sigma(T)$ is a closed subset of $\mathbb{C}$.

52. Let $(V,\langle\cdot,\cdot\rangle)$ be a Hilbert space over $\mathbb{C}$, let $T \in L(V)$, and let $\lambda,\mu \in \rho(T)$. Prove each of the following:

(a) $R(\lambda,T) - R(\mu,T) = (\lambda - \mu)R(\lambda,T)R(\mu,T)$.

(b) $R(\lambda,T)R(\mu,T) = R(\mu,T)R(\lambda,T)$.

(The first equation is sometimes called the Hilbert relation.)

53. Let $(V,\langle\cdot,\cdot\rangle)$ be a Hilbert space over $\mathbb{C}$ and let $T \in L(V)$ be an isometry, that is, $\|T(x)\| = \|x\|$, $x \in V$. If $\lambda \in P_\sigma(T)$, prove that $|\lambda| = 1$.

54. Let $(V,\langle\cdot,\cdot\rangle)$ be a Hilbert space over $\mathbb{C}$ and let $T \in L(V)$ be unitary. If $\lambda \in \sigma(T)$, prove that $|\lambda| = 1$.

*55. Let $(V,\langle\cdot,\cdot\rangle)$ be a Hilbert space over $\mathbb{C}$ and let $T \in L(V)$. Suppose $\lambda$ is on the boundary of $\sigma(T)$, that is, $\lambda \in \sigma(T) \sim \mathrm{int}[\sigma(T)]$. Prove that $\lambda \in P_\sigma(T) \cup C_\sigma(T)$.

*56. Let $(V,\langle\cdot,\cdot\rangle)$ be a Hilbert space over $\mathbb{C}$ and let $T \in L(V)$. If $p(z) = \Sigma_{n=0}^{m} a_n z^n$ is a polynomial with complex coefficients, define $p(T) = \Sigma_{n=0}^{m} a_n T^n$. Clearly $p(T) \in L(V)$. Prove that $\sigma[p(T)] = p[\sigma(T)]$; that is, prove that the spectrum of $p(T)$ consists precisely of all values that $p(z)$ assumes on the spectrum of $T$. This result is called the Polynomial Spectral Mapping Theorem.

57. Let $V = \ell_2$ and define $T \in L(\ell_2)$ by $T(\{a_k\}) = \{b_k\}$, where $b_1 = 0$ and $b_k = a_{k-1}$, $k = 2,3,4,\ldots$ . Prove that $\sigma(T) = \{z \mid z \in \mathbb{C}, |z| \leq 1\}$ and that $P_\sigma(T) = \emptyset$. Use Problem 55 to conclude that $\{z \mid z \in \mathbb{C}, |z| = 1\} \subset C_\sigma(T)$.

58. Let $V = L_2([a,b],dt)$, where $-\infty < a < b < \infty$, and define $T \in L(V)$ by $T(f)(t) = tf(t)$, $t \in [a,b]$ (see Problem 40). Prove that $\sigma(T) = [a,b]$ and that $P_\sigma(T) = R_\sigma(T) = \emptyset$.

59. (Corollary 13.14.1) Let $(V,\langle\cdot,\cdot\rangle)$ be a Hilbert space over $\mathbb{C}$. Prove that, if $T \in L(V)$ is self-adjoint, then $\sigma(T) \subset \mathbb{R}$.

60. (Theorem 13.14.3) Let $(V,\langle\cdot,\cdot\rangle)$ be a Hilbert space over $\mathbb{C}$ and let $T \in L(V)$ be self-adjoint. Prove that $M_T \in \sigma(T)$, where $M_T = \sup\{\xi \mid \xi \in \omega(T)\}$.

61. (Corollary 13.14.3) Let $(V,\langle\cdot,\cdot\rangle)$ be a Hilbert space over $\mathbb{C}$. If $T \in L(V)$ is self-adjoint, prove that

$$\|T\| = \sup\{|\lambda| \mid \lambda \in \sigma(T)\}.$$

62. Let $(V,\langle\cdot,\cdot\rangle)$ be a Hilbert space over $\mathbb{C}$ and let $T \in L(V)$ be self-adjoint. Prove that $R_\sigma(T) = \emptyset$.

63. Let $(V,\langle\cdot,\cdot\rangle$ be a Hilbert space over $\mathbb{C}$ and let $T \in L(V)$ be self-adjoint. Recall from Problem 41 that $T$ is positive; that is, $T \geq 0$ if $\langle T(x),x\rangle \geq 0$, $x \in V$. Prove each of the following:

(a) $T$ is positive if and only if $m_T \geq 0$, where

$$m_T = \inf\{\xi \mid \xi \in \omega(T)\}.$$

(b) $T$ is positive and invertible, that is, $T^{-1}$ exists and $T^{-1} \in L(V)$, if and only if $m_T > 0$.

(c) If $m_T > 0$, then $T^{-1}$ is a positive self-adjoint transformation and $m_{T^{-1}} = (M_T)^{-1}$ and $M_{T^{-1}} = (m_T)^{-1}$.

*64. Let $(V,\langle\cdot,\cdot\rangle)$ be a Hilbert space over $\mathbb{C}$ and let $T \in L(V)$. Prove that $\omega(T)$, the numerical range of $T$, is convex. This result is called the Toeplitz-Hausdorff Theorem.

*65. Let $(V,\langle\cdot,\cdot\rangle)$ be a Hilbert space over $\mathbb{C}$ and let $T \in L(V)$. Prove that $\sigma(T) \subset \mathrm{cl}[\omega(T)]$.

66. Let $(V,\langle\cdot,\cdot\rangle)$ be a Hilbert space over $\mathbb{C}$ and let $T \in L(V)$ be self-adjoint. Prove that $\mathrm{cl}[\omega(T)] = \overline{\mathrm{co}}[\sigma(T)]$; that is, the closure of the numerical range is the smallest closed convex set that contains the spectrum of $T$. This result is also true for normal transformations (see Problem 37).

67. Let $(V,\langle\cdot,\cdot\rangle)$ be a nonseparable Hilbert space over $\mathbb{C}$. If $T \in L(V)$ is compact and self-adjoint, prove that $0 \in P_\sigma(T)$.

*68. Let $(V,\langle\cdot,\cdot\rangle)$ be a Hilbert space over $\mathbb{C}$ and let $T \in L(V)$. Prove that $T$ is compact if and only if $\{T(x_n)\}$ converges in $(V,\|\cdot\|)$ whenever $\{x_n\}$ converges in $(V,T^W)$.

69. Let $(V,\langle\cdot,\cdot\rangle)$ be a Hilbert space over $\mathbb{C}$ and suppose $\{T_n\} \subset L(V)$ and $T \in L(V)$ are such that $\lim_n\|T - T_n\| = 0$. If $T_n$ is compact for $n = 1,2,3,\ldots,$ prove that $T$ is also compact.

70. Let $(V,\langle\cdot,\cdot\rangle)$ be a Hilbert space over $\mathbb{C}$ and let $T \in L(V)$. Prove that $T$ is compact if and only if for every $\varepsilon > 0$ there exists some $T_\varepsilon \in L(V)$ such that $R(T_\varepsilon)$ is finite dimensional and $\|T - T_\varepsilon\| < \varepsilon$; that is, prove that $T$ is compact if and only if $T$ is the limit in $(L(V),\|\cdot\|)$ of a sequence of transformations with finite-dimensional ranges. (Hint: Recall that $T$ can be written as $T = A + iB$, where $A$ and $B$ are compact and self-adjoint. Corollary 13.15.1 can be applied to $A$ and $B$.)

71. Let $(V,\langle\cdot,\cdot\rangle)$ be a Hilbert space over $\mathbb{C}$ and let $T \in L(V)$. If $T$ is compact, prove that $T^*$ is also compact.

72. Let $(V,\langle\cdot,\cdot\rangle)$ be a Hilbert space over $\mathbb{C}$, let $T \in L(V)$, and let $\{e_n\}$ be an orthonormal set in $V$. If $T$ is compact, prove that $\lim_n\langle T(e_n),e_n\rangle = 0$.

73. Let $(V,\langle\cdot,\cdot\rangle)$ be a Hilbert space over $\mathbb{C}$ and let $T \in L(V)$ be compact and self-adjoint. By Corollary 13.15.1 we may write $T$ as $T = \Sigma_{k=1}^{\infty}\lambda_k P_k$. If $S \in L(V)$, prove that $S$ commutes with $T$, that is, $TS = ST$, if and only if $S$ commutes with $P_k$, $k = 1,2,3,\ldots$ .

74. Let $(V,\langle\cdot,\cdot\rangle)$ be a Hilbert space over $\mathbb{C}$ and let $T \in L(V)$ be compact. Prove that, if $\lambda \neq 0$, then $\lambda \in P_\sigma(T)$ if and only if $\bar{\lambda} \in P_\sigma(T^*)$. Give an example to show that the assumption $\lambda \neq 0$ cannot be dropped.

75. Let $(V, \langle\cdot,\cdot\rangle)$ be a Hilbert space over $\mathbb{C}$ and let $T \in L(V)$ be compact. If $\lambda \in P_\sigma(T)$, $\lambda \neq 0$, prove that $\{x \mid x \in V, T(x) = \lambda x\}$ is a finite-dimensional linear subspace of $V$.

76. Let $(V, \langle\cdot,\cdot\rangle)$ be a Hilbert space over $\mathbb{C}$ and let $x_o \in V$, $x_o^* \in V^*$. If $T : V \to V$ is defined by $T(x) = x_o^*(x)x_o$, prove that $T \in L(V)$ and $T$ is compact.

77. Let $V = \ell_2$ and define $T \in L(\ell_2)$ by $T(\{a_k\}) = \{a_k/k\}$. Prove each of the following:

(a) $T$ is compact and self-adjoint.

(b) $P_\sigma(T) = \{1/n \mid n = 1,2,3,\ldots\}$ and $C_\sigma(T) = \{0\}$.

(Hint: Problem 12.6.1 is useful.)

## REFERENCES


[AkG] N. I. Akhiezer and I. M. Glazman, *Theory of Linear Operators in Hilbert Space*, Vols. I and II, Ungar, New York, 1963-1965.

[Al] G. Alexits, *Convergence Problems of Orthogonal Series*, Pergamon Press, New York, 1961.

[Ba] G. Bachman, *Elements of Abstract Harmonic Analysis*, Academic Press, New York, 1964.

[BaNr] G. Bachman and L. Narici, *Functional Analysis*, Academic Press, New York, 1966.

[Bb] N. Bourbaki, *Eléments de mathématique. Espaces vectoriels topologiques*, Actualités Scientifiques et Industrielles 1189 and 1229, Hermann, Paris, 1953-1955.

[Ca] L. Carleson, "On the convergence and growth of partial sums of Fourier series", *Acta Mathematica*, *116*, 135-157 (1966.).

[C] R. V. Churchill, *Fourier Series and Boundary Value Problems*, 2nd ed., McGraw-Hill, New York, 1969.

[CoHi] R. Courant and D. Hilbert, *Methods of Mathematical Physics*, Vol. I, Interscience, New York, 1953.

[Da] M. M. Day, *Normed Linear Spaces*, Springer, Berlin, 1958.

[de] L. de Branges, "The Stone Weierstrass Theorem", *Proceedings of the American Mathematical Society*, *10*, 822-824 (1959).

[Du] J. Dugundji, *Topology*, Allyn and Bacon, Boston, 1966.

[$DS_1$] N. Dunford and J. T. Schwartz, *Linear Operators. Part I: General Theory*, Interscience, New York, 1958.

[$DS_2$] N. Dunford and J. T. Schwartz, *Linear Operators. Part II: Spectral Theory*, Interscience, New York, 1963.

[Dn] P. L. Duren, *Theory of $H^p$ Spaces*, Academic Press, New York, 1970.

[$E_1$] R. E. Edwards, Functional Analysis: Theory and Applications, Holt, Rinehart and Winston, New York, 1965.

[$E_2$] R. E. Edwards, Fourier Series: A Modern Introduction, Vol. I, Holt, Rinehart and Winston, New York, 1967.

[$E_3$] R. E. Edwards, Fourier Series: A Modern Introduction, Vol. II, Holt, Rinehart and Winston, New York, 1967.

[Ei] M. Eisen, Introduction to Mathematical Probability Theory, Prentice-Hall, Englewood Cliffs, N.J., 1969.

[F] N. A. Friedman, Introduction to Ergodic Theory, Van Nostrand-Reinhold, New York, 1970.

[Ga] T. Gamelin, Uniform Algebras, Prentice-Hall, Englewood Cliffs, N.J., 1969.

[Gr] A. Garsia, Topics in Almost Everywhere Convergence, Markham, Chicago, 1970.

[Gn] F. P. Greenleaf, Invariant Means on Topological Groups, Van Nostrand-Reinhold, New York, 1969.

[$Ha_1$] P. R. Halmos, Lectures on Ergodic Theory, Chelsea, New York, 1956.

[$Ha_2$] P. R. Halmos, Introduction to Hilbert Space and the Theory of Spectral Multiplicity, Chelsea, New York, 1957.

[Hs] B. Harris, Theory of Probability, Addison-Wesley, Reading, Mass., 1966.

[Hy] E. Helly, "Über Systeme linearer Gleichungen mit unendlich vielen Unbekanten", Monatshefte für Mathematik und Physik, 31, 60-91 (1921).

[He] G. Helmberg, An Introduction to Spectral Theory in Hilbert Space, North-Holland, Amsterdam, 1969.

[$HR_1$] E. Hewitt and K. A. Ross, Abstract Harmonic Analysis, Vol. I, Springer, Berlin-Göttingen-Heidelberg, 1963.

[$HR_2$] E. Hewitt and K. A. Ross, Abstract Harmonic Analysis, Vol. II, New York-Heidelberg-Berlin, 1970.

[HSt] E. Hewitt and K. Stromberg, Real and Abstract Analysis, Springer, New York-Heidelberg-Berlin, 1969.

[HlP] E. Hille and R. S. Phillips, Functional Analysis and Semi-groups, Colloquium Publication XXXI, American Mathematical Society, Providence, R.I., 1957.

[Ho] K. Hoffman, Banach Spaces of Analytic Functions, Prentice-Hall, Englewood Cliffs, N.J., 1962.

[Ht] R. A. Hunt, "On the convergence of Fourier Series", Orthogonal Expansions and Their Continuous Analogs, pp. 235-255, Southern Illinois University Press, Carbondale, Ill., 1968

[Hu] W. Hurewicz, Lectures on Ordinary Differential Equations, M.I.T. Press, Cambridge, Mass., 1958.

[J] R. C. James, "Characterizations of reflexivity", Studia Mathematica, 23, 205-216 (1964).

[Kh] J. P. Kahane, Some Random Series of Functions, Heath, Lexington, Mass., 1968.

[Kt] S. Kakutani, "Weak topology, and regularity of Banach spaces", Proceedings of the Imperial Academy of Tokyo, 15, 169-173 (1939).

[Ka] Y. Katznelson, An Introduction to Harmonic Analysis, Wiley, New York, 1968.

[KeNa] J. L. Kelley, I. Namioka, et al., Linear Topological Spaces, Van Nostrand, Princeton, N.J., 1963.

[K] G. Köthe, Topological Vector Spaces, Vol. I, Springer, New York, 1969.

[Ko] A. N. Kolmogorov, "Une contribution à l'étude de la convergence des séries de Fourier," Fundamenta Mathematica, 5, 96-97 (1924).

[L] R. Larsen, An Introduction to the Theory of Multipliers, Springer, Berlin-Heidelberg-New York, 1971.

[LiSb] L. A. Liusternik and W. I. Sobolev, Elements of Functional Analysis, Ungar, New York, 1961.

[Lo] L. Loomis, An Introduction to Abstract Harmonic Analysis, Van Nostrand, Princeton, N.J., 1953.

[Lu] N. N. Lusin, The Integral and Trigonometric Series (Integral i trigonometricheskii ryad), 2nd ed., Gostekhizdat, Moscow, 1951 [Russian].

[M] C. J. Mozzochi, On the Pointwise Convergence of Fourier Series, Springer, New York-Heidelberg-Berlin, 1971.

[Nb] L. Nachbin, The Haar Integral, Van Nostrand, Princeton, N.J., 1965.

[N] M. A. Naimark, Normed Rings, Noordhoff, Groningen, The Netherlands, 1964.

[Pa] E. Parzen, Modern Probability Theory and Its Applications, Wiley, New York, 1960.

[Ph] R. R. Phelps, Lectures on Choquet's Theorem, Van Nostrand, Princeton, 1966.

[RzNg] F. Riesz and B.Sz.Nagy, Functional Analysis, Ungar, New York, 1955.

[R] K. A. Ross, Fourier Series and Integrals, Yale University Department of Mathematics, New Haven, Conn., 1965.

[Ry] H. Royden, Real Analysis, 2nd ed., Macmillan, New York, 1968.

[$Ru_1$] W. Rudin, Fourier Analysis on Groups, Interscience, New York, 1962.

[$Ru_2$] W. Rudin, Real and Complex Analysis, McGraw-Hill, New York, 1966.

[Sa] H. Sagan, Boundary and Eigenvalue Problems in Mathematical Physics, Wiley, New York, 1961.

[Sn] G. Sansone, Orthogonal Functions, Interscience, New York, 1959.

[Sh] L. Schwartz, Théorie des distributions, Actualités Scientifiques et Industrielles 1245, Hermann, Paris, 1959.

[So] M. H. Stone, Linear Transformations in Hilbert Space and Their Applications to Analysis, Colloquium Publication XV, American Mathematical Society, New York, 1932.

[Sz] G. Szegö, Orthogonal Polynomials, Colloquium Publication XXIII, American Mathematical Society, New York, 1939.

[T] A. E. Taylor, Introduction to Functional Analysis, Wiley, New York, 1958.

[Tr] F. G. Tricomi, Integral Equations, Interscience, New York, 1957.

[V] B. Z. Vulikh, Introduction to Functional Analysis, Pergamon Press, Oxford, 1963.

[We] A. Weil, L'intégration dans les groupes topologiques et ses applications, Actualités Scientifiques et Industrielles 8690-1145, Hermann, Paris, 1951.

[$W_1$] A. Wilansky, Functional Analysis, Blaisdell, New York, 1964.

[$W_2$] A. Wilansky, Topology for Analysis, Ginn, Waltham, Mass., 1970.

[Y] K. Yosida, Functional Analysis, 2nd ed., Springer, Berlin-Heidelberg-New York, 1968.

[Z] A. Zygmund, Trigonometric Series, Vol. I, 2nd ed., Cambridge University Press, Cambridge, 1968.

INDEX

L



M